Biological Diversity

Balancing Interests Through Adaptive Collaborative Management

Biological Diversity

Balancing Interests Through Adaptive Collaborative Management

Edited by
Louise E. Buck • Charles C. Geisler
John Schelhas • Eva Wollenberg

CRC

CRC Press
Boca Raton London New York Washington, D.C.

Library of Congress Cataloging-in-Publication Data

Biological diversity : balancing interests through adaptive collaborative management / Louise E. Buck ... [et al.], editors.
p. cm.
Includes bibliographical references (p.).
ISBN 0-8493-0020-7 (alk. paper)
1. Protected areas—Management. 2. Conservation of natural resources. 3. Biological diversity conservation. I. Buck, Louise.

S944.5.P78 B56 2001
333.95′16—dc21 2001018419

Catalog record is available from the Library of Congress

This book contains information obtained from authentic and highly regarded sources. Reprinted material is quoted with permission, and sources are indicated. A wide variety of references are listed. Reasonable efforts have been made to publish reliable data and information, but the author and the publisher cannot assume responsibility for the validity of all materials or for the consequences of their use.

Direct all inquiries to CRC Press LLC, 2000 N.W. Corporate Blvd., Boca Raton, Florida 33431.

Visit the CRC Press Web site at www.crcpress.com

International Standard Book Number 0-8493-0020-7
Library of Congress Card Number 2001018419
Printed in the United States of America 1 2 3 4 5 6 7 8 9 0
Printed on acid-free paper

Foreword

Norman Uphoff
Director
Cornell International Institute for Food, Agriculture and Development (CIIFAD)

Adaptive collaborative management (ACM) is a concept as well as a practical strategy whose time is coming. ACM derives both from experience and from views of the world that have their foundations in natural science and contemporary culture. This book examines these worldviews as well as some of the experience with natural resource management that has encouraged and supported their articulation.

The worldwide movement to establish conservation policies and practices that will reverse the present loss of biodiversity has gained impetus from many sources: from government agencies and legislatures, civil-society institutions, academics and researchers, community resistance to commercial encroachment on natural resource domains, and a growing conviction among people everywhere that the maintenance of environmental integrity and services can no longer be taken for granted.

Initial responses, official and nongovernmental, to threats to the survival of ecosystems and to the flora and fauna they support have tended to be regulatory or coercive. The powers of the state were mobilized to enforce the protection of natural resources, and the logic and knowledge of science were enlisted to justify and guide such actions. The role of citizens was envisioned as principally one of compliance.

But this traditional approach has been found inadequate for stemming the erosion of biological resources, because state capacities were less than anticipated, the knowledge base for effective and sustainable management was inadequate, and cooperation from various publics was not necessarily forthcoming under such circumstances. Interest in ACM arises in the first instance because there is evident need to devise strategies for natural resource conservation that can be more successful in enlisting both knowledge and public support. In a modern, democratic age, purely authority-based approaches have become anachronistic, especially when the expertise supporting them is overstated.

As governments, nongovernmental organizations (NGOs), and others have become more engaged in conservation efforts, the limitations of our knowledge for conservation management have become more evident. Mechanistic, linear projections fail to predict the future of natural systems. They can underestimate the immense recuperative power of such systems in the same way that they overlook inherent vulnerabilities that can cause quick collapse. In nature, great sensitivity and tremendous resilience coexist, which makes its nurture and preservation more difficult and less certain than when systems are simpler and more predictable.

The conjunction of needing to proceed in more flexible ways — not simply implementing preconceived plans but always working in a learning process mode — and in a more participatory manner — enlisting not just the cooperation of various stakeholders but also their contribution of knowledge — has led us and others to formulate the concepts and principles of ACM, which are laid out in the introductory chapter.

ACM is still in its initial stages, so it is a work-in-progress. No final judgment can be rendered on how sufficient or effective it will be to cope with the many complexities and great difficulties of biodiversity conservation. And like the nature it seeks to protect, we can expect that ACM will itself evolve over time as experience accumulates. So ACM is more an approach than a fixed strategy. It certainly does not offer any detailed blueprints. However, the case for proceeding with such an approach is very strong, given both the shortcomings of past efforts and the apparent merits of ACM.

This approach is consistent in the social realm with an ecological perspective in the natural world. In both domains, one needs to appreciate the complexity and relatedness of things, expecting that relationships as well as components will continue to evolve. There can and will be conflict and competition, but at the same time there is interdependence and some necessary degree of cooperation. Just as blueprints can be constructed only *post facto* in nature, we expect that efforts to conserve nature must themselves be based on an understanding of complexity and interrelatedness, with multiple objectives being served. Since nature remains full of surprises, its management needs to remain empirical.

There have been many previous efforts to make the world fit mechanistic understandings of relationships and of cause and effect. But such a clockwork ("Newtonian") worldview has been strongly challenged by other branches of (more modern) science that encourage more holistic understanding and draw on the insights coming from chaos and complexity theories (such as Capra, 1982, Prigogine and Stengers, 1984, Stacey, 1993; Gell-Mann, 1994; Uphoff, 1996).

ACM is in tune with these ways of understanding reality, not abandoning more deterministic thinking altogether, but expanding upon it with contemporary insights into reality. Biophysical and socioeconomic domains have more in common than conventional disciplinary boundaries suggest, and ACM seeks to connect these realms for the sake of biodiversity conservation and for meeting human needs.

Mechanistic, deterministic images of reality can be useful approximations. But they are always metaphorical, not real, so when we use them, we should not accept them as absolute truths. We do well to understand the ancient Greek philosopher Heraclites who said: You cannot step in the same river twice — because it is always changing. Indeed, stepping into the river changes it.

In fact, Heraclites' river was a metaphor for the real world around us, which is itself always evolving, usually slowly, but sometimes fast. And even if certain things are changing relatively little, their surroundings are not. So the significance and possibilities of even slow-changing things are continually being modified.

There are a number of implications of such a worldview for approaching ACM. These can be simply stated:

1. *We live in open systems, not closed ones.* It is tempting, for the sake of simplicity or greater certainty, to assume the latter. But this is a trick of the mind, since closed systems are more artificial than real. This appreciation makes cause-and-effect relations more complex, so that *steering* and *guiding*

are more appropriate words to describe the process of management than are *controlling* and *regulating*. Planning is a meaningful process if it is indicative and purposive, based on a *learning process* (Korten, 1980), rather than an attempt to determine the future unilaterally. Having continual feedback and evaluation is also essential.

2. *Knowledge needs to be continuously validated and revalidated.* We should not conclude that our knowledge is always or necessarily inadequate. It is possible to have useful and valid knowledge, and scientific efforts to develop universalizable principles and relationships are still important. (The extreme postmodernist position decrying scientific investigation defies the evidence of experience.) However, we should be appropriately modest about our knowledge and recognize its limits, remaining good observers and not being blinded by preconceptions. The application of knowledge to real-world situations needs to be empirical, and it is justified (or not) according to its relevance and impacts.
3. *Reality is pluralistic.* On this point, the postmodernists make a useful contribution, directing us away from monolithic and overly abstract, generalized concepts of truth. One problem is that our language tends to homogenize reality by using singular forms of nouns, rather than plurals. We should speak of *conservations* rather than *conservation*, as there are many kinds and degrees, some more desirable or sustainable than others. Indeed, the apparently plural term *natural resources* is still a monolithic abstraction, better replaced with more concrete referents like *soil*, *water*, *forests*, and *flora and fauna*. These are still rather general categories, but more closely related to reality. Few terms are more misleading than *the small farmer* or *the environment* when we should be talking about *small farmers* and *environments*, with the plural form reminding us of the diversity that exists.
4. *The world continues to contain many surprises*. We have invested hundreds of thousands of scientist-years by now in trying to understand the environment(s) with implications for how policies and practices can best maintain its (their) vitality and productivity. This productivity itself relates to human needs but also to natural processes. If we appreciate that our knowledge is imperfect and incomplete, not just because of our own intellectual limitations but because the world is changing, and that reality is pluralistic, we are better prepared for acting realistically and effectively toward the world as it is, not as we have preconceived it.

We hear at present debates over whether "markets" are good or are bad for conserving biological diversity. Any effort to generalize like this is bound to produce misleading conclusions. Leaving aside the question of whether one can meaningfully talk about biodiversity as a single phenomenon, there are surely ways in which certain market processes can be positive, and others can be negative. And outcomes that were observed at one point in time could be quite different at another time. So we should be prepared to find unexpected things. We will be better prepared for surprises, however, if we proceed in our biodiversity conservation efforts with the above four observations as ontological and epistemological premises.

We might also keep in mind an observation made by Jack Duloy of the World Bank at a Bellagio conference some years ago that "the real world is a special case." We should develop as much general knowledge as we can to be better prepared to deal with the pluralism and uncertainty of operational circumstances. But we should also remember that solving problems and meeting goals in concrete situations requires a blending of knowledge from various sources and disciplines, and a sagacious abstracting and application from experience. No two situations are the same; however, there are some recurrent or common elements that we can know and build upon. ACM does not proceed assuming a blank slate (*tabula rasa*). It explicitly draws on existing knowledge and experience, but it expects that solutions will have to be crafted, evaluated, modified, and evolved, that they are not fully determinable in advance.

This book gives substance, conceptual and empirical, to ACM. ACM is an approach that by its nature will evolve, as practitioners and policy makers as well as academics and evaluators assess the contribution it can make to the protection of natural ecosystems. Where the conservation of biodiversity is at stake, it frequently offers a state-of-the-art approach to increasing the likelihood that treasured natural resources will be sustained, not just for the 5 or 10 years of a project's life but over generations. It is urgent that we find better ways to ensure that ecosystems that have required millions of years to evolve to their present state do not deteriorate and disappear as a result of human actions — or inaction.

REFERENCES

Capra, F., 1982. *The Turning Point: Science, Society and the Rising Culture,* Simon and Schuster, New York.

Gell-Mann, M., 1994. *The Quark and the Jaguar: Adventures in the Simple and the Complex,* W. H. Freeman, New York.

Korten, D. C., 1980. Community organization and rural development: a learning process approach, *Publ. Adm. Rev.,* 40(5), 480–511.

Prigogine, I. and Stengers, I., 1984. *Order out of Chaos: Man's New Dialogue with Nature,* Bantam Books, New York.

Stacey, R., 1993. Strategy as order emerging from chaos, *Long Range Plann.,* 26, 10–17.

Uphoff, N. T., 1996. *Learning from Gal Oya: Possibilities for Participatory Development and Post-Newtonian Social Science,* Intermediate Technology Publications, London.

Editors

Louise E. Buck is a Senior Extension Associate with the Program in Ecology and Management of Landscapes in the Department of Natural Resources at Cornell University, Ithaca, New York. She is a Senior Associate Scientist with the Program on Local People, Devolution and Adaptive Co-Management at the Center for International Forestry Research (CIFOR) in Bogor, Indonesia and a program coordinator with the Cornell International Institute for Food, Agriculture and Development (CIIFAD). Dr. Buck's research, teaching, and extension interests include agroforestry, protected area management, and knowledge systems for integrated resource management. Her current research focuses on multistakeholder approaches to non-timber forest product conservation and development in northeastern North America and in southeastern Madagascar. The work seeks to understand and improve the conditions for bringing about a coordinated infrastructure for agroforestry-based practice, learning, and institutional and policy support. Prior to joining Cornell, Louise worked for 10 years in eastern and southern Africa with the International Center for Research in Agroforestry (ICRAF), CARE International, and the Beijer Institute of the Swedish Royal Academy of Sciences. She holds a Ph.D. in natural resources from Cornell with minors in development sociology and adult education.

Charles C. Geisler is Professor of Rural Sociology at Cornell University. His teaching and research interests include land use, ownership, and reform; the equity issues associated with public ownership; the environmental dimensions of land reform; agricultural land conversion; the social impacts of land protection schemes; and the evolution and cultural contests of ownership systems. He has coedited various books and authored a range of government reports, encyclopedia entries, and journal articles. His work has been supported by the Ford and MacArthur Foundations, the World Bank, the United Nations, and various state and federal agencies.

John Schelhas is a Research Forester with the Southern Research Station of the USDA Forest Service. His research is on minority and limited-resource landowners and forests in the U.S. South, including forest-based rural development, forest values, and relationships between national forests and their neighbors. Prior to coming to the Forest Service, he was a Senior Research Associate in the Department of Natural Resources at Cornell University, where he coordinated the Cornell Program on Ecological and Social Science Challenges of Conservation and taught courses in the Graduate Minor in Conservation and Sustainable Development. He has conducted research in Costa Rica since 1988, focusing on private forests adjacent to national parks and on environmental values in rural communities. He has also worked on international conservation issues for the Smithsonian Migratory Bird Center and the U.S. National Park Service. He holds a Ph.D. in renewable natural resources, with a minor in anthropology, from the University of Arizona.

Eva (Lini) Wollenberg is a researcher with the Programme on Local People, Devolution and Adaptive Co-Management at the Center for International Forestry Research (CIFOR) in Bogor, Indonesia. Her interests and experience over the last 15 years have concentrated on resource management among forest-dependent communities in Asia, especially in Indonesia and the Philippines. Her current research examines governance related to conflict and cooperation among forest stakeholders. Before joining CIFOR, Eva worked for the Ford Foundation Asia Program as a program officer in rural poverty and resources. She completed her Ph.D. and M.S. in wildland resources science and her B.S. in conservation and natural resources at the University of California, Berkeley.

Contributors

Jon Anderson
Natural Resource Policy Advisor
Africa Bureau
U.S. Agency for International Development
Washington, D.C.

Robert Bino
Research and Conservation Foundation of Papua New Guinea
Goroka, Papua New Guinea

Eckart Boege
Research Professor
National Institute of Anthropology and History
Veracruz, Mexico

Louise E. Buck
Sr. Extension Associate
Department of Natural Resources
Cornell University
Ithaca, New York

Richard S. Cahoon
Vice President
Cornell Research Foundation
Ithaca, New York

Sarah Christiansen
Conservation Specialist
Ecoregional Conservation Strategies Unit
World Wildlife Fund
Washington, D.C.

Carol J. Pierce Colfer
Principal Scientist
Center for International Forestry Research (CIFOR)
Bogor, Indonesia

Paul D. Cowles
CORE Program Director
Pact, Inc.
Nairobi, Kenya

Eric Dinerstein
Chief Scientist
Conservation Science Program
World Wildlife Fund
Washington, D.C.

David Edmunds
Senior Associate
Center for International Forest Research (CIFOR)
Bogor, Indonesia

Jenny A. Ericson
Director
Community Conservation for the Andean Southern Cone Division
The Nature Conservancy
Arlington, Virginia

Robert J. Fisher
Head, Program Development
Regional Community Forestry Training Center
Kasetsart University
Bangkok, Thailand

Richard Ford
Professor
Programs in International Development, Community Planning, and Environment
Center for Community-Based Development
Clark University
Worcester, Massachusetts

Mark S. Freudenberger
Senior Advisor
Landscape Development Initiatives
Chemonics, Inc.
Fianarantsoa, Madagascar

Charles C. Geisler
Professor
Department of Rural Sociology
Cornell University
Ithaca, New York

Hans Gregersen
Professor
College of Natural Resources
University of Minnesota
St. Paul, Minnesota

Maria Cristina S. Guerrero
Regional Coordinator Non-Timber Forest Product (NFTP)–Exchange Programme and Desk Coordinator
Philippines NTFP Task Force
Quezon City, Philippines

Carlos F. Guindon
Research Affiliate
Monteverde Institute
Monteverde, Costa Rica

Celia A. Harvey
Research Scientist
Area de Cuencas y Sistemas Agroforestales
Centro Agronómico Tropical de Investigación y Enseñanza
Turrialba, Costa Rica

Ronald J. Herring
Director
Mario Einaudi Center for International Studies
Cornell University
Ithaca, New York

Paul Hukahu
Research and Conservation Foundation of Papua New Guinea
Goroka, Papua New Guinea

Paul Igag
Research and Conservation Foundation of Papua New Guinea
Goroka, Papua New Guinea

Janice Jiggins
Professor
Human Ecology
Department of Rural Development Studies
Swedish University of Agricultural Sciences
Uppsala, Sweden

Arlyne Johnson
Wildlife Conservation Society
Bronx, New York

Godwin Kowero
Regional Coordinator
Regional Office for Eastern and Southern Africa
Center for International Forestry Research (CIFOR)
Harare, Zimbabwe

Kai N. Lee
Professor
Center for Environmental Studies
Williams College
Williamstown, Massachusetts

Maria Paz (Ipat) G. Luna
Deputy Director
Babilonia Wilner Foundation
Ermita, Manila, Philippines

Allen Lundgren
Adjunct Professor
College of Natural Resources
University of Minnesota
St. Paul, Minnesota

Richard Margoluis
Director
Analysis and Adaptive Management Program
Biodiversity Support Program
Washington, D.C.

William J. McConnell
Anthropological Center for Training and Research on Global Environmental Change
Indiana University
Bloomington, Indiana

Cynthia McDougall
Research Fellow
Center for International Forestry Research (CIFOR)
Bogor, Indonesia

Jeffrey A. McNeely
Chief Scientist
IUCN-The World Conservation Union
Gland, Switzerland

Eufemia Felisa Pinto
Regional Program Officer
Oxfam America Southeast Asia Program
Phnom Penh, Cambodia

Ravi Prabhu
Senior Scientist
Regional Office for Eastern and Southern Africa
Center for International Forestry Research (CIFOR)
Harare, Zimbabwe

Soava Rakotoarisoa
Graduate Student
International Agriculture Program
Cornell University
Ithaca, New York

Vololona Rasoaromanana
Director
National AGERAS
Antananarivo, Madagascar

Haingolalao Rasolonirinamanana
Coordinator
AGERAS
Toliara Region
Toliara, Madagascar

Niels G. Röling
Professor
Department of Communication and Innovation Studies
Wageningen University
Wageningen, The Netherlands

Nick Salafsky
Program Officer
Kellogg Foundation
Chicago, Illinois

Jeffrey A. Sayer
Director General
Center for International Forestry Research
Bogor, Indonesia

John Schelhas
Research Forester
Southern Research Station
USDA Forest Service
Tuskegee University
Tuskegee, Alabama

Glenn Smucker
Cultural Anthropologist
Milwaukee, Wisconsin

Norman Uphoff
Director
Cornell International Institute for Food, Agriculture and Development
Cornell University
Ithaca, New York

Guillermo Vargas
Agricultural Department Manager
Coope Santa Elena
Monteverde, Costa Rica

Andy White
Program Director
Forest Trends
Washington, D.C.

Eva (Lini) Wollenberg
Senior Scientist
Center for International Forestry Research (CIFOR)
Bogor, Indonesia

Table of Contents

Introduction: The Challenge of Adaptive Collaborative Management

John Schelhas, Louise E. Buck, and
Charles C. Geisler

As we enter the 21st century, the well-being and future of humans and biodiversity are more interdependent than ever before. The reach of industrial society has extended into the most remote regions of the globe and human-induced habitat conversion and species loss are arguably at record levels (Heywood, 1995; Lubchenco, 1998). Local, national, and international organizations have responded to this situation in many ways. Biodiversity conservation and research are featured prominently in media and in the activities of international organizations. New protected areas continue to be established, international agreements such as the Convention on Biological Diversity are in place, and international spending on conservation has reached record levels (Abramovitz, 1994; Heywood, 1995; UNDP/UNEP/WB/WRI, 2000). Biodiversity conservation efforts have also moved beyond reserves and protected areas to include larger human-occupied landscapes, ecoregions, and agroecosystems (Collins and Qualset, 1988; Soulé and Terborgh, 1999; Baydack et al., 1999).

In spite of all this activity, there is little evidence that the crisis in biological diversity has diminished. This may be in part because, in biodiversity conservation, successes are often temporary or tenuous and failures permanent, as measured by extinctions and habitat loss (Orlove and Brush, 1996). But the crisis also has spawned widely diverging opinions and vigorous debate about the most effective and appropriate conservation strategies (Brechin et al., 2001). This book does not seek to resolve the debate, but rather to contribute a management logic that all parties to the debate may find useful. This logic is called adaptive collaborative management (ACM) and applies to biodiversity conservation both in and beyond formal protected areas. ACM, as used here, is a strategy that (1) proceeds in a learning process mode rather than according to a universal *a priori* blueprint; (2) considers errors, mistakes, and failures not in normative terms but as normal occurrences resulting from policy experiments; (3) engages both local and nonlocal stakeholders in a participatory process of goal setting, planning, management experimentation, and evaluation; and (4) utilizes a variety of methods to generate knowledge that keeps pace with ecosystem change resulting naturally or from expanding human activity. Biodiversity conservation is often centered around protected areas such as national parks and nature reserves, but is increasingly carried out in places that are used or occupied by humans.* The managers and conservers of biodiversity are increasingly challenged to find control mechanisms that are flexible, without being ad hoc, and collaborative, without being indifferent to science.

The past century has been characterized as the century of science (Lubchenco, 1998). Massive public funding became available for ambitious research programs following World War II to achieve technical feats and address social problems. Much funding went to subduing nature and subordinating it to human ends. But as the environment moved to center stage and public sophistication about ecological services grew, scientists came face-to-face with a new "social contract" or set of obligations to meet expectations for sustainable, ecologically sound living. Implementing this social contract, however, requires more than scientists turning their attention to the important public demands of building a sustainable society through their usual methods of research and scholarship. The late 20th century also saw new attention to participatory democracy, to using and fostering citizen participation in governance. This trend, which includes both participation itself and the building of the capacity — social capital — to institutionalize participation and collaboration in decision making, was often termed building a civic society (Cortner and Moote, 1999). ACM is potentially part and parcel of the new social contract. It seeks to harness the power of both science and lay knowledge for building management models, testing them, learning from them, and — ultimately — applying them to conservation and sustainable development.

ACM contrasts sharply to traditional natural resource and environmental management, which derived from another expression of science — scientific management. Not only was nature viewed as a trove of latent commodities awaiting harvest, but environmental stability and resource permanence were vaunted management goals. Uncertainty, disorder, and unpredictability were anathemas; scientific managers had little appreciation of chaos or the dynamic uncertainties of nature. Ecosystem services were segmented and defined in anthropocentric terms. Planning was deemed successful if its prescriptions worked in perpetuity; "failures" were unwanted feedback more apt to engender denial than adaptive management.

Late in 1998, as Lubchenco's "new social contract" message was coming to light a cross section of scholars, laypeople, and nongovernmental organization (NGO) representatives was assembled to participate in seminal thinking on ACM at Cornell University.** This was not an occasion to advocate a new approach or to preach panaceas, but to employ the insights of ecosystem modelers (e.g., Holling, 1978; Walters, 1986), business visionaries (e.g., Stacey, 1993), academics concerned with questions of governance and participation (e.g., Uphoff, 1992; Western and Wright, 1994; Brechin et al., 2001) and others in relation to the management of biodiversity.

* Biodiversity is "the total diversity and variability of living things and of the systems of which they are a part," and includes ecosystem, species, and genetic diversity (Heywood, 1995, p. 9).

** Sponsoring organizations were Center for International Forestry Research (CIFOR), Cornell International Institute for Food, Agriculture and Development (CIIFAD), and World Wildlife Fund-U.S. (WWF-US).

Few of the participants thought of their work as "adaptive collaborative management" at the time.* Most emphasized either the adaptive or the collaborative component, and sometimes such views seemed to clash. This tension, and the potential for resolving it in diverse geocultural and ecological settings, motivated the present book. The editors believe that the new social contract will remain academic if not applied adaptively and collaboratively among a diverse cross section of the public.

The current context for contemplating ACM today is contentious. ACM itself is less the eye of the storm than the multiple management perspectives that over the past two decades have moved to the fore of integrated biodiversity management. Since the early 1980s pleas to balance humanity and ecology have been widely heeded. "Sustainabilty" evolved from a single to a multidimensional concept. Conservation advocates from the public, private, and nonprofit sectors struggled with human development questions and loosely regrouped around a new conservation paradigm. Its centerpiece is the hypothesis that local communities, commonly regarded as threats to biodiversity, will ally themselves with the forces of conservation if given incentives to do so. And, the paradigm suggested, the alliance is enhanced by the backlog of environmental management knowledge that local communities utilize in times of stability and abundance. The paradigm was expressed in a variety of forms, the best-known (and most generic) being the so-called integrated conservation and development programs (ICDPs).

THE "RISE AND FALL" OF ICDPs

In the 1980s, sustainable development emerged as an alternative to earlier protectionist strategies that viewed conservation and development as opposing interests and therefore sought to establish large national parks and other reserves where "natural" ecosystems could be protected from human influences. Several new ideas challenged protectionism and set the stage for the emergence of sustainable development. First, empirical research shows the complexity of relationships between people and ecosystems. Some of this research revealed the historical influence of humans on what had generally been considered "pristine" or "natural" ecosystems (Posey and Balée, 1989; Blackburn and Anderson, 1993; Boyd, 1999); other research described groups that had met their livelihood needs in ways that also maintained biodiversity (e.g., Michon and de Foresta, 1992; Brookfield and Padoch, 1994).

Second, there was a growing awareness of the potential for a mutually beneficial, pragmatic relationship between conservation and development. Conservation success began to be seen as meeting people's needs and aspirations, and successful long-term development as avoiding natural resource depletion and environmental degradation. Perhaps setting the stage for later arguments against integrated conservation and sustainable development projects, these ideas were loosely extended to new

* An exception was CIFOR, which in 1998 brought together its research programs on (1) Criteria and Indicators for Sustainable Forest Management and (2) Devolution, Community Management and Local People into a single program under the name, Local People, Devolution and Adaptive Co-Management. The evolution of this change is described in Colfer, Chapter 15, this volume.

conservation theory, arguing that local people would be most likely to conserve when they received direct benefits from that conservation (see Kramer and van Schaik, 1997; Brandon, 1998; Oates, 1999). It was hypothesized (or, more commonly, asserted) that, since economic benefits were most important to people in lesser-developed countries, conservation should be promoted through economic development. The resulting projects, often involving agroecology, ecotourism, or natural resource–based microenterprises, were the proving grounds for ICDPs (Wells and Brandon, 1992; Orlove and Brush, 1996).

These alterations in conservation thinking played favorably in many quarters. National governments and international development agencies, including many that previously showed little interest in conservation, found the ICDP approach attractive. Conservation gained new resources and attention. Interrelated developments included (1) increased funding for conservation from international lenders, such as the World Bank, and bilateral aid agencies, such as the U.S. Agency for International Development; (2) national government attention to new conservation approaches, such as extractive reserves in Brazil, ecotourism in Costa Rica, and ministry-level interest in sustainable development in many countries; and (3) a proliferation of NOGs and community groups committed to community-based conservation in and around protected areas.

Community-based conservation argued that local control, in addition to local benefits, was essential to conservation success. Conservationists saw community-based approaches as a way to make biodiversity and natural resources relevant to local communities; local communities saw them as a way to regain control over natural resources, improve their well-being, and secure their lifestyle (Western and Wright, 1994). If there were doubts about such reasoning, they were rooted in neo-Malthusian views about local population growth, the new and potentially destructive technologies at the disposal of local community inhabitants, the ongoing development subsidies the approach might necessitate, and nonlocal forces beyond community control that might undermine its local capacity or goodwill. On the other hand, there is an extensive literature affirming the promise of locally conceived common property regimes as a tried-and-true category of community-based conservation (for a summary, see Bruce, 1999).

Although ICDPs and community-based conservation have received the most attention, still other approaches were developed concurrently, including the ACM logic motivating the present work. More macrolevel, policy-oriented recommendations pointed to important structural issues. For examples, some observers urged reform in laws that offered perverse incentives for deforestation and habitat conversion (Repetto and Gillis, 1988; Southgate, 1998) or for reorienting land reform to environmental as well as social and economic goals (Geisler, Chapter 6, this volume). Others advocated using fiscal and financial incentives to promote conservation activities (Watson et al., 1998). Some researchers suggested the use of markets in wild species to promote their sustainable use and conservation (Freeman and Kreuter, 1994; Freese, 1998), while others advocated privatizing more protected areas (Anderson and Leal, 1991; Langholz, 1999).

As ICDPs and community-based conservation experiences grew, conservation biologists questioned the hypothesized link between conservation and development and resisted the shift in conservation focus from the ecological to the social. At a general level, a number of conservationists have argued that there are often fundamental conflicts between conservation and sustainable development on a number of different grounds (Redford, 1992; Robinson, 1993; Soulé, 1995; Kramer and van Schaik, 1997; Brandon, 1998; Terborgh, 1999). They argue that, while exceptions exist, in most cases humans destroy the very resources on which they depend, especially under conditions of rapid population growth and technological/social change (Soulé, 1995). Underlying this is the idea that human shortsightedness and greed make it difficult to stop even widely recognized overexploitation (Ludwig et al., 1993). Counterarguments from social scientists have rejected these generalities and instead sought to delineate the conditions under which sustainable or unsustainable resource use is most likely to occur (e.g., Bromley, 1992; Bruce, 1999; Brechin et al., 2001).

To be sure, important questions remain about the underlying premises of ICDPs and community-based conservation. There may be fundamental differences in interests between international and national conservation objectives, often based on aesthetic and moral values, and local conservation objectives, often rooted in livelihood needs and where conservation of some species, for example, large mammals, may directly conflict with livelihood interests (Schelhas and Greenberg, 1996; Sanderson and Redford, 1997). Extractive resource use inescapably brings about some ecological change and has the potential for negative biodiversity impacts (Redford, 1992; Freese, 1997; 1998; Redford and Richter, 1999). There are also questions about whether promoting conservation through personal incentives and economic development undermines the ethical and aesthetic motivations for conservation (Oates, 1999; Schelhas, Chapter 13). And community-based conservation may be founded on idealized notions of community that may not exist under contemporary conditions (Little, 1994; Li, 1996; Brosius et al., 1998; Oates, 1999). Empirical studies show evidence of resource overexploitation by local people and of the failure of ICDPs to deliver clear conservation results (for example, see Alpert, 1996; Barrett and Arcese, 1995; Langholz, 1999; Oates, 1999; Peters, 1998; Wells and Brandon, 1992).

Thus, the outset of the 21st century finds a situation where conservationists, recognizing the limitations of reserves and protected areas, are reaching beyond them to include larger landscapes and ecoregions while sustainable development advocates are seeking natural resources and nature protection schemes that benefit local people. The literature is a cacophony of competing problem definitions, approaches, interest group positions, spatial scales, time frames, and values. The result of this ongoing polemic has tended toward two polarized camps, one emphasizing rather exclusionary and expanding reserves for species and ecosystem conservation, often for global values, and the other emphasizing nature conservation as a means to human well-being, often local in orientation. Interestingly, the different views generally have more to do with emphasis than with outright disagreement. Advocates of nature reserves recognize the importance of meeting the needs of people (Kramer, 1997; Brandon, 1998), and sustainable development advocates avow strong conservation commitments (Western and Wright, 1994). Yet a vigorous and sometimes acrimonious debate continues in both the scientific and popular literature (e.g., Redford and Sanderson, 2000; Schwartzman et al., 2000; Terborgh, 2000; Brechin et al., 2001).

POSTMODERN CONSERVATION: A PLACE FOR ADAPTIVE COLLABORATIVE MANAGEMENT

> The most relevant empirical studies ... all point to a fundamentally messy, contingent, and ambiguous intermingling of knowledge, power, interests, and chance in the workings of the world. These studies nonetheless provide what we have found to be useful beginnings for efforts to understand and manipulate long-term improvements in the management of sustainable development policies. (Parsons and Clark, 1995, p. 457)

How is one to make sense of the diverse and often contradictory results and opinions in the conservation literature? A complex and changing world of diverse interests and stakeholders leaves little room for simple answers and prescriptions. Holling (1995, p. 5) suggests that when faced with a fundamental loss of certitude and an action seems more prone to costs than benefits, it is tempting to retreat to unsupported ideology, rigid beliefs, or to tightly circumscribed action. The alternative, he suggests, is to seek new understanding, often through reflective action.

One important resource for reflection is the history of conservation. Fairfax and Fortmann (1990) describe how the progressive era roots of modern conservation in the United States produced a distinctive modern conservation ideology. One component of this is the idea that natural resource decisions should be made by technically trained experts, based on scientific notions of appropriate resource use and efficiency — often in support of narrow, time-specific resource values and end uses. Local users, from this perspective, are political actors self-servingly pursuing advantages that distort "technically correct" decisions, rather than locally knowledgeable users and managers of the resource capable of acting for the civic good and larger public interest (Fairfax and Fortmann, 1990). The same perspective is used to justify rigidly standardized national-level policies that are inappropriate when applied categorically across time and place and especially to other cultures (Fairfax et al., 1998).

The past 20 years in natural resource management has seen the rise of a broad cultural and intellectual trend of postmodernism (Orr, 1992; Rudel and Gerson, 1999; McCay, 2000). Although usually implicit rather than explicit, many of the new approaches in natural resource management reflect postmodernist thinking.* The emphasis on ecosystem services noted at the outset by Lubchenco (1998), the many ways these can be construed and managed, and the claims on them by diverse groups are all examples. These, in turn, are altering traditional scientific and government management approaches to include participation, collaborative management, new community deference, and even market-based approaches (Cortner and Moote, 1999; Mazmanian and Kraft, 2000) as ecologists have shifted away from rigid ecosystem boundaries and equilibrium models and focused on ever-larger geographic units and

---

* Rudel and Gerson (1999) summarize postmodernism as having five components: (1) a rejection of comprehensive explanations and grand theories; (2) social fluidity, in which ever-changing social conditions require flexibility and adaptability in people, who respond by changing themselves; (3) in rejecting universal truth, an emphasis on the local and particular; (4) polyvocality, or the existence of multiple groups with different concerns and voices; and (5) attention to the interpretation of signs and texts, in which meaning is contingent on social relations.

interactions across spatial scales that comprehend complex histories of human–ecosystem interaction (Vayda, 1996).

Perhaps there is wisdom in these postmodern insights of use in resolving the polemic described above. Lee (1993) notes that the expanding landscape scales on which conservation must operate have by their very nature complex ecological relationships intertwined with complex social relationships. He posits that human understanding of these systems will always be incomplete, suggesting a need for continual learning and experimentation. This invites an adaptive management approach, consisting of self-conscious learning-by-doing that integrates "existing interdisciplinary experience and scientific information into dynamic models that attempt to make predictions about the impacts of alternative policies" (Walters, 1997, p. 2). Adaptive management is distinguished from trial and error by its use of modeling to develop experimental management actions, and periodic reflection on the results of these actions to begin the process again. Walters (1997) sees modeling serving three functions: (1) problem clarification and enhanced communication among scientists, managers, and other stakeholders; (2) policy screening to eliminate options that are unlikely to do much good; and (3) identification of key knowledge gaps that make model predictions suspect.

Paradoxically, adaptive management requires ongoing use of science in management at the same time that science itself is increasingly subject to questions by postmodernists. Postmodernism and postpositivism have highlighted the role of values in problem definition and science, arguing that there are multiple rationalities and that the influence of these different rationalities is related to power (Redclift, 1987; Pretty, 1995; Roling and Jiggins, Chapter 8). Although science is seldom completely free of bias, politics, and power relationships, others have argued that science itself can perhaps best sort out its own shortcomings (Harris, 1999; McKay, 2000). That is, being more scientific, with an emphasis on replicable empirical, rather than more rhetorical results is a key element in sorting out competing claims and establishing the foundation for work toward mutually beneficial solutions by diverse interest groups. Thus, although it is clear that science alone is not the solution, it is equally clear that science — with an explicit recognition of the complex relationships among science, values, and power — must be a part of the solution.

Although science is important, operating in large landscapes and in complex human-occupied ecosystems necessarily brings in different stakeholders with changing values and frames of reference (Uphoff, 1992; Cortner and Moote, 1999). Conservation decision making is no longer the exclusive domain of bureaucrats and scientists. Citizen involvement and participation has become mainstream, posing challenging questions about who should and actually does decide resource policy among landowners, governments, and NGOs (Cortner and Moote, 1999). The result is what Gunderson et al. (1995) have termed "institutional panarchies." Unlike top-down, hierarchical networks in which the flows of knowledge and power are clear, panarchies are loose associations of actors, institutions, and processes. What is good for one community is not necessarily taken to be good for another or for society at large (Woodhill and Röling, 1998). Although tensions may exist between collaboration and leadership, flexibility and consistency, inclusiveness and accountability, expert and open decision making, conflict and collaboration, and centralization and decentralization, these must be negotiated and are not choices between two mutually exclusive options (Westley, 1995; Cortner and Moote, 1999).

The idea of panarchies is consistent with Uphoff's (1992) post-Newtonian social science, where the world is not regarded as a vast mechanism but rather as a hugely complex, chaotic, and contingent system. It also conforms well with the adaptive management notion of inadequate knowledge and unpredictability in understandings of the interactions between people and ecosystems (Gunderson et al., 1995; Holling, 1995). This means that the exact result of any management action or policy cannot be predicted with certitude. It is best approached, postmodernists and adaptive managers would agree, through a social learning approach that integrates science, participatory decision making, dynamic management, and policy in a thoughtful, self-reflective process.

Lee (1993), whose recent work appears in Chapter 1 of this volume, represented this merging of science and democracy with the metaphors of compass and gyroscope in his seminal book on natural resource management in the Columbia River Basin. The "compass" is adaptive management, in which science is brought to bear on management actions and policies viewed as experiments. In adaptive management, theory and observation are blended by designing planned interventions in human/ecological systems, carefully observing, and learning from the results and responses (Gunderson et al., 1995). But science can never resolve the questions of value and distribution that underlie the management of natural resources with claims on them from diverse interest groups. Lee proposes the metaphor of the "gyroscope" to represent the democratic political process. Noting that political competition is a messy process, Lee (1993, p. 10) suggests that competition with an inclusive spirit and bounded by written and unwritten rules can, like a spinning gyroscope, stabilize efforts.

ACM builds on the ideas of Lee and others. It proposes a strategy for conserving biodiversity that uses both science in an adaptive management framework and participatory decision making through a range of collaborative processes. Relevant methods and tools evolve continuously in various contexts. Some that may be especially useful for ACM in protected-area contexts include:

1. Participatory appraisal of forest dependence, local institutions, and governance systems;
2. Stakeholder analysis;
3. Collaborative problem solving, conflict management, and resource use negotiation;
4. Spatial analysis with Geographic Information System (GIS) and other mapping techniques;
5. Socioinstitutional boundary mapping;
6. Action research;
7. Decision-support modeling;
8. Conceptual modeling;
9. Population modeling;
10. Vegetation cover modeling;

11. Future scenario analysis;
12. Testing of criteria and indicators for sustainable forest management; and
13. Participatory monitoring and evaluation, and others.

Inventing, combining, and reshaping various tools to the needs and opportunities of specific protected-area situations present creative challenges for adaptive managers.

ORIGINATING QUESTIONS AND CHAPTER OVERVIEW

As the editors of this volume planned a symposium on ACM, various questions motivated their selection of authors and subjects to be emphasized: What are the state-of-the-art programs and initiatives for integrating biodiversity conservation with development goals in and around protected areas? What new perspectives are at hand for moving beyond the quarreling that occurs between conservation biologists and social reformers, both of whom usually share an interest in long-term conservation and sustainability for the planet? How can ACM models and methods be tailored most effectively to current and future challenges of protected-area management? What are good working examples of ACM theory, models, and methods? Within current systems of protected-area management, where are the best programmatic, institutional, and policy opportunities for ACM to develop in the future?

The chapters that comprise the volume have been selected and shaped from material that was shared at the symposium and thereafter edited by each author. They are grouped to correspond with themes that emerged from the foregoing questions. Five chapters comprise the opening section on foundations of ACM, offering insight into the logic of ACM thinking, programmatic initiatives in biodiversity conservation that are giving rise to this approach to management, and strategic opportunities and constraints that are likely to affect its success.

Kai Lee sets the stage with an appraisal of adaptive management as an experimental approach to implementing policies, which he maintains is important to the search for a new meaning for conservation that is bioregional in scope, collaborative in governance, and adaptive in managerial perspective. He concludes that adaptive management has been an influential idea, that it is best used once disputing parties have agreed on an agenda of questions to be answered using the adaptive approach, and that efficient, effective social learning of the kind facilitated by adaptive management is likely to be of strategic importance in governing ecosystems as humanity searches for a sustainable economy. McNeely, in a global overview of the expansion of national protected-area systems, argues the case for collaboration in management. He contends that modern management systems must be structured to represent broad segments of civil society, and portrays a vision of protected areas as engines for new forms of rural development that ensure a better life for all, provided that many challenging requirements are met.

Christiansen and Dinerstein present the rationale for current trends in conservation toward ecoregional, landscape-scale approaches, drawing upon a range of examples to illustrate the implications for management. Their work articulates the case, from a professional conservationist perspective, for management models that are both inclusive (collaborative) and iterative (adaptive). Sayer is concerned with facilitating the transition between conventional command-and-control models of protected-area management and more desirable modern systems that are genuinely collaborative and adaptive, stressing the types of organizational change that are required. He looks to innovations in management from the corporate sector, in addition to three decades of experience with ICDPs, to suggest that conditions are ripe for bringing about a new generation of conservation approaches that can be assisted by action research, experimentation, and learning. To complete this section, Fisher draws attention to the practical challenges of inducing collaborative models of management in which local people enjoy genuine decision making and benefit sharing. He warns that issues of trust rooted in conflicted historical relationships between state agencies and local people are pervasive, suggesting that an action-research approach to facilitating community management and to documenting and publicizing success cases can help to repair the situation.

The second group of chapters deepens insight into how ACM thinking and practice can affect the design and development of institutions and policies to improve integrated conservation and sustainable development initiatives. Geisler holds that the institution of land reform can be an important way of power sharing to address inequalities in property power — an essential requirement for successful conservation comanagement. He makes a case for a contingency-based, policy-as-experiment approach to land reform in protected-area settings that can customize greener land reforms to local circumstances. Cahoon explores how advances in bioproperty are altering traditional property institutions in wild biota within and beyond protected areas. He demonstrates in several case studies how defining and using biota property rules can play a defining role in multiparty arrangements for sustainable use of natural resources.

Rölings and Jiggins draw on the concept of soft systems to argue that sustaining the ecosystem functions of the Earth requires "shared sense making," negotiated agreement and accommodation, and deliberate concerted action among the stakeholders in the system. They claim that having become a force of nature, humans must collectively learn to guide their actions by an ecological rationality, and offer diverse theoretical underpinnings for reshaping their institutions in this direction. Anderson follows with insights into the nature of complex adaptive systems from a biological and a human activity perspective. His characterization of pluralistic, integrated (ecosystem, local social) complex systems lays groundwork for crafting correspondingly dynamic and robust learning systems to support ACM.

Herring draws attention to public authority in decision making about nature protection at appropriate scales of operation, maintaining that claims to managerial authority are rooted at the intersection of politics and knowledge, or legitimacy and science. He cautions against current obsessions with localism — arguing that local, state, and global authority are needed to gain acceptance by citizens on the ground as well as scientific validation of claims to manage expanded ecosystems. Devolution policy to support local land tenure security is the focus of Luna's piece, which highlights the challenges of implementing the legal, political, and cultural dimensions of innovative protected-area policy in a coordinated fashion. An ACM perspective on managing the change process can instill confidence by demonstrating the accumulated learning value of incremental (or adaptive) progress.

Section three of the volume illuminates how an ACM perspective influences the application of science to support the management of protected landscapes. This material suggests how science can improve the understanding of relationships between human activity and ecosystem function and to generate support for collective management decisions. White and his colleagues lead off, offering a simple analytic framework for protected-area system reform, which doubles as an outline of the conditions for effective execution of the public protected-area approach. They demonstrate how the complexities and uncertainties of the system can be treated by being attentive to learning from experience, which in turn needs to employ guidelines that all research should follow for objectiveness, reproducibility of results, and representativeness of samples used. Lessons should be transferred and adopted subject to a set of rules agreed upon by the collaborators in management. Schelhas poses the challenge of how to introduce state-of-the-art scientific knowledge into collaborative management processes for regional landscapes, emphasizing the necessity of bringing together an interdisciplinary understanding of both ecosystem and human processes that increasingly shape these landscapes. He demonstrates the value of conceptual modeling to capture new trends and ongoing processes, as a complement to formal modeling and advocates the use of a suite of partial models — formal quantitative, conceptual, and folk — to develop the hypotheses on which to base adaptive management and research.

Ericson and her co-authors characterize an applied research program that is designed to monitor and manage population in-migration around a biosphere reserve, illustrating the complementary roles of dialogue between stakeholders, a low-cost monitoring system, and a participatory land use planning process that takes account of complex human population dynamics. Information gathered primarily through qualitative methods reveals the complex mosaic of causes and environmental consequences of in-migration so that coordinated action can be taken to address the phenomena. Colfer and her collaborators examine the roles of social criteria and indicators (C&I) in ACM of forests, and elaborate an iterative C&I process of conceptualizing, field trials, evaluation, revisions, and more field trials. The authors reveal how CIFOR is applying social C&I to an emergent research program in ACM for sustainable forests with applications in protected-area management.

Salafsky and Margoluis share their "measures of success" approach to monitoring biodiversity conservation projects, highlighting how it applies to common constraints in effective monitoring. They stress the importance of using outputs of monitoring systems to adapt and learn, and of integrating them into project design and management by sharing them among stakeholders. Wollenberg and her coauthors, recognizing that the new adaptive management seeks to be responsive to local needs and to facilitate collaboration among multiple stakeholders, suggest that anticipating and exchanging perspectives about the future can be a source of learning equally as important as monitoring past actions. They show how anticipatory scenario techniques can be used as practical tools of ACM to prepare forest stakeholders to adapt to change.

Case studies comprise the final group of chapters, which examine various applications of ACM in practice. The cases highlight initiatives in ACM that aim to improve protected-area policies, institutions, knowledge, and management practice. They feature biodiversity-rich settings in countries that have adopted progressive social and/or environmental protection policies designed to expand local involvement in management. Effective social learning methods, which serve to build common understandings and trust-based relationships among diverse stakeholders, are illustrated by the cases.

Johnson and coauthors analyze the effect of a project monitoring system in Papua New Guinea designed to test hypotheses that selected socioeconomic development will result in biodiversity conservation. They explain how the conceptual model that underpinned the monitoring system was an essential tool for uniting an interdisciplinary, multinational project team in analysis and planning, and how joint analysis of the outputs refocused project activities over time. Guindon and his colleagues discuss how a reserve complex managed by several public and private organizations in Costa Rica exemplifies the challenges of integrating local, regional, national, and international visions of conserving and managing biological resources using ACM strategies. They illustrate how a rich research base of biological information can be used to influence land use in private farmlands to conserve biodiversity through processes of consultation, education, and joint planning.

Ford and McConnell characterize a five-phase planning process that illustrates the use of geomatic tools in a participatory conservation initiative in Madagascar. The tools were used to engage initially reluctant villagers in ICDP activities to reduce agricultural land-use pressures on adjacent protected areas. Their case demonstrates how applying a sequence of tools and methods can serve to build trust among stakeholders, and how continual fine-tuning and amending of action plans and agreements are needed to sustain them. Drawing from a different case in the same country, Cowles and his coauthors reflect on a collaborative ecoregion-based planning process that was pilot-tested to address, in a coordinated fashion, widespread and severe biodiverity loss from a variety of sources. Highlighting the value of a three-tiered network of public task forces, a neutral facilitator, and action-research in building legitimacy, trust, and relevant knowledge, the authors consider how ongoing planning and management can address deeply rooted social conflict in an area that biologists feel is critical to protect.

Finally, Guerrero and Pinto consider how marginalized indigenous communities can employ ancestral domain claims to influence the demarcation and management of environmentally critical areas in the Philippines. The amendment of these claims is seen as a variant of ACM. As such, they consider how ACM can be valuable in catalyzing community engagement, building equitable power-sharing arrangements across levels of government, and negotiating flexible intracommunity collaboration mechanisms while promoting recognition of indigenous land rights and meeting local livelihood needs.

In concluding, the editors reiterate that it is not their intention to present ACM as a new and comprehensive approach, nor are they able to provide clear guidelines or formulas on how it can be implemented. The problem of biodiversity conservation in a world where the human population is growing, employing new technologies, and changing its needs and values at ever more rapid rates represents one of the fundamental challenges facing human society in the 21st century. As such, it will be addressed by varied and dispersed thought and action in science and governance, not simply by conservation and development professionals. This group has a critical role to play, however, in the complex processes of bringing new science and methods of governance to bear on efforts to address the widely shared but sometimes conflicting goals of biodiversity conservation and improving human well-being.

The editors thank the contributors to this book for sharing their thoughts and experiences, for their role in the development of the idea of ACM that is presented here, and for their creativity and persistent hard work in bringing the chapters to fruition. The satisfaction and payoff for this work is the opportunity to contemplate the complex relationships between science and governance in conservation, with the hope of advancing the knowledge and techniques that underlie conservation theory and practice.

REFERENCES

Abramovitz, J. N., 1994. Trends in Biodiversity Investments: U.S.-Based Funding for Research and Conservation in Developing Countries, 1987–1991. World Resources Institute, Washington, D.C.

Alpert, P., 1996. Integrated conservation and development projects: examples from Africa, *BioScience,* 46(11), 845–855.

Anderson, T. L. and Leal, O. R., 1991. *Free Market Environmentalism,* Westview Press, Boulder, CO.

Barrett, C. B. and Arcese, P., 1995. Are Integrated Conservation-Development Projects (ICDPs) sustainable? On the conservation of large mammals in sub-Saharan Africa, *World Dev.,* 23(7), 1073–1074.

Baydeck, R. K., Campa III, H., and Haufler, J. B., 1999. *Practical Approaches to the Conservation of Biological Diversity,* Island Press, Washington, D.C.

Biodiversity Conservation Network (BCN), 1999. *Final Stories from the Field: Evaluating Linkages between Business, the Environment, and Local Communities,* Biodiversity Support Program, Washington, D.C.

Blackburn, T. C. and Anderson, K., 1993. *Before the Wilderness: Environmental Management by Native Californians,* Ballena Press, Menlo Park, CA.

Boyd, R., Ed., 1999. *Indians, Fire, and the Land in the Pacific Northwest,* Oregon State University Press, Corvallis.

Brandon, K., 1998. Perils to parks: the social context of threats, in *Parks in Peril: People, Politics, and Protected Areas,* Brandon, K., Redford, K.H., and Sanderson, S.E., The Nature Conservancy and Island Press, Washington, D.C., 415–440.

Brechin, S. R., Wilshusen, P. R., Fortwanler, C. L., and West, P. C., 2001. *Reinventing the Square Wheel: A Critique of the New Protectionist Paradigm in International Biodiversity Conservation,* SUNY Press, New York.

Bromley, D. W., 1992. *Making the Commons Work: Theory, Practice, and Policy,* Institute for Contemporary Studies Press, San Francisco.

Brookfield, H. and Padoch, C., 1994. Appreciating agrodiversity: a look at the dynamism and diversity of indigenous farming practices, *Environment,* 36(5), 271–289.

Brosious, J. P., Tsing, A. L., and Zerner, C., 1998. Representing communities: histories and politics of community-based natural resource management, *Soc. Nat. Resourc.*, 11, 157–168.

Bruce, J. W., 1999. Legal Bases for the Management of Forest Resources as Common Property, Community Forest Note 14, FAO, Rome, and Land Tenure Center, Madison.

Collins, W. W. and Qualset, C.O., Eds.,1998. *Biodiversity in Agroecosystems,* CRC Press, Boca Raton, FL.

Cortner, H. J. and Moote, M. A., 1999. *The Politics of Ecosystem Management,* Island Press, Washington, D.C.

Dinerstein, E., 1995. *A Conservation Assessment of the Terrestrial Ecoregions of Latin America and the Caribbean,* World Bank and World Wildlife Fund, Washington, D.C.

Fairfax, S. K. and Fortmann, L., 1990. American forestry professionalism in the third world: some preliminary observations, *Popul. Environ.*, 11(4), 259–272.

Fairfax, S. K., Fortmann, L., Hawkins, A., Huntsinger, L., Peluso, N. L., and Wolf, S. A., 1998. The federal forests are not what they seem: formal and informal claims to federal lands, *Ecol. Law Q.*, 25(4), 630–646.

Freeman, M. R. and Kreuter, U. P., 1994. *Elephants and Whales: Resources for Whom?* Gordon and Breach Science Publishers, Basel, Switzerland.

Freese, C. H., Ed., 1997. *Harvesting Wild Species: Implications for Biodiversity Conservation,* Johns Hopkins University Press, Baltimore, MD.

Freese, C. H., 1998. *Wild Species as Commodities: Managing Markets and Ecosystems for Sustainability,* Island Press, Washington, D.C.

Grumbine, R. E., 1997. Reflections on "What is Ecosystem Management?" *Conserv. Biol.*, 11(1), 41–47.

Gunderson, L. H., Holling, C. S., and Light, S. S., 1995. Barriers broken and bridges built: a synthesis, in *Barriers and Bridges to the Renewal of Ecosystems and Institutions,* Gunderson, L. H., Holling, C. S., and Light, S. S., Eds., Columbia University Press, New York, 489–532.

Haney, A. and Power, R. L., 1996. Adaptive management for sound ecosystem management, *Environ. Manage.*, 20(6), 879–886.

Harris, M., 1999. *Theories of Culture in Postmodern Times,* AltaMira Press, Walnut Creek, CA.

Heywood, V. H., Ed., 1995. *Global Biodiversity Assessment,* Cambridge University Press, Cambridge.

Holling, C. S., 1978. *Adaptive Environmental Assessment and Management,* John Wiley, London.

Holling, C. S., 1995. What barriers? What bridges? in *Barriers and Bridges to the Renewal of Ecosystems and Institutions*, Gunderson, L. H., Holling, C. S., and Light, S. S., Eds., Columbia University Press, New York, 3–34.

Korten, D. C., 1980. Community organization and rural development: a learning process approach, *Publ. Adm. Rev.*, 40(5), 480–511.

Kramer, R. A. and van Schaik, C. P., 1997. Preservation paradigms and tropical rainforests, in *Last Stand: Protected Areas and the Defense of Tropical Biodiversity,* Kramer, R., van Schaik, C., and Johnson, J., Eds., Oxford University Press, New York, 3–14.

Langholz, J., 1999. Exploring the effects of alternative income opportunities on rainforest use: insights from Guatemala's Maya Biosphere Reserve, *Soc. Nat. Resourc.*, 12, 139–149.

Larson, P., Freudenberger, M., and Wyckoff-Baird, B., 1997. *Lessons from the Field: A Review of World Wildlife Fund's Experience with Integrated Conservation and Development Projects,* 1985–1996, World Wildlife Fund, Washington, D.C.

Lee, K. N., 1993. *Compass and Gyroscope: Integrating Science and Politics for the Environment,* Island Press, Washington, D.C.

Li, T. M., 1996. Images of community: discourse and strategy in property relations, *Dev. Change,* 27, 501–527.

Little, P. D., 1994. The link between local participation and improved conservation: a review of issues and experiences, in *Natural Connections: Perspectives in Community-based Conservation*, Western, D. and Wright, R. M., Eds., Island Press, Washington, D.C., 347–372.

Lubchenco, J., 1998. Entering the century of the environment: a new social contract for science, *Science,* 279 (January), 491–497.

Ludwig, D., Hilborn, R., and Walters, C. J., 1993. Uncertainty, resource exploitation, and conservation: lessons from history, *Science,* 260, 17, 36.

Mazmanian, D. A. and Kraft, M., Eds., 2000. *Toward Sustainable Communities: Transition and Transformations in Environmental Policy,* MIT Press, Cambridge, MA.

McCay, B. J., 2000. Post-modernism and the management of natural and common resources, *Common Property Resourc. Dig.*, 54, 1–8.

Michon, G. and de Foresta, H., 1992. Complex agroforestry systems and the conservation of biological diversity, in *Proceedings of the International Conference on Tropical Biodiversity, "In Harmony with Nature,"* Malayan Nature Society, Kuala Lumpur, Malaysia.

Oates, J. F., 1999. *Myth and Reality in the Rain Forest: How Conservation Strategies Are Failing in West Africa,* University of California Press, Berkeley.

Orlove, B. S. and Brush, S. B., 1996. Anthropology and the conservation of biodiversity, *Annu. Rev. Anthropol.*, 25, 329-352.

Orr, D. W., 1992. *Ecological Literacy,* SUNY Press, Albany, NY.

Parsons, E. A. and Clark, W. C., 1995. Sustainable development as social learning: theoretical perspectives and practical challenges for the design of a research program, in *Barriers and Bridges to the Renewal of Ecosystems and Institutions*, Gunderson, L. H., Holling, C. S., and Light, S. S., Columbia University Press, New York, 428–460.

Peters, J. 1998. Transforming the Integrated Conservation and Development Project (ICDP) approach: observations from the Ranomafana National Park Project, Madagascar, *J. Agric. Environ. Ethics*, 11, 17–47.

Posey, D. A. and Balée, W., Eds., 1989. *Resource Management in Amazonia: Indigenous and Folk Strategies,* New York Botanical Garden, Bronx, NY.

Pretty, J. N., 1995. *Regenerating Agriculture: Policies and Practice for Sustainability and Self-Reliance,* Joseph Henry Press, Washington, D.C.

Redclift, M., 1987. *Sustainable Development: Exploring the Contradictions,* Metheun, London.

Redford, K. H., 1992. *The empty forest, BioScience,* 42(6), 412–422.

Redford, K. H. and Richter, B. D., 1999. Conservation of biodiversity in a world of use, *Conserv. Biol.,* 13(6), 1246–1256.

Redford, K. H. and Sanderson, S. E., 2000. Extracting humans from nature, *Conserv. Biol.,* 14(5), 1362–1365.

Repetto, R. and Gillis, M., 1988. *Public Policies and the Misuse of Forest Resources,* Cambridge University Press, Cambridge.

Robinson, J. G., 1993. The limits for caring: sustainable living and the loss of biodiversity, *Conserv. Biol.,* 7(1), 20–28.

Rudel, T. K. and Gerson, J. M., 1999. Postmodernism, institutional change, and academic workers: a sociology of knowledge, *Soc. Sci. Q.,* 80(2), 213–228.

Sanderson, S. E. and Redford, K. H., 1997. Biodiversity politics and the contest for ownership of the world's biota, in *Last Stand: Protected Areas and the Defense of Tropical Biodiversity,* Kramer, R., van Schaik, C., and Johnson, J., Eds., Oxford University Press, New York, 115–132.

Schelhas, J. and Greenberg, R., 1996. The value of forest patches, in *Forest Patches in Tropical Landscapes,* Schelhas, J. and Greenberg, R., Eds., Island Press, Washington, D.C.

Schwartzman, S., Moriera, A., and Nepstad. D., 2000. Rethinking tropical forest conservation: perils in parks, *Conserv. Biol.,* 14(5), 1351–1357.

Smythe, K., Bernabo, J. C., Carter, T. B., and Jutro, P. R., 1996. Focusing biodiversity research on the needs of decision makers, *Environ. Manage.,* 20(6), 865–872.

Soulé, M. E., 1995. The social siege of nature, in *Reinventing Nature? Responses to Postmodern Construction,* Soulé, M. E. and Lease, G., Eds., Island Press, Washington, D.C., 137–179.

Soulé, M. E. and Terborgh, J., 1999. *Continental Conservation: Scientific Foundations of Regional Reserve Networks,* Island Press, Washington, D.C.

Southgate, D., 1998. *Tropical Forest Conservation: A Critique of the Alternatives in Latin America,* Oxford University Press, New York.

Stacey, R., 1993. Strategy as order emerging from chaos, *Long Range Plann.,* 26(1), 10–17.

Terborgh, J., 1999, *Requiem for Nature,* Island Press, Washington, D.C.

Terborgh, J., 2000. The fate of tropical forests: a matter of stewardship, *Conserv. Biol.,* 14(5), 1358–1361.

United Nations Development Program/United Nations Environment Program/World Bank/World Resources Institute (UNDP/UNEP/WB/WRI), 2000. *World Resources Report 2000–2001: People and Ecosystems: The Fraying Web of Life,* Elsevier Science, New York.

Uphoff, N. T., 1992. *Learning from Gal Oya: Possibilities for Participatory Development and Post-Newtonian Social Science,* Cornell University Press, Ithaca, NY.

Vayda, A. P., 1996. *Methods and Explanations in the Study of Human Actions and Their Environmental Effects,* CIFOR and WWF, Jakarta, Indonesia.

Walters, C., 1986. *Adaptive Management of Renewable Resources,* Macmillan, New York.

Walters, C., 1997. Challenges in adaptive management of riparian and coastal ecosystems, *Conserv. Ecol.* [online], 1(2), 1–22.

Watson, V., Cervantes, S., Castro, C., Mora, L., Solis, M., Porras, I., and Cornejo, B., 1998. *Making Space for Better Forestry,* Policy that Works for Forests and People Series 6, Centro Cientifico Tropical and International Institute for Environment and Development, San Jose, Costa Rica and London, England.

Wells, M. and Brandon, K., 1992. People and Parks: Linking Protected Areas Management with Local Communities, The World Bank/The World Wildlife Fund/U.S. Agency for International Development, Washington, D.C.

Western, D. and Wright, R. M., 1994. The background of community-based conservation, in *Natural Connections: Perspectives in Community-Based Conservation*, Western, D. and Wright, R. M., Eds., Island Press, Washington, D.C., 1–12.

Westley, F., 1995. Governing design: the management of social systems and ecosystems management, in *Barriers and Bridges to the Renewal of Ecosystems and Institutions*, Gunderson, L. H., Holling, C. S., and Light, S. S., Eds., Columbia University Press, New York, 391–427.

Woodhill, J. and Röling, N. G., 1998. The second wing of the eagle: the human dimensions in learning our way to more sustainable futures, in *Facilitating Sustainable Agriculture: Participatory Learning and Adaptive Management in Times of Uncertainty*, Röling, N. G. and Wagemakers, M. A. E., Eds., Cambridge University Press, Cambridge, 46–71.

Section I

Foundations of Adaptive Collaborative Management

1 Appraising Adaptive Management

Kai N. Lee

CONTENTS

INTRODUCTION

Adaptive management (Holling, 1978; Walters, 1986) — implementing policies as experiments — is a methodological innovation in resource management. As with any method, the adaptive approach implies revised ends as well as novel means: as its name implies, adaptive management promotes learning to high priority in stewardship. This chapter considers the difficulties of realizing the promise of adaptive management in natural resource management and biodiversity conservation. I write as a social scientist and erstwhile decision maker who sought to use adaptive management; I am an outsider to the technical practice, and my observations are meant to complement those of Walters and Holling (1990) by emphasizing the organizational and human dimensions of learning while doing. The questions proposed at the end of this chapter invite critique from insiders as well as nongovernmental organizations (NGOs), managers, and others for whom the uncertainties of the natural world imply opportunity as well as concern.

The adaptive approach is an important component of a search for a new meaning for conservation — a meaning that is *bioregional* in scope and *collaborative* in governance, as well as adaptive in managerial perspective. Conservation of this kind is emerging from two forces: the realization that highly valued ecological processes

0-8493-0020-7/01/$0.00+$1.50

and species can only be preserved in large ecosystems and (2) the recognition that many ecosystems high in biodiversity are and will continue to be inhabited by humans. These factors inform a redefinition of conservation in a way that points toward an ambitious goal: reconciling conservation biology with sustainable development — that is, bringing together two of the principal themes of environmentalism. I return to that grand aspiration below.

APPRAISAL

Adaptive management, like other policy innovations, can be appraised using a framework devised by Garry Brewer (1973). Brewer proposed that appraisal be done by considering four dimensions of a policy design:

1. Conceptual soundness: Is the idea sensible?
2. Technical: Is the idea translated into practice well?
3. Ethical: Who loses and who wins?
4. Pragmatic: Does it work?

Appraisal examines questions that are obvious — although not so obvious that they are considered automatically or even often.

Conceptual Soundness — Learning by Experimenting

Adaptive management has so far been much more influential as an idea than as a way of performing conservation. Given that influence, consider theory first: Why should one perform adaptive management at all? (see Holling, 1978).

Adaptive management is grounded in the admission that humans do not know enough to manage ecosystems. Managing is different from exploiting, which requires knowledge of how to capture or harvest. Harvest is a formidable task, but it is not management; managing is closer to cultivation or agriculture. Yet cultivating an ecosystem to foster its wild state is paradoxical. This paradox has been resolved by turning around the objective: to think of ecosystem management as managing the *people* who interact with the ecosystem. This focus for management raises questions to which there are few reliable answers, but they can be explored, among other ways, by experimentation.

Adaptive management, from this perspective, formulates management policies as experiments that probe the responses of ecosystems as people's behavior in them changes. (This experimental emphasis is called "active" adaptive management in Walters and Holling, 1990.) In conducting these experiments the aim is to learn something about the processes and structures of the ecosystem, and one seeks both to design better policies and to contrive better experiments. Note that the goal is to learn *something*. Experiments can surprise the experimenter, and one mark of a good scientist is that he or she recognizes surprise and pursues its implications. This has not been considered the mark of a good manager, however, who is rewarded instead for steadfast pursuit of objectives.

Experimentation is not the only way to learn; indeed, an adaptive approach is often not the obvious way, as shown in Table 1.1 (also see Marcot, 1998).

Many public policies are grounded in anecdotal knowledge, especially those enacted by legislatures, referenda, and general-purpose governments. From this perspective, trial and error is an unusually systematic way to learn. In that light, one can reasonably ask how much of the scientific rigor of the laboratory is attainable in the field setting of adaptive management.

The adaptive approach rests on a judgment that a scientific way of asking questions produces reliable answers at lowest cost and most rapidly; this may not be the case very often. As Walters has emphasized, adaptive management is likely to be costly and slow in many situations (e.g., Walters et al., 1993), so those involved in stewardship need to think through whether the scientific approach is worthwhile in specific cases. In particular, it is important to spell out how much difference in management might result if adaptive learning proceeds as envisioned (Walters and Green, 1997).

A research-based approach can be reasonable when one takes into consideration the complexity and subtlety of natural systems, including those that have been driven far from their undisturbed state by human utilization. The complexity suggests that even simple steps may yield surprising outcomes — and science is an efficient way of recognizing and diagnosing surprise. In principle, the scientific approach leads to reliable determination of causes; in practice, that means being able to learn over time how management does and does not affect outcomes. The complexity of the ecosystems and human behavior in the situations discussed here implies, however, that causal understanding is likely to emerge slowly, perhaps more slowly than the long struggle to understand the causal mechanisms of economic policy (Hall, 1989; Stein, 1996). The slow emergence of an economic policy paradigm reflects the fact that, to be effective, learning must become *social* — knowledge that informs public policy and collective choice (Heclo, 1974; Parson and Clark, 1995).

Reliable knowledge of natural systems used by humans is essential if a sustainable economy is to be achieved. An experimental approach may be costly and onerous in the near term, but it is probably the only way to root out *superstitious* learning — erroneous connections between cause and effect. As Walters has stressed, management of natural systems takes place against a dynamic background, and it is usually impossible to sort out the effects of management from those of concurrent changes in the natural environment (e.g., Walters and Holling, 1990). The field is profoundly different from a laboratory in this respect. Yet the designs needed to distinguish treatment from background tend to be more costly and slower than nonexperimental analyses of the past (what Walters called "passive" adaptive management).

So it is important to remember the value of explicit experimentation, which also addresses two other social misdirections of learning. The first is regression to the mean. Most environmental and resource problems come to notice in extreme situations, such as the decline of a commercial fishery. Yet, in a dynamic, mutable world, extreme situations are usually followed by less-extreme ones. There is regression to the mean, not because something has been fixed but simply because the mix of fluctuating causal factors has changed. This is fertile ground for erroneous inferences.

TABLE 1.1
Modes of Learning

Each Mode of Learning	Makes Observations...	And Combines Them...	To Inform Activities...	That Accumulate into Usable Knowledge	Example
Laboratory experimentation	Controlled observation to infer cause	Replicated to assure reliable knowledge	Enabling prediction, design, control	**Theory** (it works, but range of applicability may be narrow)	Molecular biology and biotechnology
Adaptive management (quasi-experiments in the field)	Systematic monitoring to detect surprise	Integrated assessment to build system knowledge	Informing model building to structure debate	**Strong inference** (but learning may not produce timely prediction or control)	Green Revolution agriculture
Trial and error	Problem-oriented observation	Extended to analogous instances	To solve or mitigate particular problems	**Empirical knowledge** (it works but may be inconsistent and surprising)	Learning by doing in mass production
Unmonitored experience	Casual observation	Applied anecdotally	To identify plausible solutions to intractable problems	**Models of reality** (test is political, not practical, feasibility)	Most statutory policies

Notes: Political conflict tends to weigh more heavily than scientific debate as one moves downward in the table. “Scientific uncertainty can be high so long as acceptability is high” (Walters and Holling, 1990, p. 2067).

Environmental *policy* has been formulated in response to unmonitored experience (e.g., disappearance of valued species), but environmental *management* assumes one of the other modes of learning is possible (e.g., maximum sustainable yield).

Second, as Levitt and March (1988, p. 326) pointed out, superstitious learning is also enhanced when "evaluations of success are insensitive to the actions taken." The mechanism is related to regression to the mean. In competitive situations, for example, one contestant may be slightly ahead of the others for reasons that are not within the competitors' control. Yet, as every athlete knows, contestants who are striving mightily are convinced of their own explanations for success or failure. Many of these do not stand up to scientific scrutiny. The more that resource managers are held to standards that have no grounding in ecological science, the more likely it is that accountability itself will induce superstitious learning. The rigors of experimentation provide a cure, but it is usually not an inexpensive one.

Experimentation has three components: (1) a clear hypothesis, (2) a way of controlling factors that are (thought to be) extraneous to the hypothesis, and (3) opportunities to replicate the experiment to check its reliability. These guide the selection of treatments applied to test hypotheses, as well as the selection of techniques that define what is being controlled and which measurements are being replicated. Hypothesis, controls, and replicates are all important to reliable knowledge, but none is easily achieved in conservation practice.

Adaptive management is learning while doing. Adaptive management does not postpone action until "enough" is known, but acknowledges that time and resources are too short to defer *some* action, particularly actions to address urgent problems such as human poverty and declines in the abundance of valued biota. Adaptive management emphasizes, moreover, that ignorance of ecosystems is uneven. Management policies should accordingly be chosen in light of the assumptions they test, so that the most important uncertainties are tested rigorously and early. This, too, is a criterion that managers have not valued. Management responds to problems and opportunities, and that is different from an experimental scientist's desire to explore a phenomenon systematically. Accordingly, there is no reason to think that adaptive management will work smoothly, that it will be easy to coordinate.

In theory, adaptive management recapitulates the promise that Francis Bacon articulated four centuries ago: To control nature one must understand her. Only now, what one wishes to control is not the natural world but a mixed system in which humans play a large, sometimes dominant role. Adaptive management is therefore experimentation that affects social arrangements and how people live their lives. The conflict encountered in doing that is discussed below; that conflict is a central reason that adaptive management has had more influence as an idea than as a way of performing conservation.

Technical: Cost of Information

The essence of managing adaptively is having an explicit vision or model of the ecosystem one is trying to guide (Walters, 1986). That explicit vision provides a baseline for defining surprise. Without surprise, learning does not expand the boundaries of understanding.

The technology of the geographic information system (GIS) now provides a ready template for assembling models. Into a GIS one can import physiographic and topographic databases, natural history observations, scientific measurements,

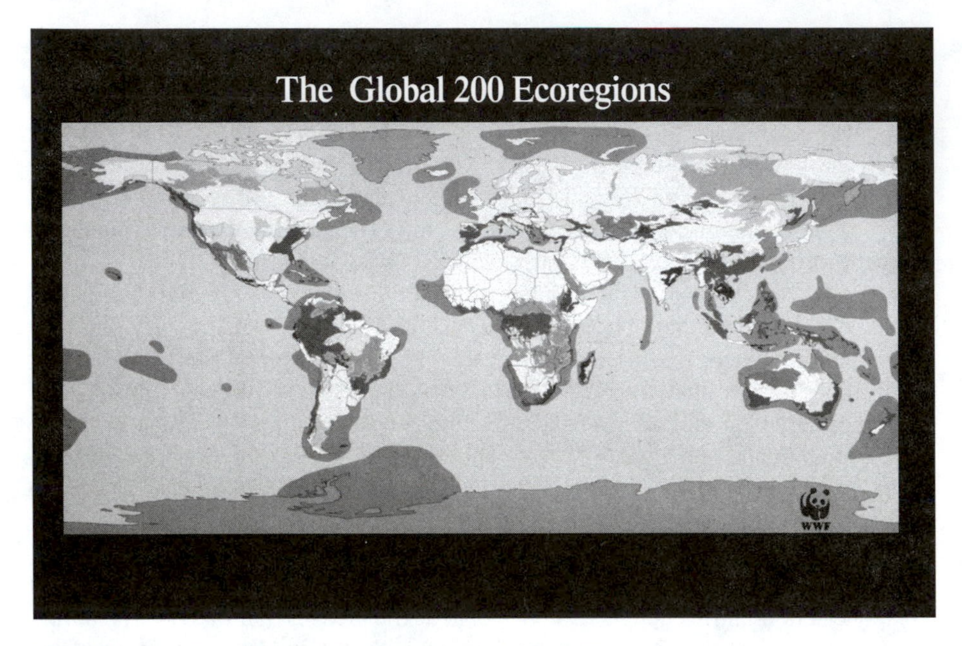

FIGURE 1.1 The Global 200 ecoregions. © 1998 WWF.

and social and economic data. Assembling information, attaching spatial coordinates and dates to the data, and preparing maps is a rapid, powerful way to create a shared view of the landscape. Figure 1.1 brings together a wide-ranging literature review by Dinerstein and colleagues at the World Wildlife Fund (Olson and Dinerstein, 1998). Its identification of 200 regions of key importance to preserving representative species and ecosystem processes has catalyzed an international commitment to conservation at large spatial scales.

Three cautions, at least, are important. First, any map emphasizes static structure rather than dynamic processes, although one may, of course, use a sequence of maps to show such changes as shrinking forest area. But a map is so engaging that one can easily forget that it is by itself static. Ecosystems are dynamic: what matters to learning is whether one can see policy-induced changes in the behavior of the ecosystem. For this, it is essential to ask whether the manager-experimenter *expects* to see a measurable difference due to manipulation of the experimental variables (Walters and Green, 1997). This is a question for which a model is needed even to speculate sensibly; a map is not a model, in this important respect. A second caution is that relatively little significant information is now available in georeferenced form. As a result, measurements made at particular locations are often attributed to much wider areas. Maps, like statistics, can easily mislead — indeed, any good map should make a clear point, which means that the mapmaker is deliberately leaving out a lot. Third and most important, having an explicit vision of an ecosystem does *not* mean having a complete or detailed or even correct baseline suite of data. Adaptive management is about urgency, acting without knowing enough, and learning. One can be surprised by one's own ignorance — and one can learn from it. The focus should be on learning, not on getting ready to learn. A map is not an end in itself but a means.*

There is a broader theme here. Information is expensive. Scientists know this well, since they work hard for each data point. But scientifically trained professionals, who know something about statistical significance and error, have been slow to face a pressing problem in adaptive management: how to get information cheaply and with as few organizational and procedural hassles as possible. Unfortunately, would-be adaptive managers have often jumped too quickly to thinking of information gathering as monitoring. That *is* what an adaptive approach leads to, but it should emerge from a skeptical appraisal of what kinds of information one can afford to collect (Rogers, 1998). The model of rapid assessment in conservation biology is worth remembering; the value of information needs to be balanced against the human and environmental values one is seeking to protect. Action guided by imperfect information is often — although not invariably — better than action guided by no information at all.**

The issue is the cost-effective testing of hypotheses. Adaptive management is not laboratory science, where the burden of proof is tilted toward highly reliable findings by rules such as $p < 0.05$, the notion that one's inferences should be reliable 95% of the time. In public policy and the world of action, the usual test is "more likely than not" — that is, $p < 0.5$.*** The findings that emerge from such roughshod hypothesis testing will not be as reliable as academic science (see Walters and Green, 1997). But that is the point: adaptive management is likely to be worthwhile when laboratory-style precision seems infeasible but trial and error seems too risky. And that is much of the time in conservation.

ETHICAL: BEAR IN MIND AMBIGUITY

Adaptive management is an unorthodox approach for people who think of management in terms of command. Learning is information intensive and requires active participation from those most likely to be affected by the policies being implemented (see Margoluis and Salafsky, 1998). Those who operate the human infrastructures of harvest — farmers, ranchers, dam operators, loggers, fishers — are usually those who know most, in a day-to-day sense, about the condition of the ecosystem. Their reports constitute much of the information that can be obtained at reasonable cost. Harvesters also see themselves as stewards of the resources upon which they rely, a claim that frequently turns out to be well founded (McCay and Acheson, 1987; Ostrom, 1990; Getz et al., 1999).

---

* To be sure, a good baseline is desirable — so much so, Walters (1997) suggests, that some resource managers are using models to *infer* parameters, as a substitute for actual measurements in the field. This is, of course, logically incorrect, compounding the errors in the information that went into the model in the first place.

** The author does not, of course, suggest replacing monitoring with rapid assessment, a family of methods aimed at taking an approximate inventory of the biodiversity of a place.

*** There is also an important difference between methods that control for Type I errors, the kind discussed here, and the problem of Type II errors, which is often more germane to adaptive management of ecosystems; see Anderson, 1998.

Those who would preserve species and ecosystems propose to alter the behavior of just these user stewards. Change is normally resisted. Moreover, when conservation is the objective of management, environmental decline has become apparent, and those who have been users, owners, or governors of an ecosystem are already under fire from critics. Under these conditions, it makes sense to have low expectations. First, adaptive management will often be resisted or sabotaged. Second, when adaptive management works, it will usually be the tool of those who want to affect how the human inhabitants of an ecosystem earn their living; moreover, inhabitants of protected areas in the developing world are often vulnerable and poorly represented in official deliberations. Win or lose, there is ethical ambiguity.

It is tempting to ignore this ambiguity but important not to do so because it usually emerges as conflict. Conflict is an essential element of governance. Differences are inevitable, and an orderly approach to resolving those differences is essential. Over the time periods needed to establish conservation practices, conflict and turbulence must be expected and should be welcomed. But conflict needs to be bounded — disputes should be conducted within the boundaries of a social process that the disputing parties perceive as legitimate. Unbounded conflict can tear apart the social fabric, thwarting learning. The difficulty is that conflict is a situation in which control of the rules of engagement are themselves contested. When the conflict is between sovereign powers, such as national governments, there is no superior authority able to impose a bounded process. In practice, even parties with little power can delay the resolution of conflicts long enough to frustrate experimentation and learning.

A surprising aspect of the ethical ambiguity is that environmentalists have often been unwilling to confess the ignorance upon which adaptive management is founded. Armed with legal mandates such as the Endangered Species Act in the United States, environmentalists often act to force reluctant authorities to obey their own laws. This is essential when environmental activists seek recognition as legitimate stakeholders. Forcing action demonstrates power. Yet when environmentalists exercise power, they often do so by denying that the natural world is uncertain. This forestalls the learning that will be needed if a sustainable policy is to be devised.

Another ethical challenge of adaptive management lies in the fact that knowledge is a public good. Once discovered, knowledge can be transferred at much lower cost than was necessary to make the initial discovery. For this reason, an agency or property-owner implementing adaptive management faces a situation of increasing but not well-controlled transparency. What is learned from the adaptive process reveals not only the way the ecosystem responds but also what the managers are doing, whether it works, and whose interests it serves. Because information needs to be gathered, usually from a variety of sources in the ecosystem, it is difficult to keep what is being learned from diffusing outward via the communication channels through which data are collected.

Undertaking an experimental approach presents the manager with two faces of learning. There are benefits from increasing understanding of the social and natural interactions — the usual justification for an adaptive approach. But there are costs, which Walters has pioneered in estimating (Walters and Green, 1997). Perhaps more important than quantifiable costs, there are risks of disclosure of activities that look inappropriate in the eyes of one or more stakeholders. The balance between the uncertain future benefits of increased understanding and the risks of inconvenient disclosure, which may come soon or by surprise, is necessarily subjective. This means that the judgments applied in initiating and sustaining adaptive management are volatile. Moreover, the scope of the cooperation needed to gather information for adaptive management means that many besides the official manager or owner need to maintain a commitment to the learning process, each weighing anticipated benefits against costs and risks. It is likely in such a setting that some members of the coalition will waver or resist participating.

These frailties underscore the importance of leadership in making adaptive management work. The obvious leader is the manager, since the manager usually controls the flow of benefits of harvest or protection of the ecosystem, a key role in motivating those whose cooperation is essential for information gathering, analysis, and diagnosis of surprises. When the manager is a public official, the balance between benefits and risks of learning is likely to be measured in political metrics. Thus, using the term *adaptive management* as a buzzword — when what is happening is much less likely to lead to disruptive disclosures than truly active adaptive learning — is a temptation hard to resist.

In Table 1.1 the Green Revolution is used as an example of successful learning by focused experimentation. It is useful to bear in mind that the benefits of increased rice, corn, and wheat harvest were visible in a short time, and that the sponsors of the learning initially were private donors who were removed from needing to please diverse stakeholders. Moreover, the controversies that came to beset the Green Revolution arose well after the process of breeding new varieties was well established as an economic strategy. For all these reasons, the Green Revolution is both a clear model and a difficult one to follow in the realm of natural resource management and conservation.*

PRAGMATIC: MAKING A DIFFERENCE

The pragmatic question is simple — does adaptive management work? The answer is not yet known. It is not known for two reasons. First, the battle for control of ecosystems is not decided in a lot of places. So the learning process that adaptive management organizes is subverted or ignored if it is even attempted. Second, the timescales for ecosystem response are typically long, and it is too early to know how or even if changes in human management policies have made an unambiguous difference. Most natural indicators yield one data point a year; even a simple trend takes patience in a world with a 24-hour news cycle, quarterly profit reports, and congressional elections every other year.

* Bruton (1997, p. 142) stimulated the observations above with a remark that the Green Revolution is also an unusual example of direct transfer from science to agrarian practice. His comment is also a caution for adaptive management: the discoveries that result from learning may not be readily applied by all or even any of the stakeholders.

The high-water mark, so far, in adaptive management practice appears to be a careful series of management experiments conducted in groundfish fisheries by Keith Sainsbury of the Australian CSIRO in Tasmania. Beginning in 1988, Sainsbury designed an adaptive management regime for a declining groundfish fishery off northwest Australia (Peterman and Peters, 1998). Using a decision analysis framework to organize hypotheses and available information, Sainsbury analyzed the value of additional information to be gathered by an experimental program. He showed in that situation that the expected value of catch could be quadrupled by carrying out one set of management experiments. This was done, with results indicating that one of the four hypotheses had strong support. Fisheries regulation has now been altered in response to these findings. The adaptive learning program took about a decade to yield practical results in fisheries management.

In the United States, adaptive management was initially adopted in 1984 by the Northwest Power Planning Council, as a way of organizing the activities of the council to protect and enhance Pacific salmon in the Columbia River basin (Lee, 1993, chap. 2). Those efforts were diverted in 1990 by litigation under the Endangered Species Act, so that the experimental phase of the Columbia basin program did not get very far (Volkman and McConnaha, 1993; National Research Council, 1996).

Adaptive management has been implemented in several other settings (see Gunderson et al., 1995; Walters and Green, 1997). Three recent instances are noteworthy. First, the U.S. Forest Service has attempted to forge a consensus management plan for its Pacific coastal forests in California, Oregon, and Washington (FEMAT, 1993). These have included the definition of Adaptive Management Areas "for land managers, researchers and communities [to] work together to explore new methods of doing business" (Olympic National Forest, 1998). The Forest Service definition of adaptive management does not emphasize experimentation but rather rational planning coupled with trial-and-error learning. Here "adaptive" management has become a buzzword, a fashionable label that means less than it seems to promise.

Second, the Plum Creek Timber Company (1998), a major landowner in Washington State, adopted a habitat conservation plan for its Cascade region lands in 1996, enabling harvest in a landscape where endangered species are found. Plum Creek has made specific commitments to experimental methods in the way it will carry out the conservation plan (see also Plum Creek, 1999). Also in 1996, the U.S. Department of the Interior sought to rebuild riparian habitat in the Grand Canyon by deliberately releasing large quantities of water from Glen Canyon Dam (Glen Canyon Environmental Studies, 1996; Grand Canyon Monitoring and Research Center, 1998; see also Barinaga, 1996). This spring flood was accompanied by a substantial monitoring effort, and it has been followed by research studies now being reviewed by a committee of the National Academy of Sciences.

None of these efforts has been as systematic as Sainsbury's decision analysis in Australia. That is probably appropriate to the state of the art: decision analysis assumes there is a single decision maker, with a rationally structured set of preferences. In an economically significant fishery, it makes sense to measure preferences via the value of catch as Sainsbury did. But when conventionally measured economic

TABLE 1.2
A Schematic Appraisal of Adaptive Management

Adaptive Management

Treats management policies as experiments that probe the *responses of ecosystems* as human behavior changes

- *Conceptual soundness*
  Hypotheses, controls, replication
  Is the idea sensible? Learning is valuable, but learning is *always* a precarious value compared with action.
- *Technical*
  Models, cost-effective monitoring, value of learning in making decisions
  Is the idea translated into practice well? Too little attention is paid to the cost and delays of gathering information.
- *Equity*
  People live in and use ecosystems
  Who loses? When conflict pushes aside learning, the ecosystem and resource-dependent communities decline while struggle continues.
- *Pragmatic*
  Does it work? Not yet known.

value conflicts with environmental values, as in the U.S. cases, the fundamental premise that a noncontroversial optimal decision can be identified is not plausible.*

In sum (Table 1.2), adaptive management is an idea highly attractive to the scientifically sophisticated, who understand how little is really known about the behavior of modified ecosystems that continue to be used by humans. Its requirements for patient record keeping and clear-headed assessment turn out to be difficult to muster where there is conflict — that is, in all the important cases. This practical reality has not seemed to dim the luster of the idea itself (e.g., National Research Council, 1996). The author is unsure whether to worry or to take comfort in that impression, but the uneven success of adaptive management is one indication of how far from realization is the "new social contract" (Lubchenco, 1998) that thoughtful leaders have urged upon the scientific community and the society it serves.

CONSERVATION IN A NEW KEY

Adaptive management and learning also play a strategic role in the emergent question of conservation at the ecosystem scale. There is growing agreement that biologically effective preservation of species, habitats, and ecological processes requires working on large spatial scales (see Figure 1.1; see also Wilson, 1993, chap. 11; Olson and Dinerstein, 1998). Such ecoregions are so large that many of those landscapes will

* Walters and Green (1997) propose a way to estimate the economic value of nonmarket attributes. Although their method is less cumbersome than the elicitation methods developed for decision analysis, its practical utility remains to be shown.

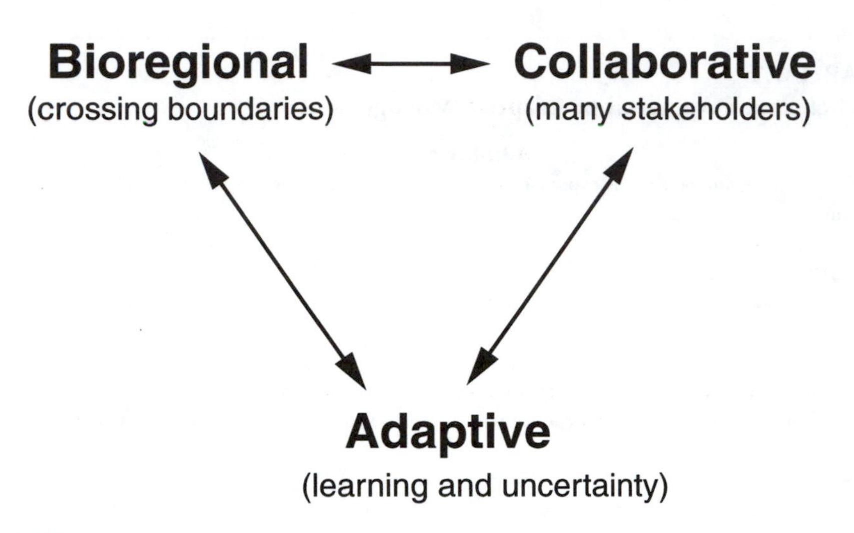

FIGURE 1.2 Ecosystem-scale conservation.

be inhabited or used by humans for the foreseeable future. Moreover, the lands and waters will not be under the control of a single owner or management agency, as is the case in existing nature preserves and parks. The formidable task of biodiversity conservation in such settings seems to require integrating at least the three themes in Figure 1.2.

1. *First*, working in the biological template of an ecosystem almost always requires acting across human boundaries. Figure 1.3 shows a portion of the Global 200 map in Figure 1.1. None of the ecoregional boundaries coincides with human jurisdictions; even the coastal zones are claimed under national sovereignty. When the jurisdictional boundaries mark off different human purposes, as does the line dividing forest from suburb, achieving coherent, coordinated action can be difficult.
2. *Second*, in inhabited landscapes there are many stakeholders. Consider those who are *always* present: governments, owners, and consumptive users of the land and waters. In addition, there are environmental activists, political insurgents, and would-be investors from outside. These groups have conflicting goals. The art of the conservationist under these conditions is to reconcile conflicting objectives, at least temporarily, to make agreements on land use and other protective actions possible.
3. *Third*, there are the challenges of adaptive management discussed above.

Trying to act in this fashion, ecosystem conservationists encounter structural challenges (Figure 1.4). The word *structural* should be highlighted. Trying to work at the ecosystem scale sets an agenda for collaboration, yet each of the elements of ecosystem conservation stirs up conflict. Cooperation is hard-won and difficult to sustain. A bioregional definition of the landscape crosses boundaries. A collaborative

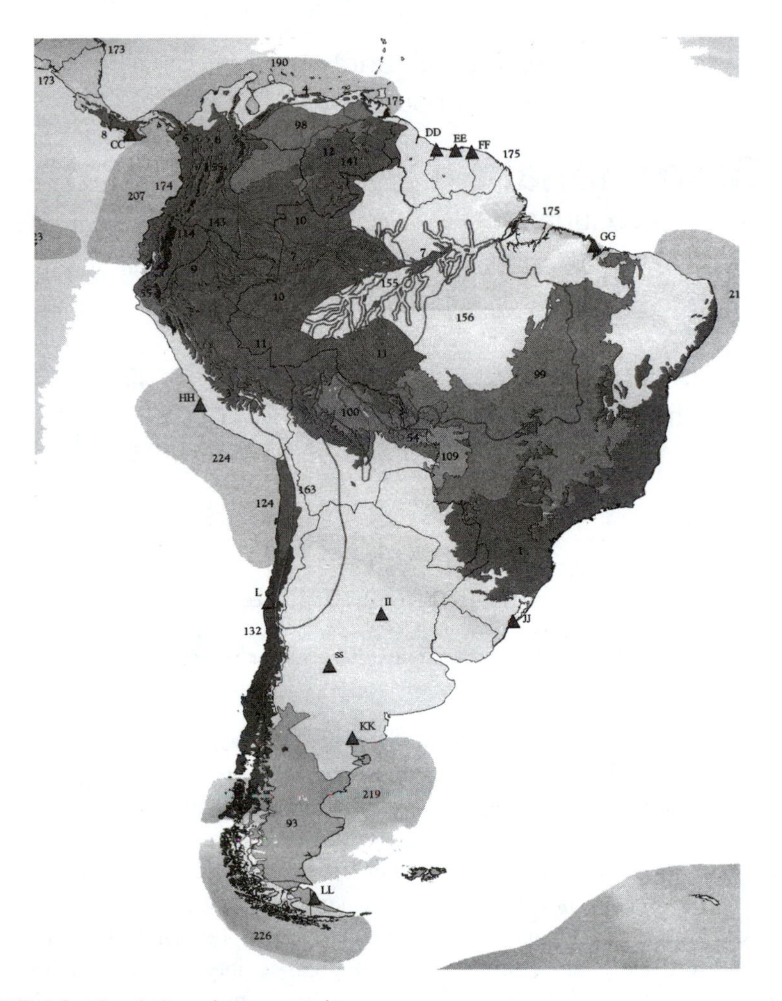

FIGURE 1.3 South American ecoregions.

approach complicates experimentation, requiring stakeholders with divergent ends to work together if learning is to produce reliable knowledge. Experimentation often takes a long time to gather significant findings — and may thus be difficult to reconcile with urgent declines in highly valued species.

The persistent theme is conflict. Stakeholders usually cannot be bypassed if they are not cooperative. So the alternative is to find ways to overcome or avoid their opposition. This is easy to say, but hard to do. The scientist's predisposition, faced with these social complications, is to find somewhere else to carry out the experiment. Why look for hassle when there is all that uncertainty to probe? This is correct from a scientific perspective. But if a landscape is of high biological value, the social complications may be unavoidable. In this important respect, a conservationist is not just a scientist.

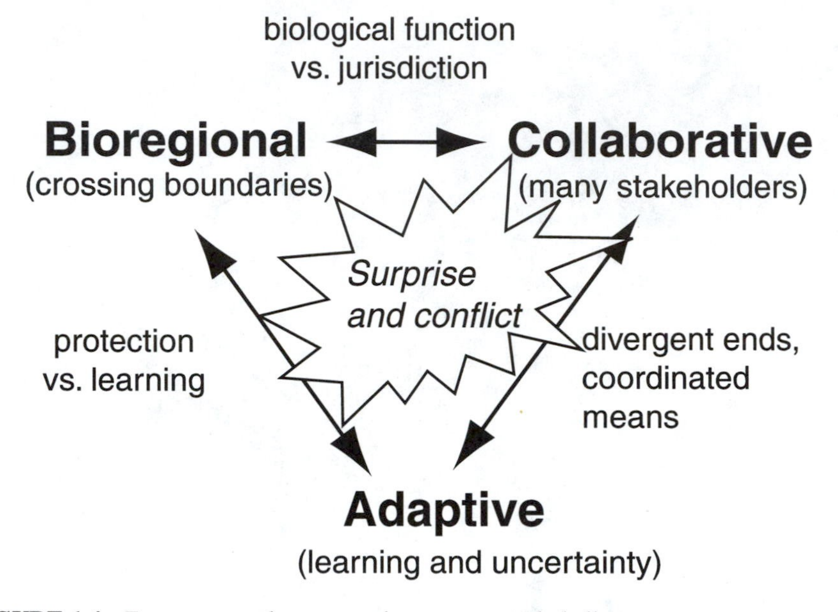

FIGURE 1.4 Ecosystem scale conservation — structural challenges.

Consider Figure 1.5, modified from an analysis advanced a generation ago by Don Price (Lee, 1993, chap. 7). Price's point is often forgotten today, but it remains valid. Scientific investigation is fundamentally different from the exercise of power. Power is about responsibility for the welfare of others; science is about determining truth or, more modestly, finding reliable knowledge. Price concluded that the goals of Truth and Power are not incompatible, exactly, but that there are usually trade-offs: a real human being cannot be for long a philosopher-king of the kind Plato described. More sadly, in a complex, technological society most people are laypersons most of the time, unable on their own to exercise much power or to determine truth without a lot of assistance. In people's work lives, Price argued, truth and power *do* work together, but in roles that constrain as well as enable knowledge and responsibility to be combined.

A question still being answered is where ecosystem conservation might fit in this chart, not only on an individual level but as an organized endeavor. After all, international corporations have shown considerable capacity for climbing toward the upper-right corner, marshalling both knowledge and power for their own ends. So it seems possible that a collaborative, bioregional, adaptive conservation strategy might be able to accomplish things that environmentalists or government officials cannot do by themselves, while opening opportunities for economic actors that developers in the traditional mold might not think to explore.

How might that happen? The author does not have a recipe, but here is a suggestion, spelled out via another diagram, shown in Figure 1.6 (Lee, 1993, chap. 4). Here, two sociologists looked at how decisions were made in different social settings. They observed that institutions and groups organized themselves differently to

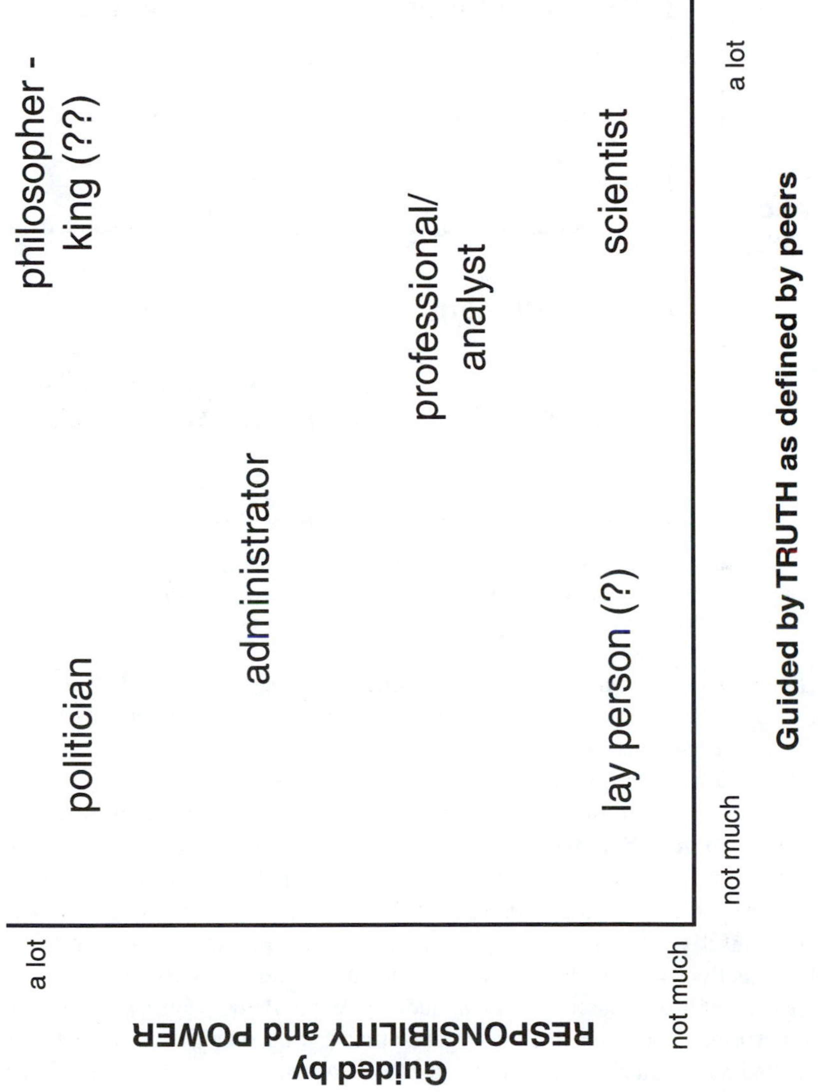

FIGURE 1.5 The spectrum from Truth to Power. (From Lee, K. N., *Compass and Gyroscope*, Island Press, Washington, D.C., 1993, chap. 7. With permission.)

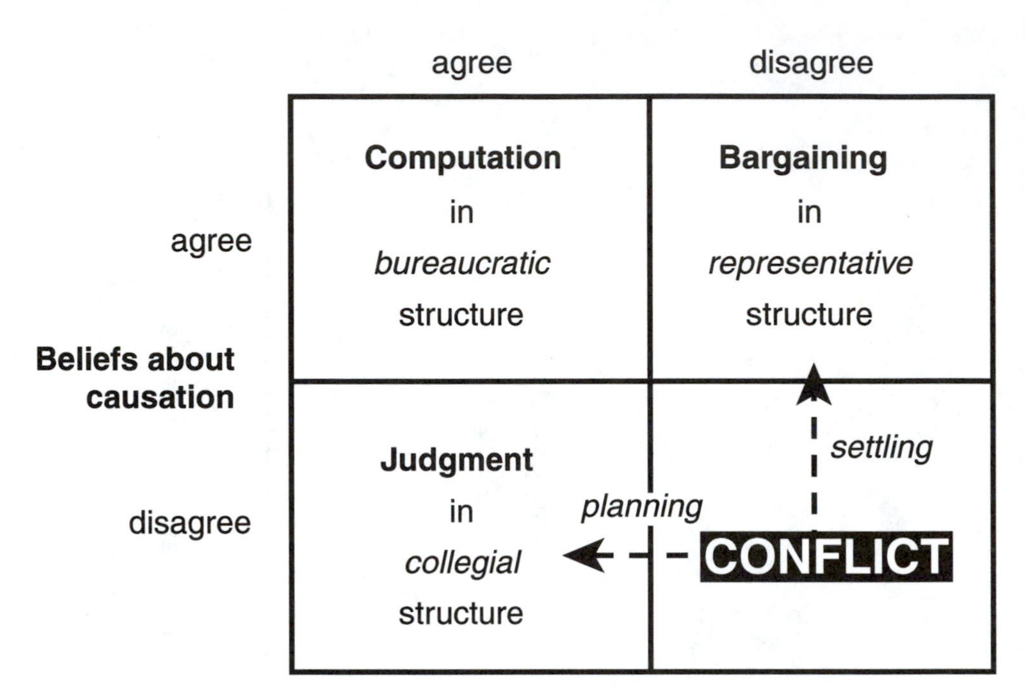

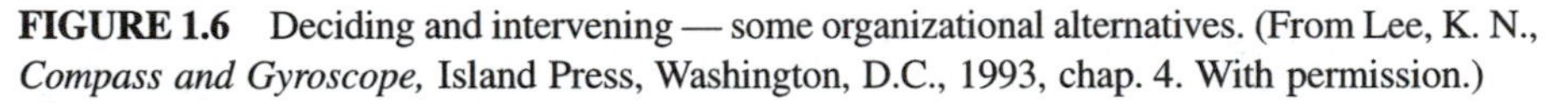

FIGURE 1.6 Deciding and intervening — some organizational alternatives. (From Lee, K. N., *Compass and Gyroscope,* Island Press, Washington, D.C., 1993, chap. 4. With permission.)

address different kinds of decision-making problems. Here, four circumstances are described, in search of answers to the question, What kinds of decision-making situations are central to ecosystem conservation?

The distinction drawn in this matrix is between outcomes and causality. That is similar to ends and means. In the upper-right box, people face the need to harmonize divergent ends (preferences about outcomes). This is what happens, for example, in assembling a budget: apples and oranges must be combined under a single spending ceiling. The way that is normally done is by bargaining — trading votes for what is often disdainfully referred to as "pork." However it is characterized, reconciling dissimilar objectives is something that all organizations do routinely.

The upper-left box is the realm of bureaucracy. When there is agreement on both preferred outcomes and causation, an organization can be set up around rules that can be applied mechanically. The judgment of whether someone can drive safely is reduced to a licensing examination — clear-cut rules that usually frighten only a 16-year-old. Bureaucracies like the Environmental Protection Agency are designed *as if* there were no arguments about the preferred outcomes they should pursue — that is, their missions. That is false, of course, which is one reason that life is often frustrating for people who work in those settings: the structure and function of the organization are ill-suited to the social environment in which it must work.

In the lower left is the realm of science, the arena in which causation is poorly understood. Peer review is one of the collegial devices familiar to academic folk: an anonymous colleague is asked to judge whether a paper is worth counting as a significant contribution to knowledge.

The lower-right box is where the hard stuff happens, including ecosystem conservation. When there is disagreement over both means and ends, over causation and desirable outcomes, there is no structural solution. Thompson and Tuden (1959) put in this box originally, "Intuitive decisions made by charismatic authority."

The author's suggestion, building on the considerable success of informal dispute resolution over the past 20 years, is to think of dispute resolution as strategies designed to move conflicts from the unresolvable to one of the cases one does know how to handle structurally. This is what one might try to do with ecosystem conservation.

It matters a lot which path one pursues. The most common cure for environmental problems is to recommend planning. Indeed, the modes of learning-while-doing listed in Table 1.1 all require planning as an initial step. The point stressed now is that interventions in established patterns of human-nature interactions typically engender dispute. Planning is often thought to be an adequate response to those tensions; even though it can do that sometimes, science is put at risk, usually inadvertently.

Planning is the arrow pointing to the left. Planning attempts to defer conflict. The planner says, "We are all in this complicated situation together — at an abstract level, our preferences about outcomes are not really different." Let a group of experts figure out some ways to get to that consensus future. Their plan will show everyone how to work together. While sorting out a complex situation usually opens up opportunities for collaboration that were obscured by conflict, planning illuminates irreconcilable differences, too. This typically leaves the experts — including scientists — vulnerable to the remaining disagreements.

In ecosystem conservation, furthermore, it is normally the scientists who are the troublemakers to begin with. They are among the advocates who want to divert the path of economic development. So, when the process for resolving conflict is planning, it is highly likely that scientific ideas like adaptive management or a bioregional template will be perceived to be the instruments of an invading power. This stiffens resistance. The questions that adaptive management seeks to answer will be seen in terms of whether or not they spell further trouble for ecosystem inhabitants. People who worry about getting the "wrong" answer can usually interfere with experiments enough that the experimenters cannot obtain a reliable answer. They are likely in any case to seek to shape the perception of what is found. Although these tensions cannot be avoided entirely, their potential to disrupt adaptive management should inform one's judgment of the worth of learning in a conflict-laden situation.

What about the other approach, which is labeled "settling"? Now, the strategy is to focus first on the fact that preferences about outcomes diverge. The bargaining searches for objectives that disputants want to pursue even though they do not agree on long-term goals. In an environmental settlement negotiation, one of the bargainers' tasks is to hammer out agreements on the questions about causality that all parties want answered. From that agreement, an adaptive approach to conservation can emerge.

A danger of the settling strategy is that there may be no common ground: in order for conservation to be accepted at all, it needs to be legitimate — something that must be done at some level, although there may be room to bargain over how much to do at a given time. In many places, the human representatives of biodiversity have yet to be given a place at the table. The rapid proliferation of protected areas may mean that there are now getting to be enough places where settling can create a viable structure for ecosystem conservation.

If conservation is legitimate, then science can play quite a different role than it does in planning. Now, the scientific questions can be answered with a shared sense of their importance. Science is not the servant of one set of interests but a tool to pursue a common agenda. And its answers, which may still be unwelcome, will have the considerable protection of a shared interest in asking the questions in the first place.

What this means, bluntly, is that a collaborative structure should be in place *before* an adaptive exploration of the landscape is under way. This is not how adaptive management has been used. Instead, experimentation has been adopted in a planning context. This may be a reason there have been few successes.

There is another frailty in the settling strategy. It lies in the word *representative*. In a jumble of jurisdictions of the kind one finds in bioregional conservation, it is not clear what counts as proper representation. What has happened in North America is that general-purpose governments like states, native American tribes, and national governments all weigh in. This complicates the bargaining because there are lots of parties and lots of arenas in which to bargain. Moreover, the institutional independence of the judiciary in parliamentary democracies means that litigation is often a parallel disputing process, available to the disgruntled. This is a consequential problem because it enlarges the social resources needed before biologically significant actions can be taken. Yet it is not easy to suggest workable alternatives. Sir Winston Churchill remarked once that democracy is the worst form of government save for all others; an ecosystem conservationist often ponders that observation.

SEARCHING FOR TRANSITIONS

Adaptive management is difficult to initiate and to sustain. However valuable the surprises it produces may turn out to be, adaptive management is unlikely to be considered affordable in many instances. But there is reason to think that this mode of learning is important, possibly essential, in the search for a durable and sustainable relationship between humans and the natural world. It is useful to recall that this is not the first generation to face the tension between human power and natural wealth.

Over the past two centuries, human population growth, technological change, and economic development have transformed the conditions of human life and altered the planet. In the coming century, there is reason to be hopeful (National Research Council, forthcoming). The long-term demographic transition, signaled by the absolute peaking of the population growth rate in the late 1960s, continues, with human numbers growing but at steadily slowing rates; a stable population size is plausible in the 21st century. It also appears that technological improvements can moderate the waste of energy and materials that pollute local and regional environments and alter the atmosphere and climate (see Ausubel, 1996). More and more places have been declared protected areas (see World Conservation Monitoring Centre, 1999), demonstrating the will to conserve biodiversity, even as the sixth great mass extinction in the history of life continues (Wilson, 1993, chap. 12).

If humanity is to achieve a transition toward a sustainable economy from the recent patterns of development, however, learning is of central importance. None of the elements of the transition sketched here, even population stabilization, is likely to be completed without the support and encouragement of societies and their leaders. Moreover, all aspects of preserving the life-support systems of Earth, particularly the protection of biological diversity and ecological processes, lie at or beyond the frontier of established practice or reliable knowledge. The history of "sustainable" exploitation of natural resources is a discouraging one (Ludwig et al., 1993; see Levin, 1993). Analyses of sustainable yield have been used not only to guide action, but also to rationalize heedlessness. This is not a reason to dismiss analysis, but rather to see that analysis is one stage of a historical process of adaptation. Analysis and learning can be influential, even when the policies they inform do not completely succeed. The U.S. Endangered Species Act, beleaguered throughout its quarter century as a legislative mandate, does seem to be bringing ecological science to bear in a small but significant set of decisions about land use (Kareiva et al., 1999).

Adaptive management is a mode of learning. It is one attractive to natural scientists, drawn to the trustworthiness of experimentation as a way to establish reliable knowledge. Practitioners of adaptive management are moving the method toward the pragmatics of trial-and-error learning, while seeking to preserve the rigor of scientific logic (Walters and Green, 1997). In doing so, the adaptive approach becomes entangled with the sociological difficulties of the Spectrum from Truth to Power (see Figure 1.5). Adaptive management has been framed to win favor with those who are nominally in charge of stewardship — typically government managers or harvest regulators and private landowners. When the legitimacy of these stewards is under attack, adaptive management has at times appeared to be a way to deflect criticism by opening the way to trying novel ideas. But unless those novel ideas turn out to command a consensus within a short time of adoption, adaptive management is no more than a way to justify trial and error in the midst of a political free-for-all. Logically, one does not need the armature of scientific methods for that purpose. If there is a lurking consensus, however, the careful thought demanded by adaptive management may be a way to elicit agreement (Walters and Green, 1997).

An alternative is proposed above, to hammer out a provisional agreement before turning to experimental methods to sort out uncertainties that are agreed to be significant by all stakeholders. Using this strategy does not assure that the initial consensus about questions to answer will hold. Indeed, since surprise is endemic to experimental learning, emergent Truth will inevitably challenge Power. That is, of course, a reflection of the necessity for conflict and debate in the process of finding a way to live in durable prosperity. Sadly, humans have difficulty distinguishing between message (conflict is healthy) and messenger (adaptive management reveals conflict rather than causing it). In an approach to learning that requires the support of large organizations, this confusion may be the limiting factor in the ability to find a sustainable future.

Conservation now requires harmonizing the insistent demands of poverty and capitalism with the quiet obligations of the biotic community. Heeding those obligations requires an uncommonly broad perspective, one that takes seriously social institutions, human needs, and biogeography. It is also a perspective that recognizes the productivity of conflict, while sustaining a patience consonant with the rhythms of the biosphere.

As in the spectrum from Truth to Power, the author does not propose that single individuals embrace all these qualities. But how to organize the skills and commitments of a diverse human community to strive for these uncommon aims is a challenge that is only beginning to be addressed. The unifying idea is the dream of humans living in harmony with nature. For the first time, humans are seeking to govern themselves in a way to preserve something they cannot see — the web of life upon which all depend. But it is far from the only time that humans have tried to govern themselves according to invisible ideals; indeed, the influence of ideas and ideals is usually thought to be a distinctive and (sometimes!) admirable trait of the species.

To paraphrase Walters (1997), the need now is not so much for better ammunition for rational debate but for creative thinking about how to make adaptive management and social learning an irresistible opportunity, rather than a threat to various established interests. That is, the need is to show that adaptive management can create feasible accommodations for those who gain their livelihood from the land and waters, property owners, public officials, and environmental activists — as well as scientists.

Where this cannot be shown, one should concede that science cannot displace power often enough to be a basis for policy. Science is at most the servant of those who act; the question is, whom should scientists seek to serve. Those who can agree, despite conflict, to search for answers bring a degree of social stability that seems little enough for the practical realities of adaptive management. Those who cannot agree furnish reason for citizens, including scientists, to act politically; that is not adaptive management but a different, sometimes higher calling.

CONCLUSIONS

Seeking to stimulate the creative thinking called for above, the chapter concludes by advancing some troublesome questions to which the author does not have answers, with an invitation to comment:

1. Are there clear-cut successes in adaptive management, other than Sainsbury's work?
2. When an adaptive management design is *not* feasible, is there experience to guide trial-and-error learning?
3. Adaptive management appears to be unsuitable *unless* there is a governmental authority that monopolizes physical access to the resources being managed. This raises the question of what can be learned under conditions of partially open access or limited enforcement of regulations. Relevant experience is invited.

4. Put another way, adaptive management appears to be a "top-down" tool, useful primarily when there is a unitary ruling interest able to choose hypotheses and test them. Markets are inherently bottom-up learning mechanisms, with messages transmitted by prices and availability of goods and services. Collaborative management works by sharing understanding of the behavior of an exploited natural system within a community of harvesters (Ostrom, 1990). Is there a broader analytical perspective on how action improves understanding that would be useful to add to this set?
5. Are there clearly articulated scientific criteria for putting short-term conservation ahead of learning? That is, are there conservation situations where enough is known not to need to worry about surprises?
6. When should near-term conservation objectives (notably, species preservation) take precedence over an ecosystem approach? When is a conservationist justified in short-circuiting participation by inhabitants and other stakeholders in decision making? Much action in behalf of endangered species implicitly answers these questions in the affirmative.
7. Efforts to design sustainable-use regimes in collaboration with established or traditional user communities have had difficulty integrating scientific perspectives, and adaptive methods have not been used to the author's knowledge. What lessons are emerging from comanagement concerning the usability of ecosystem models and explicitly adaptive processes?
8. Any learning process seems to require enough stability for the learning to take place, but environmental problems are characteristically accompanied by social change. Are there ways to record observations and assumptions so that one enhances the probability of learning by those who follow, even if they are not the same persons and may be motivated to learn for different reasons?
9. The emphasis on a transition to a sustainable material economy in the last section does not equate growth with greed (see Ludwig et al., 1993). Is this a mistake?
10. Is it possible to achieve biodiversity conservation without sustainable development … or *vice versa*? The development of national parks in the United States and other Organization for Economic Cooperation and Development (OECD) countries implicitly assumed the possibility of conservation — although not of biodiversity as it is now understood — without sustainable development.

ACKNOWLEDGMENTS

This chapter began as a talk at a September 1998 conference on "Adaptive Collaborative Management of Protected Areas," sponsored by the Cornell University International Institute for Food, Agriculture, and Development, and the Center for International Forestry Research. The author is grateful to Louise Buck for the invitation to participate in that program, and to Niels Röling for a helpful discussion there. Colleagues at the Board on Sustainable Development, National Research Council, have developed the idea of a sustainability transition discussed in the concluding section, and the author pays them the compliment of appropriating their thoughts. Four anonymous reviewers provided useful critiques; and the author is once again indebted to C. S. Holling for encouragement and inspiration.

REFERENCES

Anderson, J. L., 1998. Errors of inference, in *Statistical Methods for Adaptive Management Studies,* Sit, V. and Taylor, B., Eds., Lands Management Handbook 42, Ministry of Forests, Research Branch, Victoria, British Columbia, Canada.

Ausubel, J. H., 1996. Can technology spare the earth? *Am. Sci.,* 84, 166–178.

Barinaga, M., 1996. A recipe for river recovery? *Science,* 273, 1648–1650.

Brewer, G. D., 1973. *Politicians, Bureaucrats, and the Consultant. A Critique of Urban Problem Solving,* Basic Books, New York.

Bruton, H. J., 1997. *On the Search for Well-Being,* University of Michigan Press, Ann Arbor.

FEMAT (Forest Ecosystem Management Assessment Team), 1993. Forest Ecosystem Management: An Ecological, Economic, and Social Assessment, Appendix A of draft supplemental environmental impact statement on management of habitat for late-successional and old-growth forest related species within the range of the Northern spotted owl, Federal Interagency SEIS Team, Portland, OR, July.

Getz, W. M., Fortmann, L., Cumming, D., du Toit, J., Hilty, J., Martin, R., Murphree, M., Owen-Smith, N., Starfield, A. M., and Westphal, M. I., 1999. Sustaining natural and human capital: Villagers and scientists, *Science,* 283, 1855–1856.

Glen Canyon Environmental Studies, 1996. Floods in the Grand Canyon, available at http://www.usbr.gov/gces/rod.html; visited 9/10/98.

Grand Canyon Monitoring and Research Center, 1998. Programs and Announcements, available at http://www.usbr.gov/gces/prog.htm; visited 9/10/98.

Gunderson, L. H., Holling, C. S., and Light, S. S., Eds., 1995. *Barriers and Bridges to the Renewal of Ecosystems and Institutions,* Columbia University Press, New York.

Hall, P. A., Ed., 1989. *The Political Power of Economic Ideas: Keynesianism across Nations,* Princeton University Press, Princeton, NJ.

Heclo, H., 1974. *Modern Social Politics in Britain and Sweden,* Yale University Press, New Haven, CT.

Holling, C. S., Ed., 1978. *Adaptive Environmental Assessment and Management,* John Wiley & Sons, New York.

Kareiva, P., Andelman, S., Doak, D., Elderd, B., Groom, M., Hoekstra, J., Hood, L., James, F., Lamoreux, J., LeBuhn, G., McCulloch, C., Regetz, J., Savage, L., Ruckelshaus, M., Skelly, D., Wilbur, H., Zamudio, K., and NCEAS HCP working group, 1999. *Using Science in Habitat Conservation Plans,* National Center for Ecological Analysis and Synthesis, Santa Barbara, CA; and American Institute of Biological Sciences, Washington, D.C.

Lee, K. N., 1993. *Compass and Gyroscope, Integrating Science and Politics for the Environment*, Island Press, Washington, D.C.

Levin, S. A., Ed., 1993. Science and sustainability, *Ecol. Appl.,* 3(4).

Levitt, B. and March, J. G., 1988. Organizational learning, *Annu. Rev. Sociol.*, 14, 319–340.

Lubchenco, J., 1998. Entering the century of the environment: a new social contract for science, *Science,* 279, 491–497.

Ludwig, D., Hilborn, R., and Walters, C., 1993. Uncertainty, resource exploitation, and conservation: lessons from history, *Science,* 260, 17.

Marcot, B. G., 1998. Selecting appropriate statistical procedures and asking the right questions: a synthesis, in *Statistical Methods for Adaptive Management Studies,* Sit, V. and Taylor, B., Eds., Lands Management Handbook 42, Ministry of Forests, Research Branch, Victoria, British Columbia, Canada.

Margoluis, R. and Salafsky, N., 1998. *Measures of Success. Designing, Managing, and Monitoring Conservation and Development Projects,* Island Press, Washington, D.C.

McCay, B. J. and Acheson, J. M., Eds., 1987. *The Question of the Commons,* University of Arizona Press, Tucson.

National Research Council, 1996. Upstream: Salmon and Society in the Pacific Northwest, Report of the Committee on the Protection and Management of Pacific Northwest Anadromous Salmonids. National Academy Press, Washington, D.C.

National Research Council, 1999. *Our Common Journey: A Transition Toward Sustainability,* report of the Board on Sustainable Development, National Academy Press, Washington, D.C., http://books.nap.edu/catalog/9690.html.

Olson, D. M. and Dinerstein, E., 1998. The Global 200: a representation approach to conserving the earth's most biologically valuable ecoregions, *Conserv. Biol.,* 12, 502–515.

Olympic National Forest, 1998. The Olympic Adaptive Management Area, available at http://www.fs.fed.us/r6/olympic/ecomgt/nwfp/adaptman.htm; visited 1/5/1999.

Ostrom, E., 1990. *Governing the Commons, The Evolution of Institutions for Collective Action,* Cambridge University Press, Cambridge.

Parson, E. A. and Clark, W. C., 1995. Sustainable development as social learning: theoretical perspectives and practical challenges for the design of a research program, in *Barriers and Bridges to the Renewal of Ecosystems and Institutions,* Gunderson, L. H., Holling, C. S., and Light, S.S., Eds., Columbia University Press, New York, chap. 10, 428–460.

Peterman, R. M. and Peters, C. N., 1998. Decision analysis: taking uncertainties into account in forest resource management, in *Statistical Methods for Adaptive Management Studies,* Sit, V. and Taylor, B., Eds., Lands Management Handbook 42, Ministry of Forests, Research Branch, Victoria, British Columbia, Canada.

Plum Creek Timber Co., L.P., 1998. Cascade Region Habitat Conservation Plan (Web summary), available at http://www.plumcreek.com/eleader/initiatives01.htm; visited 9/10/98.

Plum Creek Timber Co., L.P., 1999. Plum Creek Native Fish Habitat Conservation Plan (Web summary), available at http://www.plumcreek.com/eleader/initiatives05.htm; visited 1/5/99.

Price, D. K., 1965, *The Scientific Estate,* Oxford University Press, New York.

Rogers, K., 1998. Managing science/management partnerships: a challenge of adaptive management, *Conserv. Ecol.,* 2(2), R1, available at http://www.consecol.org/vol2/iss2/resp1.

Sit, V. and Taylor, B., Eds., 1998. *Statistical Methods for Adaptive Management Studies,* Lands Management Handbook 42, Ministry of Forests, Research Branch, Victoria, British Columbia, Canada.

Stein, H., 1996. *The Fiscal Revolution in America: Policy in Pursuit of Reality.* 2nd rev. ed. of the 1969 version, AEI Press, Washington, D.C.

Thompson, J. D. and Tuden, A., 1959. Strategies, structures and processes of organizational decision, in *Comparative Studies in Administration,* University of Pittsburgh Press, Pittsburgh.

Volkman, J. M. and McConnaha, W. E., 1993. Through a glass, darkly: Columbia River salmon, the Endangered Species Act, and adaptive management, *Environ. Law,* 23, 1249–1272.

Walters, C., 1986. *Adaptive Management of Renewable Resources,* Macmillan, New York.

Walters, C., 1997. Challenges in adaptive management of riparian and coastal ecosystems, *Conserv. Ecol.,* (2):1, Available at http://www.consecol.org/vol1/iss2/art1.

Walters, C. and Green, R., 1997. Valuation of experimental management options for ecological systems, *J. Wildl. Manage.,* 61, 987–1006.

Walters, C. J. and Holling, C. S., 1990. Large-scale management experiments and learning by doing, *Ecology*, 71, 2060–2068.

Walters, C., Goruk, R. D., and Radford, D., 1993. Rivers inlet sockeye salmon: an experiment in adaptive management, *North Am. J. Fish. Manage.,* 13, 253–262.

Wilson, E. O., 1993. *The Diversity of Life,* Harvard University Press, Cambridge, MA.

World Conservation Monitoring Centre, 1999. Conservation Databases, Web site, Available at http://www.wcmc.org.uk/cis/index.html; visited 5/25/99.

2 Roles for Civil Society in Protected Area Management: A Global Perspective on Current Trends in Collaborative Management

Jeffrey A. McNeely

CONTENTS

0-8493-0020-7/01/$0.00+$1.50

INTRODUCTION

The Convention on Biological Diversity (CBD) has marked a significant shift in the perception of protected areas by governments. It has linked protected areas to larger issues of public concern, such as sustainable development, traditional knowledge, access to genetic resources, national sovereignty, equitable sharing of benefits, and intellectual property rights. Protected area managers are now sharing a larger and more important political stage with agricultural scientists, nongovernmental organizations (NGOs), anthropologists, ethnobiologists, lawyers, economists, pharmaceutical firms, farmers, foresters, tourism agencies, the oil industry, indigenous peoples, and many others. These competing groups claim resources, powers, and privileges through a political decision-making process in which biologists, local communities, the private sector, and conservationists have become inextricably embroiled (McNeely and Guruswamy, 1998). The challenge is to find ways for the various stakeholders to collaborate most effectively to achieve the conservation and development objectives of modern society. This chapter is a response to the challenge, with protected areas as the stage, drawing on work just completed for the Asian Development Bank (McNeely, 1998).

A COLLABORATIVE APPROACH TO PROTECTED AREA MANAGEMENT

Protected Areas Provide Functions Critical to Society

Protected areas carry out numerous functions that are beneficial to humans, and even essential to human welfare; ten important functions are listed below. Some of these functions can also be provided by unprotected nature, agricultural lands, or even degraded wastelands, but properly selected and managed protected areas typically will deliver more of these functions per unit area at lower cost than will most other kinds of land use in the biologically important areas that require protective management. The way these functions are transformed into benefits for people will depend on the management objectives of the protected area and how effectively these objectives are converted into action.

1. **Biodiversity**. Conserve genetic resources and biological diversity more generally, enabling evolution to continue and providing raw materials for biotechnology.
2. **Watershed protection**. Protect watersheds for downstream hydroelectric, irrigation, and water supply installations.
3. **Storm protection**. Protect coastlines against damage from storms (especially coral reefs and mangroves), and absorb heavy rainfall (especially wetlands and forests).
4. **Tourism**. Provide destinations for nature-based tourism and recreation.
5. **Local amenity**. Ameliorate local climate conditions and provide amenity values to nearby communities.

6. **Forest products**. Provide a wide range of nontimber forest products, and limited amounts of timber.
7. **Soil**. Build soils, control soil erosion, and recycle nutrients.
8. **Carbon**. Sequester carbon, thereby contributing to global efforts to address anthropogenic climate change.
9. **Research**. Provide sites for scientific research on a wide range of ecological, social, and economic topics.
10. **Cultural values**. Conserve culturally important sites and resources, and demonstrate the national interest in natural heritage.

The Major Functions of Protected Areas Deliver Different Benefits at Different Scales

Protected areas are important at many levels, including local, national, and global. Drawing on the list of the functions of protected areas presented above, Table 2.1 presents a model of the various scales at which benefits are delivered by these functions, ranging from local to global. The range of possible benefits at each scale indicates the importance of defining objectives for individual protected areas; different management approaches will provide different mixes of benefits at different levels.

TABLE 2.1
The Scale at Which Benefits are Delivered by Protected Area Functions

	Scale at Which Benefits Delivered		
Key Functions	**Local**	**National**	**Global**
1. Biodiversity	0–4	2–4	4
2. Watershed protection	4	2–4	1–3
3. Storm protection	4	2–4	1–3
4. Tourism	0–4	4	2
5. Local amenity	2–4	1–2	0–1
6. Forest products	0–4	1–2	1–2
7. Soil	0–4	1–2	1–2
8. Carbon	0–1	1–2	2–3
9. Research	0–3	2–4	2–3
10. Cultural values	0–4	2–4	1–2

Protected areas provide benefits to people at all levels. Using the ten critical functions listed in the text, this table provides a model of the scale at which benefits can be derived, from 0 (= no benefit) to 4 (= maximum benefit). More precise determinations can be made for individual protected areas or for national protected area systems.

The first step in collaborative protected area management is to determine objectives at both the system and site levels; these objectives determine who gets what benefits, and pays what costs at what scale. This is a political process that should involve dialogue with the key stakeholders, including landowners, scientists, local communities, NGOs, and the private sector. The fundamental point is that protected areas must be managed first and foremost to maintain ecological integrity, so they can continue to deliver ecosystem functions to society. The second commitment is to providing quality service to "clients" of protected areas, including visitors, the general public, people living in the surrounding areas, scientists, and the private sector. Because different objectives involve trade-offs in terms of the distribution of costs and benefits, they should to be made explicit in management terms. Further, many of the public good benefits of protected areas provide significant advantages for the global community, including conservation of biodiversity, sequestration of carbon, and the results of ecosystem research. Capturing appropriate rents at the national or local level from these global benefits remains a challenge that is only partially being met by intergovernmental processes such as the CBD, Ramsar, and World Heritage. NGOs that capture willingness to pay among consumers in wealthy countries or sectors of society play an important role in this regard.

Many Stakeholders Have Interests in Protected Areas and Important Roles to Play in Their Management

Stakeholders who can be part of collaborative management, such as the private commercial sector, NGOs, research institutions, and local communities, contain considerable variability as well as important potential to contribute to various aspects of protected area management. However, these different categories of stakeholder tend to have very different major motivations, leading to different roles that they can play in protected area management (Table 2.2). The way that the resources of a protected area are used in any particular place and time is the result of accommodation among conflicting interests between stakeholders having different objectives. Seldom does any single group dominate absolutely, and resources can be used in many different ways at the same place and time. Thus, protected area management is part of an ongoing process in which an appropriate balance is sought among the different interests of the various stakeholders. A national protected area system plan can provide the basis for this process (see below).

NGOs have been major contributors to protected areas in virtually all countries, providing funds and expertise, building public support, promoting action, and advocating conservation interests. Although NGOs can provide very practical support to protected areas, their contributions are likely to be most useful when a clear understanding has been reached between the NGO and the protected area management authority. NGOs can diversify efforts and approaches to management of protected areas, sometimes using methods very different from those adopted by government agencies. Locally based NGOs can often use their familiarity with local issues and resources to operate effectively where government agencies or national NGOs have difficulties.

None of the four major "civil society sectors" operates independently, and indeed the best results may come from some form of joint venture among local people,

TABLE 2.2
Major Motivations and Roles of Key Stakeholders

Stakeholder	Major Motivation	Major Roles
Protected area management agency	Addressing its public mandate	Setting policy; protected area management
Private commercial sector	Economic profit	Managing profitable operations; providing sponsorship
NGOs	Conserving public goods	Public information; technical advice; linkages among stakeholders; funding from public
Research institutions	Scientific curiosity	Research and monitoring; technical advice
Local communities	Sustainable livelihoods	Resource management; buffer-zone management

Five major categories of stakeholders in protected areas are listed. This chart presents a model of their motivations and roles, while taking into account that these will vary considerably from place to place. The point for government resource management agencies is to recognize the main motivations, harness the strengths of each stakeholder, and be aware of the limitations of each.

government, and the private sector, often with NGOs serving as facilitators. In the face of predictable further reductions in the resource management budgets of many governments and constantly increasing pressure on resources in virtually all countries, collaborative management of natural resources is becoming promoted more widely in many parts of the world.

However, formidable challenges face partnerships among protected area agencies and the private sector, NGOs, researchers, and local human communities. Many protected area staff and academic conservationists are concerned that this cooperative approach could ultimately reduce the quality of the protected area, and that strong legislation supported by vigorous law enforcement is the best option for long-term conservation (e.g., Kramer et al., 1997). And, indeed, experience has shown that local people often are as likely to misuse privileges under cooperative management as anyone else, and the private sector sometimes more so; enforcement of regulations must be part of management. Still, given the insufficient staff and logistics support available to most protected areas, the "strict preservationist approach" is both impossible to implement and of doubtful viability on socioeconomic grounds. The collaborative approach involving multiple stakeholders advocated in this chapter may be the only practical option in today's conditions in most of the world.

The Major Problems Facing Protected Areas Should Be Addressed by Institutions at the Appropriate Scale, With Appropriate Roles

Just as different benefits of protected areas are delivered differently at different scales, so too must the different problems faced by protected areas be addressed by the right institutions operating at the appropriate scale. The first step in determining appropriate management responses is to identify clearly the problem being addressed. In general, local people possessing secure tenure can deal with most day-to-day threats better than governments can, while governments can resist major abuses better than local people can, providing they have the technical and institutional resources and political will to do so. When the main threat to a protected area arises from cumulative overuse by too many people making too many demands on ecosystems to meet their day-to-day subsistence needs, local regulation and social control may be required, along with investments in improved agricultural practices or alternative livelihoods (Caldecott, 1997). When poaching of endangered species is a major problem, law enforcement will be a critical element. However, many of the factors leading to the loss of biodiversity and degradation of protected areas originate in national government policies far from protected area boundaries, such as national development priorities that may subsidize industrial agriculture in buffer zones, promote resettlement in remote areas, build roads or dams in protected areas, and issue timber concessions in protected areas or buffer zones. These require broader approaches, such as improved national policies on development, trade, land tenure, and land-use planning.

Institutional arrangements can also vary considerably with the category of protected area. For protected areas of global importance, such as World Heritage Sites, the central government will need to have a considerable level of involvement and usually the primary management responsibility, supported by international recognition. National governments must remain responsible for collective national conservation concerns that are best managed by government agencies at national or subnational level; thus, International Union for Conservation of Nature and Natural Resources (IUCN) Categories I and II, Strict Nature Reserves and National Parks, are managed for national conservation objectives and are properly the responsibility of the national or provincial governments. But Categories IV, V, and VI, whose objectives often are to provide benefits to local communities, may be more appropriately managed by more local levels of government, the private sector, NGOs, research institutions, and local communities to meet more local needs. Buffer zones typically are best managed at the local level by institutions whose mandates focus on development.

Involving multiple stakeholders in protected area management has many advantages. The key challenge is to specify appropriate functional roles, as suggested in Table 2.3. As indicated above, how these roles are distributed will depend on the management objectives of each individual protected area and how these are implemented. But, generally, in order for protected areas to be both economically productive and sustainably managed over the long term, governments should:

- Provide the necessary infrastructure, including research, identification, and compilation of information on biodiversity;
- Establish appropriate legislation and institutional mechanisms covering both conservation and sustainable use of biological resources;
- Implement a program to enhance public information about protected areas;
- Establish national objectives for the protected area system;
- Ensure that the various approaches to protected area management are contributing to the national system;

TABLE 2.3
Functional Roles in the Management of Protected Areas

Functional Role	Government	Private Sector	NGOs	Research Institutes	Local Communities
Systems planning	4	1	2	2	2
Site planning	4	1	2	2	3
Establishing norms	4	1	1	1	2
Maintenance of roads	4	1	0	0	1
Maintenance of trails		1–2	2	1	2
Running of hotels, lodges	0–4	0–4	0–2	0	0–4
Running of campsites	0–4	0–4	0–2	0	0–4
Habitat management	4	1–2	1–2	1–2	1–4
Wildlife management	2–4	1	1–2	1–2	1–4
Public information	2–4	1	1–4	1–2	1
Public relations	2–4	2	1–4	0–2	0
Extension	1–4	1	1–4	2–3	1
Research	0–4	1	1–4	2–4	1
Education	2–4	1	1–4	2–4	1
Monitoring	0–4	1	1–4	2–4	1–2
Bio–prospecting	0–1	4	1	2–4	2
Issuing permits	4	0	0	0	2
Funding	2–4	1–3	1–3	1	1

Although each protected area has rather different challenges, the general distribution of responsibility among the government, the private sector, NGOs, research institutions, and local communities can be assessed for each of the functional roles for protected areas. This table assesses the importance of the role for each of the five groups, scoring from 0 (no role) to 4 (lead role). These scores are indicative only, and will vary with the site.

- Support the interests of protected areas in the face of alternative land uses;
- Establish means for exchanging lessons learned from the various approaches; and
- Provide an appropriate regulatory framework to ensure quality control.

Partnerships are also required for the effective management of buffer zones, but since these typically fall outside the direct jurisdiction of protected area managers, the mix of functional roles is somewhat different (Table 2.4). Buffer zones, too, need management plans so that stakeholders are clear about rights and responsibilities.

In short, a balanced approach to protected area management will allow for many institutional players to participate in strengthening the national network of protected areas. Dangers to be avoided include governments abandoning their responsibility for the management of the system under the guise of privatization; reduced protection of core areas in the forlorn hope that buffer-zone management and local development will reduce threats to strictly protected core zones; excessive local control over nationally or internationally important resources; and excessive privatization to the detriment of public support for protected areas.

TABLE 2.4
Functional Roles in the Management of Buffer Zones

Functional Role	Government	Private Sector	NGOs	Research Institutes	Local Communities
Planning	1–3	1	1–2	1–2	2–4
Habitat management	0–4	1	1–3	1–2	2–4
Wildlife management	2–3	1	1–3	1–2	2–4
Tourism management	1–2	1–3	1–3	1–2	2–4
Plantation management	1–2	1–2	1	1–2	4
Bioprospecting	1–2	2–3	1–2	2–3	2–4
Harvesting of medicinal plants	1	2–3	1	0	2–4
Agriculture	1	1–2	1	1–2	4
Hunting management	1–2	1–2	1–2	1–2	2–4

As in Table 2.3, functional roles can also be assessed for buffer zones. This table assesses the importance of the role for each of the five groups, scoring from 0 (no role) to 4 (lead role). The government agency need not necessarily be the protected area agency.

PROTECTED AREAS ARE BEST CONCEIVED OF AS PARTS OF A NATIONAL SYSTEM OF LAND USE

As called for under the CBD, each country should treat its protected areas as a system, with different parts of the system designed to provide different kinds of benefits to different groups of stakeholders, although, of course, with considerable redundancy built into the system to ensure sustainability. Table 2.1 implied that protected areas should be conceived as a national system, with some sites designed to provide primarily national benefits, others designed primarily to meet needs of local people for watershed protection, other sites to ensure sustainable use of non-timber forest products, others designed primarily to conserve biological diversity, and some so important that they are considered of outstanding universal value and, therefore, inscribed on the World Heritage List. Within any national system, some areas will be particularly attractive for tourism; such sites can earn a considerable financial surplus that can subsidize other parts of the protected area system that do not have easily marketed values, even though they may have high economic value in terms of watershed protection, carbon sequestration, or conservation of genetic resources.

A national protected area systems plan will ensure that all major ecosystems are well protected, that the different components of the system are managed to the appropriate objectives, that each protected area in the system is assigned to one or more of the IUCN categories (or other system of applying different management approaches to achieve different major objectives), that connections between protected areas are promoted where possible, that developments in adjacent lands (buffer zones) are supportive of the protected area system, that roles for different stakeholders are identified, and that priorities for investment are specified.

A national protected area system plan should identify the fundamental sources of threat to protected areas, and help ensure that these are addressed at the national planning level. Linkages also should be established between protected areas and the many other sectors that may relate to them, ranging from international trade to climate change to agriculture, forestry, fisheries, tourism, and so forth. The protected area systems plan therefore needs to help inform the national policies governing overall development.

It is important to recognize that many large and wide-ranging animal species such as tigers, rhinos, elephants, and large birds of prey, as well as rare species of forest trees, need very large expanses of territory; establishing even 10 or 15% of a country's land area as legally protected is unlikely to be a sufficient response to their conservation needs, especially if the consequence is that the remaining land is all converted to land uses that are incompatible with these species. Many conservationists are concerned about protected areas becoming "islands of nature in a sea of cultivation," but this metaphor is oversimplified because protected areas have numerous relationships with the surrounding lands. These relationships are the basis of new approaches being taken to protected area management and systems planning, under names such as "bioregions," "landscapes," "integrated conservation and development projects," and "biosphere reserves" (e.g., Miller, 1996). All of these refer to approaches that treat the protected areas as the core zones, perhaps even "engines of development" that support appropriate and sustainable developments in the surrounding countryside and generate positive relationships among the various land uses. Such approaches can be fostered through the system plan.

FACILITATING BROADER COLLABORATION IN SUPPORT OF PROTECTED AREAS

Build a Strong Economic Foundation for the Protected Area System

Protected areas have usually been seen in primarily biological or ecological terms, but the recent studies (e.g., Costanza et al., 1997; Dixon and Sherman, 1990) are indicating the economic importance of land managed for conservation objectives. Whenever a serious examination of benefits from protected areas is carried out, it becomes apparent that the benefits of protected areas outweigh their management costs by a considerable margin, often as much as a factor of ten or even more. But even though protected areas can be significant sources of revenue for the national economy, they are deteriorating because insufficient investment is made to ensure their continued productivity.

Part of this lack of investment arises because protected areas are basically public goods, being provided to everyone, rather like education, defense, and law and order. Because many of the economic benefits of protected areas are available to all, fewer incentives exist for any one individual, community, or commercial firm to conserve the resource than would seem to make sense from the perspective of society as a whole; indeed, private benefits from converting socially valuable protected areas to other uses often are substantial. Thus managing protected areas may require some form of intervention to correct for the market failures involved, for example, offsetting the substantial subsidies that promote economically and environmentally damaging resource exploitation in agriculture, forestry, mining, and other activities that have negative impacts on protected areas (amounting to about US$1.4 trillion/year globally; Myers and Kent, 1998). Market economies typically underprovide public goods like protected areas, because their full social benefits are beyond appropriation by markets.

On the other hand, many biological resources within protected areas are potentially private goods, whose use is rival, whose control can be made exclusionary, and whose value is commercially marketable; these include, for example, genetic resources, tourism resources, and water resources. Means of capturing these economic benefits can help deliver the potential financial benefits of the protected area, especially if market prices can be charged for the full range of benefits provided by protected areas (Shah, 1995).

What activities should be funded from government allocations as opposed to activities that should be covered by various kinds of user fees? The distinction between public and private benefit is a useful guide. Generally, the taxpaying public should pay for the costs of establishing and maintaining protected areas (a public good), whereas those who derive a personal or commercial benefit from the use of such areas should pay for the associated costs. Thus, governments may decide that user fees should pay for goods and services such as camping, water, information, access, and genetic resources. The problem with this approach is that only those goods and services that can be given a monetary value will earn income, potentially leaving other values ignored, even if they are in the public interest. Further, some very valuable functions of protected areas, such as watershed protection, are politically difficult to capture through market mechanisms. Although collecting fees from the sale of hydroelectricity from a dam whose watershed is protected is feasible, few governments are yet willing to charge farmers the full costs of high-quality irrigation water emanating from a protected area.

A major issue is funding of individual sites as opposed to funding of entire protected area systems. Some protected areas will be more financially profitable than others, especially those that are very popular with international tourists or those linked to major hydroelectric projects. Others will provide significant benefits to farmers in terms of watershed protection, but the government might wish to subsidize these benefits. These subsidies should be recognized for what they are, and compensatory payments should be provided to the protected areas. But if governments decide that protected areas must be more self-reliant, then governments will need to enable protected areas to retain the funds they earn from various sorts of user fees, such as gate fees, bioprospecting fees, and even water-use charges.

One means of determining appropriate levels of support for a protected area system is to perform a thorough financial analysis of the full range of economic benefits provided by each protected area and by the national protected area system. This full financial assessment could then be compared with alternative uses of the land and the management costs of the protected areas. The mismatch between economic values of protected areas and investments being made in them would then be clearly identified, providing a strong basis for justifying additional investments in the protected area system. These investments can be generated through measures such as those outlined below.

- *Encourage investment in protected areas.* The key to the future of financing for protected areas is building a national consensus on overall environmental priorities. Protected area institutions must have a broad mandate from civil society to be able to work with other public sector agencies and civil society to set national goals for conservation. Conservation investments by governments are as essential to the welfare of society and as legitimate a public investment as defense, communications, justice, health, and education. Protected areas benefit the nation; that is the reason some of the best ones are called "national parks." They also benefit the world; that is the reason outstanding sites are recognized under the World Heritage Convention, the Wetlands Convention, and the Biosphere Reserve Programme. So both national and international sources of financial support should be tapped.
- *Promote private sector investment.* The private sector is often discouraged from investing in protected areas because of high levels of market and political risk, high initial capital costs, returns that may be earned only in the distant future, and difficulties in implementing user charges due to high exclusion costs. Private investors should be provided with appropriate incentives, such as security of tenure, appropriate contractual relations, the removal of perverse economic incentives, correction of distortionary policies, and removal of barriers to entry. Where such conditions are met, private investment in protected areas can be increased significantly.
- *Establish a national conservation trust.* New responses to insufficient or unbalanced investment in protected areas include fostering innovative funding mechanisms such as trust funds, dedicated funding of receipts from tourism, or debt-for-nature swaps. A nonprofit and transparent national conservation trust could serve as a general mechanism for mobilizing such funds. In the case of the private sector, the trust could serve as a partner of "green" business by certifying the performance of businesses operating in and around protected areas. Further, a national conservation trust would be able to play a significant role in leading the private sector down paths of sustainability by using its capital to create ventures that both make a profit and use protected area resources in a responsible way.
- *Change laws to encourage fund-raising.* The general public is often very interested in supporting protected areas but lacks any effective means of demonstrating its support, except perhaps through increased visitation. Given appropriate structures, the general public will often be extremely generous in its support of conservation, especially through conservation-related NGOs. To tap this potential, governments should examine laws and regulations governing the activities of the nonprofit private sector. Partnerships with for-profit concerns should be encouraged, and tax breaks

for charitable contributions should be instituted or enhanced. In some cases, it may be possible to establish a special protected areas fund on the basis of contributions from the energy sector, with the payment related to the benefits being provided by protected areas. A national conservation trust could also play a role here.

- *Enable protected areas to retain more of their value*. Revenues from tourism operations, fees for collecting genetic resources, income from watershed protection, and so forth generally feed into the central treasury rather than to the individual protected area or even the protected area system. If earnings from protected areas are returned to the central treasury, it is quite understandable that a protected area agency will regard such "profits" as losses because it loses control over them. Dedicated funding is a policy option that would enable a closer connection between income and expenditure. Rather than tourist concession fees going into the central treasury, if they are returned directly to the protected areas, then the concession-holder becomes much more part of the management enterprise. One option is to enable a national conservation trust to receive funds from gate fees, camping grounds, parking fees, bed fees, and various other fees, which can then be reinvested in the protected area.
- *Form independent companies to manage tourism to protected areas*. Government protected area management authorities could form independent companies to administer tourism matters, retaining shares in the company, which would pay dividends to provide a consistent flow of income to the protected area. Forming an independent company enables the necessary investment capital to be raised, harnesses the business skills of the private sector, reduces dependency on the budget of the protected area agency, fosters investment in marginal and rural areas, strengthens private sector support for the conservation effort, and improves access to loan finance. The joint venture company could be a nonprofit corporation whose income is redistributed to either tourism development or conservation projects under the auspices of a national conservation trust, and provides a mechanism for the establishment of joint projects with local communities and the private sector. In some cases, local communities also could be given shares in the company, perhaps ranging from 10 to 25%. Making the local community shareholders provides a number of benefits, including fostering a sense of ownership and accountability for the environment among the communities, improving communication between protected area managers and local communities, stimulating secondary entrepreneurial opportunities, and empowering people through their ability to participate in local decision-making processes (Pollack, 1995).

DESIGN ROBUST PROTECTED AREA INSTITUTIONS

A protected area system needs diversity in institutional approaches. Government conservation institutions in many Asia-Pacific countries claim an exclusive mandate to manage conservation areas and activities but lack the necessary human, financial, and technical resource capacities to carry out that mandate effectively. But protected areas support biological processes that often operate at small scales that vary dramatically in climate, elevation, structure, and importance from one site to the next. An overemphasis on centralized protected area agencies can undermine institutional mechanisms at smaller scales, such as traditional approaches to conservation based on local knowledge about specific complex interactions and concerns about natural capital that can be applied in daily life. This clearly is not an either–or situation, but instead calls for creating new systems of governance for protected areas, with different institutions having different responsibilities at different scales. Simply stated, large-scale, centralized governance units do not, and cannot, have the variety of response capabilities — and the incentives to use them — that large numbers of local institutions can have (Ostram, 1998).

Decentralizing both management and funding will also help cure one of the perennial problems of protected area management, namely, the lack of incentives for attracting and keeping the most energetic and qualified personnel working at the local level rather than in the cities. In seeking to find the most appropriate institutional arrangement for managing protected areas, it is useful to seek basic organizational principles that will lead to effective institutions for managing protected areas, appropriate to the national setting. The following principles should be considered:

- *Clearly identify stakeholders*. Identify who has an interest in how the area is managed, and the degree of legitimacy of the various stakeholders.
- *Clearly define boundaries*. Ensure that all stakeholders are very clear about what are the boundaries of the protected area and its surrounding buffer zone, although the outer boundary of the buffer zone may be very fuzzy, and who has the rights to what resources under what conditions within these boundaries. This helps ensure that those who invest in the resource benefit from their investments, thereby providing an incentive to invest.
- *Adapt regulations to local conditions*. Regulations are required to ensure that time, place, technology, quantities of resources to be managed, and so forth are related to the local conditions and to the rules regarding labor, materials, and funds. Such regulations should be site specific, because uniform rules established for an entire nation can seldom take into account the specific attributes of specific protected areas, the resources they contain, and the socioeconomic conditions of local stakeholders.
- *Make collective-choice arrangements*. The stakeholders affected by protected area regulations should be able to participate in modifying them. The reality for most protected areas is that the government agency seldom has sufficient presence to play a fully effective role into the day-to-day enforcement of protected area regulations. Compliance is more likely under conditions of government by consent, social pressure by the community, and enlightened self-interest, backed up by law enforcement where necessary.
- *Establish a means of monitoring*. A means must be available to audit the conditions of the protected area and its resources and the behavior of the

various stakeholders involved, to provide feedback to management. Most traditional systems of resource management are organized so that monitoring is a natural by-product of using the resource, but modern protected area managers should make an explicit effort to build monitoring into management.

- *Provide graduated sanctions*. Those who violate the regulations of protected areas should receive appropriate sanctions within the context of the managing institution. Local sanctions work best in long-enduring communities that are stable, and are unlikely to work very well in highly dynamic settings, for example, where many villagers are recent immigrants. The latter condition may require greater investments in law enforcement by the protected area agency.
- *Establish conflict-resolution mechanisms*. Both local communities and protected area managers should have rapid access to low-cost, local arenas to resolve conflict among stakeholders or between villagers and government officials (Lewis, 1996). This helps provide the feedback needed for adaptive management.

In seeking the best institutional arrangement for protected areas, governments should give careful consideration to establishing parastatal institutions, in conjunction with a national conservation trust. Such institutions could be more efficient and cost-effective than the institutions that are currently in place.

Provide Incentives to Encourage the Private Commercial Sector to Contribute to Protected Areas

A major challenge — how to link private sector operations to the benefits provided by protected areas. The private sector has proved extremely effective in avoiding costs; it is an effective "free rider." This challenge can be addressed by implementing the following guidelines:

- *Governments should provide an open, competitive market.* While some government regulation is certainly required, an appropriate level of competition will lead to a better product at the best price, enabling entrepreneurs to enter the marketplace with new ideas and new approaches that are consistent with the national protected area objectives established by the government.
- *Governments should provide a clear and stable policy framework*. The private sector should have very clear rules that apply equally to all competitors. Environmental standards should be clear and explicit, and be sufficiently powerful to ensure that the resources of protected areas are well managed, but not so strict that they serve as a disincentive to investment. More positively, tax breaks or other economic incentives for contributions to protected areas could generate greater private sector support.
- *The private sector should adopt appropriate standards*. Consumers should have confidence that the private sector is behaving in a way that is appropriate

to the public interest in the field of protected areas. The businesses involved in protected areas should be encouraged to agree voluntarily to appropriate standards of quality and environmental performance, based on guidelines provided by government.

- *The private sector should volunteer stronger support to protected areas.* It is also possible to harness greater involvement of the private sector through voluntary cooperative programs, such as those involving various energy-related companies around Kutai National Park in Indonesia's East Kalimantan Province (MacKinnon and MacKinnon, 1986). In at least some cases, these voluntary agreements may be far more effective than those forced by regulations. But in other cases, regulations, or at least the threat of them, may be an essential incentive.
- *Stakeholders should agree on realistic objectives.* Governments, the private sector, NGOs, researchers, local communities, and other stakeholders should work together to agree on objectives and set targets that recognize the realities under which business operates. These targets should encourage efficiency and cost-effectiveness, permit flexibility of responses to meet goals, allow for gradual introduction of any new regulations so that business has time to adjust, be fair and equitable across business sectors, and provide transparency of compliance to eliminate free riders (WBCSD, 1997).
- *Governments should use economic instruments to motivate the private sector.* Governments should design and implement economic instruments to encourage actions that work toward national objectives for protected areas. Numerous economic incentives for conservation are available for use by governments (OECD, 1997; McNeely, 1988). For example, governments could direct agricultural subsidies to activities by the private sector that promote behavior that is consistent with the requirements of protected areas. Tax policies can also encourage businesses to provide support to protected areas, for example, through enabling charitable donations to be deducted from corporate taxes and considering support to protected areas as a normal business operating expense. Similarly, disincentives such as fines or taxes for inappropriate corporate behavior should be part of the package.
- *Governments, NGOs, and the private sector should educate the market.* The "consumers" of protected areas need to be educated about the multiple values of protected areas. Economists argue that harnessing market forces is an important step in this process, beginning by making appropriate information available to consumers. Paying a fair price for benefits received is a basic principle that should be applied to protected areas, but the "market" must be made aware of what the fair prices are.
- *Governments, NGOs, and the private sector should promote microenterprises.* Relatively little has yet been done in the region to stimulate the creative efforts and energies of small biodiversity-based rural businesses, which will often be at the microscale where much innovation can take place. Investment, for example, through a dedicated fund, could facilitate private sector development of enterprises based on sustainable use of

> biological resources and conservation of biodiversity in and around protected areas, also helping to ensure more equitable distribution of benefits arising from such use (and thereby achieving all three objectives of the CBD). Commercialized microenterprises based on biological resources could become critical components in developing buffer zones around protected areas. If planned and developed carefully, they can help safeguard the protected areas and generate revenues for the local communities. It would seem most appropriate for the private sector to take the lead in this field, working closely with interested NGOs with connections at both the grassroots and at the marketing level, and with the local communities who will be the key producers.

If suitable incentives can be provided to enterprises to assume a certain degree of financial risk, a wide range of private investments could be secured by protected area managers. Encouraging investment will require an accessible framework for providing information, structuring negotiations, and ensuring project security. Although the market itself may be able to regulate financially viable investments, for high-risk investments some sort of claim certification must be provided. One possibility might be to use a national conservation trust as suggested above to provide this secure framework.

Seek New Approaches to Involve Local Communities in Protected Area Management

Far more needs to be done to build support from local communities for protected areas. This will require a challenging combination of incentives and disincentives, economic benefits and law enforcement, education and awareness, employment within the protected area and employment opportunities outside, enhanced land tenure and control of new immigration, especially where the buffer zones around protected areas are targeted for special development assistance. The key is to find the balance among the competing demands, and this will usually require a site-specific solution.

An important principle is first to do no harm. Relocating communities from protected areas has characterized the establishment of protected areas in the past, but the modern perspective recognizes that relocation can cause severe negative social, economic, cultural, and even ecological impacts. This approach should be used only as a last resort after careful study and planning, and only where it is clearly documented that resident peoples are truly detrimental to the objectives of the protected areas and where adequate alternatives and mitigating measures have been established for effective relocation that will improve the standard of living of those relocated (West and Brechin, 1991). Some communities may be pleased to move under such conditions.

A key factor is the stability of rural communities, implying that governments should be particularly cautious when contemplating major efforts at relocating people from one part of the countryside to another. Those people who have developed long-term relationships with particular settings, and have developed knowledge about how to manage the resources contained within those ecosystems, are likely to have very different relationships with the land and its resources than are new immigrants who have no particular linkage to local resources and often receive considerable subsidies from outside. The new arrivals frequently are responsible for more destructive land-use practices than are the long-term residents, but of course new technologies and new markets can be expected to change behavior of local villagers irrespective of their traditional conservation practices.

At a minimum, local communities should be deeply involved in buffer-zone development activities, and should be consulted on any decisions that affect them. In many cases, giving local people preferential treatment in terms of employment within the protected area, including seasonal or project-based employment, providing economic incentives to establish tourism or other income-generating activities in the buffer zone, and ensuring an appropriate flow of benefits from the protected areas to the surrounding lands can help to build a positive relationship between protected areas and local communities.

In other cases, it might be most sensible to return the full management responsibility for at least the buffer zones to the local community, leading to community-owned forests that serve at least some of the functions of protected areas. These may be managed under forest stewardship contracts between government agencies and local communities. Successful project interventions in buffer zones should address community priorities, providing incentives such as material benefits in a way that builds long-term partnerships rather than dependency. Successful projects add diversity to the development options available to the local community and build community self-reliance. While economic incentives are important to compensate local communities for opportunity costs, such compensation should be in the form of improving access to suitable productive resources, such as better agricultural land or technology, which can provide continuity between prior modes of production and opportunities for economic improvement. This is far better than simply providing cash.

Most rural communities in and around protected areas are anxious to find new ways of earning income. While local communities generally retain effective control over at least minor forest products, the fact that they are seldom legally entitled to these historical rights leads to a degree of insecurity that militates against sustainable utilization and instead fosters conflict between local people and the protected area. Because this impulse can have negative impacts on protected areas, many Integrated Conservation and Development Projects (ICDPs) are actively seeking alternatives to resource harvesting and land conversion. For-profit enterprises involving local communities around protected areas can either build dependence on local resources or seek to reduce such dependence. Generally speaking, enterprises within a protected area should be basically nonconsumptive, such as tourism or limited collection of genetic materials, while enterprises in the buffer zone can be both nonconsumptive and consumptive. An important decision is the choice between coupling and decoupling the economic interests of local people from the ecological interests of the protected area. Projects in buffer zones, such as plantations of fast-growing trees to relieve the pressure on forest timber, cash-crop initiatives, butterfly farms, investment in better farming practices and so forth, are designed to shift the economic interests of local people away from exploiting resources in the protected area.

By contrast, some activities are designed to enhance the dependence of the local communities on the natural resources or ecological services to be conserved. For example, nature-based tourism will bring revenues as long as the local environment is well preserved and attractive to tourists. Selling hunting rights to tourists is viable and lucrative only so long as the protected area sustains a sufficiently abundant wildlife population and the protected area regulations permit such use. Medicinal plants can be collected in the wild only as long as they are not overexploited, and so forth. Whether a "coupling" solution is likely to be more effective than a "decoupling" one, or whether a combination of the two is preferable, can be established only within a specific ecological and socioeconomic context. A general point here is that the greater the interaction of a community with biological resources of the forest and the greater the proportion of the community that gains or loses from that interaction, the more likely is the success of comanagement projects.

It is possible that some local communities have a limit on their perceived needs, and once their basic needs are met, then they will reduce their impact on protected area resources. But this rosy assumption is far from a generality, and most communities contain at least some individuals who happily will try to exploit more from a system than can be supported in a sustainable way, even if the social costs far outweigh the private benefits. This means that protected area management should be based on a clear understanding of rules and regulations, and effective means of enforcing them through various kinds of incentives (such as employment, clean water, various kinds of linked development, and so forth) and disincentives (such as public ostracism, fines, and jail terms).

A more positive relationship between local communities and protected area managers can be built on the following steps:

- Identify all critical interactions (physical, biological, economic, and cultural) that link the protected area to local communities, regional landscapes, and private enterprises.
- Understand the meanings and values that local communities attribute to the protected area and the region.
- Inform local populations about the national and international significance of protected area resources and strive to develop a sense of pride in the protected area and the region.
- Provide benefits that compensate for any opportunity costs paid by local communities because of the protected area, over and above the benefits the area provides to them.
- Conduct planning as an open process that provides opportunities for all stakeholders to express their opinions and views about the future of the protected area and the region.
- Avoid preconceived ideas about how things have to be done, because most problems have multiple solutions. Give priority to local solutions for local problems.
- Use advisory councils with members who can contribute to maintaining open communications with local populations and enterprises and who are sensitive to local values within the region (after Zube, 1995).

A GENERAL SOLUTION: DESIGNING AND IMPLEMENTING COLLABORATIVE MANAGEMENT OF PROTECTED AREAS

One promising overall approach to building cooperation between civil society stakeholders and protected area managers is through "collaborative management" or "comanagement" of protected areas — partnerships by which various stakeholders agree on sharing among themselves the management functions, rights, and responsibilities for a territory or set of resources under protected area status (Borrini-Feyerabend, 1996). The stakeholders can include the protected area management agencies, local residents and resource users, NGOs, business, research institutions, and others. Three main principles provide the foundation for such efforts:

1. A critical first step in any collaborative management program is a careful and realistic assessment of the social, cultural, and economic benefits and costs that accrue to the various stakeholders from the protected area. Successful collaborative management is based on an understanding of the conditions and trends of the biological resources to be managed in and around the protected area. Such information also provides a foundation for determining whether project interventions are being effective.
2. Collaborative management processes and agreements need to be tailored to fit the unique needs and opportunities of each specific context, so approaches to stakeholder participation in different protected areas should be designed to fit their specific historical and sociopolitical contexts. In some cases, conservation easements — legally binding agreements under which landowners agree to sell or rent the rights to certain uses of their land — may be feasible ways to expand the effective protected area estate.
3. Conservation initiatives may be expected to change over time, along with changes in the legal, political, socioeconomic, and ecological factors that induce consequent modifications to the institutional setting; thus, collaborative management regimes may vary both from place to place and over time within a specific location. Adaptive management approaches, which include monitoring and research, therefore are highly relevant.

Under collaborative management, the various stakeholders develop an agreement — often a signed contract, such as a forest stewardship contract — that specifies their respective roles, responsibilities, and rights in the management of the protected area. Such an agreement or contract usually identifies:

- The functions, rights, and responsibilities of each stakeholder;
- An agreed-upon set of management priorities and a management plan;
- The budgetary allocation to implement the management plan;
- Procedures for dealing with conflicts and negotiating collective decisions;
- Procedures for enforcing such decisions; and
- Specific rules for monitoring, evaluating, reviewing, and updating the agreement and the management plan.

Powersharing between stakeholders in a comanagement arrangement can be promoted through free exchange of information, clear identification of the visions of the various stakeholders, and developing a shared vision through negotiation. Communities are often the least powerful of the stakeholders, so particular efforts are needed to ensure that they can participate meaningfully. For example, they can be empowered through electing representatives of local communities to protected area management boards, other stakeholders acknowledging the level of knowledge and expertise of local communities, and training community leaders in communication and negotiation skills.

The critical advantage of collaborative management is in establishing a formal agreement that establishes a durable, verifiable, and equitable form of participation that involves all relevant and legitimate stakeholders in the management of the protected area. It establishes the conditions for a true partnership among the local community, NGOs, the private sector, research organizations, and the government, with the latter retaining at least some responsibility even if it is only the provision of an overall policy framework for managing the protected areas.

Such formal sharing of authority among stakeholders has several significant advantages:

- Providing technical and financial incentives in conjunction with certain standards of behavior on the part of the stakeholder can forge a formal link between continued economic well-being and the effective management of the protected area.
- Stakeholders are likely to have a stronger commitment to the protected area if they have a clear sense that they will be part of the management arrangements and decision-making bodies that will be established as a result.
- Decentralizing management responsibility and vesting authority in nongovernmental, private sector, and community institutions as partners in collaborative management arrangements for protected areas can result in more effective action, mobilizing resources from the local communities and relevant private-sector agencies, thereby reducing the need for regulatory measures that often are very costly.
- Through vesting of some management responsibility at the community level the government protected area management authority can help empower local communities and develop local institutions, thereby contributing to a critical social development agenda (Renard, 1997).

One useful mechanism for putting this vision into practice is through "Integrated Conservation and Development Projects" that seek to reconcile conservation and community interests through promoting social and economic development among communities in and around protected areas. Such projects should be carefully designed to ensure that the interests of the various stakeholders are well represented, in a cost-effective manner.

CONCLUSIONS

This chapter generally has been upbeat and positive, seeking to present options that could work, and indeed have worked in some situations. But how justified is such optimism? The picture on the ground is decidedly mixed; every success can be matched by a disaster, and even "success" may be illusory or temporary, ready to evaporate when external funds have been expended. People in rural areas in all parts of the world are facing human pressures on resources that are unprecedented in human history. These new pressures require innovative responses, designed to support the long-term interests of both people and the many species upon which human welfare depends. It is too late for painless solutions, and for most of parts of the world, solutions will involve trade-offs among options that have both positive and negative aspects. Progress does not necessarily depend on "grand solutions," but rather on moving ahead with the means available.

Each country has its own specific challenges in establishing and managing its protected area system to meet the needs of its society. Protected areas are created by people, so they are expressions of culture and serve as models of the relationship between people and the rest of nature. Thus, the culture of each country is reflected in its system of protected areas, so each will tend to have different characteristics.

The single, overriding issue for protected areas is how to find the right balance between the generalized desire to live harmoniously with nature and the need to exploit resources to sustain life and develop economically. The problems facing protected areas are thus intimately related to socioeconomic factors affecting communities in and around protected areas, including poverty, land tenure, and equity. They also involve national-level concerns, such as land use, tourism, development, balance of payments, energy, and resource management, and global concerns such as biodiversity, climate change, and generation of new knowledge about life.

The program for national protected area systems advocated here needs to include both firm governmental action and alliances with the other stakeholders at all levels. National governments cannot delegate their role as guarantors of the conservation of a country's cultural and natural heritage, so the appropriate authorities must build the capacity to fulfill their regulatory and management duties and responsibilities. But civil society can share certain rights and responsibilities regarding the management of protected areas after careful preparations and an adequate definition of roles and responsibilities. Given the interests of NGOs, businesses, indigenous peoples, and local communities who live within or close to protected areas, alliances can be created among stakeholders enabling each to play an appropriate role according to clear government policies and laws. Social and economic incentives can be used to reward landholders and private sector industries that contribute effectively to protected area management.

If governments and the general public recognize the many economic, social, cultural, ecological, developmental, and political values of protected areas; if appropriate institutions are established to manage protected areas in close collaboration with other stakeholders; if sustainable economic benefits are enabled to flow to protected areas and their surrounding communities; and if information from both traditional knowledge and modern science can be mobilized to enable protected areas to adapt to changing conditions, then the protected areas of the 21st century can be the engines for new forms of rural development that ensure a better life for all.

REFERENCES

Borrini-Feyerabend, G., 1996. *Collaborative Management of Protected Areas: Tailoring the Approach to the Context,* IUCN, Gland, Switzerland.

Caldecott, J., 1997. Indonesia, in *Decentralization and Biodiversity Conservation*, Lutz, E. and Caldecott, J., Eds., The World Bank, Washington, D.C., 43–53.

Costanza, R. and 12 others, 1997. The value of the world's ecosystem services and natural capital, *Nature,* 387, 253–260.

de Klemm, C., 1993. *Biodiversity Conservation and the Law: Legal Mechanisms for Conserving Species and Ecosystems,* IUCN, Gland, Switzerland.

Dixon, J. and Sherman, P. B., 1990. *Economics of Protected Areas: A New Look at the Benefits and Costs,* Earthscan, London.

Kramer, R., van Schaik, C., and Johnson, J., Eds., 1997. *Last Stand: Protected Areas and the Defence of Tropical Diversity,* Oxford University Press, London.

Lewis, C., 1996. *Managing Conflicts in Protected Areas,* IUCN, Gland, Switzerland.

MacKinnon, J. and MacKinnon, K., 1986. *Review of the Protected Areas System in the Indo-Malayan Realm,* IUCN and UNEP, Gland, Switzerland.

McNeely, J. A., 1988. *Economics and Biological Diversity: Developing and Using Economic Incentives to Conserve Biological Diversity,* IUCN, Gland, Switzerland.

McNeely, J. A., 1998. *Mobilizing Broader Support for Asia's Biodiversity: How Civil Society Can Contribute to Protected Area Management,* Asian Development Bank, Manila.

McNeely, J. A. and Guruswamy, L., 1998. Conclusion: how to save the biodiversity of Planet Earth, in *Protection of Global Diversity: Converging Strategies*, Guruswamy, L. and McNeely, J. A., Eds., Duke University Press, Durham, NC, 372–387.

Miller, K. R., 1996. *Balancing the Scales: Guidelines for Increasing Biodiversity's Chances through Bioregional Management,* World Resources Institute, Washington, D.C.

Myers, N. and Kent, J., 1998. *Perverse Subsidies: Tax Dollars Undercutting Our Economies and Environments Alike,* International Institute for Sustainable Development, Winnipeg, Canada.

OECD, 1997. *Investing in Biological Diversity: The Cairns Conference,* Organization for Economic Cooperation and Development, Paris.

Ostram, E., 1998. Scales, policentricity, and incentives: designing complexity to govern complexity, in *Protection of Global Diversity: Converging Strategies*, Guruswamy, L. and McNeely, J. A., Eds., Duke University Press, Durham, NC, 145–163.

Pollack, G., 1995. Kwazulu case study, in *African Heritage 2000: The Future of Protected Areas in Africa*, Robinson, R., Ed., IUCN, Gland, Switzerland, 49–53.

Renard, Y., 1997. Collaborative management for conservation, in *Beyond Fences: Seeking Social Sustainability in Conservation,* Borrini-Feyerabend, G., Ed., IUCN, Gland, Switzerland, 65–67.

Shah, A., 1995. *The Economics of Third World National Parks: Issues of Tourism and Environmental Management,* Edward Elgar, Aldershot, U.K.

WBCSD, 1997, *Signals of Change: Business Progress towards Sustainable Development,* World Business Council for Sustainable Development, Geneva, Switzerland.

West, P. C. and Brechin, S. R., Eds., 1991. *Resident Peoples and National Parks: Social Dilemmas and Strategies in International Conservation,* University of Arizona Press, Tucson.

Zube, E. H., 1995. No park is an island, in *Expanding Partnerships in Conservation*, McNeely, J. A., Ed., Island Press, Washington, D.C., 169–177.

3 Ecoregional Perspectives in Conservation: Recent Lessons and Future Directions

Sarah Christiansen and Eric Dinerstein

CONTENTS

INTRODUCTION

The use of adaptive collaborative management* in conservation over the past decade has tended to focus on important sites for biodiversity such as protected areas, hoping to assure the long-term stewardship of natural resources within these "localized" areas. As a result of these experiences, a great deal has been learned about the potential of local collaborative strategies for conservation (Brandon and Wells, 1992; Larson et al., 1998; Salafsky and Margolius, 1999). These experiences, including lessons about why these efforts sometimes fail, are important to the continued adaptation and refinement of local conservation actions. However, because many

* For the purposes of this chapter, adaptive collaborative management is used as the operative term to reflect collaboration as an inherent part of adaptive management employed within broad-scale conservation.

0-8493-0020-7/01/$0.00+$1.50

complex forces operate beyond local conservation sites, what has been learned from these experiences is not by itself enough to sustain conservation efforts.

Joint management of parks, enterprise-based conservation initiatives, and even supportive national policies for conservation are often too small or too short-lived to make a difference in the face of powerful and influential forces. Even where local conservation is successful, often it only creates islands in larger degraded landscapes. In these circumstances — now common around the world — local conservation gains tend to be outweighed by the irreversible loss of biological diversity, degradation of ecosystem function, and corresponding deterioration in human economic and social well-being.

Faced with these realities, conservation thinking and action has shifted toward larger scales. This shift has important implications for adaptive collaborative management (ACM). This chapter will discuss the importance of broadening the ACM process to reflect this shift. Examples from large-scale conservation efforts are used to introduce this next generation of conservation thinking and action. They will also inform exploration of the issues, challenges, and potential for defining and refining ACM in conservation.

THE NECESSITY OF SCALING UP

The evolution of conservation toward broader scales is driven by the recognition that (1) the ecological and evolutionary processes that create and sustain biodiversity occur at a much larger scale than that on which conservation has traditionally been practiced and (2) the scale of threats to the survival of biodiversity is greater and more complex than that addressed by conservation design and management in the past. Conservationists are increasingly looking comprehensively beyond sites and species to address the broader social, economic, and policy forces that can make the difference between conservation success and failure.* Although approaches that target different scales, including combinations of bottom-up with top-down approaches are not entirely new to conservation, the benefits of acting at multiple scales often are not fully realized. One critical reason is that the biological targets of conservation programs are often not clearly defined, and therefore there is no reference point for weighing threats and measuring conservation success. To address this shortcoming, practitioners are beginning to articulate their biological conservation targets within landscapes and waterscapes. Framing these targets within conservation units that better reflect how nature operates, such as ecoregions,** allows the establishment of more useful biological benchmarks.

In ecoregion conservation, these biological targets reflect a design that seeks to conserve and, where necessary, restore the biological diversity of an entire ecoregion.

---

* Many conservation organizations are developing broad-scale conservation principles and approaches under a variety of names. See "Conservation in the 21st Century" in the appendix to this chapter for a joint vision statement developed collaboratively with CI, IUCN, TNC, WRI, and WWF.

** An ecoregion is defined as a large area of land or water containing geographically distinct assemblages of species, natural communities, and environmental conditions. Worldwide, 868 terrestrial ecoregions are identified with 104 marine and freshwater ecoregions (Olson and Dinerstein, 1998).

This does not mean that every individual of every species must be protected; it does mean, however, (1) that species, communities, and habitats must be represented and maintained within the ecoregions; (2) that the areas maintained must be large enough or connected enough that they are resilient to disturbances and change; (3) that viable populations of key species are guarded; and (4) that key ecological processes are also maintained. Developing conservation targets grounded on these principles across large-enough areas such as ecoregions is an important evolution in thinking to help develop strategies that can work across the multiple scales of influences that impact biodiversity.

Scaling up spatially to larger landscape units must be accompanied by temporal scaling up to much longer timescales. While the role and level of investment of a conservation advocate can, and probably should, evolve over time, there is a need to make longer-term commitments than project designs or funding cycles have typically allowed in the past. Scaling up means facing more uncertainties and becoming more vulnerable to mistakes. As such, one must be prepared to learn and adjust the approach over time. One way this can be done is by using the principles of ACM. Because ACM can be a time- and resource-consuming strategy, it is important that one learns from current and past ACM efforts.

SCALES FOR ADAPTIVE COLLABORATIVE MANAGEMENT

One reason that locally targeted collaborative management schemes have failed to achieve conservation results is that investment is often targeted at too small an area — for example, a cluster of village communities — to produce results across the landscape in which a resource management problem has arisen. The reality is that conservation successes will be short-lived if they do not address the driving forces behind the problem. In other words, an integrated set of conservation actions that will positively impact the state of the entire landscape under threat is needed, with implementation focusing on certain key, or catalytic parts, of that area.

To address this challenge of scale and focus, conservationists have been developing analytical tools that allow them to understand the key threat associated with identified conservation targets. Based on the scale and pathways of these threats, conservationists are better able to identify appropriate points of intervention and measures of success over time (TNC, 2000; Salafsky and Margolius, 1999). Some tools are helpful for understanding immediate threats, while others trace relationships back to the driving forces behind biodiversity loss (e.g., WWF MPO, 1999). The challenge now facing conservationists is the application of these tools at broader spatial and temporal scales.

The conversion of rich habitats by local farmers provides a good example of this challenge. On the surface, habitat conversion may appear to be a result of the farmers' efforts to meet immediate livelihood needs. However, by reviewing this occurrence in larger context, it may become clear that it is driven by more spatially and temporally distant factors, such as regional agricultural policies, international trade policies, or the changing global economy. Thinking across this broader context allows interventions to be more specifically targeted. It should also help practitioners evaluate whether or not adaptive collaborative management strategies should be applied at the farmer level or at higher levels, such as key junctures in the chain of agribusiness activities (Clay, 2000).

The idea of operating at multiple scales with diverse and often nontraditional actors and stakeholders is not entirely new to conservation thinking. However, this approach has rarely been used in combination with strategically defined landscape and ecoregional conservation targets.

MONITORING AND EVALUATION FOR ADAPTIVE COLLABORATIVE MANAGEMENT

The World Wildlife Fund (WWF) ecoregional conservation program is using a "Driver-State-Impact-Response" model to understand better how to measure progress over time and at broader scales, and to facilitate testing and learning.* This model seeks to understand the forces that *drive* human activities impacting biodiversity, the different *states* of biodiversity that can result, the *impacts* of different states of biodiversity on society, and, finally, how key segments of society *respond* to these states and impacts to influence or adjust the drivers of biodiversity loss. Table 3.1 outlines how the simpler version of this model, known as the "Pressure-State-Response" model, was used in the Gulf of California Ecoregion. To determine whether or not different conservation strategies and approaches are making a difference requires knowing how changes along the pathways of the model are induced by conservation efforts and other factors. There are difficult challenges in distinguishing between ongoing environmental and ecological changes, for example, periodic droughts that may cause human pressure on forests and desertification, and the effects of conservation actions over long time frames. In many cases, the level of complexity at larger spatial and timescales make it difficult to disaggregate these factors. Only by applying keen biological and socioeconomic insight throughout the design, monitoring, evaluation, and adaptation of implementation stages can this complexity begin to be addressed. An effective way of achieving this is to develop tracking indicators that help highlight trends and patterns in conservation status over time.

Although there are a number of environmental "report cards"** that help to understand global conservation status, there have not been any measures available for ecoregions or other conservation units. Data are needed, in particular visual data such as maps, tables, and graphs, using coherent indicators that reflect monitoring at several distinct levels. These should include sensitive species of special concern, critical habitats and landscapes, and ecological processes as criteria from which a monitoring system can be tailored and adapted over time.

---

* This model is derived from a more common framework called the "Pressure-State-Response" model with the addition of socioeconomic impacts to the package of indicators tracked over time (Freese, 2000).

** For example, the Living Planet Report is based on two indices: (1) the Living Planet Index which seeks to quantify changes in the state of the Earth's natural ecosystems over time and (2) the Ecological Footprint, which measures human pressures from consumption of renewable resources and pollution.

Another way monitoring systems for these units is being explored is in the Klamath–Siskiyou ecoregion, which is using simple and measurable biological and social indicators to show the changing state of an ecoregion over time. It provides the ecoregion with an important monitoring tool and a resource for informing and engaging stakeholders (PNP, 2000). Although developing these models is important, it is equally important for practitioners to embrace the principles of adaptive management in their thinking. This is a state of mind, a disposition of learning by doing that is critical to developing adaptive and collaborative conservation efforts. There are many examples of integrated conservation and development projects that dissolved in the absence of skilled capacity for systematic rethinking and learning (Brandon and Wells, 1992). ACM requires adjustments and changes, which often can be a source of conflict. Learning how to manage these in ways that do not threaten collaboration is critical.

Although the conflicts generated by multiple interests present a key challenge in ACM at the ecoregional level, it is important to understand that, if productively harnessed, they can be the "engine" that drives participants forward to achieve their goals (Lee, 1993). Conflict resolution and management skills, including understanding and knowing how to engage stakeholders, are now almost mandatory for conservation practitioners. However, three conditions must be in place if practitioners are to use conflict as a constructive dynamic within ecoregion conservation: (1) a framework for structured debate between conservation partners and stakeholders, (2) the capacity for sound monitoring and analysis at the ecoregional level, and (3) strong links or feedback loops between these two to promote learning (Razanatahina, 2000).

Although ACM at the ecoregional level involves many challenges, ongoing efforts are already providing important lessons and experiences. The remainder of this chapter will describe the experiences of three ecoregional initiatives, showing how conservation practitioners are using ACM as an integrating tool for the full suite of conservation strategies employed across an ecoregion.

LEARNING BY DOING — EXAMPLES FROM ECOREGIONS

WORKING WITH INSTITUTIONAL FRAMEWORKS IN THE MADAGASCAR SPINY FOREST ECOREGION*

The Madagascar Spiny Forest Ecoregion is a mosaic of spiny forests, riparian and gallery forests, freshwater ecosystems, and coastal mangroves. The spectacular plant communities — dominated by the endemic family of the Didieraceae — have developed unique ways to adapt to low rainfall, poor soil conditions, and periodic droughts. It is estimated that over 75% of all medicinal plants used in Madagascar come from this area. The ecoregion is also home to some of Madagascar's outstanding wildlife, including the ring-tailed lemur, Verreaux'sifaka, two terrestrial tortoises,

* Information on this example was adapted from documents from the WWF Madagascar Spiny Forest Ecoregion Program and personal communication with Anitry N. Razanatahina, WWF-Madagascar.

TABLE 3.1
Pressure-State-Response Model for the Gulf of California

Project/Issue	Goal	Pressure Indicators	State Indicators	Response Indicators
Fishing practices				
Shrimp bottom trawlers	Maintain and improve population, diversity, and size of benthic species	• Area trawled (km^2) in shrimping grounds • Trawlable area under use (percent)	• Relative by-catch species abundance (tons or populations) • Size of by-catch (cm) • By-catch/shrimp catch ratio (tons by-catch per year/tons shrimp catch per year)	• Number of permits or other restrictions • Use and efficiency of by-catch reduction technology • Reduction of trawling effort (percent of 2000 level) • Area of exclusion zones (km^2) • Market price of shrimp ($/kg) and percent exported
Artisinal fisheries	Sustainable use of marine resources by artisinal fisheries	• Number of boats and number and type of fishing gear • Intensity of fishing effort by location • By-catch of marine mammals (number/year) • Price of target species, e.g., shark fin, sea cucumber, lobster, scallops, jaiba, at market ($/kg)	• Stock of target species, e.g., snapper, octopus, swimming crab, shrimp, etc. (tons or populations) • Age structure of populations (life tables of three to four target species)	• Evidence of out-of-season fishing • Number of fishing concessions • Fish certified (tons and percent of production) by species • Area of no-fishing zones (km^2) • Number of permits per location per fishery • Number of products per fishery • Number of staff, budget, and technical capacity of regional fisheries authorities
Other commercial fisheries	Sustainable use of marine resources by commercial fisheries	• Size of sardine catch (tons/year) • Size of squid catch (tons/year)	• Stocks of sardine (tons) • Stocks of squid (tons)	—
Sports fishing	Sustainable use of marine resources by sports fishers	• Size of sialfish and marlin catch (number of tons/year) • Number of trips (per year per state/location)	• Abundance of sports fishing species, especially marlin and sailfish (population or tons) • Size of fish caught (cm)	• Reports from sport fishers (number/year)

Conserve regional wetlands	Conservation of critical habitat and freshwater inflows to maintain ecological processes	• Conversion of wetlands to shrimp aquaculture, marinas, etc. (ha/year and total area) • Pollution/water quality, e.g., industrial effluent, agricultural runoff, organic compounds (BCOD), heavy metals, or inorganics (mg/l) • Freshwater withdrawals for human purposes (km^3/year) and inflow to gulf (km^3/year) by catchment area • Number of households using mangrove wood	• Area, fragmentation, and isolatedness of mangroves (hectares and fragmentation index) • Area of other coastal ecosystems, e.g., mudflats, coastal lagoons, wetland vegetation, estuarine habitat, etc. (ha) • Mangrove quality (tree species diversity) • Abundance of waterfowl and shorebirds (populations or breeding pairs) • Presence of larvae of breeding fish and other key species (e.g., *Leptocephalus* sp.) • Presence of corvin breeding near estuaries	• Area under active management regime (hectares and percent of original area) • Capacity of responsible agencies (number of staff, funding, and technical capacity) • Level of public participation in management planning • Freshwater flow assigned to environmental use (percent of total flow) per catchment area • Freshwater allocation for human use (percent of total flow) per catchment area
Management capacity for protected areas	Maintain healthy island ecosystems, including terrestrial and marine components	• Fishing pressure (number of fishing boats) • Presence of exotic species (rodents, cats, goats, buffel grass)	• Abundance of key species, e.g., sea lions, seabirds, especially endemics (populations, breeding pairs) • Composition of benthic biota (populations) • Presence of sea bass larvae • Aggregation of reproductive fish (presence/absence)	• Establishment of marine protected areas (MPAs) • Management plan implementation (e.g., eradication programs) • Management capacity (number of staff, equipment, mandate/authority, funding) • Public support for MPAs
Cetacean conservation and awareness	Maintain Gulf of California as globally critical cetacean habitat	• Squid catch (tons/year) • Red tides (number/year)	• Abundance of selected whale and dolphin species, especially top predators (populations)	• Number of employees and revenue of whale-watching industry ($/year)
	Maintain viable vaquita population	• Vaquita by-catch (number/year)	• Abundance of vaquita (populations)	• Fraction of range protected (percent of total) • Proportion of fishing boats with excluder devices (percent fishing within vaquita range)

BCOD = Biochemical Oxygen Demand; MPA = Marine Protected Area.

and endemic species of birds, reptiles, and amphibians. Less than 3% of the total spiny forest area is under legal protection, and therefore most of the unique biodiversity of the ecoregion is found outside formal protected areas.

The social and economic dynamics of the ecoregion are characterized by, on the one hand, a rural agropastoral population that is heavily dependent on forest resources, and, on the other hand, a fast-growing urban population in the cities of Tulear, Ampanihy, Ambovombe, and Betioky. Slash-and-burn agriculture is a long-standing way of life in this area, as in other parts of Madagascar; however, repeated droughts coupled with a general rural crisis have accentuated the unsustainability of this practice during the last 15 years. Moreover, due to recent opening to regional markets in Mauritius and Reunion, slash-and-burn agriculture of maize for export has become a major source of deforestation. Maize is both exported to neighboring countries to produce animal feed and consumed locally during periods of drought. Extensive cattle grazing and uncontrolled fires are closely linked to the agropastoral life of the Mahafaly and Tandroy ethnic groups that inhabit most of the southern parts of the ecoregion, where the zebu is an integral of every important event of one's life. Unsustainable fuelwood harvesting to meet the demands of urban populations is a serious threat. The spiny forest is also being cleared for commercial agriculture of cotton and sisal. Whereas cotton is mainly processed for the local market, recent expansion of sisal plantations is partly a response to increasing demand for biodegradable packaging in European countries. Although the Mahafaly and Tandroy people have traditions that respect forests through taboos and stewardship of sacred forest areas, these are jeopardized by increasing in-migration of jobless, landless people driven by a pervasive economic crisis.

The long-term conservation vision for the Spiny Forest Ecoregion is to conserve a network of representative, sufficiently large, natural habitat blocks that can sustain viable populations of species and maintain ecological processes, and also to maintain the services and products from natural habitats that are important to people's well-being. This requires employment of innovative conservation strategies outside protected area boundaries through collaboration with key stakeholders. To achieve this, ecoregional staff are working with partners from the national environmental program, and local and regional nongovernmental organizations (NGOs).

Since 1996, the government of Madagascar, with funding from USAID, has been experimenting with an "ecoregional" approach to conservation that is oriented toward natural resources management. "Ecoregions" within this program are defined around major protected areas and the corridors connecting them. The WWF Madagascar Spiny Forest Ecoregion partly overlaps with one of the Madagascar government's ecoregions. The Malagasy Environmental Program has implemented a regional planning process based on the use of spatially derived data (AGERAS).* AGERAS is a process that allows for participation and collaboration of stakeholders in regional planning that incorporates conservation concerns. The AGERAS process is designed to promote an adaptive management approach as a foundation for the following processes within each of their priority areas:

* Appui à la Gestion Régionalisée et l'Approche Spatiale.

1. Raise awareness about the medium- and long-term benefits of managing natural resources in a sustainable way, with emphasis on the relationship between the environment and lifestyle and livelihoods of people who rely upon biological resources.
2. Facilitate the creation of a consultative structure that represents stakeholder interests.
3. Facilitate a participatory analysis of the threats to, and opportunities for, natural resource management and conservation.
4. Facilitate discussion on major strategies and priority actions of natural resource management through the consultative structure.
5. Facilitate the identification of funding sources for priority activities.
6. Implement activities.
7. Monitor, evaluate, and adjust strategies to learn along the way.

The ecoregion conservation process for the Spiny Forest Ecoregion is developing in close collaboration with these regional and local planning efforts. In the northern part of the ecoregion, the Comité de Régional de Programmation (GRP) plans, coordinates, and facilitates conservation action for the areas around Tulear. There are three major local/intercommunal-level consultation and planning structures under the CRP: (1) the FIMAMI,* a civil society association for the protection of the Mikea forests, (2) the AICPCM,** an association of the mayors of the towns bordering the calcareous Mahafaly plateau, and (3) the FMKB,*** an association for the protection of the Belomotra forests in the central western part of the ecoregion (WWF-Madagascar, 2000). These structures comprise community members, local elected officials, representatives of government agencies, local businesses, and traditional leaders. The WWF ecoregional strategy is to catalyze the creation of such structures in the remaining areas of the ecoregion and to reinforce the effectiveness of existing structures in integrating conservation objectives. This involves systematically reviewing these objectives and the impacts of their implementation and making necessary adjustments.

At the other extreme of the ecoregion, the CRP in the Fort Dauphin area, provides another notable forum for collaborative management. This committee was established in 1998 by a regional development planning body**** as a condition for the settlement of a large mining project. The objective of creating the committee was to promote participatory development planning in the area. The World Bank and USAID have provided considerable financial support to the implementation of a strategic environmental assessment (SEA) process that should help prioritize development options for this area. The objective of the SEA process is to integrate environmental concerns as early on as possible. Thus, it is an avenue for consultation

* Fikambanana Miaro ny Alan'i Mikea.

** Association Inter-Communale pour la Protection du Plateau Calcaire Mahafaly.

*** Fikambanana Miaro ny Alan'i Belomotra.

**** Commissariat Général au Développement Intégré du Sud (CGDIS) coordinates development activities throughout the southern region. It was created by the government in the late 1980s to organize food and water supply during the famine due to heavy drought in the south and continues as a forum for major government and industry actors.

and negotiation among powerful interests in government and industry and the less powerful communities and city-dwellers of the Fort Dauphin area. Injecting clear biodiversity targets within the process of this structure provides an opportunity for engaging actors at a broader scale who otherwise rarely enter into community-oriented efforts for collaborative management.

The ecoregional process acts as a catalyst and facilitator for conservation by providing critical supporting activities, such as research on ecological processes and socioeconomic drivers of resource use decisions in local areas and across the landscape. For example, at the request of CRP in Tulear, WWF has undertaken biological inventories and research to support the planning processes around Tulear and the Plateau Mahafaly. The information on the state of the biological resources of the area will serve to orient the prioritization of natural resource management projects in these two areas and contribute to raising the environmental awareness of communities and other stakeholders. Similarly, in the Fort Dauphin area, spatial biological information and biodiversity valuation methods are used to define conservation priorities and opportunities for the Anosy region. These will be included in the regional development plan for the area to be submitted to various donors. Other forms of support include the facilitation of linkages with other economic sectors, development agencies, and donors, as well as investment in the development of conservation management, organizational, and monitoring capacity. The ecoregion program facilitated the initiation of a health program that integrates environmental messages in basic primary health and family planning interventions in towns surrounding conservation zones in the southern part of the ecoregion by linking a local nongovernmental organization (NGO) health partner to the Summit Foundation. An initiative to link conservation activities to rural credit is currently under consideration between several governmental and nongovernmental development and conservation agencies. Since its beginning in 1997, the WWF ecoregional program has supported the University of Tulear in providing graduate-level courses in conservation biology and resource management. Other training initiatives consist of strengthening the organizational capacities of local communities to enable them to manage their resources, facilitating their access to a legal transfer of resource management from the government.

A law passed in 1996 that enables the transfer of natural resource management to community groups offers a promising avenue for collaborative management.* Under this law, communities wanting to take responsibility for natural resource management must formally organize themselves into an entity to design sustainable management policies for the resource. A contract is then entered between the community and the state, with the state devolving management responsibility to the community, thereby providing tenure security for the land or resource. This is an important step toward "legalizing" new, collaborative management structures. The Spiny Forest Ecoregion program is working with local communities around the forests of Ankodida to help them obtain such a management transfer. Major constraints experienced are suspicion about the new law and government commitment

* The GELOSE (Gestion Locale Sécurisée) law was issued in 1996 and its implementation decreed in January 2000. The first two management transfer contracts signed in May and June 2000 — Zombitse and Marojejy ICDPs — have both been facilitated by WWF.

in general, particularly as the informal traditional agreements were created to make up for government failure. There is also a lack of flexibility in the law, requiring long administrative procedures and the intervention of a government-accredited environmental mediator, sometimes resulting in discouraged communities that are no longer willing to engage in the process.

The AGERAS process itself presents many challenges. For example, during the first year of implementation, major constraints encountered by AGERAS included the lack of coherence in national-level policies and the difficulty of reconciling bottom-up and top-down aspects of the process (AGERAS, 1999). Since the structures created under the AGERAS process are all informal — lacking a legal mandate — the long-term prospects are unclear. The process could be vulnerable to factors such as policy changes of the national government, most specifically the decentralization process now under way in Madagascar

Generating Coordination Across the Klamath–Siskiyou Ecoregion

One of the most powerful aspects of broad-scale conservation is the potential to coordinate conservation across landscapes that reflect a mosaic of stakeholder interests. This is particularly true in the U.S. Pacific Northwest in the Klamath–Siskiyou Ecoregion. Recognized as among the top in diversity on Earth for temperate coniferous forests, the Klamath–Siskiyou straddles the Oregon–California border near the Pacific Ocean and stretches over 10 million acres. The region is home to more than 3500 plant species, 281 of which are found only in the region, and an extraordinary diversity of habitat types. More than 22% of the region's forests is late-successional/old growth, mostly on public lands. Within the last 100 years, bighorn sheep, California condor, gray wolf, grizzly bear, and McCloud River bull trout have all been extirpated and numerous mollusks are on the brink of being lost. Almost half of the watersheds are degraded, requiring restoration and improved management. Since 1994, logging in national forests and federal lands has dropped by 80% while harvesting on private lands has increased. Logging, combined with road building and other cumulative impacts, continues to degrade habitats and threaten watershed values and important indicator species. In spite of the imperiled state of much of the biodiversity of this region, there is significant potential for conservation.

In the last few decades, this region has experienced a shift from an economy that relied on an extractive industry, primarily of timber, to one now predominantly reliant on jobs in the service and retail sector. Throughout this transition there has been a polarization of business and environmental interests that has left a legacy of conflict that continues today. These deeply embedded political interests tend to hinder collaborative efforts. The association of an organization with "green" interests can, because of historic tensions and conflicts, impede efforts to negotiate a balance between development and conservation. This reality poses a significant challenge to organizations like WWF who want to work in the region, and necessitates the consideration of new collaboration approaches. In learning how and when to engage different stakeholders, organizations like WWF are forced to reflect anew upon their own role and function as a stakeholder in the region.

There are several existing initiatives in the ecoregion that seek to integrate conservation with the resource uses of communities, business, and government, such as the Healthy and Sustainable Communities Project* and the Humbolt County Planning Initiative.** However, in spite of these many and wide ranging efforts, they tend to be of limited geographic scope and there has been little coordination across the Klamath–Siskiyou Ecoregion to support integrated management for long-term biodiversity conservation. In response, WWF collaborating with local environmental groups in the Klamath–Siskiyou Alliance, supported the initiation of the People and Nature Partnership (PNP). While the PNP offered an opportunity for diverse stakeholders to gather, it became apparent that WWF acting as a primary convenor in this partnership limited the role of WWF and also limited the diversity of stakeholders involved. From the process of addressing these challenges at a conference of the PNP, the Jefferson Sustainable Development Initiative (JSDI) emerged as an effort to reach out to a broader spectrum of stakeholders and to function as a forum for communication and coordination of collaborative efforts across the ecoregion. Unlike previous coalitions, which tended to have narrow and specific interests, JSDI was designed to reach out to the broad spectrum of stakeholders: from private landowners, to commercial timber companies, to government representatives.

While WWF still provides important support for the JSDI, this kind of neutral body allows for a broader range of stakeholder voices at the table and for WWF to take on a stronger advocacy role for certain processes appropriate to conservation interests. For example, WWF can play an instrumental role in developing and monitoring conservation efforts to better ensure that conservation goals are addressed and met. Although it is still early in the process of working with JSDI, there is strong potential, for example, for WWF to promote the targets outlined in an ecoregional biodiversity vision for the Klamath–Siskiyou to inform what kind of activities can be promoted across the landscape. The success of this entity will depend on the ability to maintain a forum for multiple interests rather than being swayed toward any single objective or advocacy role. This requires linking regional awareness and shared commitment to an expanding network of individuals and organizations working together in on-the-ground projects that enhance the health of both communities and biodiversity.

Power of an Ecoregional Vision in the Wadden Sea Ecoregion***

The Wadden Sea is one of the world's most important tidal wetlands, extending along the North Sea coasts of Denmark, Germany, and The Netherlands. It is a

* The Healthy and Sustainable Communities Project is associated with the Southern Oregon Regional Services Institute (at Southern Oregon University), which developed livability indicators through a community-based process. Within the two counties — Jackson and Josephine — in which this initiative developed, the current challenge is moving from indicators to actions on the ground.

** The Humbolt County Planning Initiative offers an opportunity for citizens to take part in an effort to improve data collection and planning processes that help stimulate economic development, protect the environment, and enhance quality of life.

*** Information in this example was adapted from documents and communication from the WWF Wadden Sea Ecoregion Program and a case study written by Curt Freese, Senior Fellow, WWF.

dynamic ecosystem of 13,500 km^2 with tidal channels, mudflats, sands, beaches, salt marshes, dunes, estuaries, and a transition zone to the offshore area. The high biological productivity of the coastal habitats of the Wadden Sea is of extraordinary importance for birds, seals, shellfish, and fish. For example, more than 10 million coastal waterbirds breed in the region or stop over during migration — originating from a breeding area covering half the Arctic. It is also important habitat for 10,000 harbor seals and a nursery area for numerous fish species, many of commercial importance.

The negative effects of human activities on the Wadden Sea are primarily the high level of contamination by pollution, and habitat disturbance and destruction by recreation, hunting, commercial fisheries, and human infrastructure. Global warming may result in rising sea levels, and more frequent and intense storms. Protection efforts began in the 1970s, in particular with the formalization of engaging senior environment officials in all three countries in trilateral Wadden Sea conferences. Even with this strengthening of intergovernmental cooperation, economic necessities and other factors have contributed to declining public support for conservation efforts. Further erosion of public participation is magnified by the additional regulations imposed on many stakeholders by European Community law. Because the public has largely been left out of this process and is uncertain of the purposes and consequences of new developments, important conservation initiatives lack public support, or worse, are vehemently opposed.

The Wadden Sea team, comprising NGOs from across the three countries, worked together to develop a common vision in 1991. This vision is still active, resulting in important progress for actions within countries. At the same time, grassroots conservation groups and local stakeholders have tended to concentrate on local or national problems and needs without relating these to the entire ecoregion. In this context, the potential for collaboration among this coalition has not been fully realized.

Although "ecoregion conservation" is not the shared common approach across the various constituencies, WWF has used this framework to help augment and revitalize efforts beyond the political or institutional boundaries traditionally followed. This broadened scope is being harnessed to revisit the vision process to integrate better socioeconomic aspects. In particular, the emphasis on socioeconomic issues and public participation within the context of an ecoregionwide vision has served to reunify and refocus the efforts of the Wadden Sea team. An important breakthrough for the Wadden Sea team was attaining observer status on the Trilateral Working Group (TWG) that coordinates the three-country intergovernmental effort. The TWG is the implementation body of the agreements between the three ministers of environment from each country and is therefore responsible for the management plan developed in 1997. By obtaining observer status, the NGOs of the Wadden Sea team have more direct access to information and discussions within the ministerial conferences. For an organization such as WWF, this is critical to raising the profile of conservation issues as well as helping WWF and other NGOs to plan and act more strategically across the entire ecoregion.

There are many challenges in the future. Although there have been several national-level public participation forums for Wadden Sea planning, a regionwide plan for involving the public has yet to be approved or implemented. WWF ecoregional staff have proposed an innovative plan to rectify this situation that, if adopted in whole or large part by the TWG, would fundamentally change public participation and greatly improve the prospects for Wadden Sea conservation. This plan was acknowledged by senior officials in 2000 and the crucial opportunity for progress in adoption will be at the 2001 ministerial conference. In preparation for this, WWF staff in the ecoregion are working collaboratively with various NGOs, such as the Dutch Wadden Society, to develop contributions for this next conference. By stressing the importance of public participation in the overall ecoregional plan, conservation NGOs are building an important bridge between local stakeholders and high-level policy work. Local businesses, politicians, and others are interested in and supportive of this effort, and such local interest can quickly stimulate responses by local government. For example, the Environment Ministry of Lower Saxony, which is historically reluctant to engage with conservation NGOs, is now working with the Wadden Sea team on a collaborative project.

The experience of the Wadden Sea highlights the reality that, even though most stakeholders involved are not aware of or familiar with the term *ecoregion conservation*, there are initiatives such as this throughout the world that are already operating at larger scales and provide well-poised opportunities for promoting greater conservation. In this example, working to revitalize existing structures for collaborative management across the ecoregion is making progress toward more fully realizing the potential for transborder collaboration. Coupled with the already ongoing work within countries for broader stakeholder participation, the ecoregional approach is leveraging efforts toward a broader vision for conservation. This broader perspective is essential when many threats and opportunities for restoration and conservation are ecoregionwide.

IMPLICATIONS FOR CONSERVATION PRACTITIONERS

Approaching conservation challenges at the ecoregional scale, and therefore, by necessity, acknowledging and responding to a range of stakeholder interests and influences, has interesting implications for conservation organizations such as WWF. As shown by the European and U.S. examples, taking on conservation challenges at the ecoregional scale will expose conservation organizations to a far wider spectrum of structures, fora, and tensions than traditionally encountered or dealt with at the site level. As a result, conservation at the ecoregional scale will often be part of a larger regional equation — driven by political, private sector, or civil interests. This can present conservationists with the dual role of being a stakeholder as well as initiating or supporting facilitation of a participatory process that convenes a broader array of agendas and interests. In such circumstances, it is important that organizations like WWF identify their role and function relative to other stakeholder groups and the goals that they are advocating. The scale and complexity of ecoregional conservation, and the ambitious goals that it sets, mean that the "leadership" of conservation efforts may or may not be conservation organizations. Rather, the leadership of ecoregional efforts may fall to those best placed to mitigate the actions or attitudes that threaten the state of biodiversity in an ecoregion. As such, the role of the conservation organization can vary immensely across ecoregions — from participation in a coalition of NGOs such as in the Wadden Sea, to the provider of funding support for the creation of a neutral entity designed to overcome historical conflicts as in the Klamath–Siskiyou Forests, to the facilitator of partnership between government and community as in the Madagascar Spiny Forests.

CONCLUSIONS

Driven by the realities of global biodiversity loss, conservation has undergone an important evolution in recent years toward larger geographic and spatial scales. ACM, a key concept in the conservation toolbox, must respond by scaling up both spatially and temporally. As examples from Africa, Europe, and North America attest, broadening our thinking and design does not resolve the many complex issues inherent within ACM at any scale. It does, however, provide practitioners with a compelling challenge to clarify biological goals at broader ecoregional scales because these can act as a benchmark for both the identification and weighing of threats, and determination of the most appropriate scale for intervention. ACM also requires that practitioners pay more attention to monitoring and evaluation. This focus should encourage more thoughtful tracking of actions and impacts at multiple scales while also building a mind-set for learning by doing. Finally, embedded in this evolution of conservation paradigms are the implications for how WWF or other conservation organizations identify their role within the larger realm of interests that influence decisions on how to use natural resources. These early lessons in ecoregion conservation highlight both the challenges and the potential for better mitigating the pressures that threaten to dismantle the ecological systems upon which the range of stakeholders depends.

REFERENCES

AGERAS/MIRAY USAID, 1999.

Brandon, K. and Wells, M., 1992. Planning for people and parks: design dilemmas, *World Dev.,* 20(4), 557–570.

Clay, J., 2000. Extractive Industries and the Environment, unpublished document, World Wildlife Fund, Washington, D.C.

Freese, C., 2000. Monitoring and Evaluation of Ecoregion Conservation, unpublished document, World Wildlife Fund, Washington, D.C.

Larson, P., Freudenberger, M., and Wycoff-Baird, B., 1998. *WWF Integrated Conservation and Development Projects: Ten Lessons from the Field 1985–1996,* World Wildlife Fund, Washington, D.C.

Lee, K. N., 1993. *Compass and Gyroscope,* Island Press, Corvallo, CA.

Olson, D. and Dinerstein, E., 1998. The Global 200: a representation approach to conserving the Earth's most biologically valuable ecoregions, *Conserv. Biol.,* 12(3), 502–515.

Razanatahina, A. N., 2000. *Adaptive Management: General Guidelines for WWF Ecoregion-based Conservation Practitioners,* World Wildlife Fund, Washington, D.C.

Salafsky, N. and Margolius, R., 1999. *Greater Than the Sum of Their Parts. Designing Conservation and Development Programs to Maximize Results and Learning,* Biodiversity Support Program (BSP), Washington, D.C.

(TNC) The Nature Conservancy, 2000. *The Five-S Framework for Site Conservation,* Vol. 1, 2nd ed., TNC, Washington, D.C.

(PNP) People for Nature Partnership, 2000. *State of the Klamath–Siskiyou Region,* Ashland, Oregon.

(WWF) World Wildlife Fund, 2000a. *The Living Planet Report,* Gland, Switzerland.

(WWF) World Wildlife Fund, 2000b. Ecoregion Conservation unpublished paper, Ecoregional Conservation Strategies Unit, R&D, Washington, D.C.

(WWF) World Wildlife Fund, 2000c. *Stakeholder Collaboration — Building Bridges for Conservation,* Ecoregional Conservation Strategies Unit, R&D, WWF, Washington, D.C.

(WWF) World Wildlife Fund, Antananarivo, Madagascar, 2000. Draft Vision for the Madagascar Spiny Forest Ecoregion, unpublished document.

(WWF) World Wildlife Fund, MPO, 1999. *Root Causes of Biodiversity Loss,* Macroeconomics and Policy Office, Washington, D.C.

APPENDIX

Conservation in the 21st Century

Joint Statement by Conservation International, IUCN, The Nature Conservancy, The World Resources Institute, and WWF — World Wide Fund for Nature

September 1999

Over the past several years our organizations have dramatically changed the way we think about, and seek to implement, conservation. Several terms have been developed for these new approaches, including the ecosystem approach (Convention on Biological Diversity), ecosystem-based management (IUCN), ecosystem conservation (CI), bioregional planning (WRI, IUCN), and ecoregion-based conservation (TNC, WWF). Although there are differences in methodology and application reflecting the distinctions in our organizational missions and strategies, the overall guiding principles are in each case the same. These principles represent a shared vision and goals that we believe should point the way forward for conservation in the 21st century, and they are set out below.

Our vision is of a world in which both the full diversity of life and the richness and well-being of human cultures are secured for future generations. This will involve a balance between built and cultivated areas, a strong emphasis on networks of fully implemented protected areas, corridors linking these core areas and buffer zones for restricted use — a balance that will secure the future both for humans and for the millions of other species with which we share the planet. The size and interrelationships of these areas will depend on the dynamics of the species, ecosystems and human populations in the regions concerned. But the result will be the same wherever our vision is achieved: regions and communities in which people can realize their potential and live with dignity while ensuring that the full range of species and ecosystem diversity is maintained, both for their own sakes and for the vital ecosystem services they provide. The intrinsic value of biodiversity and its critical importance to human welfare mean that the aim must be zero loss of species due to human intervention. In practice this means planning and implementing integrated conservation and development programs on a larger scale than has been attempted so far. We are broadening our focus to encompass landscape, seascape, and regional scales, working closely with the key stakeholders and using the best scientific information to help ensure that we conserve the most important and representative terrestrial, freshwater and marine ecosystems both within and between countries. Moving to larger scales will also help us to address more effectively the broader social, economic, and policy factors that are critical to sustainable livelihoods and ecosystems. This is important because some of the impacts upon a particular region may originate in other parts of the world, for example, international demand for a particular commodity. Achieving our vision will be challenging because the pressures on the natural world generated by the sheer scale of human activities worldwide are greater than ever. But the alternative is impoverishment — not only of the natural world, but of our own children and of theirs. As we all know, the conservation of nature and natural processes is not an optional extra in achieving sustainable development, but the essential foundation of human welfare.

Guiding Principles

1. **Conserving — and where necessary restoring — the full range of biodiversity:** Conserving genes, species, communities, ecosystems and ecological phenomena on a scale that ensures their integrity is critical both for their own sake and because of the key role they play in providing the ecosystem and other services upon which we all ultimately depend. Achieving this means recognizing that the ecological processes that sustain biodiversity operate at large scales. Hence, larger scales are needed to accommodate migratory patterns, anticipate nature's time cycles, and absorb the impacts of global change. It is becoming clear that, albeit other components of the landscape matrix will be vital to ensure that our common goals are met, an extensive network of protected areas will be necessary to maintain the full range of biodiversity. Furthermore, for the highly threatened, top priority biodiversity hotspots and ecoregions, the preservation of all remaining habitat will be of paramount importance.

2. **Planning conservation and development at landscape or regional scales:** Moving to larger scales of planning and action requires, more than ever, that we be sensitive to the interaction of the social and economic as well as the ecological factors that shape the threats to, and opportunities for, conservation and development. A lasting reconciliation between the needs of human development and the conservation of natural systems depends critically on the engagement and commitment of key stakeholders, from local people to society at large, corporations, governments and donor institutions. Although the processes for participation will vary from one region to the next, the approach is a multi-stakeholder, collaborative endeavor, building

consensus by bringing together more diverse groups than ever before to determine the best means of ensuring the future both for the natural environment and its human inhabitants. Integrated conservation and development at the landscape or regional scale requires coordinated action by many actors. Government or non-government agencies can be catalysts or in some cases leaders, but diverse partnerships — with communities, NGOs, universities, governments, corporations, and others — will be essential to each phase of the design and implementation of conservation and development plans at regional scales.

3. **Investing in good science:** Stakeholder discussions — and the decisions and actions they lead to — need to be informed by the best available scientific guidance on all the key issues. Without good data and information, the quality of decision making will be flawed. For example, such information is critical for assessing threats to biodiversity and developing strategies to abate them, and in addressing the question of what size protected areas need to be to maintain biodiversity and ecological functions. The combination of rigorous science, traditional knowledge and practical politics are the foundation stones for success.

4. **National sovereignty and international cooperation:** It is important to establish cooperative arrangements among central, regional, and local government entities as well as private interests, that respect national and institutional sovereignty and existing mandates. Likewise, many ecosystems cross international boundaries, or multilateral cooperation is necessary to address such issues as the illegal trade in endangered species. Therefore international cooperation agreements for consultation, joint research, information management etc. are often important components of programs bringing together biodiversity conservation and sustainable development.

5. **Long-term commitment:** Cooperation among stakeholders and partners to bring about the social, economic, and ecological changes required to secure genuine conservation and sustainable development for a region is a long-term endeavor, with new challenges constantly arising to replace the old. Holistic and lasting solutions require long-term commitment.

In conclusion, the traditional focus of our organizations on species, protected areas, environmental policy, and public information has generated significant conservation action. But the challenges are growing, and it is clear that we need to work together with all the stakeholders at larger geographical scales to address these challenges, learning and adapting as we go, without reducing emphasis on the traditional conservation activities that remain critically important. The various terms that are being applied to such approaches are far less important than their shared principles, based on a long-term commitment to the welfare of species, ecosystems and human societies.

4 Learning and Adaptation for Forest Conservation

Jeffrey A. Sayer

CONTENTS

INTRODUCTION

The past three decades have seen significant investments in programs to protect forests in developing countries. As a result of these efforts the number and aggregate area of National Parks and equivalent protected areas have grown considerably. But a recent report released by the World Wildlife Fund — World Bank Forest Alliance (World Bank/WWF, 1999) has concluded that only a small proportion of these areas is under secure and effective conservation management. There is a clear mismatch between the high-level global commitment to nature conservation, as manifested in the 160 nations adhering to the Convention for Biological Diversity, and the lack of concrete conservation achievement in the field. We are in the odd situation of having apparently uniform agreement among the informed public that nature conservation is a desirable component of development. Almost nobody seriously argues for the destruction of tropical nature. But the success rate of on-the-ground investments in conservation remains disappointing (Wells and Brandon, 1992; Kramer et al., 1997; Wells et al., 1999).

This chapter draws upon the author's personal experience as a manager of conservation programs in tropical Africa and Asia. My involvement in conservation began in the days of the paramilitary command and control culture of postcolonial Africa. My first project in the late 1960s in Zambia may have been the first to use the term *integrated conservation and development*. But at that time it was still considered normal for our staff to parade every morning so that we could ensure that their boots were polished and their rifles properly greased. Monthly reports

0-8493-0020-7/01/$0.00+$1.50

focused on the number of transgressors that we imprisoned and the number of casualties that we incurred in the process. As we enter the 21st century, approaches to conservation have come to be based on engagement with, and participation by, people living in and around protected areas. We understand much more about the constraints to which conservation is subject. We recognize that there is no single "best outcome" for conservation. Biodiversity and ecosystem services are valued differently by different people. Conservation has to be based upon negotiation between the different interested parties, mediated by enforced regulation of the agreed objectives. Conservation has to adapt continuously as people's conditions and aspirations change and as nature itself evolves. The hypothesis of this chapter is that, although the conceptual framework for achieving conservation is now stronger, the institutions responsible for management are still dominated by a command-and-control culture that they have inherited from the past. The chapter therefore explores some of the organizational changes that are required if management of conservation forests is to be genuinely collaborative and truly adaptive.

LEARNING FROM MISTAKES

Much of the analysis of conservation failures has focused on the design of internationally funded conservation projects (e.g., Caldicott, 1996). This is perhaps understandable given that such projects have been the main mechanism through which international organizations have invested in developing-country conservation programs. Perhaps because of the accountability culture of the aid agencies that have supported most of these projects, the analysis of their impact has focused heavily on the intricacies of their design and implementation. There has been a failure to evaluate critically some of the fundamental assumptions underlying the projects. The incentive structures of most development assistance agencies provide disincentives to recognizing, or learning from, mistakes. Project managers in multilateral development banks or development assistance agencies will almost always attempt to rationalize failures in the performance of their projects. Failure is attributed to technical obstacles, a lack of capacity, or a lack of political commitment. It is unusual to challenge the assumptions that underlie the project, for example, to acknowledge that the outcomes being sought were not priorities for the major stakeholders involved. Many of the shortcomings of development assistance to conservation projects are indistinguishable from those that have characterized the failures of development assistance in general (Edwards, 1999).

Development assistance agencies always aspire to respond to demand from recipient countries, but in reality donors have only managed to maintain high levels of investment in conservation projects by allocating special budget lines for this purpose. Donors who have been rigorous in allowing countries to set the priorities have seen the share of their resources allocated to conservation decline rapidly. Conservation projects generally have been tolerated by host governments and local people rather than actively solicited. The outcomes that the projects sought to achieve were rarely of high priority to the people they affected. Project designers have been driven by the assumption that education and rationality would eventually lead people to support conservation. There was a touching faith on the part of conservation advocates in the ability to find win — win outcomes whereby local people's development needs would be met and conservation would be achieved. But this flew in the face of the compelling logic that long-term environmental gains rarely outweigh short-term local benefits among poor people struggling to survive. When the win — win outcomes did not emerge, it was attributed to a lack of understanding or awareness among people or an absence of technologies (agroforestry, buffer zones, accurate mapping, boundary demarcation, etc.), which would enable production and conservation to be reconciled.

It is interesting to compare the development of protected areas and their management agencies in developed and developing countries. In developed economies with greater levels of transparency and more secure land rights, the trade-offs between national conservation values and local needs were usually addressed when the protected area was established. This resulted in more finely tuned networks of smaller, more strategically located parks and reserves, complemented by extensive multiple-use areas. The need for *ex ante* negotiations has meant that in developed economies conservation objectives have had to be explicit and measurable.

In less-developed countries, planners in capital cities or in international conservation or development organizations set extravagant conservation goals based largely on the perception that it was possible and desirable to maintain large, apparently pristine areas in their existing state. This ignored the reality that these areas rarely were pristine, that they were always subject to some use by local people. Planners also made extravagant assumptions about the willingness of poor people to forgo future development benefits. Planners drew lines around empty areas on maps and worried later about sorting out the concerns of any local people or any conflicting claims upon resources. The past two decades have seen far more efforts to involve local people in protected area planning, but the relations between conservationists and local people may still be too heavy on convincing or coercing and light on genuine negotiation and accommodation (e.g., Saxena, 1997).

The proliferation of international initiatives to support conservation may also have shifted resources away from dealing with practical realities on the ground and toward a "mechanistic" one-size-fits-all view of conservation. Recognition, rewards, and professional advancement for developing country conservationists have gone to those who perform well in international meetings, not to those whose talents lie in achieving conservation outcomes in the field. A void has developed between conservation theorists and practitioners. Edwards (1999) has made the case for economic development requiring 10% international inspiration and 90% national perspiration, and this must certainly apply to achieving conservation.

International conservation projects still have a role to play, but to succeed they must be rooted in local realities. They must genuinely recognize the costs that are being imposed upon local people, and the need for negotiations with these people must be conducted on a "level playing field" (Edmunds and Wollenberg, in press). They must recognize that win–win situations are the exceptions, not the rule. In those situations where win–win solutions do really exist, one might expect the "invisible hand" of the marketplace to find them. For meaningful negotiations to occur it is necessary for conservation objectives to be defined in terms of more precise and measurable characteristics of the fauna, flora, or ecology. Similarly, development objectives must also be amenable to objective measurement. Negotiations must be based upon agreed criteria and these must be backed up by measurable indicators. This will force conservation agencies to be realistic about the difficulty of defending areas when strict protection clearly does not represent the optimum land use as viewed from a local perspective.

The failure of conservation projects in the developing world must be seen in this context of false underlying assumptions. Integrated conservation and development projects (ICDPs) and various forms of local, participatory management have all emerged as responses to heightened recognition of the need for genuine local buy-in for conservation. However, notwithstanding the considerable resources invested in them, their record of achievement has been disappointing (e.g., Sayer, 1991; Wells and Brandon, 1992; Wells et al., 1999).

The weaknesses of many local conservation projects, particularly the so-called ICDPs, lie in the persistence of the fundamental assumption that relatively short-term technical assistance projects could cause people to change their behavior in ways favorable to conservation. It was assumed that the provision of new technological options or other incentives to conserve natural areas would lead people to forgo the economic and social advancement sought by other members of societies and to pursue more ecological, locally sustainable lifestyles (Sayer, 1999). They have tried to persuade people living around protected areas to opt out of the world of globalization and market-driven economies and live in islands where environmental values predominate (Bossel, 1998).

Figures 4.1 and 4.2 characterize the behavioral changes that ICDPs have sought to impose on the subjects of their conservation projects. ICDPs often try to make people adopt a development trajectory that differs from that which would occur if market forces and normal government policies alone were to determine their behavior. People living in the areas of ICDPs are expected to adopt lifestyles different from those of the rest of the population. In the case of some of the larger ICDPs this is tantamount to attempting social engineering on a quite large scale. Some recent ICDPs in Indonesia and China have attempted to change the course of development of geographic areas with populations and economies as big as a moderate-sized country. Success would require significant changes in behavior among millions of people, many of whom are struggling for material advancement and who stand to gain little or nothing from the protected area.

Notwithstanding this gloomy portrayal of the recent history of conservation projects, there are signs that progress is being made. There are increasing numbers of examples where approaches to local conservation management have shown genuine respect for local people and have encouraged meaningful participation by local people. There has been a greater willingness of conservation managers to accept trade-offs and compromises. The conservation literature is replete with failures (Kramer et al., 1997), but it is perhaps useful to examine some of the elements of success that do undoubtedly exist. Success appears to be linked to the adoption of concepts derived from what have been variously described as "ecosystem," "collaborative," and "adaptive" approaches to management (Lee, 1993; Gunderson et al., 1995; Szaro and Johnston, 1996; Maltby et al., 1999).

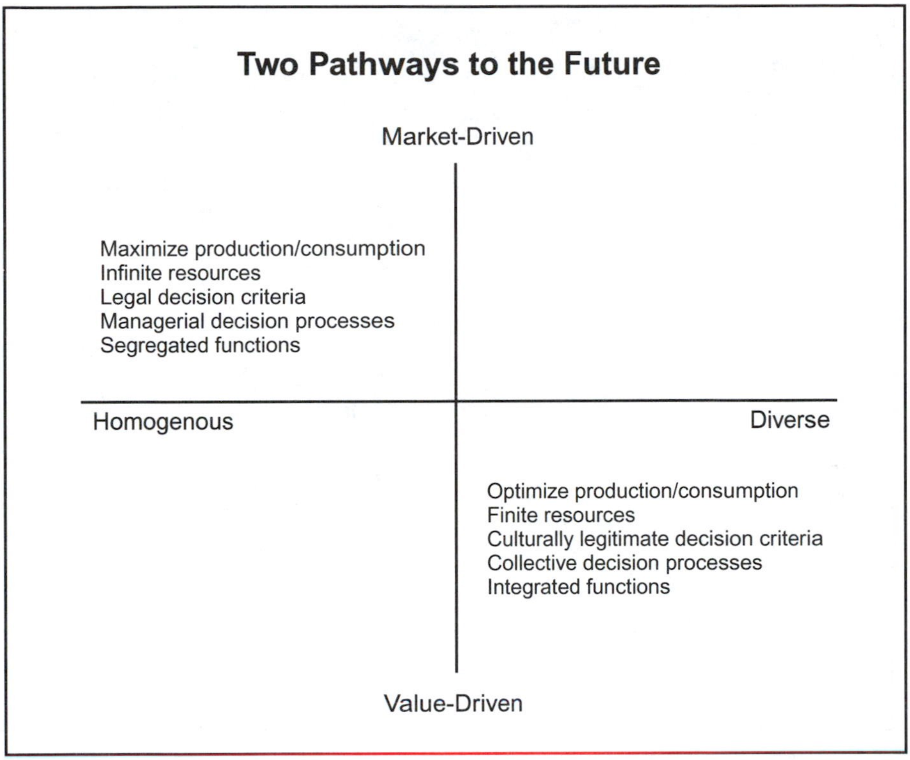

FIGURE 4.1 Market- and value-driven futures.

The fundamental characteristics of these approaches are that they are explicit about trade-offs and they involve genuine and fair processes of negotiation. Adaptive collaborative management (ACM) approaches are less driven by preconceptions and advocacy and are characterized by humility and willingness to learn on the part of their advocates. Significantly, these approaches are driven by the pursuit of agreed-upon outcomes with checks in place so that management can be modified to ensure that these outcomes are, to the extent possible, attained.

THE WAY FORWARD

The intellectual underpinnings for the management systems that are likely to work are now well established. A number of significant books have reviewed the understanding of integrated and holistic approaches to resource management that have been shown to be effective in developed economies (Lee, 1993; Szaro and Johnston, 1996). Ecosystem management was originally used to describe approaches to management of the full complexity of ecological systems in the Pacific Northwest of the United States, but people were still treated as exogenous to the system. The recent debate on the use of this term by the conference of the parties to the Convention on Biological Diversity (2000) has moved to a definition in which people

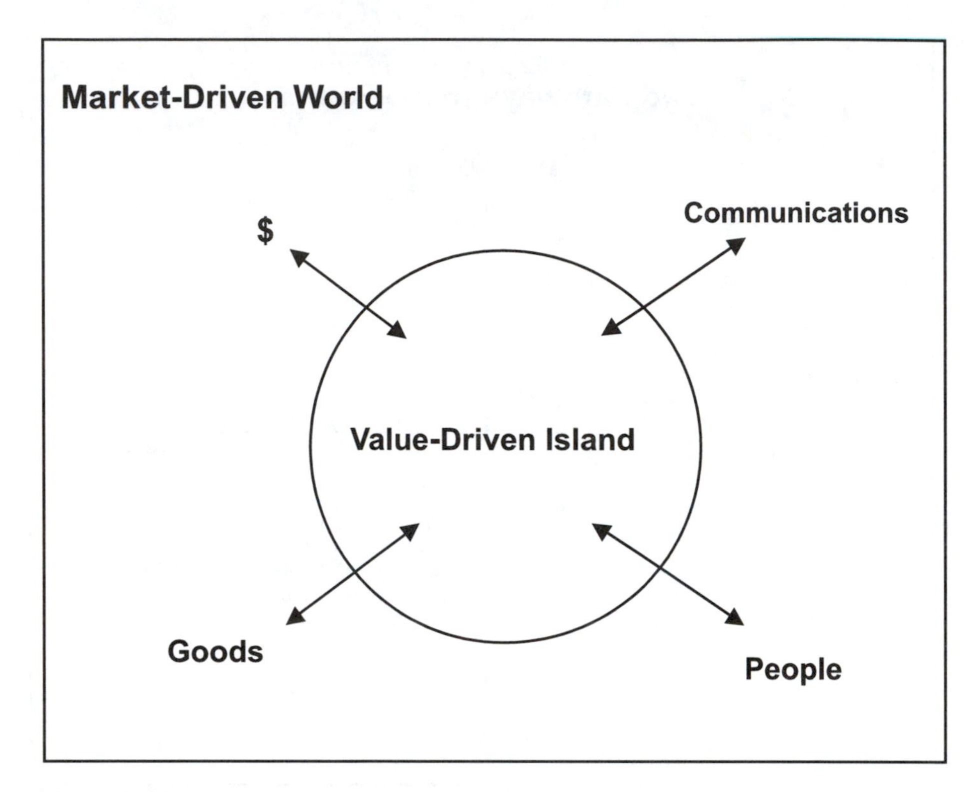

FIGURE 4.2 The myth: ecodevelopment in a value-driven world.

are endogenous. This is embodied in the so-called Malawi Principles for ecosystem management, which provide a compelling and concise summary of current thinking on collaborative and adaptive approaches to management. See the appendix to this chapter.

It is instructive to consider more carefully why so little progress has been made in the practical application of these principles in tropical developing countries.

Institutional status quo: Experience in the developed world would suggest that part of the problem lies in the conservatism and inertia within resource management institutions. The influence of a powerful civil society was needed to bring about the reforms in the U.S. Forest Service that led to the adoption of ecosystem management approaches in the Pacific Northwest. Forest services in other industrial countries have been slow to adopt broad-based and participatory approaches to management. The fact that aid donors work predominantly through recipient country institutions means that there is a built-in avoidance of any real challenge to the legitimacy of the institutional status quo.

Land rights and legal processes: Uncertainties about land rights and a lack of legal redress in support of land claims in developing countries is undoubtedly an obstacle to the attainment of ecosystem management objectives. Good fences make for good neighbors, but many conservation projects are operating in a quasi-common property context with uncertainty and insecurity surrounding access to resources.

There are often overlapping or conflicting claims on land and resources compounded by population growth and migration and rapid development of infrastructure and corporate activity. In these dynamic situations, sorting out land rights may take decades and little is likely to be achieved within the lifetime of a typical 3- to 5-year project. Similarly, the great power differentials that still exist between officials and ordinary people in many countries mean that genuinely equitable negotiations between stakeholders are the exception rather than the rule.

Conservation project culture: The procedures of the development assistance agencies that finance natural resource management projects are also an obstacle to progress. Projects are still locked into the rigidity of logical frameworks, are conceived and funded for short time spans, and are firmly rooted in a command-and-control institutional culture. In these circumstances they are unlikely to respond to local needs and perspectives. Where then might one turn for guidance on the organizational change that will be required to support management that is truly adaptive and collaborative?

EMERGING MANAGEMENT PRINCIPLES

In many respects, the problems of designing effective natural resource management agencies are similar to those confronting managers of any complex and unpredictable system. There is much in the present generation of management literature that can be applied to conservation. The same problems that confront conservation managers in developing countries are the subject of much debate and study among students of the management of all the complex systems that constitute the modern corporate world (Sherman and Schultz, 1998).

The challenges of managing modern "knowledge-based" businesses have much in common with those confronting natural resource managers. The business literature is replete with advocacy of new management paradigms (e.g., Stacey, 1993). The parallels between much modern business thinking and the challenges of managing protected areas are striking. Both deal with uncertainty, unpredictability, connectivity (to distant influences), compromise, judgment, etc. A number of fundamental principles recur in both the business literature and in experience with conservation practice. The following are examples of principles that should underpin effective approaches to management.

- *"One size does not fit all."* Management has a strong tendency to place conservation activities into categories; to have international definitions of protected areas; to speak of demonstration projects and model projects; to assess success by aggregate figures for the areas under any particular management regime. But in the real world every site is different. If management is truly attuned to local realities, then the management of every site will be unique. So frameworks for assessing overall progress are useful, but they should be secondary to the need to define local management in terms of local needs, constraints, and opportunities.
- *Common conceptual frameworks*. A fundamental ingredient of successful management is to have agreement and clarity about the outcomes being

pursued. Many conservation projects have very fuzzy, imprecise, difficult-to-measure goals. There must be a framework against which performance can be judged. Ironically, some of the recent attempts to develop criteria and indicators for assessing biodiversity have probably done more to obfuscate than to add clarity to management. Biodiversity criteria and indicators have often been developed by taxonomists and are based upon value sets that may not be shared by local people. To the taxonomist all species may have equal value, whereas for local people different components of biodiversity will have different values. Practical management-oriented indicators are needed for quick assessments of whether management is succeeding or not. Are boundaries delineated, are the staff patrolling the area, is the natural vegetation still present?

- *Consistent competency and reward structure*s. Good local management needs good local managers. Reward systems, professional advancement, and recognition must be based upon concrete conservation outcomes. Criteria against which managers are assessed must be objective, transparent, and linked directly to the agreed-upon conservation objectives for the site. Incentives must offset the difficulties of working in remote locations and the various social disadvantages that site managers will often experience. The scarcest commodity in international conservation at present is the competent field practitioner. This is true both among the professionals of developing-country conservation agencies and among the droves of international experts who come to help and advise them. It is sobering to observe how few of the latter have ever had any practical experience upon which to base the advice that they so generously dispense.
- *Realistic objectives*. Objectives must be realistic. Many international projects make hugely optimistic assumptions and, in the words of Beinert, are based upon the international advisers' "technical imagination of what might be possible" (cited in Scott, 1998). With hindsight it is easy to see examples throughout the developing world where grandiose projects have failed and where more modest, pragmatic interventions might have succeeded. Systems of smaller, more strategically located protected areas, realistically located in terms of competing land uses have a far higher likelihood of success than many of the overly ambitious targets that tend to dominate international and national conservation action plans (CIFOR, 1999, Sayer et al., 2000).
- *Knowledge systems thinking*. Business thinking on knowledge management, learning organizations, systems approaches, and teamwork offers much of relevance to the protected area manager. Conceptualizing knowledge systems as configurations of social actors (Checkland, 1990; Roling and Wagemakers, 1998) with stakes in the outcome of learning and negotiation processes helps legitimate local knowledge and thus level the playing field among parties at interest. The Learning and Innovation Loans (LILS) and other flexible delivery mechanisms of the World Bank may provide useful models for protected area projects. Many other significant

contributors of conservation funding are still locked into very rigid delivery frameworks that are not conducive to locally adapted management. Progress will require profound changes in management cultures, with greater authority and discretion given to local managers.

- *Property rights in biota.* Innovations must be sought in terms of ownership of biological resources and of management interventions. Ownership often predicates good management, but there are few examples of either local people or managers being able to capture any of the global values of the biodiversity that they are expected to conserve. This is an area where a great deal of research will be needed to develop mechanisms for capturing benefits, without which long-term success is unlikely to be achieved.

CONCLUSIONS

The knowledge and understanding derived from three decades of experience with ICDPs combined with innovations in management from the corporate sector should provide the basis for a new generation of conservation approaches. It should now be possible to achieve far more with the resources available for conservation. Every project, in fact, all management, should be seen as "action research," a process of experimentation and learning. All good management must adapt constantly to changing circumstances and in response to the impact of its own activities. Good management must have clear objectives, a framework for negotiating desired outcomes has to be established, and the performance of the manager assessed within this framework. People will only tolerate protected natural areas to the extent that they perceive some rationale for them and value from them. They will more readily tolerate, and one hopes welcome, conservation programs if they perceive justice and fairness in the way the program is developed and implemented.

REFERENCES

Bossel, H., 1998. *Earth at a Crossroads: Pathways to a Sustainable Future,* Cambridge University Press, Cambridge, U.K.

Caldecott, J., 1996. *Designing Conservation Projects,* Cambridge University Press, Cambridge, U.K., 1–312.

Checkland, P. and Scholes, J., 1990. *Soft Systems Methodology in Action,* John Wiley, Chichester, U.K.

CIFOR, UNESCO, Government of Indonesia, 1999. World Heritage Forests: The World Heritage Convention as a mechanism for Conserving Tropical Forest Biodiversity, CIFOR, Bogor, Indonesia, 1–54.

Edmunds, D. and Wollenberg, E. (in press). A strategic approach to multistakeholder negotiations. *Dev. Change.*

Edwards, M., 1999. *Future Positive: International Co-operation in the 21st Century,* Earthscan, London, 1–292.

Gunderson, L. H., Holling, C. S., and Light, S. S., Eds., 1995. *Barriers and Bridges to the Renewal of Ecosystems and Institutions,* Columbia University Press, New York.

Kramer, R., van Schaik, C., and Johnson, J. Eds., 1997. *Last Stand: Protected Areas and the Defense of Tropical Biodiversity,* Oxford University Press, New York, 242 pp.

Lee, K. N., 1993. *Compass and Gyroscope: Integrating Science and Politics for the Environment,* Island Press, Washington, D.C.

Maltby, E., Holdgate, M., Acreman, M., and Weir, A., 1999. *Ecosystem Management, Questions for Science and Society,* Royal Holloway Institute for Environmental Research, Royal Holloway, University of London, Egham, U.K.

Roling, N. G. and Wagemaker, M. A. E., 1998. *Facilitating Sustainable Agriculture: Participatory Learning and Adaptive Management in Times of Environmental Uncertainty,* Cambridge University Press, Cambridge, U.K., 318 pp.

Saxena, N. C., 1997. The Saga of Participatory Forest Management in India, CIFOR, Bogor, Indonesia.

Sayer, J. A., 1991. *Rainforest Buffer Zones, Guidelines for Protected Area Managers*, IUCN, Gland, Switzerland.

Sayer, J. A., 1995. Science and international nature conservation, CIFOR Occasional paper 4, Bogor, Indonesia, 1–14.

Sayer, J. A., 1999. Globalisation, localisation and protected areas, in *Imagine Tomorrow's World*, McNeely, J. A., Ed., IUCN, Gland, Switzerland, 1–258.

Sayer, J. A., Ishwaran, J., Thorsell, J. W., and Sigaty, T., 2000. The World Heritage Convention and the conservation of tropical forest biodiversity, *Ambio,* 29(6), 302–309.

Scott, J. C., 1998. *Seeing Like a State: How Certain Schemes to Improve Human Condition Have Failed,* Yale University Press, New Haven, CT, 1–445.

Sherman, H. and Schultz, R., 1998. *Open Boundaries: Creating Business Innovation through Complexity,* Perseus Books, Reading, MA, 232 pp.

Stacey, R., 1993. Strategy as order emerging from chaos. *Long Range Plann.,* 26(1), 10–17.

Szaro, R. C. and Johnston, D. W., Eds., 1996. *Biodiversity in Managed Landscapes*, Oxford University Press, New York.

Wells, M. and Brandon, K., 1992. *People and Parks: Linking Protected Area Management with Local Communities*, The World Bank, Washington, D.C.

Wells, M., Guggenheim, S., Khan, A., Wardojo, W., and Jepson, P., 1999. *Investing in Biodiversity,* The World Bank, Washington, D.C.

World Bank/World Wildlife Fund Forest Alliance, 1999. Threats to Forest Protected Areas, World Bank/WWF, Washington, D.C.

APPENDIX

ECOSYSTEM MANAGEMENT PRINCIPLES

From the Web site of the Convention on Biological Diversity:

http://www.biodiv.org/Decisions/COP5/html/COP-5-Dec-06-e.htm

The following 12 principles are complementary and interlinked.

Principle 1: *The objectives of management of land, water and living resources are a matter of societal choices.* Different sectors of society view ecosystems in terms of their own economic, cultural, and society needs. Indigenous peoples and other local communities living on the land are important stakeholders and their rights and interests should be recognized. Both cultural and biological diversity are central components of the ecosystem approach, and management should take this into account. Societal choices should be expressed as clearly as possible. Ecosystems should be managed for their intrinsic values and for the tangible or intangible benefits for humans, in a fair and equitable way.

Principle 2: *Management should be decentralized to the lowest appropriate level.* Decentralized systems may lead to greater efficiency, effectiveness, and equity. Management should involve all stakeholders and balance local interests with the wider public interest. The closer management is to the ecosystem, the greater the responsibility, ownership, accountability, participation, and use of local knowledge.

Principle 3: *Ecosystem managers should consider the effects (actual or potential) of their activities on adjacent and other ecosystems.* Management interventions in ecosystems often have unknown or unpredictable effects on other ecosystems; therefore, possible impacts need careful consideration and analysis. This may require new arrangements or ways of organization for institutions involved in decision making to make, if necessary, appropriate compromises.

Principle 4: *Recognizing potential gains from management, there is usually a need to understand and manage the ecosystem in an economic context. Any such ecosystem-management programme should:*

a. Reduce those market distortions that adversely affect biological diversity;
b. Align incentives to promote biodiversity conservation and sustainable use;
c. Internalize costs and benefits in the given ecosystem to the extent feasible.

The greatest threat to biological diversity lies in its replacement by alternative systems of land use. This often arises through market distortions, which undervalue natural systems and populations and provide perverse incentives and subsidies to favor the conversion of land to less diverse systems.

Often those who benefit from conservation do not pay the costs associated with conservation and, similarly, those who generate environmental costs (e.g., pollution) escape responsibility. Alignment of incentives allows those who control the resource to benefit and ensures that those who generate environmental costs will pay.

Principle 5: *Conservation of ecosystem structure and functioning, in order to maintain ecosystem services, should be a priority target of the ecosystem approach.* Ecosystem functioning and resilience depend on a dynamic relationship within species, among species, and between species and their abiotic environment, as well as the physical and chemical interactions within the environment. The conservation and, where appropriate, restoration of these interactions and processes is of greater significance for the long-term maintained conditions, and, accordingly, management should be appropriately cautious.

Principle 6: *Ecosystems must be managed within the limits of their functioning.* In considering the likelihood or ease of attaining the management objectives, attention should be given to the environmental conditions that limit natural productivity, ecosystem structure, functioning, and diversity. The limits to ecosystem functioning may be affected to different degrees by temporary, unpredictable, or artificially maintained conditions, and, accordingly, management should be appropriately cautious.

Principle 7: *The ecosystem approach should be undertaken at the appropriate spatial and temporal scales.* The approach should be bounded by spatial and temporal scales that are appropriate to the objectives. Boundaries for management will be defined operationally by users, managers, scientists, and indigenous and local peoples. Connectivity between areas should be promoted where necessary. The ecosystem approach is based upon the hierarchical nature of biological diversity characterized by the interaction and integration of genes, species, and ecosystems.

Principle 8: *Recognizing the varying temporal scales and lag effects that characterize ecosystem processes, objectives for ecosystem management should be set for the long term.* Ecosystem processes are characterized by varying temporal scales and lag effects. This inherently conflicts with the tendency of humans to favor short-term gains and immediate benefits over future ones.

Principle 9: *Management must recognize that change is inevitable.* Ecosystems change, including species composition and population abundance. Hence, management should adapt to the changes. Apart from their inherent dynamics of change, ecosystems are beset by a complex of uncertainties and potential "surprises" in the human, biological, and environmental realms. Traditional disturbance regimes may be important for ecosystem structure and functioning, and may need to be maintained or restored. The ecosystem approach must utilize adaptive management to anticipate and prepare for such changes and events and should be cautious in making any decision that may foreclose options, but, at the same time, consider mitigating actions to cope with long-term changes such as climate change.

Principle 10: *The ecosystem approach should seek the appropriate balance between, and integration of, conservation and use of biological diversity.* Biological diversity is critical both for its intrinsic value and because of the key role it plays in providing the ecosystem and other services upon which everyone ultimately depends. There has been a tendency in the past to manage components of biological diversity either as protected or nonprotected. There is a need for a shift to more flexible situations, where conservation and use are seen in context and the full range of measures is applied in a continuum from strictly protected to human-made ecosystems.

Principle 11: *The ecosystem approach should consider all forms of relevant information, including scientific and indigenous and local knowledge, innovations, and practices.* Information from all sources is critical to arriving at effective ecosystem management strategies. A much better knowledge of ecosystem functions and the impact of human use is desirable. All relevant information from any concerned area should be shared with all stakeholders and actors, taking into account, *inter alia*, any decision to be taken under Article 8(j) of the Convention on Biological Diversity. Assumptions behind proposed management decisions should be made explicit and checked against available knowledge and views of stakeholders.

Principle 12: *The ecosystem approach should involve all relevant sectors of society and scientific disciplines.* Most problems of biological diversity management are complex, with many interactions, side effects, and implications, and therefore should involve the necessary expertise of stakeholders at the local, national, regional, and international level, as appropriate.

5 Experiences, Challenges, and Prospects for Collaborative Management of Protected Areas: An International Perspective

Robert J. Fisher

CONTENTS

INTRODUCTION

The idea of collaborative management (CM) of protected areas, under a variety of different names, including joint management and comanagement, is now established as a major element of protected area (PA) management policies and rhetoric in much of the world. This stems from several factors, including (1) pragmatic recognition that active local support of PAs is important to the effectiveness and cost-efficiency of PA management; (2) recognition, from the point of view of social justice and equity, of the legitimate needs of people who have been deprived of access to

0-8493-0020-7/01/$0.00+$1.50

resources when their homelands have been declared to be PAs; and (3) the linking of funding from international donors to less authoritarian models of PA management.

A superficial glance at the burgeoning literature on CM would suggest that the approach really has taken off, particularly in much of Asia and Africa. More careful examination of the literature shows that much of it advocates CM and discusses how it should work, but very little of it documents successful examples. The author believes that this analysis of the literature pretty well reflects what is happening on the ground. There is a lot of enthusiasm, but very few cases of CM that go beyond token attempts to "consult" with local people. Little real progress has been made on meeting the needs of local people who are negatively affected by PAs.

Although this may sound pessimistic, there are some very promising initiatives and a great deal of good will and enthusiasm that can be harnessed. CM may evolve into a dominant model for PA management,* but there are some major obstacles to be overcome if the collaboration is to be meaningful. At present, most of what is described as CM is little more than tokenism.

The aim of this chapter is to present an international perspective on experiences, challenges, and prospects for CM of PAs. The chapter focuses mainly on an Asian regional perspective, where situations vary within individual countries in Asia, and more so between countries. Several themes that are crucial to CM in the Asian region and that are generally relevant throughout the developing world are explored.

In May 1998, the International Union for Conservation of Nature and Natural Resources (IUCN) Nepal hosted an international workshop in Chitwan National Park, on "Collaborative Management of Protected Areas in the Asian Region" (Oli, 1999). One of the activities involved identifying the status of CM, as well as constraints and opportunities facing CM in the countries represented. Papers presented summaries of the situation in each country. There was serious debate at the meeting about the role of CM and to what extent it should involve real control and real involvement in decision making by local people. Participation at this meeting and subsequent discussions provide some confidence that the perspective in this chapter reflects the concerns of many practitioners about the tokenism generally evident in the application of CM in Asia.**

This chapter focuses on the CM element of adaptive collaborative management (ACM), rather than on experiences in the application of the ACM methodology itself. This emphasis stems from the view that ACM is unlikely to work unless the collaborative element is meaningful, which the author has been forced to conclude it usually is not.

---

* Although CM has not yet been implemented widely, it is important to note that conventional PA management (i.e., lacking CM) has not been particularly effective in much of Asia. Yet, paradoxically, laws, policies, and training remained geared to the conventional approach (W.J. Jackson, personal communication, 1998).

** The author thanks Andrew Ingles, Ashish Kothari, Krishna Oli, Scott Perkin, and Javed Ahmed for their contributions to a brainstorming session on issues relating to CM in PAs while planning this chapter. Also thanked are Bill Jackson and Don Gilmour for helpful comments on early drafts.

COLLABORATIVE MANAGEMENT OF PROTECTED AREAS IN SELECTED ASIAN COUNTRIES

In all, 11 country papers were prepared for the Chitwan workshop. Several of these reviewed the status of CM as related to protected areas in the particular country. The Appendix to this chapter presents a summary of these accounts, supplemented from other sources. Two key themes predominate in accounts of the status of collaborative management of protected areas in Asia:

1. Although enabling legislation and policies are often present, implementation is in a preliminary stage.
2. Even when progressive enabling legislation exists, the possibilities for CM are usually interpreted very conservatively by officials.

Only in Nepal, where CM of parts of PAs is supported by legislation, regulation, and policy, is it seriously practiced. Even in this case, benefit sharing and shared decision making are not options in core zones of PAs.

WHAT DOES COLLABORATIVE MANAGEMENT MEAN?

An initial difficulty in discussing CM in Asia is delineating the range of activities covered by the term. The term is used to describe activities that are often quite different from each other. This is complicated by the fact that rather different terms are used in different countries.

Borrini-Feyerabend (1996, p. 12) gives the following definition:

> The term "collaborative management" (also referred to as co-management, participatory management, joint management, shared-management, multistakeholder management or round-table agreement) is used to describe a situation in which *some or all of the relevant stakeholders in a protected area are involved in a substantial way in management activities.* Specifically, in a collaborative management process, the agency with jurisdiction over the PA (usually a state agency) develops a *partnership* with other relevant stakeholders (primarily including local residents and resource users) which specifies and guarantees their *respective functions, rights and responsibilities* with regard to the PA.

The key element in this definition is the collaboration of stakeholders, with particular reference to the participation of local residents. The author regards this as a minimum definition.* Within this definition, CM can include cases on a continuum from

* The primacy of the involvement of local residents seems to be an absolute minimum condition. Yet, not all conservation agencies and officials see it that way. In Uganda, the author heard a senior park official make it clear that he thought collaboration between the Uganda Wildlife Authority and the Forest Department was collaborative management. That this is not a unique perspective is clear from Idrus and Nais (1999, p. 197), who find it necessary to specify that they restrict their discussion to activity "that includes local community collaboration."

partnerships involving local people in genuine decision making about PA management to token consultation.

The author has some reservations about including formal partnership agreements as a *necessary* part of the definition. It is possible, even where formal agreements do not exist, and may not be legally possible, for informal arrangements to operate that enable genuine and substantial involvement. It can be argued that it is precisely the existence of such informal arrangements that can lead to the mutual trust, which is the first step in policy change.

If CM is such a broad field of activities, is there any way of classifying broad types of CM-related approaches? Here, it is suggested that there are two ways of classifying CM approaches. One is in terms of the nature of local "participation" along a continuum from token consultation to genuine decision making (after Arnstein, 1969). The second way to classify CM approaches is in terms of the ways they attempt to meet the need for resources normally (or previously) obtained from PAs. Four broad approaches can be identified, which can be variously combined.

1. *Substitution.* This approach provides alternative means of obtaining products previously obtained from PAs. This may be done either by direct substitution of similar products produced outside the PA, such as fuelwood grown on woodlots, or alternative products that meet a similar need such as subsidized kerosene.
2. *Poverty reduction.* This approach emphasizes income generation. It is assumed that resource degradation is a result of poverty and that increased income will enable people to purchase products rather than obtain them from PAs. This assumption is often flawed. People may choose to continue to collect necessary resources from PAs and use the income for other purposes.
3. *Income generation from nonextractive activities in the PA itself.* Examples of such activities might be employment of local people as rangers, or involvement in tourism.
4. *Sustainable management model.* In this approach the emphasis is on involving local people in the sustainable management, including utilization, of the PA itself, through an agreed-upon system of regulated use and access to resources. In Asia, apart from minor concessions for the regulated extraction of "minor" forest products, this is not a common option. The approach in the form of "extractive reserves" seems to be practiced as a serious option in some parts of South America (Murrieta and Pinzón Rueda, 1995).

The first two approaches attempt to provide alternatives to exploitation of PA resources. The poverty reduction or income-generation approach is the basis for many projects either in formal buffer zones or in areas adjoining PAs. It is sometimes naively assumed that economic development will take pressure off PA resources. Underlying this is another assumption — that poor people will not need to get resources from PAs if they have increased income. What is ignored is the possibility that new sources of income will complement rather than replace income obtained

TABLE 5.1
Conditions under Which Development Activities can Contribute to the Conservation Goals of PAs

- When they provide alternative forms of income in place of existing forest-based income. The new forms of income must be attractive in terms of relative value and inputs of time and labor. There is also a scheduling factor. If the new forms of income leave adequate periods of time (or seasons) when people can earn supplementary income from PAs, then the people are likely to continue to exploit the resources in the PAs.
- When they provide alternative ways to meet the requirements for desired products normally obtained from PAs.
- When they allow communities to utilize PA resources under managed (planned or restricted) conditions.

from PAs.* A problem with integrated development and conservation projects is that they are often not integrated. The linkages between conservation objectives and development activities need to be made explicit (Gilmour, 1994). Table 5.1 suggests some conditions under which development activities can contribute to the conservation goals of PAs.

Challenges to the Progress of Collaborative Management

The discussion so far suggests that the progress of implementation of CM, whether defined in the minimalist sense or in a more genuine form, is relatively slow. There are success stories, documented in collections such as Western and Wright (1994) and several books and reports by Kothari and colleagues. But, in general, those specifically related to PAs are about "work in progress" rather than cases of mature working models.** The author believes that there are three broad reasons collaborative management is not moving as fast as many would hope.

The Lack of Real, Documented Success Stories

On the one hand, the lack of documented success stories is an indication of relatively slow progress. But it is also a contributing factor, because it is difficult to sell CM to skeptical PA officials and policy makers in the absence of convincing examples. It is also relatively easy to dismiss CM as irrrelevant to a particular country when all the examples relate to apparently quite different circumstances. This leaves project implementors or stakeholders with a dilemma: they need success stories to suggest an approach to use and the confidence to start. Because each situation has highly

* Gilmour (personal communication, 1999) points out that decreased poverty can actually lead to *increased* demand for resources. For example, people may decide to build a larger house, which requires more, not less, wood.

** There has been more clearly documented progress in various forms of CM of natural resources outside PAs, particularly in the literature on various forms of CM of forests (see, for example, Fisher, 1995; Poffenberger, 1990). The fact that forest management outside PAs is more likely to allow serious benefit sharing may account for the relatively higher levels of local participation in CM of forests.

specific conditions, however, case studies of success elsewhere do not seem immediately applicable.

This problem is compounded by the fact that projects tend to be "overdesigned" and target driven, even when they presumably emphasize a learning process approach. There is one example in northeastern Thailand where an Integrated Conservation and Development Project (ICDP) in the buffer zone of a wildlife sanctuary had unrealistic targets for carrying out Participatory Rural Appraisal (PRA) exercises, forming interest groups, and commencing income-generating activities. There was no time for the complex negotiations necessary among all the partners. The project manager recognized this, then sought and obtained a reduction in targeted activity, but nevertheless was forced to move more quickly than he wished.

The author suggests that the key to breaking this cycle may lie in employing flexible methods for enabling action in the context of uncertainty. Action research offers a way of thinking about this. It enables one to begin without knowing exactly what to do (Fisher and Jackson, 1999). Using action research to generate case studies in selected pilot sites could illustrate the value of adaptive and exploratory management. Such case studies would focus on process documentation in a particular situation.

Lack of Trust between Parties

Lack of trust expresses itself in two important ways. PA officials tend not to trust local people, who they regard as being generally ignorant and incapable of regulating behavior in relation to resources. This is often exacerbated by differences in ethnicity between the officials and the local people. On the other hand, local people often see officials as unwilling to consider their concerns.

The attitude of local people to PAs and the officials who manage them is an extremely important factor in Asia. This issue is often glossed over, and officials may see local people as ungrateful for the concessions the officials make, without really appreciating the genuine basis of their hostility.

The first major source of hostility arises for people who have been forced out of areas where they live, or had access to resources cut off when PAs were declared. PAs in Asia are almost invariably imposed in areas that are essential to the livelihoods of resident and indigenous people. Forced relocation and denial of access to resources is a continuing process associated with the development of protected areas in Asia and compensation (if any) and provision of alternative homes and livelihoods often are not adequate. Two examples of the way resident people are affected by the declaration of PAs come from Thailand.*

Buffer zone of Ang-rue-nai Wildlife Sanctuary (Eastern Thailand)

In the early 1990s the Ang-nai-rue Wildlife Sanctuary was enlarged and people living in the newly demarcated areas were forced to move out. Some people were resettled in the buffer zone and given small allocations of land (Kijtewachakul, 1998). Initially, there was considerable hostility between the people and the Royal Forest Department and other agencies. However, due largely to the efforts of the author's colleagues at RECOFT, through an action research project, relationships have improved and a

* For a discussion of the political economy of PA management in Thailand, see Sato (1998).

community forest, providing some access to resources, has been established in a forest in the buffer zone. Although this is promising, a recent field trip to one village made it clear that not everyone is content. A serious problem is that the land provided to resettled families has no secure title and cannot be sold legally. Consequently, people cannot borrow to invest in land improvement. Land is sold informally, but, because sale is illegal and because continued occupancy is subject to administrative whim, prices are very low. Many of the original settlers are now landless and have nowhere to go.

Proposed National Park at Phu Kha (Northern Thailand)

Within the likely boundaries of the proposed Phu Kha national park are Hmong tribal people who would probably be resettled when the park is declared. Apart from uncertainty about where alternative land might be available, there is an additional injustice here: these people were forcibly *relocated into the area* by the army during the communist insurgency. They were subsequently forced to give up opium farming and shifting cultivation. They are now regarded as encroachers in an area into which they were forcibly relocated.

Buffer-zone development projects do not always adequately compensate people for loss of access to resources. Such projects often are not taken very seriously by officials. They tend to be top-down and funds are frequently scarce. Even in Nepal, where there are regulations that ensure a significant proportion of PA revenue will be used for community development in the buffer zones or conservation areas for the benefit of residents, the amounts involved are small in all except a few PAs, such as Sagarmartha (Mt. Everest), the Annapurna Conservation Area, and Chitwan National Park. In other more remote and less popular areas, revenue from visitors is small and opportunities for income from activities such as tourism are very limited.

In cases such as these, where PA income and associated business opportunities are significant, or cases where substantial donor-funded projects support buffer-zone development, compensation may be adequate and net benefits may accrue to local communities from the presence of a PA. However, the author has not seen convincing evidence of cases where there are net benefits and where these benefits are equitably distributed. An interesting potential scenario relates to buffer zones that are so successful that conditions are better than in neighboring areas outside the buffer zone, possibly attracting new immigrants into the area (Andrew Ingles, personal communication, 1998).

With the exception of a few examples where compensation or alternative livelihoods may have occurred, experience suggests that most local people affected by PA formation have been severely disadvantaged. Their hostility is understandable.

The causes of hostility do not stop with past dispossession and its effects. Governments, while pleading the need for sacrifice in the national interest, continue to send messages that confirm local people are not the major players. Kothari et al. (1997, p. 10) mention the "denotification of protected areas to make way for mining, hotels, industries, and other commercial activities." In Thailand there is an ongoing debate about concessions to tourism and other interests in PAs, when local people are excluded or under continuing threat of relocation. Even in the case of Kakadu National Park in Australia, which seems to go farther than almost any other case in devolving decision-making authority and benefit sharing to indigenous people, the current Australian government has recently approved mining operations at Jabiluka.

This is an area that has always been an excision from the park but that has not previously been mined. Outrageously, during a demonstration at the site in late 1998, at least one traditional Aboriginal landowner was arrested for trespass. Such actions by governments and officials reinforce this hostility and suspicion.

Reluctance to Share Decision-Making and Benefits

Involvement of local people in decision making about PA management has been strongly advocated by some activists. Publications by Kothari and various co-workers are particularly relevant here. The argument in favor of this is a pragmatic one that people are more likely to follow management practices and regulations if they have had a say in developing them and if these regulations take some account of their needs and objectives. Other arguments include the value of indigenous knowledge to PA management.

There are some PAs where local/indigenous people are genuine, even dominant partners in decision making, notably Kakadu National Park in Australia. There are some cases where legislation or policy requires representation by local people on PA management boards, for example, the National Integrated Protected Areas System (NIPAS) legislation in the Philippines. These cases where local people have a genuine role in decision making for the PAs themselves are rare. In Nepal, encouragement of community participation in planning and management of buffer zones is policy, but community involvement in core zone management is essentially restricted to "consultation."

Rao and Sharma (1999, p. 105) have specifically rejected the view that "co-management must go beyond mere consultation and participatory planning and should entail a conscious and official distribution of responsibility, with formal vesting of some authority." It is not uncommon for wildlife or PA officials to reject this approach to CM, although it is rare for the opposition to be expressed so clearly and explicitly. Rao and Sharma state clearly that it cannot be disputed that it is necessary to integrate "local peoples' interests into the policies and programmes of PA planning and management" (p. 105). However, they object to the argument that many advocates of co-management view it as a part of wider social agenda, and raise related issues of social justice, self-determination, and democratization. In pursuing an oversocialized viewpoint, the essential need for integrating environmental and human concerns is frequently lost sight of.

As an alternative to this "oversocialized" approach to CM, Rao and Sharma describe the Government of India's policy for promoting "ecodevelopment" around PAs. The program supports the development of sustainable livelihood practices around, but not within, PAs. They claim that the program has been accepted without qualification everywhere it has been introduced.

While pointing out that PA management in India often provides for some direct benefits from PAs, such as grazing rights, Rao and Sharma specifically argue against the view that more resources should be provided directly from PAs. One of their key points is that encouraging such use merely "reinforces the vicious cycle of resource dependence and resulting degradation … [contributing] to the continuing impoverishment of these peoples. It is ironic that some social policy people advocate the perpetuation of such precarious existence and keeping the people alienated from the mainstream, ostensibly in the interest of preserving traditional knowledge and lifestyles" (1999, pp. 105-6).

The author believes Rao and Sharma are correct in pointing out the importance of providing opportunities for people to move beyond dependency on forest resources for subsistence. But in using this as an argument for maintaining strict protection of PAs, they miss the point that moving beyond dependency is meaningful only if alternatives are provided. By their own admission, the ecodevelopment program has "been extended to some 80 PAs, but due to resource constraints, the interventions have been on a relatively small scale" (1999, p. 107). Indeed, to avoid spreading resources too thinly, the focus is to shift to more intensive coverage of fewer areas. In other words, "alternatives" are provided to only a small proportion of the people negatively affected by PAs. In the absence of alternatives, it makes no sense to maintain a "people out" policy for the peoples' own good.* A policy of direct benefit sharing may be unnecessary where a program exists to provide direct compensation for forgone benefits in the form of substitution of products or enhanced income generation. Where such benefits cannot be provided, there is a social justice or equity issue involved in providing direct benefits from PAs. Support for a PA from local people can hardly be expected in the absence of benefit sharing.

Given that local people are not often involved in decision making about PA management and that the existence of PAs often involves costs in terms of forgone access to resources, reluctance to become involved in CM projects is hardly surprising. What is surprising is that some local residents do become involved in purely protective activities, i.e., when there are no direct benefits even in buffer zones, where some benefits could normally be expected. Further, community-initiated protection oriented activities are common. For examples from India, see the various papers by Kothari and colleagues. These initiatives indicate that there is a potential resource for conservation that could be tapped through a genuine willingness to share decision making about PAs through collaborative management.**

Prospects

The author has concluded so far that progress toward CM in the fuller sense involving real sharing of decision making, and therefore power sharing, has been slow in Asia and the developing world in general. The situation need not lead to pessimism, however. Although it is useful to examine each situation critically to acknowledge the serious obstacles that need to be addressed, it is not suggested that it is necessary to address problems such as the reluctance of officials to "let go the strings" as a *precondition* for any progress. An adaptive, exploratory process can build trust and lead to major changes in attitude.

---

* A recent paper by Byron and Arnold (1997) stresses the importance of deconstructing the category "forest dependent peoples" and providing economic choices appropriate to the type of economic relationships between particular people and forests.

** One needs to be a little careful about how to interpret community willingness to be involved in supporting official initiatives that provide no direct benefits. Frequently, participation in such activities is limited to local elites, whose participation makes the activities conspicuous, but masks the fact that there is little popular support.

Some key elements are in place for development of CM. Although there is suspicion on both sides, there is also good will on both sides, and people with quite divergent opinions about CM and devolution of authority hold these opinions sincerely. Good will can be generated when people actually work together. Even senior officials who are very suspicious of CM and the motives of those who advocate it are often open to change. They must, however, be engaged in constructive debate and they must see results.

Another positive element is the presence of significant international backing for CM, from organizations regarded as credible by PA managers. For example, IUCN has made a serious commitment to CM. Its Social Policy Group has made major contributions to the conceptual development of CM of PAs through two important publications (Borrini-Feyerabend, 1996; 1997).

Below are very brief examinations of two examples of PAs that qualify, the author believes, as CM success stories, although not necessarily unqualified successes.

Annapurna Conservation Area, Nepal

The Annapurna Conservation Area Project (ACAP) is widely recognized for its innovation and commitment to participatory management. ACAP is a project of an NGO, the King Mahendra Trust for Nature Conservation (KMTNC). The Annapurna Conservation Area is one of the most popular trekking routes in Nepal and attracts large numbers of tourists, a major source of cash income for the people who live in the area.

A basic innovation in the project, which formally started in 1986, is that a new category of PA was introduced on the urging of the KMTNC. It is a Conservation Area (CA), not a National Park (Gurung, 1994). The key feature is that the CA residents are allowed to remain in the area and to practice their traditional livelihood activities with improved practices promoted by the project. The ACAP integrated approach involves conservation activities (alternative sources of energy, plantation, and management of existing forests), community development activities (drinking water, schools, improved trails, bridges, irrigation, soil conservation), and tourism management (Gurung, 1994).

ACAP has shown a strong commitment to participatory planning and implementation. Local communities reputedly have a substantial role in decision making.

Kakadu National Park, Australia

The second example of a promising approach to collaborative management comes from outside the Asian region. However, Kakadu National Park in Northern Australia is so unusual for its management arrangements that it deserves to be mentioned. Kakadu consists of an area of about 20,000 km^2. It is listed as a world heritage site for both its natural and its cultural values (Wellings, 1994) and consists of extensive savannah woodlands with large areas of wetlands, which, in the wet season, host huge numbers of birds. There are at least 5000 sites with Aboriginal rock art in the park. The oldest of these are at least 20,000 and possibly as much as 60,000 years old (Brockwell et al., 1995).

For the purposes here the park is most interesting for its unique management arrangements. Approximately half of the land is Aboriginal land under land rights legislation. This land has been leased to the Australian Conservation Agency to be managed as a national park. The other half is subject to an unresolved land rights claim by the traditional owners, but the Australian Nature Conservation Agency has a strong commitment to work with Aboriginals over the whole park (Wellings, 1994). The result is that the entire park is treated as if it were Aboriginal land.

The management of the park is a joint activity of the traditional Aboriginal owners and the Australian Nature Conservation Agency. Remarkably, 10 of the 14 members of the Board of Management are Aboriginal nominees of the traditional owners of the land (Press and Lawrence, 1995). This level of commitment to joint management is very unusual, and perhaps unprecedented.*

The Potential Role of Action Research and Adaptive Collaborative Management

These cases and, perhaps, a few like them, provide hope for the emergence of genuine CM in the future. The author would argue, however, that the spread of genuine CM will require increased willingness on the part of governments and agencies to devolve decision making and benefit sharing to affected communities. The reluctance to do so partly results from reluctance to give up institutional and personal power. There is no simple way to resolve this issue. But the reluctance is also partly the result of a lack of trust in local communities and fear of the consequences to conservation of relinquishing control. Efforts to develop real examples of working CM can lead to increased trust and, therefore, to greater willingness to relinquish control. ACM and action research, a broadly similar methodology, have great potential here.

Action research and ACM methodologies are at least broadly similar in their concern with incremental change, exploration, and participation of stakeholders. The claims made below for the relevance of action research also apply to ACM.

Action research can be defined as a collaborative process through which "a group of people with a shared issue or concern collaboratively, systematically and deliberately plan, implement and evaluate actions" (Fisher and Jackson, 1999).

Action research applied to CM seems to offer a number of advantages (Fisher and Jackson, 1999):

- It provides a learning process approach to implementation of CM in a situation where there are many unknowns and where the process must be developed through action. In other words, it provides an opportunity to act without having all the unknowns worked out in advance.
- By seeking the involvement of a range of stakeholders and giving them a chance to work together, it can contribute to increased trust and cooperation.
- Various stakeholders can develop a sense of ownership by being involved in decision making.

Of these factors, it is the possibility of building trust and confidence through successful collaborative activities that is potentially most important.

---

* It may also be under threat, as evidenced by the action of the current Australian government in overriding the desires of traditional owners in allowing the opening of a uranium mine at Jabiluka (see above).

CONCLUSIONS

There is much rhetoric about CM of PAs in Asia and the rest of the developing world. There also are numerous donor-funded projects that aim for the integration of conservation and development and espouse participation. When one looks at the quantity and quality of collaborative activities, however, the situation is not encouraging. There is a shortage of good examples of what CM could be if PA authorities were to "let go the strings" and confer real decision-making authority. These limitations are compounded by a lack of trust.

On the positive side, there are initiatives that are providing opportunities for development of trust and each successful example of collaboration helps to build some confidence that collaboration can work. Legislation and policy may assist the development of collaboration, but they are not enough. It may be that the incremental, exploratory, and experimental steps that ACM embodies will be more valuable in the long run.

REFERENCES

Ali, S. S. and Ghulab Habib, Md., 1999. Country paper — Bangladesh, in *Collaborative Management of Protected Areas in the Asian Region*, Oli, K. P., Ed., IUCN, Kathmandu, Nepal, 77–104.

Arnstein, S. R., 1969. Citizen participation, *J. Am. Inst. Planners,* 35, 216–224.

Borrini-Feyerabend, G., 1996. *Collaborative Management of Protected Areas: Tailoring the Approach to the Context*, Issues in Social Policy, IUCN, Gland, Switzerland.

Borrini-Feyerabend, G., Ed., 1997. *Beyond Fences: Seeking Social Sustainability in Conservation*, 2 vols., IUCN, Gland, Switzerland.

Brockwell, S., Levitus, R., Russell-Smith, J., and Forrest, P., 1995. Aboriginal heritage, in *Kakadu: Natural and Cultural Heritage and Management*, Press, T., Lea, D., Webb, A., and Graham, A., Eds., Australian Nature Conservation Agency and North Australia Research Unit, The Australian National University, Darwin, 15–63.

Byron, N. and Arnold, M., 1997. What Futures for the People of the Tropical Forests? CIFOR Working Paper 19, Bogor.

Chantaviphone, I. and Berkmuller, K., 1999. Country paper — Lao PDR, in *Collaborative Management of Protected Areas in the Asian Region*, Oli, K. P., Ed., IUCN, Kathmandu, Nepal, 177–195.

Custodio, C. C. and Tabaranza, B. R., Jr., 1999. Country paper — Philippines, in *Collaborative Management of Protected Areas in the Asian Region*, Oli, K. P., Ed., IUCN, Kathmandu, Nepal, 213–224.

Fisher, R. J. and Jackson, W. J., 1999. Action research for collaborative management of protected areas, in *Collaborative Management of Protected Areas in the Asian Region,* Oli, K. P., Ed., IUCN, Kathmandu, Nepal, 235–243.

Gilmour, D. A., 1994. Conservation and development: seeking the linkages, in *Community Development and Conservation of Forest Biodiversity Through Community Forestry,* Wood, H., McDaniel, M., and Warner, K., Eds., Proceedings of an International Seminar held in Bangkok, Thailand, Oct. 26–28, RECOFTC Report 12, Regional Community Forestry Training Center, Bangkok, 7–20.

Gurung, C. P., 1994. Linking biodiversity conservation to community development: Annapurna Conservation Area Project approach to protected area management, in *Community Development and Conservation of Forest Biodiversity through Community Forestry,* Wood, H., McDaniel, M., and Warner, K., Eds., Proceedings of an International Seminar held in Bangkok, Thailand, Oct. 26–28, 1994, RECOFTC Report 12, Regional Community Forestry Training Center, Bangkok, 54–66.

Hassan, F., 1999. Country paper — Pakistan, in *Collaborative Management of Protected Areas in the Asian Region,* Oli, K. P., Ed., IUCN, Kathmandu, Nepal, 61–76.

Idrus, R. and Nais, J., 1999. Country paper — Malaysia, in *Collaborative Management of Protected Areas in the Asian Region*, Oli, K. P., Ed., IUCN, Kathmandu, Nepal, 197–212.

Kijtewachakul, N., 1998. Some findings and outstanding issues regarding community forestry in bufferzone management, in *Community Forestry at a Crossroads: Reflections and Future Directions in the Development of Community Forestry,* Victor, M., Lang, C., and Bornemeier, J., Eds., Proceedings of an International Seminar held in Bangkok, Thailand, 17–19 July 1997, RECOFTC Report 16. Regional Community Forestry Training Center, Bangkok, 275–284.

Kothari, A., 1996. Is joint management of protected areas desirable and possible? in *People and Protected Areas: Towards Participatory Conservation in India,* Kothari, A., Singh, N., and Suri, S., Eds., Sage, New Delhi, 17–49.

Kothari, A., 1997. Key issues in joint protected area management, in *Building Bridges for Conservation: Towards Joint Management of Protected Areas in India,* Kothari, A., Vania, F., Das, P., Christopher, K., and Jha, S., Eds., Indian Institute of Public Administration, New Delhi, 15–28.

Kothari, A., Singh, N., and Suri, S., Eds., 1996. *People and Protected Areas: Towards Participatory Conservation in India,* Sage Publications, New Delhi.

Kothari, A., Vania, F., Das, P., Christopher, K., and Jha, S., Eds., 1997. *Building Bridges for Conservation: Towards Joint Management of Protected Areas in India,* Indian Institute of Public Administration, New Delhi.

Murrieta, J. R. and Pinzón Rueda, R., Eds., 1995. *Extractive Reserves.* IUCN, Gland Switzerland.

Oli, K. P., Ed., 1999. *Collaborative Management of Protected Areas in the Asian Region,* IUCN, Kathmandu, Nepal.

Poffenberger, M., Ed., 1990. *Keepers of the Forest: Land Management Alternatives in Southeast Asia,* Ateneo de Manila University Press, Manila.

Press, T. and Lawrence, D., 1995. Kakadu National Park: reconciling competing interests, in *Kakadu: Natural and Cultural Heritage and Management,* Press, T., Lea, D., Webb, A., and Graham, A., Eds., Australian Nature Conservation Agency and North Australia Research Unit, The Australian National University, Darwin, 1–14.

Rao, K. and Sharma, S. C., 1999. Country paper — India, in *Collaborative Management of Protected Areas in the Asian Region,* Oli, K. P., Ed., IUCN, Kathmandu, Nepal, 105–116.

Sato, J., 1998. The political economy of bufferzone management: a case from western Thailand, in *Community Forestry at a Crossroads: Reflections and Future Directions in the Development of Community Forestry,* Victor, M., Lang, C., and Bornemeier, J., Eds., Proceedings of an International Seminar held in Bangkok, Thailand, 17–19 July 1997, RECOFTC Report 16, Regional Community Forestry Training Center, Bangkok, 87–99.

Sharma, U. R., 1999. Country paper — Nepal, in *Collaborative Management of Protected Areas in the Asian Region,* Oli, K. P., Ed., IUCN, Kathmandu, Nepal, 49–59.

Wellings, P., 1994. Biodiversity conservation and community development in Kakadu National Park, Australia, in *Community Development and Conservation of Forest Biodiversity through Community Forestry,* Wood, H., McDaniel, M., and Warner, K., Eds., Proceedings of an International Seminar held in Bangkok, Thailand, Oct. 26–28, RECOFTC Report 12, Regional Community Forestry Training Center, Bangkok, 157–166.

Western, D. and Wright, R. M., Eds., Strum, S. C., Assoc. Ed., 1994. *Natural Connections: Perspectives in Community-Based Conservation,* Island Press, Washington, D.C.

APPENDIX

Status of Collaborative Management of Protected Areas in Selected Countries in Asia

Malaysia

Idrus and Nais (1999) found no specific legislation in either peninsular Malaysia or Sabah explicitly supporting CM. There is a provision in Sarawak for the appointment of honorary rangers from community members living around PAs. In general, PA management is based on a "people out" approach, although there are some exceptions to this and there are some small initiatives for CM.

Pakistan

Hassan (1999, p. 64) states, "There are currently no protected areas in Pakistan where collaborative management is practiced in its true spirit. The existing laws have no provisions to support joint management of protected areas." He mentions several initiatives to promote CM "both inside and outside PAs."

Lao PDR

Chantaviphone and Berkmuller (1999, p. 177) state, "Collaborative management within protected areas ranges from the merely consultative to the truly participatory. However, all efforts are still in their early stages." They also point out that "although the law has basic provisions in support of collaborative management, local interpretation of the law stresses state control and detailed regulations are absent" (p. 177).

Philippines

The National Integrated Protected Areas System (NIPAS) Act of 1992 provides that a Protected Area Management Board (PAMB) be created and that it should include representatives from each affected village and other NGO and community organization representatives (Custodia and Tabaranza, 1999). In fact few PAMBs are in operation because of shortages of funds. According to Custodia and Tabaranza (1999, p. 219) CM "is relatively new" and implementation of various programs and strategies is "still in the trial and error stage."

Bangladesh

Ali and Habib (1999, p. 78) state, "Conventional management methods in protected areas have been in practice for a long time. However, national management principles have been given topmost priority, with the interests of local people and the adjacent community attracting little or no attention. In many cases, local people and communities sacrificed their self-interest, and lost access to resources." While there are provisions for CM under the new Forestry Sector Project (1997–98 to 2003–2004), these have "yet to be put in use in the field" (Ali and Habib, 1999, p. 79).

Nepal

Sharma (1999, p. 49) describes the main features of legislation relating to PAs in Nepal. Nepal's approach "calls for strict control of forests within ... the park or reserve ... combined with intensified agriculture and forestry on public and private properties outside the park." The National Parks and Wildlife Conservation Act, 1973 makes "provision for financing community development activities in buffer zones and conservation areas by ploughing back royalties accumulated from park-generated businesses such as tourism. Buffer zones receive 30–50% of the royalties, with 100% going to conservation areas" (Sharma, 1999, p. 51). Sharma describes a continuum of CM models (p. 50). At one extreme ranging from PAs (Conservation Areas or buffer zones) where people are "guardians of the land" and manage resources and generate income "through the formation of self-made user groups." In the lowland national parks and reserves "people's participation is limited to seeking opinions before preparing any management plans or strategies," but people have been allowed to enter these parks for a few days annually to collect such products as grasses. A middle category includes Himalayan parks, which have "provision for the people living in the park enclaves to collect natural resources of daily necessities such as fuelwood, leaf litter, small wood and fodder."

India

In India, "consumptive use of wildlife resources is prohibited" within PAs, but regulated grazing is permitted (Rao and Sharma, 1999, p. 106). Kothari (1996, p. 29) points out that the Wild Life (Protection) Act which is, on the one hand, invoked "to stop local community activities," also contains a provision "which allows for the continuation of local rights." In India, Joint Protected Area Management (JPAM) between those who argue for genuine local involvement in decision making and benefit sharing in and around PAs (Kothari et al., 1996; 1997) and those (such as Rao and Sharma, 1999) who reject this approach in favor of ecodevelopment (development of the multiuse surrounds of PAs). While the legal status and official support for JPAM may be ambiguous and contested, the social movement supporting various CM approaches is significant and real (if often partial) success stories do exist.

Thailand

In Thailand, a regime of strict protection within PAs exists in theory, although there are large numbers of people actually living in PAs including hill tribes with histories of occupation predating any demarcation of PAs. Relocation (both forced and strongly encouraged voluntary relocation) has been common. The proposed Community Forestry Bill, which has been hotly contested by elements in the Royal Forestry Department (RFD) and the environmental movement, would provide for some tenurial security for people with long histories of residency in what are now PAs, but this would apply only to certain categories of PA. The bill, if passed, would also allow management plans to be developed collaboratively. However, at present, there is no legal basis for collaborative management in protected areas. Local community forestry initiatives that do exist in PAs are entirely protection oriented. Collaborative management of buffer zones, usually involving NGOs, is an emerging trend, but is very protection oriented, as use of forest products is not permitted by the RFD (although minor use is sometimes tolerated). There are some new ICDPs concerned with income generation in buffer zones.

Section II

Institutions and Policies

6 Adapting Land Reform to Protected Area Management in the Dominican Republic

Charles C. Geisler

CONTENTS

INTRODUCTION

The commonsense aspects of adaptive comanagement (ACM) seem all too apparent. Adaptive management keys into the basic wisdom of evolutionary biology: there are no evolutionary "mistakes," only experiments. Certainly sustainability will be served, as Lee (1994) states, by viewing "failure" as the opportunity to begin again with better knowledge. One is more apt to fix what is broken knowing that it will break again and require new fixes. And at the very practical level of integrated conservation and development project (ICDP) work, although problematic to some (e.g., Wells and Brandon, 1992; Larson et al., 1997; Oates, 1999), it is more than comforting to steady the course with Lee's "compass and gyroscope" of adaptive comanagement. Those placing faith in integrated approaches to ecoregionalism,

0-8493-0020-7/01/$0.00+$1.50

ecosystem management, and protected area growth have surely gained resiliency as ACM has diffused over the policy landscape.

As a sociologist, however, the author feels a worm turning in the weight-bearing beams of conservation. The concern is with power, manifest in the growth of protected areas but, as Lee reminds, too often ignored in discussions of comanagement and collaborative planning.* Differential power, like a rogue magnetic force besieging the Earth's poles, can distort both compass and gyroscope. It can corrupt and coopt. It can impose solutions that in themselves are problems. So, in embracing ACM, it is important not to turn a blind eye to everyday power realities operating in conservation work. This chapter is about ACM in the real world of unequal power and what, from one standpoint, might be done to rectify this situation.

The chapter begins by highlighting the frequent disparity in power between those who seek to expand the conservation estate and those whose livelihood and sense of place, home, identity, and security come from its internal frontiers. It is a disparity that has many facets, including land use, access, and ownership rights. In many countries of the world where protected areas are growing, relatively affluent people enjoy net gains in land tenure while marginal populations experience net losses and bear the social opportunity costs of conservation (Geisler, 1993; 2000; Inamdar et al., 1999; Geisler and de Sousa, 2000). To ignore inequalities in property power in designing conservation schemes is to ignore land and resource dispossession among stakeholders and to think that the privileged and the precarious alike are willing ACM collaborators.

The formulation here is this. Despite good intentions and rhetorical appeal, ACM is largely academic without power sharing, and property sharing is an important way of accomplishing this. Marshall Murphree (1994, p. 405–407) underscores this in reflecting on community-based conservation and collaboration in Zimbabwe:

> What is required to make the concept of participation viable is proprietorship, which means sanctioned use rights, including the right to determine the mode and extent of management and use, rights of access and inclusion, and the right to benefit fully from use and management. Proprietorship provides the necessary tenurial component for an adequate institutional framework.

In settings where ICDPs respect local tenure (Murphree is open to individual or collective private ownership, communal tenure or other), there is less chance of disenfranchizing and alienating — materially and politically — local conservation partners. He recognizes, as one must, that local property empowerment means a reduction in state and other nonlocal authority systems. The risks herein for conservation interests are matched by the risks of conservation mutinies from below.

This chapter poses what might be called "Murphree's law": successful conservation comanagement entails local empowerment, and the redistribution of property rights, as Dorner (1992) notes, is essential to this process. Community-based

* Power is only one of several problems which plague adaptive management, according to Lee. He holds out faith that democracy ("gyroscope") is contentious yet stable. It marks the boundaries of acceptable force and threat and through debate it channels discrepant views into a policy area that is constantly amended and refined (adpated).

resource management is more likely to align with conservation objectives if local people, at a minimum, enjoy tenure security that shields them against eviction, if they are fairly compensated, if they are resettled should eviction occur, and if those who remain have access and use rights to resources that they help determine. Such property-based power sharing already exists, of course, and is called land reform. This chapter will attempt to show, using the Dominican Republic as a case in point, that land reform is a reasonable way to operationalize Murphree's law on property and power. And where land reform fails or falls short, as it sometimes does, an adaptive approach to its rehabilitation is advised.

LAND REFORM AND POWER

Land reform is, almost by definition, large-scale experimentation in restructuring property relations with both social and environmental implications (Sobhan, 1993). Where frontier lands abound, land reform is often used as a means of colonization to secure borders or provide outlets for population surges. Where land is scarce and/or monopolized, it can be a symbolic gesture at partitioning resources by the state or a revolutionary action from below (Powelson, 1987; Sobhan, 1993). Two major inducements to land reform in the latter decades of the 20th century have been the United Nation's support for land reform as a sectoral development strategy and the Cold War climate of the 1960s. In the latter setting, the United States and the Soviet Union advanced competing models of ownership and the role of the farm sector in modernization. In the post–Cold War era, land reform has to some degree been appropriated by disciples of privatization, despite a legacy that spans centuries, cultures, and many forms of tenure.

What conservation advocates often forget in contemplating regional landscape approaches to ecosystem protection is something land reformers take for granted. Landscapes are proprietary and the ownership therein is rarely equitable. Chronic landlessness and near landlessness are inversely related to land concentration. Many countries distinguished for their biologically diverse ecosystems such as Costa Rica, Kenya, Madagascar, and the Philippines have high land concentration rates.* In such settings, land consolidation for natural protection has severe cumulative impacts for those with few or no property rights, and proposals to set aside from 12 for 50% of such countries aside for exclusive protection are socially catastrophic (Inamdar et al., 1999; Geisler, 2000; Brechin et al., 2001).** Where large owners are enlisted to protect biodiversity (Langholz, 1996), they are compensated or their lands become private parks and reserves, which garner income, status, or tax benefits or land swaps.

* Both Eckholm (1979) and Sobhan (1993) provide data on landlessness and land concentration. The latter notes that in some so-called developing countries land availability for cultivation per agricultural household increased in recent decades, but this "good news" is deceptive where such expansion comes from marginal lands with relatively high biodiversity.

** There are happy exceptions to this generalization, including the recently legislated ancestral domain rights of local people in the Philippines (see Chapter 22 in this volume for more detail), Australia's comanaged (with Aboriginies) protected areas (Birckhead et al., 1992) and cases cited in Western and Wright (1994).

It is one thing to work aggressively to defend nature against development; it is another to impose the costs of protection on the least secure members of society and then seek their collaboration in the end product. There is an analogue in Garret Hardin's (1968) much-repeated argument about the commons. Many concur that the commons tends to be overexploited but not because commoners are self-serving. Instead, forces beyond their control compel their overexploitive behaviors (e.g., Ostrom, 1990; Bromley, 1992). These forces include power disparities which, when factored into the Hardin model, recast the tragedy of the commons as a "tragedy of the commoners" (McCay and Acheson, 1988). The solution lies not in banishing commoners but in helping them limit access to the commons, rewarding them for long-term sustainable behaviors, and assisting their collective action that equalizes power with distant stakeholders of the resources in question. Land reformers can include similar objectives in their work and, as the examples below suggest, have previously done so.

There is a double consolation in combining land reform with protected area management. Not only does it potentially restructure power relations by rearranging property rights, but it shares the ICDP social welfare agenda. Both are integrative forms of large-scale land use planning. And, when one moves beyond bare-bones land reform (land redistribution and conferral of title) to fuller agrarian reform, locals become the beneficiaries of technical and marketing assistance, subsidized credit and loans incentives, infrastructure and services, conservation training and education, and employment diversification (World Bank, 1975). Before moving to actual experience with land reform and conservation, one would do well to revisit the logic of how land tenure and conservation behaviors are related — a subject central to Murphrees' law.

TENURE AND CONSERVATION

There is a widely held belief that land stewardship of many kinds increases as proprietary interests become secure through actual title transfer (e.g., Dotzauer, 1993).* For this and other reasons, western land reformers often make title-transfer the backbone of their efforts. Exclusive rights are believed to give rise to incentives (such as dependable future income, enhancement of land value, creditworthiness, etc.) for maintaining and investing in conservation. Absent these, planning horizons will be short term and maximize immediate profitability (Nelson, 1938) and/or consumption (Feder, 1998). By extension, tenure insecurity caused by lack of clearly defined land ownership promotes unsustainable uses by discouraging long-term investments (Wachter, 1992). These investments have environmental implications — crop choice (annuals vs. perennials), the amount of land cleared, installation of shelter belts, windbreaks, terraces, selection of labor–intensive organic practices, etc. Long-term tenure security is also thought to influences one's social standing and willingness to adopt/innovate/experiment.

* This perspective perhaps goes back to the classical economists such as Smith and Mill who believed that improvements would follow on the heals of secure ownership (Hite, 1979). For a current overview of tenure and conservation behaviors (see Beaumont and Walker, 1995).

This logic is subject to numerous qualifications, however, about which advocates of ACM should be informed. First, research indicates that the relationship between tenure and conservation is highly situational (Lee, 1980; Ferraro, 1993). Even where tenure is secure, poverty tends to raise discount rates and complicate land investment behaviors (Wachter, 1992). Nor is tenure security synonymous with individual private ownership (Findley, 1988; Larson and Bromley, 1990; Lynch and Alcorn, 1994).* Levin (1999, p. 203) points out risks attached to privatization: "Assigning property rights to some individuals may thereby deny them to others, in extreme cases creating destitutes as well as fomenting rebellion." Moreover, culture colors the meaning of security that land reform seeks to expedite (South Pacific Commission, 1985; Hirtz, 1998). And finally, off-farm labor opportunities, variable discount rates, and other conditions may induce untitled farmers to behave in ways less degrading to the land than those with titles, and property privilege can pillage land as well as protect it, depending on the ethics of the owners.

Although land reform/agrarian reform is not a panacea, it speaks directly to the matter of tenure and sets the stage for the long-term behaviors sought by ICDP and ACM proponents. Land reform is not predicated on any particular form of tenure and can be used to encourage tenure choice and diversity. It does not guarantee the comanagement of natural resources and ecosystems but creates an opportunity structure supportive of such results. For example, land reform titles (collective or individual) often remain provisional for as long as 20 years, giving comanagers time to experiment with alternative mixes of property rights and responsibilities. Simply put, land reform can experiment with title conditions that induce not only equity and efficiency, but environmental sustainability as well.

GREENER LAND REFORM

The case for utilizing land reform to advance adaptive comanagement rests not only on its ability to restructure power relations nor on the conservation potential of greater title security. It also builds on the theory and practice of land reform as deployed by economists, sociologists, and demographers, suggesting its conservation potential.

THEORY

Much theory on economic efficiency, a traditional pillar of land reform, has environmental implications. Land reform recipients typically receive small land allotments varying greatly in fertility, water availability, slope, and other critical production factors. Holding tenure security constant, are small farmers better stewards of their land and resources than their large-scale counterparts? The latter frequently have higher levels of education and exposure to environmental management information from extension and other sources. Small farmers are often seen as a threat

* The current privatization bias often ridicules common property systems even though Garret Hardin himself has recently (1994) modified his classic statement.

to biodiversity as a result of swidden cultivation, poaching activities, and their need for fuelwood, timber, and other forest resources (e.g., Southgate, 1990).*

There is some evidence, however, that small, limited-resource operators at times approach farming with conservation in mind. Small household production systems often have lower opportunity costs than do large capitalist farms and tend to be more labor absorbing (attractive under capital-poor, labor-abundant conditions) than larger ones (Ashe, 1978).** Second, size of operation typically varies inversely with output per unit of land (Dorner, 1992), although many other factors color this finding (e.g., excessive fragmentation may weaken the size–productivity relationship). Thus, productivity per unit of land rises when compared with larger landholdings (World Bank, 1975; Berry and Cline, 1979; Dorner, 1992), bolstering the theoretical case for economic sustainability of land reform recipients.

Third, economies of scale — an argument often used to dismiss land reform — are easily misapplied in agriculture (Hayami, 1991; Binswanger and Deininger, 1993). In fact, there are diseconomies of scale that accompany large-scale agricultural production, over and above social costs (low labor absorption per unit of land) and environmental costs (monocultures and the side effects on soil, water, and biodiversity of maintaining them). Scale advantages in large farming units such as input purchasing, credit, marketing and joint production strategies, where real, are available to smaller producers through cooperatives, labor federations, and other producer associations. Kaiser (1988) has used this argument in defense of small holders in the Dominican Republic.

Turning to sociological theory, particularly where small holders are drawn from rural as opposed to urban settings, land reform can benefit from capturing valuable indigenous knowledge retained by such producers (Kloppenburg, 1988). Smallholding traditional farmers are often culturally equipped to incorporate forest species, structure, and nutrient cycling knowledge into their farming operations (Wilken, 1987; de Ceara, 1985; Jacobs, 1997). It is typically smallholders who have developed conservation-prone slash-mulch and green manuring systems and applied these to small plots in lieu of slash-and-burn cultivation (Thurston et al., 1994). It is frequently traditional producers who retain sophisticated home garden techniques and other time-tested intercropping strategies, which lead to high-yielding, environmentally compatible production systems (Altieri and Macera, 1993).

Land reform also has demographic facets bearing on conservation concerns. Social equity, an intended by-product of most land reforms, is believed to correlate with low human fertility rates (Kocker, 1973; National Academy of Sciences, 1986), a critical component in conservation planning. A study of 561 land reform beneficiaries in Egypt by Stokes and colleagues (1986) found a positive relationship between size of holding and fertility and an inverse relationship between fertility and security of title. Like other authors cited above, Kirk (1971) argues that land redistribution and titling improve economic security and social status. Demographers

* Recent interest in the "Kuznets environmental curve" is indirectly relevant here. Large (more affluent) interests are thought to make the environment better in the long run (e.g., see Levin's summary, 1999). But the validity of these claims is under much dispute (Environmental Economics).

** For further research on small vs. large holdings in an environmental context, see Collins (1986) and Foy and Daly (1989).

associate both with reduced reproductive pressure. There is room for guarded optimism, therefore, with respect to the effects of land reform on the environment via population dynamics.

In terms of both social and economic theory, then, small farmers warrant serious consideration as conservation allies.* Particularly when drawn from traditional farming systems in biologically rich areas, their farms compare favorably with those of larger farmers in terms of the genetic diversity (Wilken, 1987; Thurston et al., 1994). Not only do they lack the capital to compete in the market of commercial monocultures, but their aversion to risk often prevents them from doing so. Their modest holdings yield food, fodder, shelter, and medicinal remedies in a web of local ecological knowledge. Land reform in support of small farmers — still the majority of the world's producers — seems theoretically predisposed to ecologically acceptable practices.

EXPERIENCE

Environmentally sensitive land reform is neither new nor merely theoretical. Some land reforms have incorporated soil conservation as a *sine qua non* of long-term productivity gains by their beneficiaries. The controversial Salvadorean land reform and that of the former Soviet Union employed land reform to check soil conservation as a cornerstone of national food security (Frost, 1996; McReynolds et al., 2000). Krylatykh (1991) identifies three problems (natural resource restoration, preservation, and improvement) that must be solved to attain agrarian reform in today's Russia. Brazil's land reform agency for a period provided leadership over the expansion of extractive reserves in that country, a model of extensive land reform intended to conserve biodiversity within the Amazon rain forests (Geisler and Silberling, 1992). South Africa's post-independence land reform manifesto devotes a lengthy chapter to natural resource management and conservation (ANC, 1994). Dorner (1992), Forster (1992), Theisenhusen (1991), and Ericson et al. (1997) explore environmental protection options via land reform and emphasize the imperative for sustainable agriculture on future land reform settlements.**

The occasions in which land reform has consciously been integrated into park management and planning nonetheless appear to be few. The case of Brazil's extractive reserves, a special case of protected area management in themselves, was mentioned above. A decade ago, the Costa Rican government entered an agreement with an international nongovernmental organization (NGO), granting title to untitled farmers surrounding a natural reserve on that country's Atlantic coast in exchange for farmers limiting their own entry to the reserve for hunting or logging (ANAI,

* The argument that locals are "appropriate allies" in conservation has been made before (e.g., Brownrigg, 1985; Western and Wright, 1994; among others).

** Land reform can be double-edged in its environmental implications, especially if it is defined as parellation and titling with few integrative aspects of agrarian reform. Worse, land reform is sometimes confused with poorly planned colonization in frontier areas with high biodiversity (Redclift, 1987). Thus conceived, land reform is apt to be the antithesis of conservation. Finally, land-to-the-tiller almost invariably means intensification of resource use. If this causes large numbers of people to move onto less land with superior and sustainable farming techniques, then conservation at the regional level has been served, at least in the short run. (Intensification per se can, of course, have a long list of unwanted environmental externalities.)

1988). The same government resettled farmers displaced through the expansion of Guanacaste National Park on land reform parcels in northwestern Costa Rica. In Guatemala's Sierra de las Minas Biosphere Reserve, a Nature Conservancy partner organization (Defensores de la Naturaleza) arranged a communal "land reform" in 1999 in which an indigenous community agree to a land swap for fertile lands outside the reserve on which to farm (Lehnhoff, 1999). Cook (1994) summarizes the World Bank Conference on Environment and Settlement Issues in Africa, emphasizing the importance of land reform as a way of organizing resettlement and bridging the concerns of tenure and protected area management. What follows is an experiment in land reform–assisted ICDP development in the Dominican Republic.

THE CASE OF THE DOMINICAN REPUBLIC

The Dominican Republic is of interest to the present discussion both because of a 40-year commitment to land reform and because roughly 30% of the country has recently attained protected area status of one kind or another (Geisler et al., 1997; Geisler, 2000). It offers fertile ground, so to speak, for land reform experimentation in conjunction with large-scale biodiversity and habitat protection. Of particular interest is the Los Haitises National Park (LHNP) on the country's north-central coast (Figure 6.1), a park surrounded with national land reform resettlements, some of which have been created as part of a buffer-zone strategy for PNLH (Figure 6.2).*

Table 6.1 summarizes changes in major land use categories in the Dominican Republic for almost five decades. Arable land and permanent cropland have almost doubled and currently occupy 30% of all land.** This, combined with permanent pastureland (which has multiplied nearly four times in the same period), extends over nearly three quarters of the country, large portions of which are mountainous or arid. It is thus not surprising that forest and woodland have diminished dramatically, from over 70% earlier in the century to 13% in 1991, nor that environmentalists are highly motivated to create protected areas out of what forests remain. The Dominican Republic is one of the leading countries in the region in deforestation and is second only to El Salvador in its diminutive per capita forest cover.

Thus, the Dominican Republic faces severely competing land uses. Moreover, between 1950 and 1993 (the country's most recent national census), its population grew from just over 2 million to over 7 million (Ramirez, 1988), despite massive outmigration to the United States. As population grows, landlessness and poverty have expanded as well (Dotzauer, 1993). Table 6.2 shows the highly skewed distribution of landownership among those who own land for the country as a whole. Very large landowners (over 500 ha) were 1.4% of all owners in 1971 and owned 40% of all farm and pastureland; they diminished somewhat by 1981 to just under 1% but expanded their hold to 47% of these lands. Average size of holdings for farmers with less than 500 ha remained more or less constant despite the land reform, probably because the number of holdings grew by 26%.

* The physical layout of these land reform projects in relationship to the original park boundaries is discussed in Geisler et al. (1997).

** As of this writing, revised legislation is under consideration that would reduce the area dedicated to formal conservation an undisclosed amount.

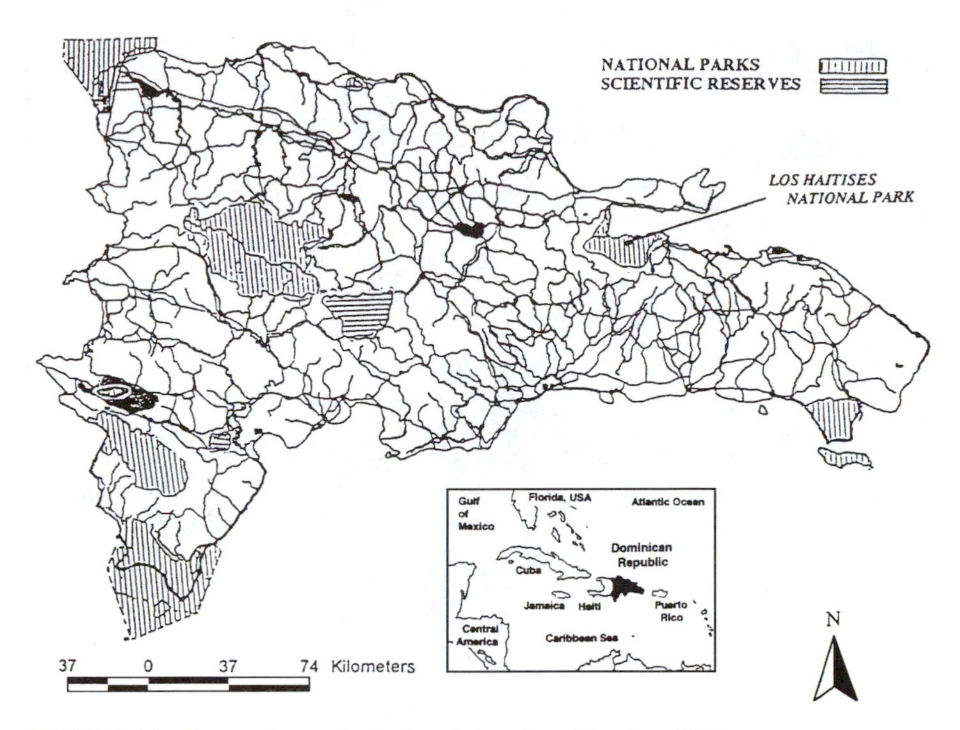

FIGURE 6.1 Protected areas in the Dominican Republic, circa 1993.

In the Dominican Republic as elsewhere, the tendency is for the disenfranchised to seek subsistence in relatively remote places where conservation groups are now focusing their protection efforts. Dominican land reformers anticipated this problem over a generation ago when the competition between subsistence agriculture and conservation was still in its infancy. The Introduction of Agrarian Reform Law 5879 makes explicit reference to the fact that excessive land concentration has driven large numbers of landless people to remote, fragile lands where their efforts to survive lead to critical losses in soil and forest resources (Gutierrez-San Martin, 1989, p. 259). The same preamble anticipates comanagement possibilities in the future, calling for "an ever greater direct participation by them in the study, discussions and solutions of their problems."

The Dominican government has used land reform to buffer LHNP since its creation. Northwest of the park in the delta of the Bajo Yuna and Barrecote Rivers, it established a far-flung series of land reform rice cooperatives, which conveyed title and economic livelihood to many thousands of families (Meyer, 1989). In 1990, it purchased lands from the State Sugar Corporation in an effort to relocate 300 families from the park on a pilot land reform project outside the park (Hughett, 1992).* And during the 1990s, the government added six additional Institute of Agricultural Developments (IAD) resettlements, which absorbed an additional 2400

* This earlier strategy has not prevented local rice farmers from hunting, fishing, and making charcoal in the park, although it has surely reduced the pressure of these behaviors.

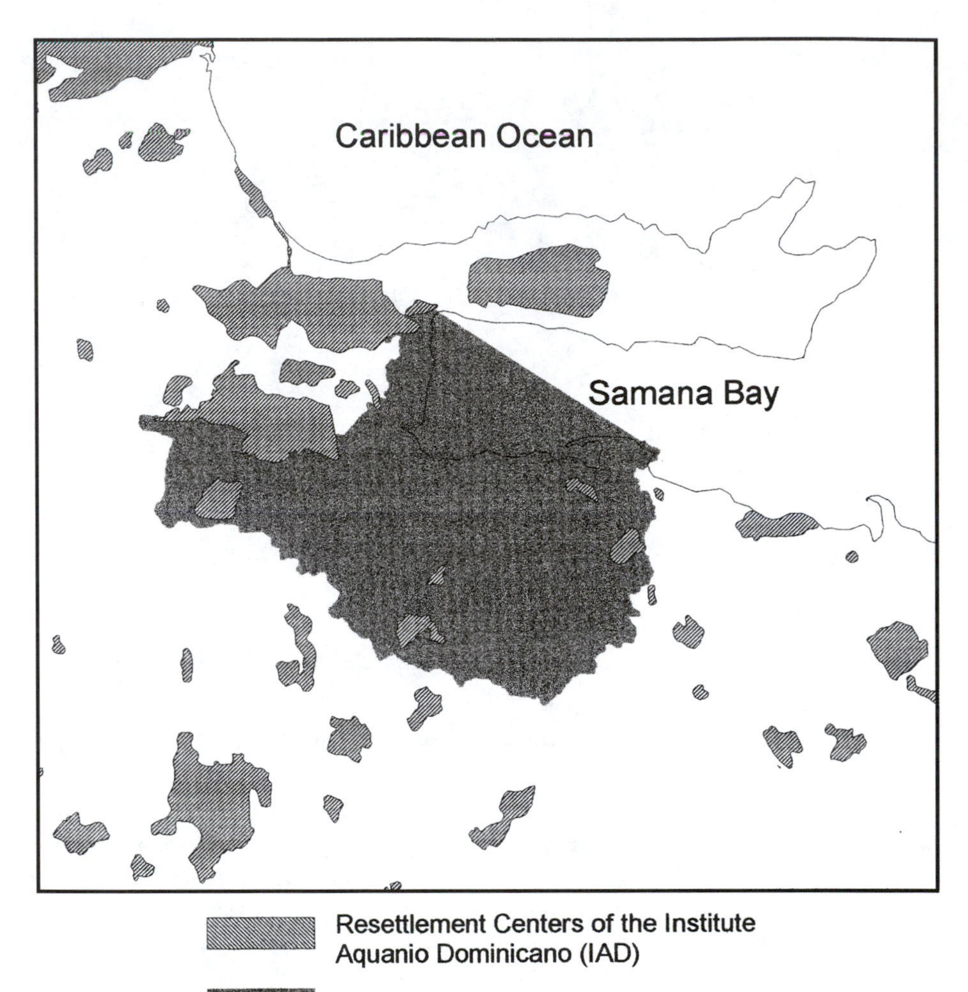

FIGURE 6.2 Land reform resettlements (1990) and Los Haitises National Park Boundaries (1999). (From Agencia Espanola de Cooperacion Internacional, 1991, p. 86.)

families (12,000 people) from relying on the park for subsistence. This expensive infrastructural development was in part the result of a detailed census jointly performed by the Dominican military and a social action agency of the Catholic Church in the region — an unusual collaboration.

Besides physically relocating former park residents, several additional conservation benefits in the region have resulted from the land reform. Partially due to the demonstration effects of collaborative work performed in the 1990s by researchers from Cornell University and the National University Pedro Henríquez Ureña (UNPHU), the governmental land reform IAD now offers short courses in organic agriculture and agroecology to encourage ecologically sound farming adjacent to

TABLE 6.1
Macro Land Use Change, 1956–1991, for the Dominican Republic (1000 ha)

	1956	1961	1966	1971	1976	1981	1986	1991
Total area	4873	4873	4873	4873	4873	4873	4873	4873
Land area	4838	4838	4838	4838	4838	4838	4838	4838
Arable land and permanent cropland	680	860	895[a]	972	1310[a]	1425[a]	1437	1446[a]
Arable land	?	580	600[a]	635[b]	960	1075[a]	1075	1000[a]
Permanent cropland	?	280	295[a]	330[a]	350[a]	350[a]	362	446[a]
Permament pasture	580	1020	1170[a]	1436	2092[a]	2092[a]	2092	2092[a]
Forest and woodland	3440	1100	1100	2225	643[b]	633	623	613[a]
Other	173	1858	1673	240	793	688	686	687

[a] Fao estimate.
[b] Unofficial.

Source: Fao Agricultural Yearbook.

TABLE 6.2
Number of Holdings, Surface, and Average Size According to the Size of the Holdings

	National Agricultural Census, 1971			National Agricultural Census, 1981		
Size of Holdings (ha)	**Number of Holdings**	**Total Surface Area (ha)**	**Average Size of Holdings (ha)**	**Number of Holdings**	**Total Surface Area (ha)**	**Average Size of Holdings**
Total	304,820	2,719,305.6	9.0	385,060	2,659,977.4	6.9
Less than 0.5	49,651	12,132.0	0.25	61,670	12,465.0	0.18
0.5–5	185,292	337,516.8	1.8	252,995	311,660.0	1.25
5–10	33,803	229,930.1	6.8	32,543	230,383.0	7.0
10–50	28,987	584,165.0	20.1	30,815	636,484.2	20.6
50–100	3,974	266,841.0	67.1	4,081	270,154.0	66.1
100–200	1,791	247,245.2	138.1	1,825	249,629.0	136.8
200–500	884	266,350.1	301.3	786	230,175.3	292.8
500–1000	222	146,861.0	661.5	184	120,574.9	655.3
1000 or more	216	628,264.4	2,908.6	161	598,452.0	3,717.1

Source: National Agricultural Census 1971 and 1981, Oficina Nacional de Estadistica.

the park (GEF, 1997).* Second, through joint efforts with the same universities, the Dominican government now has the digital mapping capacity to cross-reference soils, slope, and groundwater information with areas within the buffer zone of the park that might be opened to farming activity. This Geographic Information System

* The survey work and on-site observation suggest that large numbers of families formerly residing in the park have retained their parcels therein as a hedge against failure or termination of the resettlement projects.

(GIS) resource should reduce the conflicts that arise between ecosystem and social system activities and lay the basis for both a permitting system and a resource-driven approach to environmental education in the future.

The Dominican case offers empirical evidence that land reform can go hand-in-hand with large-scale conservation. It is too early to determine whether it has redressed major power imbalance among collaborating partners in the initiative. Despite better housing and provisional title to land in exchange for insecure property rights in the park, farmers who had relied on LHNP for all or part of their livelihood in the past were of mixed mind when surveyed in 1996 (GEF, 1997). These 600 surveys were executed in a politically charged climate of presidential elections, changing park boundaries (which expanded repeatedly, engulfing adjacent communities), and interventions by the Dominican Armed Forces (Geisler et al., 1997).* So the following results are provisional and should be compared to future restudies.

The year 1992 began a farmer diaspora from LHNP because of the decrees and military actions just noted. The author and colleagues located farmers in their original communities ("unrelocated communities"), outside the park ("urban barrios"), on land reform ("IAD") resettlements, and still others who had moved in with friends or families elsewhere. Almost 90% of the farmers interviewed relied on the park for a spectrum of subsistence activities such as grazing animals, swidden (*conuco*) agriculture, harvesting fruit, lumber, charcoal production, hunting, and gathering plants and medicinal materials. Those sampled were asked how much their families had suffered as a consequence of the forced removal. All four categories of respondents reported considerable trauma, with those on resettlements and those not forced to move reporting the least (but still 89%), probably because relocation was involuntary and entailed loss of economic improvements, family, community, and cultural ties, as well as self-sufficiency and self-worth.

Nearly all of persons interviewed "held" land in the park (with or without title, which is normal elsewhere in the country) before the forced relocation (Table 6.3). This drops to an average of just below 75% following the government intervention. Those who self-relocated experienced the most dramatic land loss; after military intervention, only 54% have property. When asked to compare the amount of land before and after relocation, 86% of the entire sample reported less land, and 94% of those on IAD resettlements reported the same. Thus, although the IAD group appeared best off in terms of current access to land, they now have less land with which to support their families.

Moreover, respondents "complained" that the postrelocation land allotments were not as productive as those held previously. Only 5% of the sample preferred their new lands and there was virtually no difference between land reform recipients and those who remained in their villages, relocated at their own expense, or moved to urban neighborhoods, often with small garden options. Thus, 91% of land reform recipients reported having more productive lands before being resettled. The identical

* In 1992, the Dominican government took a strong stand on LHNP, occupied the entire park with military forces, initiated a census of people, and proceeded to remove cattle, fencing, and human residents (Geisler et al., 1997). Eviction and relocation dragged out over 5 years because of construction and siting delays for resettlement communities and because LHNP boundaries changed several times, further complicating who was "in" and who was "outside" of the actual park and buffer zone.

TABLE 6.3
Comparisons of Tenure before and after Relocation

	IAD Resettlements	Unrelocated Communities	Self-Relocated Communities	Urban "Barrios"	Total
Before	99.5	99.4	100	100	99.7
Now	84.0	81.4	54.3	69.4	74.2
Total	100	100	100	100	100

Source: IEPD/Cornell Survey: Encuesta sobre Poblacion y Medio Ambiente en el PNLH, March, 1996.

TABLE 6.4
Comparisons of Security of before and after Relocation

	IAD Resettlements	Unrelocated Communities	Self-Relocated Communities	Urban "Barrios"	Total
More	51.2	5.7	4.7	9.8	26.0
Less	31.4	54.7	75.0	66.7	49.4
Same	14.5	32.1	20.3	19.6	20.9
Total	100	100	100	100	100

Source: IEPD/Cornell Survey: Encuesta sobre Poblacion y Medio Ambiente en el PNLH, March, 1996.

percentage among those who did not move from their park communities can be explained by the fact that they are now cut off from fertile lands within the park, the principal source of "productivity" in their overall scheme of survival.

More troubling was the response to questions of security (Table 6.4). Here, land reform recipients lead all other categories in reporting superior property security over what they had before, 51% compared with less than 10%. Yet a surprising 31% of the IAD recipients consider their land claims less secure now than previously, although the other three categories exceed this rather dramatically. These combined results suggest that postrelocation land tenure is only somewhat more secure (for IAD recipients, not for others) and that both quantity and quality of land has been reduced. Matters of productivity and security are functions of technical assistance, provided to both the land reform recipients and other farmers in the region. Curiously, 65% of the former reported this assistance as "good" or "regular," whereas 88% of the latter gave these responses (Table 6.5).

In anticipation of comanagement proposals for the park, respondents were asked where control for PNLH should be vested. Only 23% wanted the government to assume this role, whereas 72% said their own communities should do it. Roughly 80% of all respondents surveyed said their communities needed more environmental education to protect nature. Just under 60% felt farmers in the park, should they be allowed to continue farming in the park, needed to conserve their soils. Over 91%

TABLE 6.5
Evaluation of Technical Assistance to Farmers

	IAD Resettlements	Unrelocated Communities	Self-Relocated Communities	Urban "Barrios"	Total
Good	30.0	41.2	—	25.0	31.7
Regular	35.0	47.1	—	50.0	38.1
Bad	27.5	11.8	—	—	20.6
Total	100	100	100	100	100

Source: IEPD/Cornell Survey: Encuesta sobre Poblacion y Medio Ambiente en el PNLH, March, 1996.

said they would sign a contract, overseen by their community, to conserve the park in exchange for access to the buffer zone of the park. Interestingly, the land reform farmers were the least supportive of this option (78%), perhaps because the meaning of "community" was in flux.

Of ongoing concern to the Dominican government is recidivism among former park users and inhabitants, that is, the inclination of farmers relocated from LHNP to return to their former homes and *conucos* in the park despite government prohibitions. Residents in the pilot resettlement project (1991) were frustrated with waiting for housing, services, and potable water. On-site observation there by the author suggests that resident turnover was nearly 100%. Many had (illegally) sold their parcels and returned to LHNP, and others reported that approximately half of the residents were using government credit not to farm their new parcels but to improve their "abandoned" parcels in the park. The 1996 survey results indicated that newer resettlements had improved, however, and that only 51% of the land reform respondents were considering in-park work in the future compared with between 65 and 75% for other categories (Table 6.6).

Some readers will be alarmed at how many farmers, even those benefiting from land reform opportunities, contemplated returning to LHNP. At the time of the survey, a relatively large buffer zone had been added to "the park" and the government was unclear whether or not farmers would be allowed to farm it. So farmer interest in returning was less a sign of defiance than a statement that they wanted access to buffer-zone resources.

DISCUSSION

The proposition that land reform, including comprehensive or agrarian reform, can be of use in redistributing power in the form of property and hasten cooperation between conservation advocates and resident populations receives qualified support in the present analysis. Despite strong theoretical underpinnings, environmentally sophisticated land reform that might lubricate ACM and ICDP work remains sluggish. Many problems are inherent in land reform, precisely because power is at stake and political/institutional difficulties are many. Many rounds of adaptation and

TABLE 6.6
Thinking of Working in the Park in the Future

	IAD Resettlements	Unrelocated Communities	Self-Relocated Communities	Urban "Barrios"	Total
Yes	51.5	70.2	74.6	64.7	63.9
No	47.1	24.2	23.9	32.9	33.4
Total	100	100	100	100	100

Source: IEPD/Cornell Survey: Encuesta sobre Poblacion y Medio Ambiente en el PNLH, March, 1996.

collaborative work with affected communities lie ahead before anything like "success" can be recorded. It is and will be for many generations a work in progress.

It would be an oversimplification to conclude that Dominican farmers wish to return to the park because they oppose conservation. What lures them to the park is a valuable tuber crop called *yautia* (*Xanthosoma sagittifolia*). This cash crop, as any Dominican farmer will testify, is green gold, often selling for exhorbitent prices in Puerto Rico and the United States (Moreno, 1996). This episodic bonanza, well adapted to the peculiar climatic conditions of the park, makes recidivism worth the risk. If land reform is to work as a brake on such behavior, it is perhaps wise to begin experimentation with the crop in the buffer zone or even on resettlement projects. Initial observations suggest that this can be achieved with green manuring and other moisture-preserving soil management practices (GEF, 1997). Perhaps it is time for the Dominican land reform agency to begin an experiment station run by local farmers and underwritten by environmental organizations, the government, or both.

A second lesson upon which land reformers in the Dominican Republic might dwell is that the farmers displaced from the park are part of a migrant labor stream. Unlike migrant labor that is employed by agribusinesses in Europe and North America, however, this migrant labor is self-employed. For as long as the region has been cultivated, farmers have inhabited communities surrounding what is today LHNP and sojourned in it during planting and harvesting. Settling them in resettlement projects runs against a culturally engrained pattern of survival, which needs further study. Land reform has a mixed track record in the Dominican Republic. To function in the service of ICDPs it must offer farmers attractive alternatives to this entrenched migration pattern.

Third, the Dominican land reform probably erred strategically in putting resettlement projects in zones with poor soils and uncertain water supplies. The adaptation here would seem to be fuller use of the GIS system referred to earlier, which includes basic soil foundation data. Cornell and UNPHU researchers acquired recent census data and overlaid it on the soil classification system. The conformity between human settlement and good soils was striking, suggesting that farmers knew from years of trial and error where to cultivate in the buffer zone and larger region. Thus, if IAD heeds what farmers appear to know about soils, it will not have to arm-twist to get them to abide by its digital soil information and related settlement implications.

A fourth lesson is that food security and conservation security are fundamentally connected. Conversely, a precarious food system will eventually erode what appears to be a secure conservation system. Working closely with local NGOs and a social action arm of the Catholic Church, Cornell researchers have had encouraging results assisting local farmers, on and off land reform resettlements, using mulch-based agriculture, green manures, and multipurpose cover crops. Such techniques promote food security through single-farm production, as opposed to either migration or migratory *conuco* agriculture. And their initial success has built trust among a network of 200 families who, despite high *yautia* prices, have chosen to farm in one place, confident that they can improve their soils and their productivity.

A final point has to do with buffer-zone configuration. The Dominican government decreed in 1993 that the buffer zone attached to LHNP would be within the offical park boundary and limited to research and ecotourism. It later retreated from this position and is now contemplating opening the entire buffer zone to farmers whom it has spent the last 6 years evicting. Local communities are understandably confused by the location and intent of the buffer zone. Cornell has proposed a series of conditional use regimes and land trust models that might be instated in the current buffer zone, enriched with farmer incentives for compliance and opportunities for input (GEF, 1996). But it is property redistribution that addresses the power differential between both the state and local parishes (*parajes)* and between the conservation community and farmer groups. It is time to experiment with a noncontiguous buffer-zone concept, which could be constituted from surrounding land reform/agrarian reform settlements.

There are many additional land reform lessons that, learned after years of experience in many settings, could serve large-scale conservation efforts well, which aspire to ICDP goals and objectives. These range from the importance of gender equality, to optimum parcel size, to equitable lease agreements, to effective communal and cooperative arrangements, and to a wide range of supplemental services without which small holders tend to fail as farmers.

CONCLUSIONS

At one level, this chapter is about the integration of the sociosphere, which includes power dynamics, and the biosphere employing the underutilized institution of land reform. Land reform has numerous attributes to recommend it for ICDP development and strengthening. Foremost among these are its transfer and potential equalization of power, its close parallel with ICDP philosophy (integrated development), and the numerous ways it can be customized to fit local circumstances. Land reform itself is an institution in constant flux, change, and adaptation. The author has tried to argue the case, using the mixed success of land reform in and around a park in the Dominican Republic, that one can pick up the pieces and move closer to a resilient ACM mode. Murphree's law insists that this be tried in earnest, with full awareness of the imperfections of the instrument in hand. If one takes a contingency-based, policy-as-experiment approach, land reform is apt to be a store of riches, not because it always delivers equity, efficiency, and environmental integrity but because, mindful of its many failures, it has recreated itself untold times in the past in response to new priorities.

ACKNOWLEDGMENTS

This work stemmed from a paper presented at the CIIFAD/CIFOR Adaptive Comanagement Symposium at Cornell University, September 15–17, 1998. The author acknowledges support for this research from CIIFAD, the Global Environmental Facility (GEF) of the Dominican Republic, Project Number DOM/94/G31, and the helpful comments of John Schelhas, among others.

REFERENCES

AID, 1978. Agricultural Development Policy Paper, June, Agency for International Development, Washington, D.C.

Altieri, M. A. and Masera, O. 1993. Sustainable rural development in Latin America: building from the bottom-up, *Ecol. Econ.,* 7, 93–121.

ANAI, 1988. Campesino Land Tilling and Wildlife Preservation, Project 3506. Report to the World Wildlife Fund, Puerto Limón, Costa Rica.

ANC, 1994. ANC Agricultural Policy, African National Party, Westro Reproductions, Johannesburg.

Ashe, J. 1978. Rural development in Costa Rica. Interbook, Inc., New York.

Beaumont, P. M. and Walker, R. T., 1995. Land degradation and property regimes, *Ecol. Econ.,* 18, 55–66.

Berry, R. A. and Cline, W. R., 1979. *Agrarian Structure and Productivity in Developing Countries,* Johns Hopkins University Press, Baltimore, MD.

Binswanger, H. P. and Deininger, K., 1993. South African land policy: the legacy of history and current options, *World Dev.,* 21, 1451–1476.

Birckhead, J., DeLacy, T., and Smith, L., 1992. *Aboriginal Involvement in Parks and Protected Areas,* Aboriginal Studies Press, Canberra.

Brechin, S. R., Wilshusen, P. R., Fortwangler, C. F., and West, P. C., Eds., 2001. *Contested Nature: Power, Protected Areas and the Dispossessed — Promoting International Biodiversity Conservation with Social Justice in the 21st Century,* SUNY, New York.

Bromley, D., 1992. *Making the Commons Work: Theory, Practice, and Policy,* Institute of Contemporary Studies Press, San Francisco.

Brookings Institution, 1942. *Refugee Settlement in the Dominican Republic,* The Brookings Institution, Washington, D.C.

Brownrigg, L. A., 1985. Native cultures and protected areas: management options, in *Culture and Conservation: The Human Dimensions in Environmental Planning,* McNeely, J. A. and Pitt, D., Eds., Croom Helm, New York, 33–44.

Collins, J. L., 1986. Smallholder settlement in tropical America: the social causes of ecological destruction, *Hum. Organ.,* 45, 1–60.

Cook, C. C., 1994. Environment and settlement issues in Africa: towards a policy agenda, in *Involuntary Resettlement in Africa,* Cook, C. C., Ed., Africa Technical Department Series, Technical Paper 227, World Bank, Washington, D.C., 193–198.

de Ceara, I. A., 1985. Land tenure and agroforestry in the dominican republic, in *Land, Trees and Tenure, Proceedings of an International Workshop on Tenure Issues in Agroforestry,* International Council for Research in Agro-Forestry, Nairobi, 301–322.

Dorner, P., 1992. *Latin American Land Reforms in Theory and Practice,* University of Wisconsin Press, Madison.

Dorner, P. and Theisenhusen, W. C., 1992. Land tenure and deforestation: interactions and environmental implications, Discussion paper, United Nations Research Institute for Social Development 34, Geneva.

Dotzauer, H., 1993. The political and socio-economic factors causing forest degradation in the Dominican Republic, Network Paper 16d, ODI, The Rural Development Forestry Network.

Eckholm, E., 1979. The dispossessed of the Earth: land reform and sustainable development, Worldwatch Paper 30, Worldwatch Institute, Washington, D.C.

Ericson, J. E., Boege, R., and Freudenberger, M. S., 1997. Population dynamics, migration, and the future of the Calakmul Biosphere Reserve, paper presented to the Latin American Regional Meeting on Population and the Envirronment, Mérida, Yucután, Mexico, March 1.

Feder, G., Onchan, T., Chalamwon, Y., and Hongladaron, C., 1998. *Land Policies and Farm Productivey in Thailand,* The Johns Hopkins University Press, Baltimore, MD.

Ferraro, P., 1993. Mixing Property Rights Regimes: Effects on Secondary Forest Ecosystem Management, Report prepared for the Ranomafana National Park Project, Antananarivo, Madagascar.

Findley, S. E., 1988. Colonist constraints, strategies, and mobility: recent trends in Latin American frontier zones, in *Land Settlement Policies and Population Redistribution in Developing Countries*, Oberai, A.S., Ed., Praeger, New York, 271–316.

Forster, N. R., 1992. Protecting fragile lands: new reasons to tackle problems, *World Dev.,* 20, 571–585.

Foy, G. and Daly, H., 1989. Allocation, Distribution and Scale as Determinants of Environmental Degradation: Case Studies of Haiti, El Salvador, and Costa Rica, Environmental Department Working Paper 1, World Bank, Washington, D.C.

Frost, J., 1996. Land reform and soil conservation in the Soviet Union, paper prepared for graduate seminar 743, Land Reform Old and New, Spring Semester, Cornell University, Ithaca, NY.

GEF, 1996. Annual Report for GEF Project Conservation and Management of Biodiversity in the Coastal Zone of the Dominican Republic, DOM/94/G31, December.

GEF, 1997. Final Report for GEF Project Conservation and Management of Biodiversity in the Coastal Zone of the Dominican Republic, DOM/94/G31, December.

Geisler, C., 1993. Adapting social impact assessment to protected area development, in *The Social Challenge of Biodiversity Conservation*, Davis, S. H., Ed., Global Environmental Facility, World Bank, Washington, D.C., 25–43.

Geisler, C., 2000. Your Park, my poverty: using impact assessment tools to counter the displacement effects of environmental greenlining, *Protected Natural Areas and the Dispossessed*, Brechin, S. and West, P., Eds., SUNY Press, Albany, chap. 11.

Geisler, C. and de Sousa, R., 2001. From refuge to refugee: the African case, *J. Publ. Admin. Dev.,* 21, 1–12.

Geisler, C. and Silberling, L., 1992. Extractive Reserves: Appalachia and Amazonia compared, *Agric. Hum. Values,* 9, 58–80.

Geisler, C., Warne, R., and Barton A., 1997. The wandering commons, *Agric. Hum. Values,* 14, 325–334.

Gutierrez-San Martin, A. T., 1989. *Agrarian Reform Policy in the Dominican Republic,* University Press of America, Lanham, MD.

Hardin, G., 1968. The tragedy of the commons, *Science,* 162, 1243–1248.

Hardin, G., 1994. The tragedy of the commons revisited, *Ecol. Evol.,* 9, 199–206.

Hayami, Y. 1991. *Toward an Alternative Land Reform Paradigm,* Ateneo de Manila University Press, Manila.

Hirtz, F., 1998. The discourse that silences beneficiaries' ambivalence toward redistributive land reform in the Philippines, *Dev. Chang.,* 29, 247–275.

Hite, J., 1979. *Room and Situation,* Nelson-Hall, Chicago.

Hughett, C., 1992. From Access to Success: Women and Land Reform in the Los Haitises Region of the Dominican Republic, Master's thesis, Department of Rural Sociology, Cornell University, Ithaca, NY.

IEPD/Cornell Survey, 1996. Encuesta sobre Poblacion y Medio Ambiente en el PNLH, Survey completed in March and reported in Semi-Annual Summary Report to Dominican GEF in June, 1996 (Project DOM/94/G31).

Inamdar, A., de Jode, H., Lindsay, K., and Cobb, S., 1999. Capitalizing on nature: protected area management, *Science,* 283, 1856–1857.

Jacobs, T., 1997. From forest clearing to forest replacement: the political ecology of conservation in the Dominican Republic, *Cult. Agric.,* 18, 58–67.

Kaiser, H. M., 1988. Relative efficiencies of size and implications for land redistribution programs in the Dominican Republic, *Appl. Agric. Resourc.,* 3, 144–152.

Kirk, D., 1971. A new demographic transition, in *Study Committee of National Academy of Sciences, Rapid Population Growth: Consequences and Policy Implications,* Johns Hopkins University Press, Baltimore, MD.

Kloppenberg, J., Jr., 1988. *First the Seed,* Cambridge University Press, New York.

Kocker, H. E., 1973. *Rural Development, Income Distribution, and Fertility Decline,* Key Book Service, Bridgeport, CT.

Krylatykh, E. N., 1991. Agrarian reform in the USSR, in *Putting Food on What Was the Soviet Table,* Clauden, M. P. and Gutner, T. L., Eds., New York University Press, New York.

Langholz, J., 1996. Economics, objectives, and success of private nature reserves in sub-Saharan Africa and Latin America, *Conserv. Biol.,* 10, 271–280.

Larson, B. A. and Bromley, D. W., 1990. Property rights, externalities and resource degradation, *J. Dev. Econ.,* 33, 235–262.

Larson, P., Freudenberger, M., and Wyckoff-Baird, B., 1997. Lessons from the field: A Review of World Wildlife Fund's Experience with Integrated Conservation and Development Projects 1985–1996, World Wildlife Fund, Washington, D.C.

Lee, K. N., 1994. *Compass and Gyroscope,* Island Press, Covelo, CA.

Lee, L., 1980. The impact of landownership factors on soil conservation, *Am. J. Agric. Econ.,* 62, 1070–1076.

Lehnhoff, A., 1999. Swapping farmland for wildland in Guatemala, *Nat. Conserv.,* (January/February), 29.

Levin, S., 1999. *Fragile Dominion,* Perseus Books, Reading, MA.

Lynch, O. J. and Alcorn, J. B., 1994. Tenurial rights and community-based conservation, in *Natural Connections,* Western, D. and Wright, R. M., Eds., Island Press, Covelo, CA, 373–392.

McCay, B. and Acheson, J., Eds., 1988. *The Question of the Commons,* University of Arizona Press, Tempe.

Meyer, C. A., 1989. Ararian reform in the Dominican Republic: an associate solution to the collective/individual problem, *World Dev.,* 17, 1255–1267.

Moreno, R., 1996. Analisis del Uso Actual de la Tierra y Posibilidades de Produccion Agropecuaria Sostenible en los Haitises y Hatillos de la Republica Dominicana, Proyecto IICA-GTZ Costa Rica y Proyecto Manejo del Bosque Seco, Dominican Republic.

Murphree, M. W., 1994. The role of institutions in community-based conservation, in *Natural Connections,* Western, D. and Wright, R. M., Eds., Island Press, Covelo, CA, 403–427.

National Academy of Sciences, 1986. Population Growth and Economic Development Questions. Working Group in Population Growth and Economic Development. National Academy Press, Washington, D.C.

Nelson, P., 1938. Tenancy — a major factor in soil conservation, *Land Econ.,* 14, 88–91.

Oates, J. F., 1999. *Myth and Reality in the Rain Forest; How Conservation Strategies Are Failing in West Africa,* University of California Press, Berkeley, CA.

Ostrom, E., 1990. *Governing the Commons: The Evolution of Institution for Collective Action,* Cambridge University Press, New York.
Powelson, J., 1987. Land tenure and land reform: past and present, *Land Use Policy,* 4, 111–121.
Ramirez, N., 1988. *Republica Dominicana, Poblacion y Desarrollo 1950–1985,* CEPAL, San Jose, Costa Rica.
Redclift, M., 1987. *Sustainable Development: Exploring the Contradictions,* Methuen, London.
Sobhan, R., 1993. *Agrarian Reform and Social Transformation,* ZED Books, London.
South Pacific Commission, 1985. Land Tenure and Conservation: Protected Areas in the South Pacific, Topic Review 17, South Pacific Regional Environmental Program, Noumea, New Caledonia.
Southgate, D., 1990. Conservation and Development: What Are the Policy Choices Facing Ecuador? Instituto de Estrategias Agropecuarias, Quito.
Stokes, T. S., Schutjer, W. A., and Bulatao, R. A., 1986. Is the relationship between landholding and fertility spurious? A response to Cain, *Popul. Stud.,* 40, 305–311.
Theisenhusen, W. C., 1991. Implications of the rural land tenure system for the environmental debate: three scenarios, *J. Dev. Areas,* 26, 1–23.
Thiam, B., 1992. The Demographic Consequences of Environmental Degradation: Impact on Migratory Flows and on the Spatial Redistribution of the Population, United Nations Expert Group Meeting on Population, Environment, and Development.
Thurston, H. D., Smith, M., Abawi, G., and Kearl, S., Eds., 1994. *Frijol Tapado,* CATIE, Turrialba, Costa Rica and CIIFAD, Ithaca, NY.
Wachter, D., 1992. Farmland Degradation in Developing Countries: The Role of Property Rights and an Assessment of Land Titling, Land Tenure Center Paper 145, Madison.
Wells, M. and Brandon, K., 1992. *People and Parks: Linking Protected Areas Management with Local Communities,* World Bank, World Wildlife Fund, and U.S. Agency for International Development, Washington, D.C.
Western, D. and Wright, R. M., Eds., 1994. *Natural Connections,* Island Press, Covelo, CA.
Wilken, G. C., 1987. *Good Farmers,* University of California Press, Berkeley.
Wood, R. T. and Schmink, M., 1979. Blaming the victim: small farmer production in an Amazon colonization project, *Stud. in Third World Soc.,* 7, 77–93.
World Bank, 1975. Land Reform, Sector Policy Paper (May), Washington, D.C.
WRI, 1994. *World Resources 1994–95,* Oxford Press, New York.

APPENDIX

Development Assistance Committee Guidelines on Environment and Aid Guidelines for Aid Agencies on Involuntary Displacement and Resettlement in Development Projects

Organization for Economic Co-operation and Development

Paris 1991

ELEMENTS OF THE RESETTLEMENT PLAN

Resettlement policy and objectives should be embodied in resettlement action plans. Below are the basic elements that should be considered during the preparation and formulation of such an action plan.

1. *Organizational responsibilities*. The organizational framework for managing resettlement must be developed during project preparation and adequate resources provided to the agencies in charge. There may be considerable scope for involving non-governmental organizations in planning, implementing, and monitoring resettlement.
2. *Socio-economic survey.* Resettlement plans should be based on recent information about size, cultural, economic, and ecological characteristics of the population and the likely impact of displacement. Socio-economic surveys should describe i) the scale of displacement; ii) the standard household characteristics and full resource base of the affected population, including income derived from informal sector and non-farm activities, and from common property; iii) the extent to which groups will experience total or partial loss of assets, including control over resources, knowledge, and skills; iv) public infrastructure and social services that will be affected; v) formal and informal institutions that can assist with designing and implementing the resettlement programs; and vi) attitudes on resettlement options. Socio-economic surveys, recording the names of affected families, should be conducted as early as possible to help prevent inflows of population ineligible for compensation.
3. *Community participation and interaction with host populations.* The cultural and psychological acceptability of a settlement plan can be increased by moving people in groups, reducing dispersion, sustaining existing patterns of group organization, and retaining access to cultural property (temples, pilgrimage centers, etc.), if necessary, through the relocation of that property.
4. *The involvement of involuntary resettlers and hosts in planning* prior to the move is critical. To obtain effective participation, the affected hosts and resettlers need to be informed about their entitlements and systematically consulted during preparation of the resettlement plan about their options and preferences. Local leadership must be encouraged to assume responsibility for environmental management and infrastructure maintenance. Particular attention must be given to ensure that women and vulnerable groups such as indigenous people, ethnic minorities and the landless are represented and actively involved in such arrangements.
5. *Conditions and services in host communities should improve*, or at least not deteriorate. Providing improved education, water, health, and production services to both groups fosters a better social climate for their integration.
6. *Legal framework.* An analysis should be made of the local legal framework relevant to resettlement operations, including i) the scope of eminent domain power and regulations for the valuation of lost assets; ii) applicable legal and administrative procedures, including access of those affected to the grievance process; iii) land titling and registration procedures; and iv) laws and regulations relating to the agencies responsible for implementing resettlement.
7. *Compensation for lost assets.* Valuation of lost assets should be made at their replacement cost and in a transparent and openly publicized manner. Cash compensation alone should generally be avoided, except in well justified instances, as it typically leads to impoverishment. Some types of

loss ... e.g., loss of access to i) public services; ii) customers and suppliers; and iii) fishing, grazing, or forest areas etc. ... cannot easily be compensated for in monetary terms and access must be sought to equivalent and culturally acceptable resources or earning opportunities. Customary land ownership and usufruct rights must be recognized for compensation purposes to avoid the destitution of former users.

8. *Land acquisition and productive re-establishment.* Resettlement plans should provide for the fair acquisition of condemned land, the conservation of cultural properties, as well as the identification, acquisition, or allocation of land at the new resettlement sites. Land-based productive strategies are commonly the most reliable options for the socioeconomic re-establishment of agricultural families; they may include small-scale irrigation development, tree planting schemes, etc. Particular attention needs to be given to lands held under common property regimes and to the needs of the poorest groups.
9. *Access to training and employment.* General economic growth cannot be relied upon to protect the welfare of the project-affected population. For non-agricultural displaced families, or where the land is not sufficient to accommodate all former farmers, alternative employment and vocational training could be incorporated in the resettlement plan. The resettlement plan should exploit new economic opportunities created by the main investment (e.g., fisheries and aquaculture development in a new reservoir).
10. *Shelter, infrastructure, and social services.* To ensure the economic and social viability of the relocated communities, adequate resources should be allocated to provide shelter, infrastructure (e.g., water supply, feeder roads), and social services (e.g., schools, health care centres). Since community or self-built houses are often better accepted and more tailored to the resettlers' needs than contractor-built housing, provision of a building site with suitable infrastructure, model house plans, building materials, technical assistance, and "construction allowance" (for income forgone while resettlers build their houses) are options communities should be offered. Planning for shelter, infrastructure, and services should take into account population growth.
11. *Environmental protection and management.* The environmental assessment of the main investment requiring the resettlement should include the potential environmental impacts of the resettlement. In rural resettlement, if the incoming resettled population is large in relation to the host population, such environmental issues as deforestation, overgrazing, soil erosion, sanitation, and pollution are likely to become serious and plans should include appropriate mitigating measures. Urban resettlement raises other density-related issues (e.g., transportation capacity, access to potable water, sanitation systems, health facilities, etc.). If the likely impacts on the environment and the population are unacceptable, alternative and/or additional relocation sites must be found.

12. *Implementation timetable, monitoring, and evaluation.* The timing of resettlement should be coordinated with the implementation of the main investment component of the project requiring the resettlement. All resettlement plans should include an implementation timetable for each activity that covers initial preparation, actual relocation, and post-relocation economic and social activities. The plan should include a target date when the anticipated benefits to resettlers and hosts are expected to be achieved.
13. *Arrangements for monitoring implementation* of resettlement and evaluating its impact should be developed by the aid recipient agency during project preparation and used during supervision. Monitoring and evaluation units should be adequately funded and staffed by specialists in resettlement and provisions should be made to ensure a participatory approach.

7 Property in Wild Biota and Adaptive Collaborative Management

Richard S. Cahoon

CONTENTS

INTRODUCTION

Property is a fundamental organizing principle in all social systems. It defines the right to control access, possession, transfer, and use. Property rights and liabilities, and who holds them and how they use them, are a cornerstone of natural resource governance. Property rights are one of the defining principles of the design, implementation, and

0-8493-0020-7/01/$0.00+$1.50

outcome of an adaptive collaboration management (ACM) scheme or any other natural resource management system. Entities that hold property rights can be key participants in natural resource management — those who do not, have significantly less power to control or influence such management. Real property (land) and traditional personal property (e.g., crops, livestock, and manufactured goods) are usually considered in natural resource management systems, while property in wild biota is a relatively unexplored factor. Property in wild biota has recently become an important, but little understood, factor in natural resource management, in part, because of advances in biotechnology that create novel economic values and property from biota. While the role of traditional wild biota property in such management has been studied (Lueck, 1995; 1998), new values and property creates new problems and opportunities in conservation.

This chapter hopes to show that both traditional and novel property mechanisms in wild biota may, by accident or design, be a basic factor in ACM systems that encompass wild biota. To understand this wild biota property role, a property taxonomy is defined and these categories are used to describe the evolution of the wild biota property regime in the United States. This evolution has produced a complex mix of property types and rights in wild biota that includes government domains, private property holders, and biota with no property connection. Details of personal property in biota further emphasize this complexity.

Against this background, the chapter examines how biotechnology has altered traditional institutional property structures in wild biota to create important new natural resource problems and opportunities. The interplay of these rights, their various outcomes, and their role in ACM are illuminated in two case studies. One, involving biotechnological properties from microbes in Yellowstone National Park, demonstrates a failure by a conservation organization to assert its property rights in wild biota and the related failure to use such rights in affecting sustainable conservation. The other, a fungal bioprospecting partnership of the Finger Lakes Land Trust, exemplifies how defining and using biota property rules can play a role in structuring a multiparty arrangement for sustainable use of a natural resource. These case studies demonstrate the opportunity and the challenge of wild biota property regimes and rules for future adaptive collaborative management schemes.

EVOLUTION OF PROPERTY IN WILD BIOTA

Understanding property in wild biota requires a taxonomy of fundamental property regimes:

Res nullius: No property (e.g., ocean fisheries)
Res privatae: Private property (e.g., livestock, crops on private land)
Res publicae: Government ("public") property (e.g., species listed in the Endangered Species Act, fish and game)
Res communes: Private property held by a group (e.g., communal alpine pastures)

Property in wild biota depends on the biota taxon, its location, and whether the biota is "fixed" to the land or transient. For example, no property exists in free-roaming populations and individuals of almost all arthropods (i.e., *res nullius*). All nonendangered wild plants on private land are private property (*res privatae*). Animals listed in federal and state law as Endangered Species are government property (*res publicae*). Some marine fisheries have attributes of communal property (*res communes*).

In early hunter/gatherer cultures, wild biota was *res nullius,* owned by no one until captured. Tribal hunting grounds represent a form of *res communes* in wild biota (Laveleye, 1878; Larfargue, 1894). These early regimes were lost in the transition to Roman law (Lafargue, 1894). Roman law, according to the Institutes of Justinian (Epstein, 1994), considered wild biota as *ferae naturae*, a *res nullius* resource characterized as "wild and untamable" and belonging to no one but in which personal property could be acquired through first possession (e.g., killing or caging). Roman law was the basis for the evolution of wild biota property in England, the direct antecedent of U.S. law in wild biota. In England, as in Rome, free-roaming wildlife were *ferae naturae* and *res nullius* until captured or killed. However, in England there arose the notion of the sovereign's property in wild biota and the beginnings of a grand debate that continues to this day. The debate centered on the public's right in wild biota vs. the rights in such biota on a private landholder's property. Lund (1980) describes how the English legal philosopher Blackstone argued that *ferae naturae* was not *res nullius*, but rather the property of the sovereign (*res publicae*). An alternative view, espoused by Edward Christian, held that while free-roaming wild biota are *res nullius*, they are part of realty and, therefore, a landowner's rights extended to the timber, crops, and transient wild game and fish on the real property.

Today, Christian's view of wild biota as part of realty dominates in the United Kingdom while Blackstone's view became the dominant legal philosophy in the early United States; because of the new republican's rejection of old-world rights of aristocracy and because, in part, Blackstone's legal commentaries could fit in the saddlebag of itinerant, frontier lawyers and judges (Lund, 1980). The notion that free-roaming wild biota was a God-given resource, freely available to all, was also held by Christian theologians, which further cemented the idea in the fledgling United States of wild biota as a *res nullius* resource freely available for the public to take and use wherever and whenever it chose.

Before the rise of trespass law and the primacy of a landowner's possessary right to the land, many states in the early United States forbade landowners from prohibiting hunters and fishers from their pursuits on private land (Lund, 1980; Lueck, 1998). An overview of critical U.S. legal verdicts from colonial to recent times that define the jurisprudence history of property in wild biota can be found in Cahoon (2001).

DEFINING *RES PUBLICAE* BIOTA

Today, many wild biota species are *res publicae,* the property of the government in its role as trustee for the public. The federal government asserts a property interest in all endangered species, migratory birds, marine mammals, eagles, wild horses and burros, biota on federal lands, and any biota obtained contrary to state or foreign law and entered into interstate commerce. States variously assert property interests in certain wildlife including, primarily fish and game but also a variety of nongame creatures. These biota are state *res publi*cae. Looking out across the landscape of biodiversity reveals, then, a mix of state and federal biota property (federal *res publicae* and state *res publicae*).

It is appropriate to speak of a government property interest in biota. But is the government the owner of this biota? Although numerous courts have upheld the right and obligation of government to assert sovereign control over wild biota, it is clear that the government does not own the biota, as in ownership of private property (Dukeminier and Krier, 1993). *Res publicae* property is of the character of *parens patriae* in which the state holds the sovereign right to control access and use of a public good as a trustee for the public. The *parens patriae* right is not the same as a private property right. In particular, the state has little or no liability related to the property.

Over the past 200 years the legal doctrine of government property in biota in its role as *parens partriae* has emerged. That is, government has all the rights necessary to accomplish its obligation as the guardian of the public trust in the common good of wild biota. So, while government does not have all the rights in biota that would come with ownership, it does have the right and obligation to control access to, possession and use of all *res publicae* biota. Furthermore, while government does have the *parens partriae* obligation and power to protect the public good in wild *res publicae* biota, it does not have the property liability for damages caused by that biota, which would normally attend ownership.

It is convenient to think of *res publicae* property in a simple, comprehensive, and monolithic fashion. However, this ignores the great variety of types and degrees of control asserted by the many federal and state agencies that are the enforcers of government biota property. For example, the federal government asserts absolute control over the possession and use of Endangered Species, so that specimens (whole or in part) cannot become *res privatae*, and a modicum of less stringent control over migratory birds, allowing *res privatae* in some instances.

States control fish and game with significantly more latitude even to the point that free-roaming *res publicae* biota can be transformed into a *res privatae* carcass through legalized hunting, fishing, and trapping law.

Within the domain of state *res publicae*, for example, there can be a great variety of government control over possession and, particularly, use. For example, most state biota law will strictly control commercial uses of biota carcasses, which have otherwise become *res privatae*.

Federal *res publicae* control also exhibits significant variety. For example, while any possession of a specimen of an Endangered Species is prohibited, there are a plethora of rules on falconry that permit possession of raptors by private entities but retain the *res publicae* property character and impose highly restrictive controls on possession and use. Thus, state and federal *res publicae* property in wild biota is a patchwork quilt of different types and extentions of property rights of different biota taxa and locations.

DEFINING *RES PRIVATAE* BIOTA

Ownership is an imprecise term that may be best considered as the sum total of several distinct and separable property rights held; the "bundle" of distinct rights that comprise the classical concept of ownership. Becker (1977) describes Honoré's delineation of the rights and liabilities in this classical concept:

1. *The right to possess*; exclusive physical control; the right to exclude others from use or benefits
2. *The right to use* (including the right to consume, waste, modify or destroy)
3. *The right to manage* (how and by whom a thing shall be used)
4. *The right to the capital* (the power to alienate)
5. *The right to security* (immunity from expropriation)
6. *The right of transmission* (the right to bequeath)
7. *The absence of term* (indeterminate length of ownership)
8. *Prohibition against harmful use* (to others)
9. *Liability to execution* (liability to having the thing taken for repayment of a debt)
10. *Residuary character* (existence of rules governing the reversion of lapsed ownership rights)

These rights may be separately held in the same property by different parties (see Geisler, 2000). For example, holding a right of usufruct gives the right to use but not the right to possess. In other cases, one may have the right to possess but not to any and all uses. Thus, a property can be thought of in "degrees" of ownership — with degree being a function of the number of separate rights held. Few owned properties are wholly owned, but rather are a mixture of private and public. Table 7.1 depicts a classification of ownership as a function of the number of separate rights along a continuum of private to public property.

While "complete" ownership is characterized by the unfettered holding of all the rights described above, it is rare that all such rights are simultaneously and fully held by a single party. For example, the most personal of ownership rights, one's own body, is even fettered by government control over alienability and use. Even ownership of one's clothes has limits in that although people thoroughly own them, they cannot use them to hurt others and cannot remove them in public.

In wild biota property rights, the right of the state to control access and possession of a free-roaming game animal may be relinquished to a legal taker, but that taker's rights to use or alienate may remain controlled by the state. The distinction between the right to possess and the right to use wild biota becomes an important factor in the relationship of property rights in wild biota and biotechnology. Biota property that is *res privatae* is of the personal property domain and is subject to those property rules. As a result, personal property definitions are important to the overall framework of biota property.

BIOTA AS PERSONAL PROPERTY

Generally, biota is personal property (also called "Personalty") and falls under the rules of chattel. However, a few key characteristics distinguish biological chattel from nonbiological chattel. First, living biological chattel is capable of autoreproducibility and this reproductive capacity creates unique property attributes. Also,

TABLE 7.1
Classification of Ownership as a Function of the Rights Held

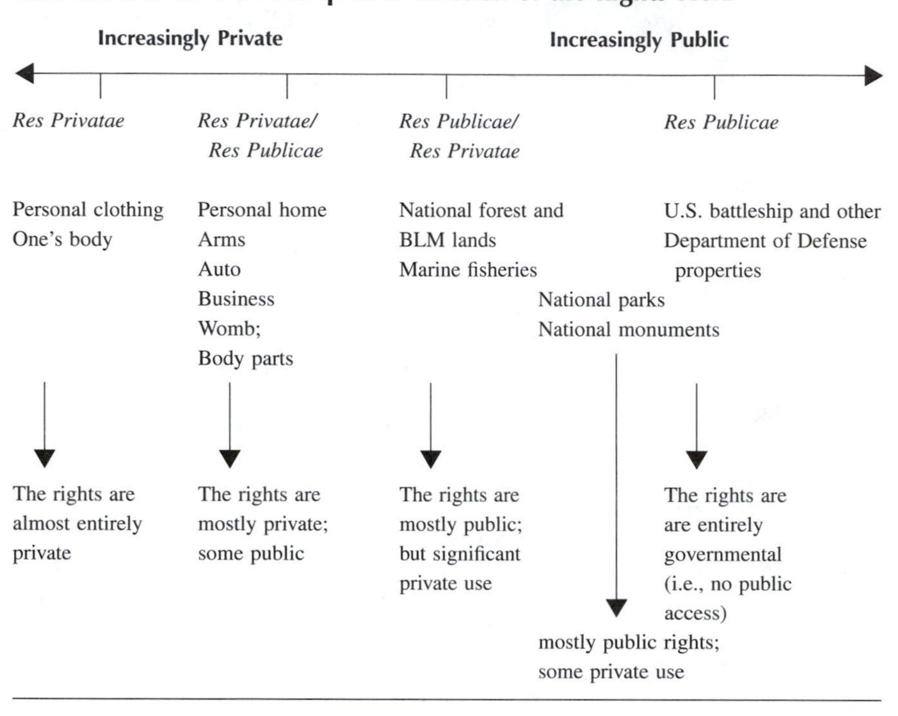

whole organisms are frequently self-locomotive and capable of self-directive and related behaviors.

The first distinction is whether the biota is transient or fixed. In the instance of obtaining a property right in a transient wild animal, the doctrine of Pierson v. Post allows a party to acquire a property right in a free-roaming, *res nullius* organism if and when a certain level of control is asserted over the creature.* This first possession requires actual or constructive control with an intention to possess.** Whereas mere chasing is an insufficient level of control to assert a possessary right, mortal wounding is.*** A captured, living *res privatae* creature that was *res nullius* prior to capture becomes *res nullius ferae naturae* upon escape.**** However, by virtue of the doctrine of *animus revertendi* a captured wild animal that escapes and is conditioned (through training or otherwise) to return to the capturer is the property of the capturer even when it is out of the sphere of control of the original captor.*****

* *State v. Shaw,* 65 NJ 875 (1902).
** *Kenon v. Cashman,* 33A 1055 (1860).
*** *Leisner v. Wamie,* 156 Wisc. (1914).
**** *Muller v. Bradley,* 53 NY Supp. 781 (1898).
***** *Ulery v. Jones,* 81 Ill., 403 (1876).

According to Hogan (1955), the assertion of *res privatae* property rights in a *res nullius* wild creature is affected by overarching public policy. In the early years of the United States, in which there was little or no industry in wild animal husbandry, courts viewed the assertion of a property right in a wild animal as strictly determined by actual physical control. Later, as such industry developed (e.g., fox pelts), courts extended property rights with a looser definition of control. The term *mansuetae naturae* arose to define *ferae naturae* that was rendered *res privatae*, through a variety of controls, as a kind of stock or merchandise. Thus, under *mansuetae naturae*, the issue is not whether the free-roaming *ferae naturae* is *animus revertendi*, but rather if the creature has value to the original captor as stock or merchandise.*

At present, the courts will generally support the property rights of an original captor that has done all that is reasonably expected to control the biota (Burke, 1983). Such rights are conversely true of liabilities. Wild biota, which has been reduced to *res privatae* through capture and which then escapes onto another's property and causes damage, remains the property and liability of the original captor. The damaged property owner may possess the wild animal until the damages are compensated (Hogan, 1955). In a related scenario, if a trespasser kills or captures a *ferae naturae* on another's property, the *ferae naturae* is the property of the landowner (Webb and Bianco, 1970).

Fixed biota property, including trees, shrubs, and herbaceous plants follows different rules. Wild plants are *fructus naturales* and are fixtures on the land and part of realty property rules unless severed from the land. *Fructus industriales* are cultivated plants and their ownership is dictated by separate rules. Generally, ripened fruit is personalty. However, fruit growing on *fructus naturales* is part of realty and, therefore, the property of the landowner. Fruit on *fructus industriales* is personal property that may be owned by the sower. Such property rights in the "fruits" of such sowing are termed *enablements* (Burby, 1961). Nursery stock is always considered as personalty, and property rights in that stock are not coincident with rights in the land.

Personal property rules govern possessed wild biota and domesticated animals. Domesticated animals are defined as "animals raised generation after generation, by the owner(s); generally of docile and manageable temperament or trait and typically requiring human intervention for survival" (St. Julian, 1995).

A bailee (one who receives physical possession of personal property but not title from the owner) of an animal is not entitled to property rights in the progeny. However, the holder of a life estate is entitled to such offspring during the period of the estate. In general, the owner of the mother is the owner of all progeny, regardless of whether the owner is in possession of the mother.** If contracts dictate ownership of progeny, the courts tend to follow. In the absence of a contract to the contrary, the owner of an animal also owns the animal by-products (e.g., milk, eggs). Pigs, cows, horses, oxen, mules, burros, yaks, and other large domesticated animals are typically owned and sold through contract. A registration or certificate may act as title (e.g., Simmental cattle, Arabian horses, Holstein dairy cows). Herds and

* *Stephens v. Albers,* 81 Colo. 488, 256 Pac15 (1927).

** *Arkansas Valley Land & Cattle Co. v. Mann,* 130 U.S.S. 69 (1888).

individual animals are typically owned through possession and contracts. A brand is used to mark and assert ownership. Poultry is typically owned as personal property and exchanged via contract, for large numbers of animals, and simple possessary exchange, for small transfers.

So, the great sea of ownerless wild biota in preindustrial eras has been divided by modern society into the domains of government property, private property, and the still-large realm of property-less biota. Far less appreciated is the domain of community property (*res communes*) in which well-defined property in wild biota is managed by a community defined by a governance structure. It is an arena in which ACM may play an important role in the sustainable management of wild biota.

Each of these property domains rests on the fundamental act of possession of the biota specimen. The property rules of *res publicae*, *res privatae*, *res nullius*, and *res communes* are tethered by the right of physical possession and its corollary, the right to use. The preceding sections have described the complex quilt of possessary rights in wild biota and these rights largely revolve on such possession. In turn, possession and use rights in wild biota factor heavily into ACM schemes.

Although the relation of traditional property in wild biota to ACM is fertile with theoretical and practical implications, the advance of biotechnology presents quantum changes in the complexity and opportunity of that relation. The challenge is to comprehend such complexity in wild biota property for ACM systems conservation purposes.

BIOTECHNOLOGY ALTERS THE *STATUS QUO* IN WILD BIOTA PROPERTY

Biotechnology creates new utilities (i.e., materials and methods useful in the marketplace of human utility) and properties from parts of biota that have not been considered valuable before and also from previously economically trivial biota. Because of this, biotechnology significantly alters the *status quo* of property right regimes in wild biota. This new economic value — including new value in wild biota previously considered economically trivial — upsets traditional relationships of wild biota, property, and economics. In addition, biotechnology creates a significant dichotomy between tangible and intangible property, even in the same wild biota specimen. All this dictates a fresh look at fundamental questions that affect biota property, natural resource economics, and management.

The tangible property in whole individuals or populations of wild biota species, their parts (e.g., meat, skin, hair, organs), and by-products (e.g., eggs, milk, embryos, semen) have been the purview of traditional analyses of political economy. Biotechnology expands that envelope to include "micro" tangible parts such as tissues, cell cultures, antibodies, enzymes, plasmids, cDNAs, and other molecular species. Most significantly, biotechnology permits the creation of intangible property in the form of intellectual property (i.e., primarily patents but also copyrights and plant breeder's rights) derived from tangible property (i.e., the biota specimen or its parts per se).

Because of these characteristics, biotechnology has created a "failure" of the market system in the use and management of wild biota. Market failures can result from ineffective property right definition and assertion in any natural resource. Here, biotechnology has exacerbated existing wild biota property voids and other flaws to produce a clear separation between that wild biota and technology-based property derived therefrom. This failure results in part from a disconnect between the value of biotechnologically derived tangible property (e.g., cell cultures, venom samples), intellectual property (e.g., patents on genes) from wild biota and the cost of maintaining and providing that wild biota.

However, while biotechnology has created these market failure problems, it is a two-edged sword that has also created potential solutions. By harnessing the power of biotechnology to create novel economic value in wild biota and to create new property rights, new institutional mechanisms may be created to link potential, biotechnologically derived value of wild biota to the cost of sustainable wild biota maintenance.

Biotechnological materials and methods permit the creation of valuable utilities in microcomponents of biota. Whereas traditional wildlife uses have included harvest of parts (e.g., skins, meat, teeth) and by-products (e.g., eggs, guano), biotechnology creates values in tissues, cell cultures, enzymes, antibodies, DNA sequences, and other cytological structures and molecules. For example, *in vitro* cultures of Japanese quail (*Coturnix japonica*) cells are key to certain modern pharmaceutical production (U.S. Patent 5,789,231).

The Biota Property Problems of Biotechnology

Several wild biota property problems emerge from biotechnology. It can create valuable *res privatae* intellectual property that is disconnected from federal and state wildlife law that asserts a *res publicae* property interest in select wildlife. Biotechnology also creates valuable property from wild biota that is *res nullius.* For example, there are numerous U.S. patents on useful chemical formulations based on spider venom discoveries, and free-roaming spiders in the U.S. are essentially *res nullius.* Biotechnology then creates a situation in which a property right and its attendant value requires capture. This situation is ripe for a biotechnologically driven "tragedy of the commons" caused by increasing scarcity of *res nullius* wild biota.

Biotechnology creates another new biota property predicament by lending itself to the creation of two distinct, but usually related, forms of biota property. It readily creates personal tangible property in biological matter such as cell lines, DNA sequences, plasmids, and other cellular components, which follows personal property rules. Concomitantly it creates intangible intellectual property. Thousands of patents on a great variety of materials and methods obtained from biota are issued by the U.S. Patent and Trademark Office every year. Patents* are issued by the federal government to inventors of novel, useful, and unobvious inventions and creations. They are the property of the inventor (unless the inventor has voluntarily contracted away this right), for a limited period of 20 years from the date of first patent application. Patents are a negative right in that they do not inherently grant the right to make, use, or sell the patented invention, but rather the right to stop others from doing

* Patents have their legal basis in the U.S. Constitution Article I, Section 8 "Powers of Congress" Paragraph 8, which states exclusive right to their respective writings and discoveries, and the Patent Act of 1790.

so. Like the physical delineation of real property, their "metes and bounds" are defined by sentences (i.e., "claims") that delineate the scope of the patented property. Patents from biota may cover materials (e.g., "the particular DNA sequence of gene Y") or methods (e.g., "the use of gene Y in producing a human therapeutic drug").

Biotechnology not only can create these two types of distinct property, but it also creates complex interactions between them. There is not necessarily a one-to-one match between these two types of property even for a particular biotechnology that arises from the same biota. For example, a biotechnology may be covered by a patent only, by personal property only, or by both. However, even if the two types are present, they may not precisely overlap. For example, a patent on the use of a specific peptide in human therapeutics is owned independently of the ownership of the personal property of the peptide in a laboratory flask. These independent property rights may be held by the same or different parties. Given the great variety of wild biota species, the biotechnological properties, and variety of patentable subject matter, the related property complexity is manifest.

There is some order in this apparent property chaos. In order for an inventor to make a patentable invention from biota, physical possession of that biota at some point is necessary. Thus, any holder of property rights in the tangible biota specimen per se can potentially control an inventor's ability to create intellectual property from that biota by controlling access and/or use. This can include inventing and perfecting intellectual property rights by satisfying the legal requirements for patentability. The crux of the relation between property rights in tangible biota and intellectual property obtained from that biota is the extent to which property rights in the wild biota specimen *per se* are defined and enforced. Since wild biota such as most arthropods are *res nullius*, there are no property rights in these specimens outside of *res privatae* through first capture. Potential inventors are free to possess, use, and perfect *res privatae* in all tangible and intangible property derived from the creature that was *res nullius* when free roaming.

Although state laws assert a certain level of *res publicae* over possession and use of state-controlled wild specimens and their parts, these same state laws do not place any *res publicae* property constraints on the potential *res privatae* of intellectual property obtained through legal possession of the biota corpus (Musgrave and Stein, 1993).

Under federal wildlife law, access, possession, and use of biota specimens are strictly controlled. For example, the Endangered Species Act (ESA) asserts federal property rights over all tangible matter of these biota. However, *res privatae* possession is permitted for certain uses, including scientific use and propagation for species survival. There is no provision in the ESA that would preclude obtaining *res privatae* intellectual property obtained from such legally held ESA biota. This is because the protocol for acquiring a patent on a material or method discovered from an Endangered Species specimen does not require any interstate commerce in the specimen or its parts per se. The Migratory Bird Treaty Act has a glaring loophole in which *res privatae* rights may be legally obtained in the tissue samples taken from a *res publicae* bird under the terms of a banding permit. In this way, federal property (the whole bird) yields biotechnologically valuable tissue that represents viable, omnipotent cells and the bird's genome, which becomes *res privatae* through first capture.

Another clear disconnect between government property interest in wild biota and biotechnologically based private intellectual property can be found in the federal Lacey Act. This law extends federal enforcement over state-controlled wild biota property when the biota "or a part thereof" is entered into interstate commerce. In a patent application on a gene gleaned from a wild biota specimen, the gene per se (i.e., the oligonucleotide macromolecule) is not entered into commerce. The patent application contains only the information obtained from the gene (i.e., the DNA sequence pattern based on the symbols of "A," "T," "G," and "C"). Thus, a patent filed by a private entity on a gene sequence from an illegally obtained wild biota of state-controlled *res publicae* would not invoke Lacey Act enforcement and would apparently leave the patent owner with unfettered benefit of *res privatae* rights in the patent.

What happens to federal *res publicae* rights if a gene sequence is patented from an illegally possessed endangered species? There is nothing in the U.S. Patent Act of 1790 or its implementing regulation (U.S. CFR Title 37) that affects the validity of a patent and/or the inventor's ownership in that valid patent except actions that directly affect the criteria of patentability. Those criteria primarily include requisites for novelty, nonobviousness, and utility. Thus, while the inventor may perfect *res privatae* in the patent on the gene sequence, it is an open question whether the federal government may assert *res publicae* rights in the biota corpus or parts to somehow mitigate the inventor's rights in the patent.

In summary, wild biota laws that assert a *res publicae* (state or federal) property interest exist entirely independently from laws that permit assertion of intellectual property rights, even when the two types of property arise from the same biota specimen.

The independence of these different types of property right is a very important principle. Property rights in a patent on some biotechnological material or method from a wild biota specimen gives the holder of those intellectual property rights absolutely no rights in the personal property of the specimen per se or other specimens in its taxon.

The failure to assert *res publicae* rights over biotechnologically based property is exemplified by the pronouncement by the U.S. Fish and Wildlife Service (USFWS), concerning genetic material of federally controlled species (Bowen and Avise, 1994). In response to a researcher's request to allow import of DNA samples from whale tissue to check Japanese compliance with the International Whaling Convention, the USFWS determined that the DNA from such whale species constitutes a "part" of a species listed under the Convention on the International Trade of Endangered Species (CITES) and under ESA and could not be imported into the United States since "parts" of such (*res publicae*) species cannot be imported. Conversely and illogically, the USFWS did allow copies of the whale DNA sequences (i.e., made by polymerase chain reaction) to be imported since, the USFWS reasoned, such DNA "copies" are facsimiles of the biota part, not unlike photographs or photocopies.

The extent of government property interest in specimens of wild biota is relatively well defined for whole organisms of *res publicae* species under federal and state control, and for *res privatae* species such as domesticated animals, plants owned as land fixtures, and wild biota that is captured and/or tamed. However, biotechnologically derived property rights in parts or intellectual property *res publicae* wild biota are not entirely clear but appear to be subject to *res privatae* control. One can imagine a scenario in which the entity responsible for maintaining a viable population of a wild biota species, such as a state government, would have no share in benefits derived from a new property right obtained from that species. Free-riders could benefit from the exploitation of such property rights without obligation to share in the cost of maintaining the wild biota. Regarding the many species that are currently *res nullius*, it seems only a matter of time before biotechnology renders some of them valuable and scarce, at which time a market failure appears likely, barring some form of property right–based intervention.

Tisdell (1991) points out that resources of mixed good character can exhibit an effective market in some goods and a simultaneous market failure in others. For example, trophy hunters who pay a large hunting fee participate in the market in big game "hunting goods" but will simultaneously participate in the failure to supply an "existence value" of charismatic megafauna. Biotechnologically derived utilities in wild biota and their underlying properties are highly susceptible to the mixed-goods problem.

The Biotechnology Wrinkle in Biota Property Taxonomy

Biotechnology modifies the property taxonomy described earlier and yields the following summary of the wild biota property milieu:

Res nullius — All transient free-roaming wild biota populations, individuals, progeny, their parts and by-products that are not covered by federal or state law. Examples include spiders, most insects, protozoans, starlings, porcupines, many reptiles and amphibians.

Res publicae — All transient wild biota populations and individuals, progeny, their parts and by-products explicitly listed in federal and state legislation; for ESA, *res publicae* extends to all tangible matter; Migratory Bird Treaty Act *res publicae* extends to all tangible matter except that legally acquired as *res privatae* under certain use permit.

Res privatae — All tangible matter obtained legally from *res publicae* biota in which there are no government restrictions on possession or use; all fixed biota on private land including plants and perhaps soil organisms; all intellectual property obtained from any wild biota.

Res communes — Private property held and cooperatively managed by a group of private entities; for example, some fisheries.

This new wild biota property taxonomy enriched by the advent of biotechnology presents a new set of property tools and potential for new institutional mechanisms that can serve as a framework for ACM systems. On the other hand, this new biotechnologically produced layer of biota property rights may, when unmanaged through neglect, unawareness, or ideological bent, thwart otherwise effective ACM schemes.

Biotechnology-Based Property and Sustainable Resource Use

If the goal is to link economic value of wild biota to its conservation, careful analysis of the property rights milieu involving potential biotechnological use of wild biota can establish the basis for effective market exchanges. When appropriate property right assertions are applied by the holder of wild biota property rights, sustainable mechanisms may be created that connect the potential value of biotechnological discoveries in wild biota to the cost of maintaining that biota.

Below, two case studies demonstrate the role that intellectual property can play in conservation and, by inference, in ACM. The first case involves microbes in Yellowstone National Park and characterizes a failure to use biota property rights in the service of conservation. The second involves a fungal bioprospecting scheme in which biota property rights are employed with a conservation focus. Each is illustrative of the promises and pitfalls of property in biota as biotechnology forces change.

YELLOWSTONE'S VALUABLE MICROBES: A CASE STUDY IN FAILURE OF WILD BIOTA PROPERTY RIGHTS

In the summer of 1966, Dr. Thomas Brock and his student, H. Freeye, collected microbial mat samples at Mushroom Hot Springs in Yellowstone Park under the terms of a National Park scientific collection permit. The permit placed essentially no restrictions on Brock and Freeye's use of the collected biological materials. Freeye isolated a bacterium (they termed it "YT-1"), and 3 years later they published a paper on *Thermus aquaticus*, a novel thermophilic bacterium. Brock and Freeye deposited living axenic culture specimens of the species in the public repository, American Type Culture Collection (ATCC) as ATCC culture number 25104. The meager restrictions on public access to ATCC 25104 placed by the ATCC were simply the requirement to pay a nominal fee and standard "hold harmless" clauses for liability.

Obtaining a sample from ATCC gave recipients *res privatae* in their particular sample. Several years later, Cetus Corporation paid the ATCC $35 fee and Kary Mullis, employed by Cetus, obtained ATCC 25104 and used it to invent the polymerase chain reaction (PCR) method. The PCR method is a powerful technique for making copies (i.e., "amplifying") of specific sequences of DNA. It is a tool that has proved to be essential for biotechnology, is used globally in basic research, and has a huge variety of commercial uses. Its discovery was so profound, *Science* declared polymerase as its "molecule of the decade" and won Mullis a Nobel prize.

Cetus filed its first U.S. patent application on the PCR technology with K. Mullis as sole inventor, on October 25, 1985. A subsequent U.S. patent application was filed by Cetus, with Mullis and five other inventors, on February 7, 1986. The October 1985 and the February 1986 applications ultimately issued on July 28, 1987 as U.S. Patent 4,683,202 and 4.683.195, respectively. U.S. Patent 4,683,202 (K. Mullis inventor; Cetus Corporation assignee) entitled "Process for Amplifying Nucleic Acid" issued with 21 claims that cover "a process for amplifying any desired specific nucleic acid sequence contained in a nucleic acid mixture thereof." In these two patents, Cetus controlled the making, using, and selling of PCR technology. In the late 1980s, Cetus and Perkin Elmer Corporation commercialized PCR through licenses and sales of PCR reagents and equipment. Hoffman-LaRoche purchased the Cetus' PCR patent estate for $300 million. PCR has continued to comprise a significant flow of commercial activity.

Yellowstone National Park, the National Park Service, and the Department of the Interior have not received any direct share of any of the commercial revenue streams resulting from the PCR property. Thus, other than normal business taxes, there is no connection between the financial beneficiaries of PCR (particularly Cetus and Hoffman-LaRoche) and those who have invested in the maintenance of the resource (the public via the Department of the Interior and the National Park Service).

This market externality appears to have arisen from the failure of the Yellowstone Park and the National Park Service (i.e., the Department of Interior) to define, assert, and maintain a *res publical* property right (i.e., ineffective collection permits and restrictions on ATTC distribution rights) and the failure to negotiate for a sharing of benefits from exploitation of biotechnologically derived properties from the tangible microbial property. This failure to assert a government property interest denied Yellowstone Park (and the public) a role in the control over the use of this park biota. An opportunity to establish an ACM scheme for the sustainable use of this biota was lost. So, a failure in wild biota property assertion can distort or eliminate an ACM process.

BIOTA PROPERTY RIGHTS IN THE SERVICE OF CONSERVATION: FUNGAL BIOPROSPECTING AND THE FINGER LAKES LAND TRUST

The inter- and intrainstitutional arrangements surrounding the fungal bioprospecting project on the lands of the Finger Lakes Land Trust (FLLT) represent the use of novel biota property assertions for conservation. In this arrangement, the FLLT perfected property rights over wild biota on its land, extended those rights to include control of possession and use of that biota by other parties even after removal from FLLT land, and asserted these property rights into the realm of intellectual property derived from FLLT biota.

Understanding how the property rights assertion by the FLLT over wild biota on its land extends to benefit-sharing from derived intellectual property sheds some light on the mechanisms and institutional evolution necessary to optimize the connection of biotechnologically derived value of biota with the cost of its maintenance. It also illuminates how wild biota property can play an important role in ACM.

FLLT was established in 1989 as a public membership, nonprofit corporation for the purpose of protecting the unique natural habitat in the Finger Lakes region of upstate New York. FLLT activities include natural resource inventories, land acquisition through purchase and donation, management of its proprietary properties (including nature preserves and trade-lands), acquisition and management of conservation easements on other's land, fund-raising, land and natural resource evaluations, financial management, member recruitment, and organizational governance.

In 1997, FLLT purchased the "Biodiversity Preserve," a natural area near Cornell University and Ithaca, New York, valued for its biological diversity. One of the reasons to acquire the preserve was to develop its bioprospecting potential. Developing that potential required a property framework that could create economic value of the biota in the preserve. That framework was dependent on FLLT perfecting and enforcing of its property rights in FLLT biota. A first step was to establish control over access to any biota on FLLT land. This was accomplished through the following functions:

1. **Control of access to FLLT land** — Using a basis in New York State trespass law, FLLT established a coherent policy and administrative capacity to control access to FLLT lands;
2. **Prohibit unauthorized taking of biota on FLLT land** — In conjunction with the establishment of controlled access, FLLT established a policy of strict prohibition against the taking of any biota on its land except through an explicit collection or hunting permit process;
3. **Regulate taking of biota** — FLLT established a biological material collection and possession permit process, which allows only permittees to take and possess FLLT biota under strictly prescribed conditions and protocols; these included justifications for collection, definition of species to be collected, number/quantity, location, and nature of collection;
4. **Assertion of property in collected biota** — The FLLT collection permit, a bailment contract defining the conditions of possession and use by the collector, places the following restrictions on the permittee's possession and use of collected biological material:
   a. Required labeling of all samples collected, regardless of form, such as whole caged organism, culture tube, petri plate, deep freeze, etc.;
   b. Prohibition of transfers to third parties without prior approval by FLLT;
   c. No commercial use (by FLLT definition) without license from FLLT;
   d. Required acknowledgment of permittee to hold collected materials as a bailment;
5. **Ownership Vigilance** — Follow-up of permittee's adherence to permit conditions, for example, biological property accounting).

With these functions, FLLT established a reasonably effective level of proprietary control over the biota on its land and was then in a position to enter into a bioprospecting partnership. In the first step, FLLT effectively asserted control over rights of access to possession to all biota on its property, fixed and transient, through its rights under trespass law. In the next two steps, FLLT began to establish property rights in its biota and finally in the last two steps, FLLT asserted property rights in the biota collected from its land.

The structure of this situation was dependent on the taxa selected for bioprospecting. Selecting fungi was convenient as no state or federal law asserts *res publical* in fungi. A different scenario would have resulted if another biota such as state-" *res publical* bears had been selected.

The Fungal Property

FLLT lands comprise a variety of unique habitats in the Finger Lakes region. Although a variety of wild biota taxa could have been chosen as a bioprospecting target, fungi were selected as a particularly promising source of useful discoveries. Although there were biological reasons for this selection, the lack of conflicting and potentially confounding government law and regulation was fortuitous. There are no Federal or New York State laws that give the government explicit control over fungal species. And, although there is ambiguity over the ownership of microbial species, it could be argued that the organisms of the soil (the majority of collected fungi from FLLT land) are, by *ratione soli* (the Roman doctrine of ownership "by reason of the soil"), the property of the landowner.

For assurance of its rights in biota, FLLT investigated the extent to which the State of New York asserted control over wild biota in the state. The state law that governs wild biota is the N.Y. State Environmental Conservation Law 90 (ECL-90). This law explicitly states at the outset that "all wildlife is owned by the State," in direct contradiction of Supreme Court decisions. Through ECL-90, New York State has established prohibitions against unauthorized taking, possessing, or using any native wildlife that is defined and listed as controlled. Such controlled species, as defined in ECL-90, include "Endangered and Threatened Species," "Fish," "Big Game," "Small Game," "Protected Wildlife," "Species of Special Concern," and "Species of Status Undetermined."

Species listed as endangered or threatened cannot be taken, possessed, transported, sold, or offered for sale except by permit from New York State Department of Environmental Conservation. Such permits may be provided for certain reasons, such as justifiable, scientific purpose. Species of "Status Undetermined" may be taken without a permit. Fish may be taken only during open season according to precise and complex rules of species, size, location, and time of year. Game, defined as either Large Game (as listed) or Small Game (as listed), may be taken only during open season dates prescribed according to precise rules of species, age, sex, and location. No wild plants may be taken (uprooted and collected) from state land without a permit.

New York State asserts no control over possessing any "uncontrolled species" defined as all wild biota not described as protected. This includes all insects (except aquatic insects) and any listed endangered or threatened or species of special concern, species of status undetermined; all plants on private property except those listed as endangered, threatened, or species of special concern; all protozoans and microbes such as yeast, bacteria, cyanobacteria, algae, fungi, and viruses that are not explicitly listed as wildlife. Microbes are not explicitly mentioned in the ECL-90 except in an oblique reference in the definition of wildlife ("wildlife shall mean … arthropods and all other invertebrates").

Thus, according to New York State law, a landowner may (or may allow others), without any state control, to take and possess any unprotected organism on the landowner's land. Prior to any collection of any protected organism, the landowner must obtain federal or state approval to take and possess either through a collection permit or hunting/fishing/trapping license. No fungi are listed in any federal or New York State law and are, therefore, either *res nullius* or (by *ratione soli*), *res privatae*. FLLT is, therefore, the unfettered "owner" of all fungi on its lands.

FLLT Collection Permit: A Bailment Contract for Retention of Property Rights

Having established the basis for control over wild biota on its land, FLLT established a protocol for authorized taking of its biota under the following terms:

1. Select purposes only (e.g., research, approved bioprospecting, conservation, education, nature study);
2. Collecting application with full disclosure of purpose and details of scope and methods required;
3. Signature of collection permit by permittee required.

Although the FLLT property rights in the fungi growing on its lands could be clearly established, maintaining such property claims can become tenuous when the specimens are physically removed from FLLT land. To assure extension of property rights after removal, FLLT transfers possessary rights, but not title to its biological materials, through a bailment (transfer of right to possess but not title). This bailment takes the form of the FLLT Collection Permit. The key terms of the FLLT Collection Permit include:

1. All collected biological material must be clearly labeled as the property of FLLT and such labeling perpetually maintained throughout any laboratory manipulation culture, transport, storage;
2. The permittee cannot distribute the materials to any third party without FLLT permission;
3. The permittee may not use the material for any commercial purpose without FLLT permission;
4. All collected biological material and any derivative (clones, propagules, etc.) materials must be destroyed or returned on request by FLLT.

The Bioprospecting Collaboration: Cornell University and FLLT

FLLT does not have the staff or expertise to systematically search, identify, characterize, catalog, and store the diverse wild biota on its lands. However, Cornell University, neighbor and bioprospecting partner of FLLT, does have this capability. Cornell and FLLT agreed to allow Cornell mycologists to collect FLLT fungi under the terms of the FLLT Collection Permit. This allowed Cornell University to collect and conduct research and to store FLLT fungi, but prohibited commercial use. It also obligated Cornell to maintain any fungal samples as a bailment for FLLT. To complete the market exchange of FLLT fungal biota, a mechanism for appropriate commercialization was required.

The Commercialization Partnership: FLLT and CRF

In the early 1930s, Cornell University established Cornell Research Foundation, Inc. (CRF), a nonprofit corporation, as its mechanism for owning and managing the transfer of Cornell's intellectual property and related proprietary technology to the commercial sector. CRF was the logical party to transform the FLLT's fungal biota held by Cornell for research purposes only, into a bioprospecting opportunity and the subject of a commercial transaction. CRF and FLLT consummated an agreement in which CRF agreed to manage the FLLT fungal specimen property for the sole purpose of CRF entering into a future bioprospecting contract with a commercial partner. Any such contract was to be in accordance with the rights of FLLT as a bailor (the fungal property owner). CRF and FLLT agreed to an equal sharing of any returns from commercial activity with the FLLT fungal property. CRF and FLLT agreed to transfer a number of the fungal specimens to the commercial partner under specified conditions of an agreement between CRF and the commercial partner, and FLLT agreed to enter into no other agreements that would be contrary to agreements between CRF and the commercial partner.

The Transfer of Possession to the Commercial Partner

FLLT, Cornell University, and CRF do not have the capability to screen large numbers of samples for potential pharmaceutical applications. Nor do they possess the capability to develop, register, and market any such pharmaceutical. Of necessity, a pharmaceutical company partner was attracted by the partners to the idea of screening the FLLT fungi held by Cornell.

CRF, acting as agent for FLLT and as manager of the commercial applications of FLLT fungi, negotiated a biological material screening agreement with the pharmaceutical company in which Cornell/CRF provided a select number of fungal samples to the pharmaceutical company. In exchange, the pharmaceutical company provides funding to support Cornell mycologists to carry out the searching, cataloging, storing, and sampling. Any discoveries or inventions, patentable or not, made using the FLLT fungal specimens that generate commercial return are subject to a royalty to be paid to CRF, which is equally shared between CRF (Cornell) and FLLT. In addition, the pharmaceutical company paid an "access fee" directly to FLLT.

The web of several contracts between FLLT and Cornell, Cornell and CRF, CRF and FLLT, and CRF and the pharmaceutical company all maintain the chain of FLLT property rights in fungal specimens, which began with FLLT "enclosing" its lands to unauthorized taking. It is important to note that ownership of any intellectual property derived from FLLT fungi throughout the chain of possession is not a requisite for sharing of commercial return. This demonstrates the power of personal property rights. A holder of property in wild biota need not be an inventor or link its biota to financial return on intellectual property exploitation.

This arrangement focuses solely on the field of human therapeutics and diagnostics, which leaves FLLT free to enter into similar arrangements for other uses of their fungi and also other arrangements for other taxa, such as insects, plants, microbes, or higher animals. However, it is likely that other taxa may be controlled by either federal or state law, which would bring either of these governmental entities into such arrangements.

The FLLT property right assertion in the fungi held by any third party exists as long as the fungi exists. The fungi must be collected nondestructively and in such a way to maintain the original population intact. These factors combine to produce the potentially sustainable use of a wild biota, which uses property rights in that biota to link the resource conservation system (i.e., the Land Trust) with the process for commercial exploitation and resultant benefit stream.

BIOTA PROPERTY AND ADAPTIVE COLLABORATIVE MANAGEMENT

What are the implications of property in wild biota to ACM concepts and practices? Will the complex mix of existing, traditional, novel, and potential property rights in wild biota affect the design or implementation of such management schemes?

Considering biota property as a distinct institutional realm provides an organizing principle that can range from the traditional to the very innovative. Situations such as a state government in its role as *parens partriae* of free-roaming game animals as a participant in an ACM system are familiar. Assertion of rights in obscure biota or in intellectual property derived from wild biota presents a new arena in which novel ACM systems may be developed.

The rules of biota property — some well established and others at the cutting edge of technology and law — provide a set of tools that may be used by the participants in such management systems to further their own interests, the interests of clusters of participants, and/or the interests of the management scheme itself. Understanding these rules and facility with their practice can provide property power to certain parties who may have been previously powerless. Holding property rights in wild biota that have previously not existed or have been unasserted can give stakeholders and actors new roles in such management schemes. The exercise of biota property rules has the potential to alter significantly the status quo of existing management systems or the design of new ones.

The complex, mosaic-like character and dynamic nature of wild biota property also have the potential to play an important role, if not deciding factors, in such systems. The intermixing of overlapping rights of land and biota, biota populations and individuals, carcasses and parts, and intellectual property necessitates a significant level of inter- and intrainstitutional negotiation and partnering.

Any ACM scheme that deals with biota, even if only indirectly, is a candidate for application of biota property theory and practice. Used intelligently, such property could provide a powerful new set of structures and tools within which such schemes can accomplish their goals. Used otherwise, it will likely waste this potentially powerful, institutional mechanism.

CONCLUSIONS

Numerous analysts of the political economy of natural resources have pondered the means to manage such systems most effectively to accomplish simultaneous social goals of sustainability, equity, and efficiency. The grand debate over the public good vs. the private right is particularly relevant to natural resource existence, supply, and use. The property taxonomy plays an important part in this large question. Should natural resources be left as *res nullius* and potentially subject to the "tragedy of the commons" or carved into a property domain that is either dominated by private or government control? The widespread belief that lack of resource ownership is inefficient and inequitable leads to the natural bipolar debate of private property vs. government control — and each side has its dedicated and vigorous advocates. It is easy to understand why each side in this debate feels strongly, given the strengths and weaknesses of both the *res privatae* and *res publicae* regimes. Although *res privatae* appears efficient from short-term, utilitarian perspectives, it suffers from problems of sustainability and inequity. *Res publicae* approaches seem bureaucratic and too susceptible to the corruption of the political process and unfair, intragenerational power plays.

Confronted by this Gordian knot, some have reached the conclusion that private property held and managed by a community (i.e., *res communes*) may provide an effective alternative to this conundrum. Ostrom (1990) has made it clear that such communal property regimes have a long history in many cultures of successful and sustainable natural resource management. Although the conventional political economy of natural resources in the United States has largely ignored *res communes*, recent institutional experiments with fisheries management in the United States suggest a promising alternative for future resource management.

Ostrom (1990) points out the importance of property rights and governance structure to the *res communes* system. Biota property mechanisms provide an obvious framework for creating and managing such communities. The example of the collaborative and sustainable use of fungi by the "community" of Cornell, CRF, FLLT, the pharmaceutical partner, as well as the public looking for health enhancement, suggests the potential.

"Biota conservation communities" that are focused on the conservation of select biota or their habitat can be designed around the thoughtful use of wild biota property mechanisms, and managed using the framework of these rules. Such communities are likely to include a variety of different stakeholders and actors, each with potentially disparate goals and objectives that must be reconciled within the governance structure. ACM processes are an obvious means of keeping these communities intact and viable. The types of such communities are limited only by the extent of creative use of biota property rules. Based on early experience with pilots of biota conservation communities like the FLLT fungi, ACM seems destined to play an important role.

REFERENCES

Becker, L. C., 1977. *Property Rights: Philosophic Foundations,* Rutledge & Kegan, St. Paul, MN.

Bowen, B. and Avise, J., 1994. Conservation research and the legal status of PCR products, *Science,* 5, 266.

Burby, W. E., 1961. *Law Refresher — Personal Property.* 3rd ed., West, St. Paul, MN.

Burke, D. B.. Jr., 1983. *Personal Property in a Nutshell,* West, St. Paul, MN.

Cahoon, R. S. 2001. Analysis of the relation between intellectual property and biological conservation law and policy in the United States, doctoral dissertation, Cornell University.

Dukeminier, J. and Krier, J., 1993. *Property,* 3rd ed., Little, Brown, Boston.

Epstein, R. A., 1994. On the optimal mix of private and common property, in *Property Rights,* Paul, E. F., Millen, F., and Paul, J., Eds., Cambridge University Press, Cambridge.

Geisler, C., 2000. Property pluralism, in *Property and Value,* Geisler, C. and Daneker, G., Eds., Island Press, Washington, D.C., 65–86.

Hogan, J., 1955. The distraint of animals *ferae naturae, Oreg. Law Rev.,* 34(4).

Larfargue, P., 1894. *The Evolution of Property Rights from Savagery to Civilization,* Swan, Sonnen, Schien & Co., New York.

Lavaleye, E., 1878. *Primitive Property,* Macmillan, London.

Lueck, D., 1995. Property rights and the economic logic of wildlife institutes, *Nat. Resourc. J.,* 35(4).

Lueck, D., 1998. Wildlife law in the U.S., in *The New Palgrave Dictionary of Economics and the Law,* Vol. 3. Peter Newman, Ed., Macmillan, New York.

Lund, T., 1980. *American Wildlife Law,* University of California Press, Berkeley, CA.

Musgrave, M. and Stein, A., 1993. *State Wildlife Laws Handbook,* Government Institutes, Inc., Rockville, MD.

Ostrom, E., 1990. *Governing the Commons: The Evolution of Institutions for Collective Action,* Cambridge University Press, Oxford.

St. Julian, R., 1995. Animals, in *American Jurisprudence,* 2nd ed., Vol. 4, Lawyers Coop.

Tisdell, C. A., 1991. Economics of Environmental Conservation: Economics for Environmental and Sociological Management, Elsevier, New York.

Webb, G. and Bianco, T., 1970. *Personal Property and Bailments,* Holt, Rinehart & Winston, New York.

8 Agents in Adaptive Collaborative Management: The Logic of Collective Cognition

Niels G. Röling and Janice Jiggins

CONTENTS

INTRODUCTION

People have become a major force of nature. Jane Lubchenco (1998) drew this conclusion in her 1997 Presidential Address to the American Association for the Advancement of Science based on the following observations:

0-8493-0020-7/01/$0.00+$1.50

- Between one third and one half of the Earth's land surface has been transformed by human action;
- The carbon dioxide concentration in the biosphere has increased by nearly 30% since the beginning of the industrial revolution (and in its third assessment report due for publication in 2001, the International Panel on Climate Change will unequivocally point to humans as the cause of climate change; *New Scientist,* 1999);
- More atmospheric nitrogen is fixed annually by humans than by all natural terrestrial sources combined (especially through fertilizer production);
- More than half of all accessible surface water is put to use by humans;
- Humans are the direct cause of the fifth largest extinction event (e.g., about one quarter of bird species have been driven to extinction);
- Approximately two thirds of marine fisheries are fully exploited, overexploited, or depleted;
- And notwithstanding global treaties banning chlorofluorocarbons (CFCs) and other atmospheric ozone-depleting chemicals, the largest-ever hole in the ozone layer over the Northern Hemisphere appeared in the first spring of the new millennium, exposing ecosystems to solar radiation with unknown effect (*New Scientist,* 2000).

Recognition of human beings as a major force of nature has three important implications.

The *first implication* is that we cannot continue to consider the Earth a substrate for human activity or use it as a resource for only human ends, and then rely on the resilience of the web of life to deliver the ecological services on which humans as biological agents depend. People have *de facto* taken responsibility for the direction in which the Earth evolves. In this sense we can say that the Earth has become a soft system: whether by default or intention, its future state emerges from human interaction (Checkland, 1981). Yellowstone, the world's oldest nature park, provides a powerful illustration of a soft system (Table 8.1).

Sustaining a soft system such as Yellowstone, or the biosphere, requires shared sense making, negotiated agreement and accommodation, and deliberate concerted action among the stakeholders in the system. Now that people *collectively* have become a major force of nature, releasing and safeguarding human opportunity increasingly means managing our own collective impact and hence our own activities. Our survival as a species is no longer only a question of learning about our environment, but increasingly also of being able to learn collectively about and control our own collective behavior. Beck (1994) calls this reflexive modernization. We shall speak of "social learning." While social science understanding has been a marginal influence over science and public policy during the era in which we have imposed increasing technical control over the biophysical world, the understanding of ourselves as a uniquely intentional species is becoming vitally important to our own survival.

The *second implication* is that people's activities must increasingly be guided by ecological rationality. Although by no means an unknown consideration in human experience, it will require a major transformation of the current rather single-minded

TABLE 8.1
Yellowstone Park as a Soft System

Yellowstone is a symbol of pristine wilderness guided by natural law, free from human interference. But this image is an illusion. The fires that occasionally burn substantial sections of the Park had been stopped for many years but are now deliberately allowed to ensure rejuvenation of sequoias. These fires are contested by the timber companies which operate outside the Park. Wolves have been eliminated from the Park since the 1930s. Their reintroduction is contested by ranching interests, who already resent the out-migrating buffalo which allegedly infect their cattle with Brucellosis. The aquifers feeding the geysers for which the Park is famous are threatened by tin mining activities from which the State of Wyoming derives a great deal of its revenue. The most ubiquitous animal in the Park is *Homo sapiens*, to whose education the wilderness has been dedicated. To deal with the fact that the Park's boundaries do not coincide with the messy, more complex human system that determines what happens to the Park, the Greater Yellowstone Co-ordination Committee has been formed to negotiate accommodations among the conflicting human interests that impinge on the Park. The Park clearly is not an operational natural ecosystem, but a soft system determined by human ends, negotiated under collaborative management.

Source: Keiter, R. B. and Boyce, M. S., *The Greater Yellowstone Ecosystem: Redefining America's Wilderness Heritage,* Yale University Press, New Haven, CT, 1991. With permisison.

TABLE 8.2
The Ifugao Rice Terraces as a Monument to the "Soft Side" of Land

The rice terraces developed by the Ifugaos during the course of 2000 years have been declared a Heritage Site by UNESCO. The sight of entire mountainsides covered by terraces awes the visitor, not only because of their beauty or the enormous effort that must have been involved in their construction, but also because of the ingenuity, organisation and collective management that such a structure requires. Unlike the pyramids and other world wonders built by tyrants who used slaves for their own glory, the Ifugao terraces are due to voluntary collaboration and organisation. Careful study reveals that the "hard" terrace system of irrigation channels, walls, protective forests and so on, has its counterpart in complex social institutions and human cognition involving spirits and gods, rituals, work organisation, discipline, leadership, shared experiential knowledge and values. The fact that the hard system now is collapsing can be traced to the erosion of the *social* system that ensured its upkeep.

Source: Gonzalez, R., Ph.D. dissertation, Wageningen University, Wageningen, 2000. With permission.

pursuits of instrumental control and economic growth, and of the institutions dedicated to fostering those enthusiasms.

The *third implication*, therefore, is that people must deliberately develop a "soft side" to the sustainable management of the biosphere (Röling, 1997; Jiggins and Röling, 1999). The soft side of land, for example, refers to the human knowledge, technology, institutions, resource allocation, and so forth from which land use emerges. Ifugao in the Northern Philippines illustrates this concept (Table 8.2).

People have always created their own context; that is, they lived in their environment according to their shared enthusiasms. During the age of religion, they built cathedrals, mosques, and temples and organized according to commands construed as given by God. In the industrial age, they built industries, distribution networks, and markets in the pursuit of human well-being and profit. In today's world, science has impact through the emergence of actor networks that replicate laboratory findings on a societal scale (Callon and Law, 1989).

The current enthusiasm seems to be to transform the whole world into a global, competitive marketplace dedicated to satisfying consumptive needs. The "battle of Seattle" in November 1999 forced the reconsideration of the agenda proposed as the basis for the next round of world trade talks. Perhaps it is an indication that significant numbers of people are beginning to see the need for a different, "soft side" of the globe, one that is more dedicated to a sustainable society. The current chapter takes this need for granted.

The chapter is ambitious. It takes seriously the three awesome implications of humans having become a force of nature. The purpose is to explore our chances of building an ecologically rational society. To this end, we first examine what ecologists and biologists have to say about the three implications. On the basis of extensive studies of complex ecosystems in which humans play major roles, the ecologist Holling and colleagues come to the conclusion that adaptive management and social learning are essential for a sustainable society. But they do not give much guidance on how to go from there. This is where the Santiago theory of cognition comes in. The Chilean biologists Maturana and Varela provide a powerful perspective on cognition that allows us to do two things. First, it allows us to define rationality from an ecological perspective, a prerequisite for elaborating the second implication of having become a major force of nature. Second, it allows us to translate the other two implications into a startling insight: a sustainable biosphere implies that the people on this Earth begin to act as a collective cognitive agent. This admittedly is a far-fetched idea. Nevertheless, it seems an inescapably logical conclusion and hence a good reason to engage in thinking the unthinkable.

The remainder of the chapter examines selected social science theories from the perspective of collective cognitive agency. Is it a pipe dream or a realistic opportunity? Answering this question requires an examination of the progress made so far in developing theory that could inform a new intersubjective awareness of humans in interaction with our environment. We suggest that such a theory could begin to replace the 19th century strategic rationality that underpins economics as the currently dominant design framework for global society.

SOCIAL LEARNING: FROM ECOLOGICAL TO SOCIAL PROCESS

On the basis of their extensive studies of complex ecosystems, the ecologist Holling and colleagues (Gunderson et al., 1995) have established that societies are bound to meet "surprises" emanating from the dynamic and cyclic nature of ecosystems, but that societies are not equipped to deal with these surprises. Moreover, "success in managing a target variable for sustained production of food or fibre apparently leads to an ultimate pathology of less resilient and more vulnerable ecosystems, more rigid and unresponsive management agencies and more dependent societies" (Holling, 1995; p. 29). As a consequence, "the essential point is that evolving systems require policies and actions that not only satisfy social objectives but also achieve continually modified understanding of the evolving conditions and provide flexibility for adapting to surprises" (Holling, 1995; p. 14).

This ecological analysis leads to an important implication for *human* behavior, that Holling calls "adaptive management" (AM). He defines AM in the following terms:

> The release of human opportunity requires flexible, diverse and redundant regulation, monitoring that leads to corrective action, and experimental probing of the continually changing reality of the external world (Holling, 1995; p. 30).

AM is a powerful guiding principle for the design of the interface between society and biosphere, between community and ecosystem. It has generated a great deal of interest worldwide. The focus of this book on collaborative AM is directly inspired by the work of Holling and colleagues.

Social learning, in turn, has been suggested as the key ingredient of AM (Parson and Clark, 1995). Social learning can be defined as that which a person learns from others rather than discovers for himself or herself. Taking this farther leads to an emphasis on collective learning, i.e., learning by a group or some other collective agent. The concept of social learning has potential as a basis for structuring *shared, intersubjective* learning at scales that make a difference to societal outcomes (e.g., Roe, 1999; Maarleveld, 2000; Jiggins and Röling, 2000). It draws attention to the procedures and incentives that encourage people to learn *together with an ecosystem* what is sensible to do to sustain the ecological capacity that supports human life. Social learning in natural resource management maintains dialogue among stakeholders, scientists, planners, managers, and others, and their environment (Maarleveld and Dangbégnon, 1999).

Scientists often seem to take a view of social learning that emphasizes helping especially policy makers acquire a better *scientific* understanding of how humans affect ecosystems and how system feedbacks affect human well-being. Social learning becomes a dialogue *about* the environment, rather than a dialogue *with* the environment. Holling's students and colleagues Stephen Light and Lance Gunderson (Light et al., 1995; 1998), for example, have invested considerable time and energy in developing, and interactively using with policy makers, simulation models of complex ecosystems. The goal is to facilitate interactive learning between scientists and policy makers about how the ecosystem is affected and reacts to the collective impact of human activity. This approach seems similar to Lubchenco's call for scientists "to tell people what is out there" and to school themselves in science communication (personal communication, 1998).

The solution to the ecological challenge here is seen in what Giddens (1984) calls "the double hermeneutic": people can learn to make sense of the world based on how others, in this case scientists, have made sense of the world. We believe this to be too narrow a view of social learning. However important scientific understanding of what is happening to our biosphere, it is only part of the story. Defining social learning only as disseminating scientific understanding traps the concept in the positivist paradigm: scientists discover the truth about the "world out there" and share this knowledge with ultimate users. This paradigm ignores (1) the need to understand the *human activities* that lead to the problems in the first place, and (2) the need for humans *collectively to take responsibility for an ecological society.* As Funtowicz and Ravetz (1993) have put it, puzzle-solving science is unlikely to get us out of the predicament that has been created by the impacts that the application of knowledge is having on our world. The irreducible uncertainties of risks manufactured by knowledge application are the hallmarks of a risk society. As Funtowicz and Ravetz argue, we need a "post-normal science," in the Kuhnian sense, a science that is marked by *widely shared searching* for ways out of the mess.

The approach characterized by the double hermeneutic is comparable with health educators showing smokers pictures of lungs that have been blackened by smoking. To be sure, the educators do well not to ignore this approach to bringing about change in behavior. But, as thousands of doctors who smoke demonstrate, scientific understanding is perhaps a necessary but definitely not a sufficient basis for behavior change. Effective AM requires not only understanding how ecosystems work, but widely shared, self-reflective understanding of how people's individual and collective actions lead to survival-threatening outcomes.

The juxtaposition of environment and people allows a second observation. Ecologists maintain a dualism, of two autonomous spheres of existence, when they position AM as an interface or bridge between people and their environment. Social learning implies more: the fundamental *duality* of people and biosphere as one structurally coupled system.

We propose that the Santiago biological theory of cognition can help develop understanding of the fundamental distinction between dualism and duality. It helps us examine the potential for creating collective cognitive agency, i.e., the power to act purposefully in concert with our environment on the basis of shared learning. It also allows us to define ecological rationality as an alternative to economic rationality. Finally, it provides an appealing, widely sharable, conceptual framework for self-reflection.

THE SANTIAGO THEORY OF COGNITION

The Santiago theory of cognition, developed by two Chilean biologists, is summarized succinctly by Capra (1996, p. 257) in the following words:

> In the emerging theory of living systems, mind is not a thing, but a process. It is cognition, the process of knowing, and it is identified with the process of life itself. This is the essence of the Santiago theory of cognition, proposed by Humberto Maturana and Francisco Varela (1992).

Their starting point was the question: how do organisms perceive? Take a frog looking at a fly. Their research showed that the image of the fly cannot be projected on the central nervous system of the frog. In fact, the physical processes that govern the image of the fly (light waves) are totally different from the neurological processes that determine the image created in the central nervous system of the frog. One could say that the central nervous system is informationally closed. There is no way that the fly can be "objectively" projected. But the presence of a fly can trigger change in the central nervous system of the frog. The frog "does not bring forth *the* fly, but *a* fly." The construction of reality is not a human prerogative but a quality of all living organisms.

But, say Maturana and Varela, the frog does not bring forth *any* fly (as pure relativists would contend). It brings forth a fly the frog can catch and eat. Organisms and their environment are structurally coupled. They maintain this coupling through mutual perturbation. The way organisms bring forth *a* world allows them to maintain structural coupling with their environment. This leads Maturana and Varela to a startling and powerful definition of *knowledge as effective action in the domain of existence*.

This definition is startling, not only because so many people think of knowledge as the prerogative of *Homo sapiens*, but also because we are taught to believe that as scientists we are building a store of ever-expanding objective knowledge. A doctoral dissertation is supposed to add to the store, and so we build our store bigger and bigger. Maturana and Varela change that metaphor. A store of knowledge developed in an old context can be an insight that blinds, and a downright barrier to taking effective action in a changing context or a new domain. We indeed have developed enormous amounts of knowledge. We appear to have very little *effective knowledge*, in the sense of guiding effective action in this new domain of existence — the domain marked by anthropogenic destruction of the conditions for life.

THE COGNITIVE SYSTEM

According to Maturana and Varela's definition, cognition is the very process of life. Mind is immanent in matter at all levels of life. There is no organism that is not capable of cognitive action, that is, of assessing experience according to some emotion and taking action accordingly. "The new concept of cognition, according to the Santiago theory, is much broader than that of thinking. It involves perception, emotion, and action — the entire process of life" (Capra, 1996, p. 170). The cognitive system as a coevolving *duality* of the perceiving organism and its environment is sketched in Figure 8.1. We observe that the system includes (1) an organism or *agent* that can perceive the environment and take action in it; and (2) the domain of existence with which the agent is structurally coupled. We further distinguish (3) an ecosystem, i.e., a space in which multiple agents interact and mutually determine each other's domain of existence. It is important to note here that Maturana in particular does not accept the identification of "emotion" with "intentionality." Nor, further, does he accept a definition of "intentionality" as implying some *a priori* setting of an objective to be attained, that motivates action (as, for example, Searle, 1984, would argue in his discussion of consciousness and the perception of freely willed action). The biological basis of cognitive systems implies, on the contrary, that the triggered response of a perceiving organism's cognitive processes to its environment is necessarily something that occurs "in the moment." Learning, that is, occurs in the continuous present and is necessarily adaptive (Jiggins et al., 2000).*

* The authors are grateful to Professor Ray Ison of the Open University, Milton Keynes, U.K., for pointing this out. See also Damasio (1999).

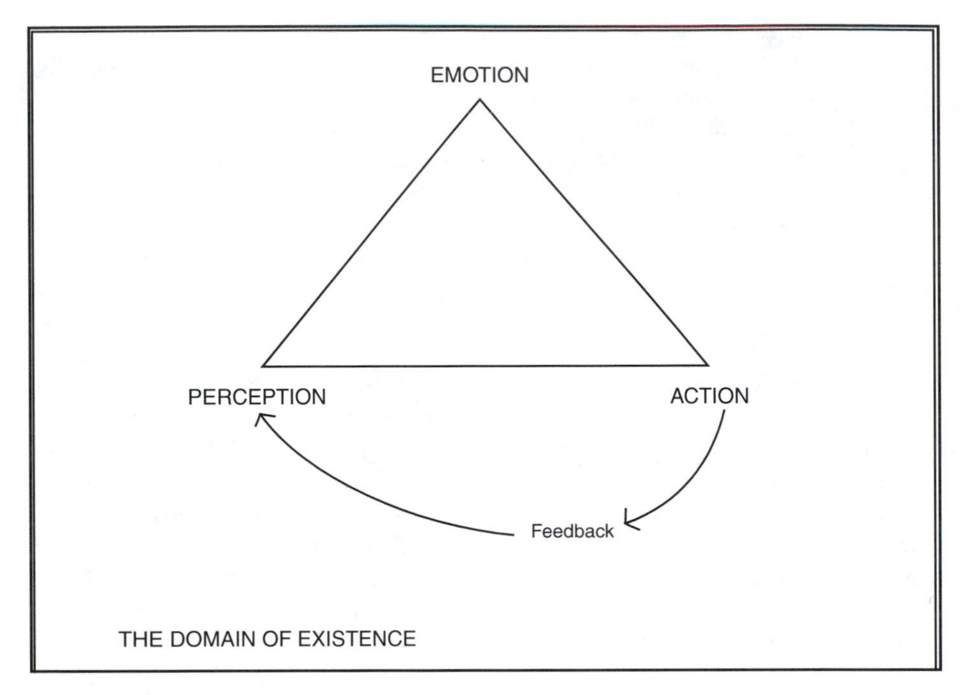

FIGURE 8.1 The cognitive system. (After Maturana and Varela, 1992; Capra, 1996.)

Ecological Rationality

The process of mutual adaptation among coevolving elements allows us to discuss *ecological rationality.* Rationality is crucially important: it is the driving force that lends dynamism to models of human cognition. The rationality implied by the Santiago theory of cognition is different from the rationality ascribed to humans by economists (which has strongly influenced the way people see themselves; Röling and Maarleveld, 1999). We begin our discussion with the Belgian institutional economist Jean Philippe Platteau, who makes a distinction between:

1. *Parametric rationality,* also called "an orientation toward individualism." Each person is assumed to maximize his or her own benefit or utility (i.e., an assumption of "Economic (Wo)man");
2. *Strategic rationality,* also called "an orientation toward competition." Each person is *also* assumed to take into account (anticipate) the behavior of others in furthering his or her private interest.

Platteau (1996; 1998) explicitly bases his theory about the evolution of land rights in sub-Saharan Africa on an assumption of *strategic* rationality. When challenged, he claims that he would like his ideas to be disproved, but has so far seen no reason to abandon them. His anthropological economic studies of fishermen in Senegal have convinced him that institutions are virtually incapable of overriding strategic rationality (Platteau, personal communication, 1999). Notice that economic rationality, as defined by Platteau, implies a *given* desire that is satisfied (1) by controlling the environment so as to make outcomes fit wants and (2) by winning from others.

The ecological rationality implied by the Santiago theory of cognition is based on a very different reasoning. Here, the ultimate rationality of the cognitive agent is to maintain structural coupling with its domain of existence. For the agent, this implies that it has the capacity to take effective action in its domain by reducing discrepancy among perception, emotion, and action. This includes the ability of the agent to adapt to changes in a domain shaped by other agents. The taking of effective action in a domain of existence therefore implies the following capacities:

- *Control:* Act to make outcomes satisfy emotion-based purpose;
- *Adaptation:* Adapt emotion-based purpose to the opportunities (perceived to be) offered by the environment, or to the outcomes that can be elicited (feedback);
- *Learning:* Develop perception to fit the opportunities or threats in the environment and adapt action and purposeful behavior to changed perception;
- *Evolution or innovation:* Adapt the ability to take effective action to the perceived and/or experienced threats and opportunities in the domain of existence.

Rationality, therefore, implies first of all that the cognitive agent tends toward correspondence with whatever opportunities there are for maintaining structural coupling. Secondly, it implies that the three elements — perception, emotion, and action — tend toward (cognitive) consistency. Third, it implies that action evolves capability to make use of the opportunities and, fourth, that perception becomes able to interpret the domain of existence selectively to maintain and feed effective action and keep purpose attuned to what is possible. A fifth and key implication is that, since the domain of existence is never stable, the agent must learn and evolve or innovate, lest it lose structural coupling and die. It is in this sense that we argue for an appreciation of the Earth as a complex adaptive system of coevolving cognitive agents.

In all, we agree with Gigerenzer and Todd (1999:18) who, when analysing ecological rationality, make a distinction between *coherence* criteria for rationality (as in cognitive consistency, or logical consistency) and *correspondence* criteria, which measure performance against the "real" world and require more domain-specific solutions. If consistency criteria override correspondence criteria, the organism becomes maladapted to its domain of existence. If correspondence criteria dominate, the organism does not "catch up" or becomes disorganised. When applied to (human) collective cognitive agency, the need to be rational in terms of both sets of criteria promises rich ground for analysis.

The concept of rationality implied by the Santiago theory not only seems much richer than the simple goal-seeking rational choice theory used by economics, it also seems a sounder guide for a self-interested "major force of nature."

Collective Cognitive Agency

The Santiago theory of cognition provides a simple but intriguing discourse that we consider useful for stimulating widely shared self-reflection about ourselves as a major force of nature. This discourse becomes even more powerful if we consider *collective* cognitive agency. As we have seen, the biosphere must be considered a soft system: what happens emerges from human interaction. Regenerating the biosphere and building opportunity means *purposefully and collectively redesigning human interaction.* This is the key point made in this chapter. To become a responsible force of nature through AM we must develop *collective cognitive agency in pursuit of ecological rationality.* The Santiago theory of cognition provides a fundamental vantage point from which to view such an enterprise. The remainder of this chapter examines selected social science perspectives to flesh out the powerful but rudimentary skeleton of the theory, and to determine whether the development of collective cognitive agency is a realistic enterprise.

RATIONAL CHOICE

Economics undoubtedly is the most widely shared perspective that people use today as a frame for thinking about, and for organizing, their relationship with each other and with their environment. We consider this a rather disturbing state of affairs. It is like a huge capital investment in a machine that can produce only goods for which there is no longer a need. We elaborate this bold statement by drawing on the work of prominent economists. First, however, we need to make explicit what we perceive to be a widely shared economic narrative (albeit one rejected by many institutional, evolutionary, and ecological economists).

We have discussed already the explicit assumptions that neoclassical economists such as Platteau make with respect to human rationality. We can take this farther by showing the fundamental assumptions of neoclassical economics with respect to human cognition (Figure 8.2).

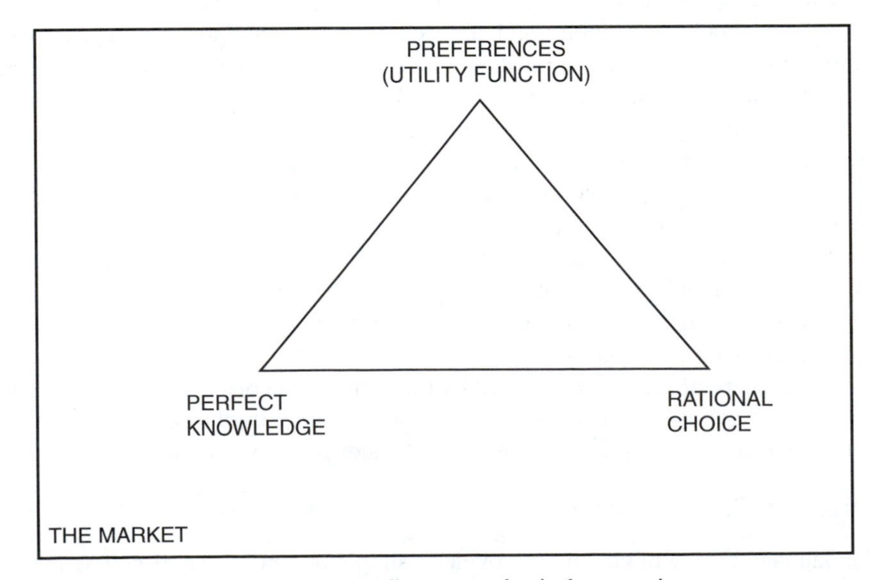

FIGURE 8.2 Human cognition according to neoclassical economics.

The preferences are generated outside the economic system; they are a *given* with which economics works. Humans are assumed to have given goals, and economics to be concerned with goal attainment. In this sense, economics is an axiomatic science. It builds knowledge about people's rational choices to achieve given goals in situations of scarcity, assuming that they are perfectly informed. At the level of the ecosystem ("market"), neoclassical economics models the equilibrium mechanisms that emerge from the interaction among rational individual agents (as we shall see later, one can call this "methodological individualism"). This market, the interaction among selfish agents in pursuit of their goals, is expected to lead to "the greatest good for the greatest number." The market acts as a force of nature.

Although this kind of economic thinking dominates much of the popular (and professional) discourse, many observers of the human condition have moved on. Recent challenges from within the economic discipline itself are exciting exactly because they have begun to question the *cognitive* assumptions of neoclassical economics.

> The knowledge of human knowledge claims a place of its own in economics. Beyond the walls of our discipline, spectacular progress is taking place in the field of empirical research into human knowledge — the so-called cognitive sciences. In the light of such advances, the old and classical axiom that nothing scientific can be said beyond the axioms of substantive rationality now looks very much like the protective belt of a degenerating programme (Tamborini, 1997).

The formulation of "bounded rationality" by Simon (1969) and the identification of institutions as mediators of information by North (1990) have led recently to numerous new perspectives with respect to the cognitive basis of economics. In an attack on methodological individualism, Arrow (1994) concludes that "social variables, not attached to particular individuals, are essential for studying the economy or any social system, and that, in particular, knowledge and technical information have an irremovably social component, of increasing importance over time."

Arthur (1994) puts it like this:

> The type of rationality assumed in economics, perfect, logical, deductive rationality, is extremely useful in generating solutions to theoretical problems. But it demands much of human behavior, which is more, in fact, than it can usually deliver.

He uses the findings of modern psychology to show that, in complicated situations, humans use characteristic and predictable methods of reasoning that are *inductive*. In the case of complex problems, we look for patterns to simplify and make tractable the problem, and we construct temporary internal models, hypotheses, or schemata to work with. As feedback comes in, we strengthen or weaken, or reject and replace them. A belief or a model is clung to, not because it is correct, but because it has worked in the past, and must cumulate a record of failure before it is discarded. This perspective is evolutionary; hypotheses must prove themselves by competing and being adapted within an environment created by other agents' hypotheses.

In a distinguished lecture on economics in government, Aaron (1994) chastises his discipline for failure to take the formation of preferences seriously ("the recalcitrant refusal of economists to venture beyond a model of human behavior others see as seriously incomplete"). He also laments the reliance on a model of utility that has no relation to current psychology. He draws attention to the following claims as directly relevant to economics:

- People never know the full consequences of their actions.
- The human brain does not contain a central processing unit, a giant server supervising many workstations. A more useful metaphor is of the brain as a massive parallel processing unit (see also Clark, 1997).
- People have a capacity for self-reference through which they can judge their own lives and relationships.
- Humans derive satisfaction from helping each other.
- People care about others as ends, not only as means.
- People derive enormous satisfaction from interpersonal relationships.
- The satisfaction people take from setting goals and achieving them has erroneously been singled out as the most important.
- The most palpable reality of all lives is internal conflict.

Aaron argues that each person operates more than one utility function, including self-respect, profit, others' regard, and social capital. The trade-offs that are made among them, and the utility function that is determinant at any time, are an empirical, contextual outcome of contingent history, opening to the economist "a vast scope for theoretical imagination."

Satz and Ferejohn (1994) conclude their social theoretical analysis of rational choice theory by pointing out that that theory is most credible under conditions of scarcity where human choice is constrained. Without constraints, agents will not behave as the theory predicts. In the conditions in which it is a strong predictive theory, the theory relies on features of the agents' environment. "We need a background theory to identify in just which contexts a psychological interpretation of rational choice theory makes sense."

This overview of a portion of recent economic theorizing allows us to draw an important conclusion. Modern economics is struggling to move away from the axioms about cognitive agency on which much of its theory is based, by incorporating insights from other sciences. However, the older "strategic narratives" of economics, that guide people's expectations of human nature and that are embedded in powerful institutions throughout society, remain determinant (Röling and Maarleveld, 1999). By creating environments that encourage people to act selfishly and strategically, selfish and strategic rationality is indeed a "blinding insight" that stands in the way of a shift to an ecological rationality.

A widely shared self-reflective understanding of ourselves and other life-forms as cognitive agents, in a domain of existence inhabited by other cognitive agents, seems a useful, perhaps necessary, if not sufficient, basis for any shift toward ecological rationality. Worded more strongly, a shift would seem to require that people are encouraged to take concerted action at different levels of nested hierarchies to manage ecosystems, i.e., that they are able to engage in a "shared learning environment" that is structurally coupled to its ecology. We see this as the most important challenge. The following sections selectively explore what different social sciences might have to offer as guidance.

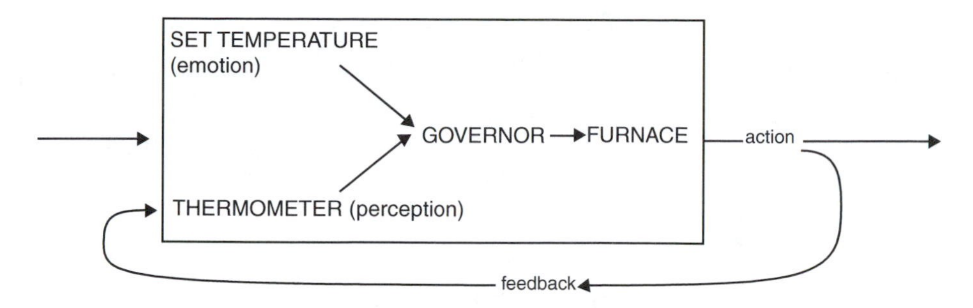

FIGURE 8.3 The simplest "cognitive agent:" the thermostat. (After Von Bertalanffy, 1968.)

GENERAL SYSTEMS THEORY

General systems theory (Boulding, 1968) distinguishes seven levels of theoretical discourse: (1) static structures (frameworks); (2) simple dynamic systems (clockworks); (3) self-regulating, cybernetic systems (thermostats); (4) self-maintaining living structures (cells); and (5, 6, and 7) more complex living and self-organizing adaptive systems. Each higher level presumes the lower one. Adequate theoretical models extend to the fourth level and not much beyond. The level of the clockwork is the level of classical natural science.

Although nonliving, the cybernetic system (level 3) clearly has a cognitive structure and provides a recognizable starting point for a discussion of human (collective) cognition (Figure 8.3). The thermometer *perceives* what is happening in the *domain of existence*; the set temperature represents *emotion (what is desired)*; and the furnace allows *action* that affects the domain and is again perceived by the thermometer (*feedback*). The simple example of the thermostat is helpful because it highlights the role of a governor, which allows communication and comparison among the elements of the system and which can trigger action. Further requisites are an apparatus for taking action and a throughput of energy. The simple thermostat makes us aware of what it *cannot* do, compared with even the simplest cellular organism. It cannot adapt or learn. There is limited dynamic interplay with the domain of existence. If a fire broke out next to the thermostat, the mechanism would stop working because it would be "warm enough."

The simple thermostat is consistent with the Santiago cognitive theory. All higher system levels, cells and more complex living systems, presume its elements. Below we examine a number of social science theories to determine whether collective human cognitive agency, that we derived as a necessary condition for a sustainable biosphere, is a realistic possibility.

We speak of a collective cognitive agent when people (1) *perceive* their domain of existence in a similar way, perhaps because they share a monitoring system; (2) have similar *emotioning* in that they have negotiated shared goals or subscribe to the same goals because they are part of a community; (3) engage in concerted *action*, based on (4) a shared *knowledge* about what is expected to be effective action in the domain of existence; and (5) construct the *domain of existence* according to a shared design. In the sense defined here, a collective cognitive agent acts as if it were one cognitive agent.

Mode of Adaptation	New Cultural Goal	Institutionalized Means
1. Innovation	+	–
2. Ritualism	–	+
3. Retreatism	–	–
4. Rebellion	±	±
5. Conformity	+	+

Note: + means acceptance; – means rejection; ± means rejection of prevailing values and the substitution of new ones.

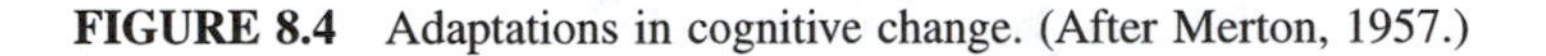

FIGURE 8.4 Adaptations in cognitive change. (After Merton, 1957.)

COLLECTIVE COGNITIVE AGENTS: THE CONTRIBUTION OF SOCIAL PSYCHOLOGY

ADAPTATIONS TRIGGERED BY COGNITIVE INCONSISTENCY

The sociologist Merton (1957) devised a classic typology of adaptation in situations in which cultural goals have changed and institutionalized means have not (Figure 8.4). Merton originally devised his scheme to explain deviant behavior. It can also be read as a typology of effort to maintain cognitive consistency.

In terms of cognitive agents, we can describe the responses as follows (Röling, 1970):

Innovation is the adaptation of action or of the ability to act to satisfy changed goals. Earlier, we called this control. Innovation might express itself in the adoption of new technologies, emigration, legal redistribution of access to assets and power, or even magic, developing supernatural means to achieve new goals.

Ritualism is the rejection of new goals to comply with existing ability to act. Old forms of action are fixated. This fixation is considered a substitute goal response. Development can be marked by extreme "traditionalism," often as an expression of frustration.

Retreatism is the rejection of the new goal *and* of the existing ability to act. It represents withdrawal resulting in apathy. A typical response is fatalism, the belief that external forces determine one's outcomes, but other responses include voluntary isolation, and escapism into alcohol, cults, or religious extremism. Such responses are often seen as the best adaptation people can make to a hopeless situation.

Rebellion is the rejection of the institutional arrangements within which the ability to act is defined (e.g., access to resources). In a way, rebellion is an innovative response.

Conformity is adherence to both the new goal and the existing ability to act. This seeming paradox can be explained by the fact that people do not allow themselves to be motivated (and frustrated) by all possible goals (e.g., standards of living to which they are exposed). Goals are limited to those perceived as attainable in one's own situation. One feels poor only in relation to the outcomes "relevant" to others' experience (the so-called reference group).

Merton's typology highlights some interesting aspects of collective cognition.

1. The "responses" detailed above are the result of individual learning. They are ways of dealing with cognitive inconsistency. They can lead to common responses (e.g., rebellion, culture of fatalism), when individuals share an experience, that are perceived by others as deviant because the responses do not achieve desired control through innovation. The common response here is merely an aggregation of individual responses within a shared domain of existence.
2. "Cultural goals" (based on what we have called emotion) are seen as more prone to change than the opportunities to satisfy them. The tension raises the salience of the human mechanisms that regulate motivation in view of realizable opportunity.
3. If we consider current global interconnectedness to be increasing, yet inequality in access to resources, capital, and other opportunities, and enjoyment of benefits, also to be increasing, then we may anticipate widespread cognitive inconsistency and "deviant" adaptations that threaten the achievement of a global system based on ecological rationality.
4. To the extent that ecological rationality means adapting cultural goals to limitations in outcomes, i.e., taking less and/or giving more, it requires (self)-management of cognitive inconsistency. This in turn implies reshaping criteria for status and achievement, new enthusiasms and new ideas about what is worthwhile, and perhaps new social institutional devices that replace the religion of old.

Reasoned Action

One of the most influential social psychological theories is Fishbein and Ajzen's (1975) theory of reasoned action. It is widely used to guide persuasive interventions (advertising, political propaganda, health education, agricultural extension). According to this theory, the intention to engage in action is determined by the interplay of three factors: the attitude toward the intended action, the social norms (or expected social control) with respect to the action, and self-perceived efficacy (one's belief in one's own ability to undertake the action) (Figure 8.5).

Persuasive efforts can try to affect each of these determinants of contemplated action. For example, the beliefs about the outcome of the action (anticipated feedback) can be the subject of intense persuasion by the health educator when he or she tries to raise the perceived likelihood of cancer as an outcome of smoking. But also the desirability of the outcome can be a target of persuasion, for example, when an advertisement claims that using toothpaste *X* increases one's sex appeal. Efforts to affect self-perceived efficacy are usually called empowerment.

The theory of reasoned action establishes levers for changing the human cognitive agent, such as anticipated feedback, "desirability," and self-confidence. The

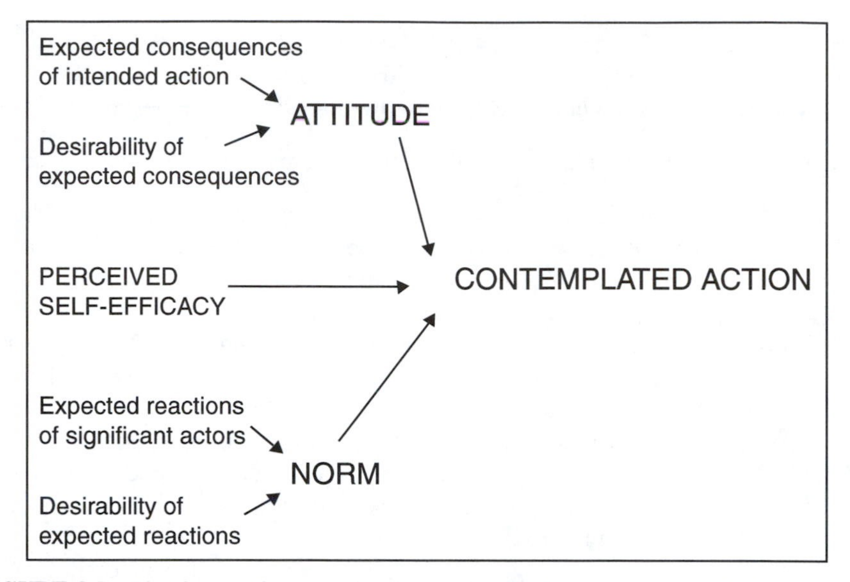

FIGURE 8.5 The theory of reasoned action. (After Fishbein and Ajzen, 1975.)

theory emphasizes the malleability of the cognitive agent in terms of the information on which he or she bases contemplated action, and the extent to which his or her reasoning is subject to social influence. This social influence does not hold only for the "norms," but also for the "attitude." Perceived self-efficacy is largely determined by the opportunities or limitations of the domain of existence. In that sense, the theory emphasizes the socially constructed nature of the human cognitive agent.

The theory also draws attention to the nature of its application. It assumes a strategic decision maker who has the means to influence people as reasoning actors. Hence, it evokes the image of a centralized power that can decide how human cognitive agents can be influenced. In the global competitive market that we are constructing, this power is increasingly concentrated in companies that want us to consume and spend. Nearly all our media for shaping collective cognition are dedicated to advertisement, sport, pornography, and to consolidation of consumer markets. How in this perspective might a groundswell of counteraction gain sufficient power and momentum to force reconsideration of the global agenda? The emergence of the groundswell might be explained by two factors:

1. The threats to the maintenance of structural coupling within our domain of existence are becoming daily more evident. In line with Kuhn's (1970) argument that "normal science" continues until the accumulating counterevidence can no longer be ignored, new paradigms are emerging to take account of the threats (e.g., Funtowicz and Ravetz (1993; 1998).
2. In his history of the idea of communication, Durham Peters (1999, p. 35 a.f.) raises "a debate between *dialogue*, for which Socrates is the greatest proponent, and *dissemination*, for which Jesus is the most enduring voice." A focus on persuasion implies that dissemination is the main instrument

TABLE 8.3
Variations of Truth in Teak Plantation Forestry

A Teakwood Investment Programme offered the Dutch public the opportunity to invest in a teak plantation in Costa Rica. Pioneered in 1989, the programme gained momentum in 1993 when it was joined by a Dutch insurance company, Ohra, and the World Wildlife Fund (WWF). Thousands in The Netherlands alone invested millions of Euro. Teakwood offered "green gold." From November 1995 onward, Teakwood increasingly met with opposition. The study describes the erosion of the credibility of the claims of the Teakwood partners and of the certification of the plantations by the Rainforest Alliance, itself accredited for certification by the Forest Stewardship Council (FSC). The Dutch branch of WWF not only supports FSC as founder and funder, but also benefited directly from the proceeds of selling Teakwood timber.

"Variations of the truth" emerged in two ways. For one, it turned out that the teak plantations did not comply with several FSC principles and criteria (agro-chemicals were used in the plantations, for example). Secondly, the production projections used by Ohra to underpin the profitability of the investments proved higher than anything described in the scientific literature. These revelations proved a sound legal basis for dissolving Teakwood contracts. Ohra pulled out of Teakwood in 1999.

The demise of Teakwood was brought about by publication, by a self-appointed activist, of the statements made by the major partners involved in Teakwood over the internet in networks of forestry professionals between January and July of 1996. These internet circulars were found to produce increasing pressure on the Teakwood partners and associated organisations (marked by court cases and television publicity). This pressure led to additional statements and justifications, which often only reinforced the inconsistency, lack of transparency and non-accountability of the original statements. Also the additional statements were published on the internet.

Source: Romeijn, P., *Green Gold. On Variations of Truth in Plantation Forestry,* Treemail, Heelsum, The Netherlands, 1999. With permission.

for change, but interactive networks are emerging that are built on dialogue, or communicative rationality (Habermas, 1984; 1987) (Table 8.3).

Social Impact Theory

Social impact theory is developing with the aid of computer simulation. There is considerable debate about the usefulness of computer-assisted simulation as a representation of reality or as a way of perceiving the world. Nonetheless, the theory is generating research that is relevant and insightful for our purposes. Latané (1997), for example, has developed a perspective that focuses on the social outcomes that emerge from interaction among agents. The theory has four main elements:

1. Individual human beings, varying in strength and other attributes, are distributed in social space. "Strength has to do with how much power people have to influence others, how wise and articulate they are, how much they are listened to and imitated."
2. Each person is influenced by his or her own individual experience and by others in proportion to a multiplicative function of their strength, immediacy, and number.

3. Unless total persuasive impact (the pressure to change to a different position) outweighs supportive impact plus individual experience (the pressure to stay where one is), the individual will not change.
4. The computer can be used to calculate dynamic social impact as the cumulative effect of the iterative, recursive influence of interacting people on each other.

Simulations of interactions among 400 agents satisfying these criteria and varying on two attributes showed the following ways in which such systems self-organize after multiple runs. *Consolidation:* The size of the minority on each attribute decreases. There is some loss of diversity. *Clustering*: As a result of immediacy (people are more influenced by neighbors than by strangers), people who are similar with respect to an attribute spatially cluster together. *Correlation:* The two originally independent attributes become correlated. "Partly, correlation emerges as a result of a perpetuation of the strength structure, with highly influential people being able to impose their views on several different issues. Primarily, however, correlation is caused by the reduction in degrees of freedom resulting from clustering. Instead of being a population of 400 independent actors, the system now acts as if there were a greatly reduced number of independent agents." *Diversity is maintained through time by clusters that preserve minorities from extinction.* People who live inside minorities are shielded from majority influence.

Social impact theory is relevant because it provides a first glimpse of a collectivity of human agents that has properties that do not emerge only from individual action. In that sense, social impact theory goes beyond methodological individualism and shows that the human "ecosystem" has a dynamic of its own, which, in turn, affects the individual agents. Interestingly, such ecosystems self-organize and begin to act as if there were a greatly reduced number of independent agents.

COLLECTIVE COGNITIVE AGENTS: SOME IDEAS FROM SOCIOLOGY AND ANTHROPOLOGY

Structuration

For the sociologist Giddens (1984), too, social structure is more than an emergent property of the interaction of individual agents. "The rules and resources drawn upon in the production and reproduction of social action are at the same time the means of system reproduction (the duality of structure)" (p. 19). He identifies the elements of structuration as *signification* (shared interpretative schemes interactively established through communication), *domination* (allocation of access to resources through power based in economic or political institutions), and *legitimation* (criteria for acceptable behavior or norms reinforced through sanctions by legal institutions). These elements of structuration appear to mirror the language of the Santiago school of cognition.

They are presented in sociological theory as necessary but not sufficient for the emergence of collective cognitive agents within an ecological rationality. The emergence of such agents also seems to require *shared monitoring* that provides evidence to counter the self-referentiality of socially constructed perceptions. They further require *capacity for concerted action*, minimally a capacity for negotiating and making collective decisions. This not only assumes leadership or policy, but also *social capital* in the sense of known, trusted, and dependable checks and balances, and procedures for conflict resolution and for maintaining justice. We explore the concept of social capital further in the next section.

Governing the Commons

Ostrom (1992; 1998) has identified conditions for establishing collective cognitive systems with respect to common pool resources, i.e., resources for which the use by one user subtracts benefits from another, and for which exclusion involves high transaction costs (Steins, 1999, p. 3). These resources are of particular interest because they include biodiversity, groundwater, the atmosphere, the oceans, silence, genetic integrity, landscape, the ozone layer, a stable climate, and other resources and ecological services that are threatened by human actions.

On the basis of experiments and studies of situations where the "tragedy of the commons" did *not* occur, Ostrom established the following conditions for what we would call a collective cognitive agent:

1. Definition of the group that can use the resource;
2. Communication, i.e., the group must interact to discuss and negotiate;
3. Agreement on the activities that maintain structural coupling with the domain of existence (e.g., with respect to the allowed off-take from a communal forest);
4. A system for monitoring compliance and its upkeep (e.g., a forest warden and his salary);
5. Sanctions.

Such conditions admittedly assume fairly small local groups and a sense of place, seemingly a far cry from the global issues that are at stake. Ostrom (1998) believes, however, that, even when the resource is huge, say the U.S. Pacific Coast and the specific marine system that has developed along it, the problematic situation is rendered manageable by acting through many small groups. Each is responsible for only a section, or a hierarchical space, in the overall system. Such parallel management allows for the force of contingent history to be expressed, for adaptability, and for corrective response (resilience). This optimism is supported by game theoretical studies that demonstrate at a minimum that distributed and contextualized shared learning and management is "better" with respect to a range of desired outcomes than centralized, top-down management.

Institutions Think

Mary Douglas (1986) has initiated another line of research and action that feeds into the development of ecologically rational behavior. A taste of the work presented in her influential book, *How Institutions Think,* is the following comment on the prevailing rational choice theory (p. 128).

> Only by deliberate bias and by extraordinarily disciplined effort has it been possible to erect a theory of human behavior whose formal account of reasoning only considers the self-regarding motives, and a theory that has no possible way of including community-mindedness or altruism, still less heroism, except as an aberration.

And she concludes as follows: "For better or for worse, individuals really do share their thoughts and they do to some extent harmonise their preferences, and they have no other way to make the big decisions except within the scope of the institutions they build" (Douglas, 1986, p. 128).

Just as Maturana and Varela's work challenges theories of cognition that insufficiently take heed of the mind as embodied matter, so she sets out to amend "unsociological views" of human cognition. She traces resistance to "the idea of a supra-personal cognitive system" to the enthusiasm of our society for individualism. "The offence taken [to the idea of a superpersonal cognitive system] in itself is evidence that above the level of the individual human another hierarchy of 'individuals' is influencing lower level members to react violently against this idea or that" (1986). Douglas builds on Durkheim who believed that the utilitarian could never account for the foundations of civil society. He was convinced that the Benthamite model, by which social order is produced automatically out of the self-interested actions of rational individuals, was too limited because it gives no explanation for group solidarity. But his ideas encountered considerable opposition and have remained underdeveloped until this day. But the commitment that subordinates individual interests to a larger social whole must be explained. Douglas considers therefore "the role of cognition in forming the social bond. ... In Durkheim's work the whole system of knowledge is seen as a collective good that the community is jointly constructing." It is this process that is the center of Douglas's book. "Half of our task is to demonstrate this cognitive process at the foundation of social order. The other half of our task is to demonstrate that the individual's most elementary cognitive process depends on social institutions" (p. 45).

In her analysis, Douglas comes to a number of important conclusions:

> Institutions systematically direct individual memory and channel our perceptions into forms compatible with the relations they authorise (p. 92).
>
> Institutions have the pathetic megalomania of the computer whose whole vision of the world is its own programme. For us, the hope of intellectual independence is to resist, and the necessary first step in resistance is to discover how the institutional grip is laid upon our mind (p. 92).
>
> The high triumph of institutional thinking is to make the institutions completely invisible. ... Since all social relations can be analysed as market transactions, the pervasiveness of the market successfully feeds us the conviction that we have escaped from the old non-market institutional controls into a dangerous new liberty (p. 99).
>
> In all, Douglas provides convincing arguments that the "cumulative experience of the world should explicitly incorporate the social nature of cognition and judgment (p. 122). ... So let no one take comfort in the thought that primitives think through their institutions while moderns take big decisions individually. That very thought is an example of letting institutions do the thinking" (p. 124).

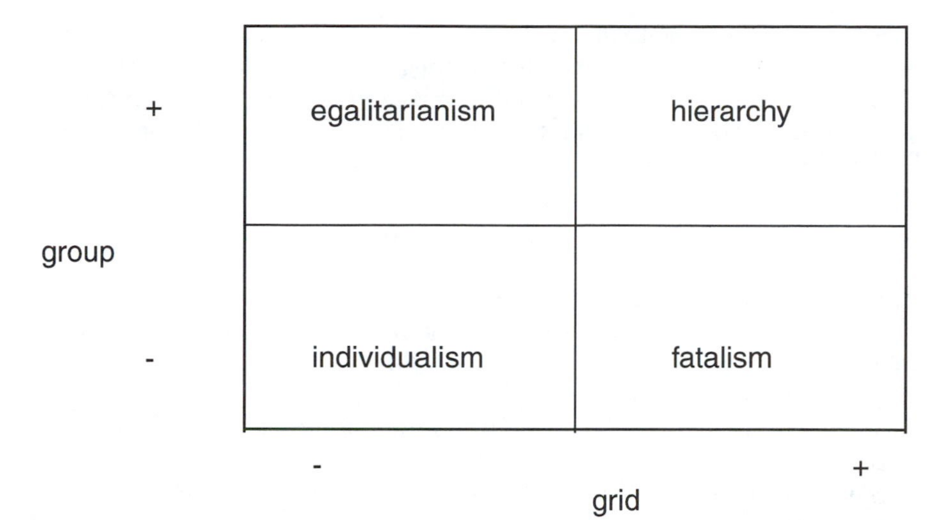

FIGURE 8.6 Typology of forms of social life. (Based on Oversloot, 1998, and Douglas, 1996.)

Douglas has been very influential through her "cultural theory" that makes more concrete the convictions presented above (reviewed by Oversloot, 1998). In brief, Douglas argues that our preferences are largely the product of our social relations. Social relations are embedded in what Douglas calls "forms of social life that recur: the individualist, sectarian (or egalitarian) and hierarchical" (p. 7). These "forms of social life" can be seen as a typology formed by two dimensions: group and grid. The former described the extent to which individuals form part of bounded units, and the latter the extent to which the life of the individual is determined by rules. The resulting typology is shown in Figure 8.6. For Douglas, the typology has tremendous predictive value. For example, "the competitive (individualist) society celebrates its heroes, the hierarchy celebrates its patriarchs and the sect its martyrs" (p. 80).

Of interest to us is the analysis of Thompson et al. (1990), quoted by Oversloot (1998), which links the typology in Figure 8.6 to "myths of nature." Thus, a perspective on nature as benign, robust, and tolerant fits with individualism. Because it is robust, nature does not need protection by others. Hence "nature benign" is consistent with individualism. A view of nature as fairly tolerant, but perverse (if you treat it badly, the damage cannot be repaired) fits with hierarchy. Because nature must be handled with care, control by the group over individuals is required. Nature is extremely vulnerable (ephemeral) for egalitarians. It must be treated with extreme care. But egalitarians do not have the means to prescribe others how they should behave. The only thing they can do is to proselytize and lead an exemplary life. Finally, the idea that nature is capricious is consistent with fatalism. There is little one can do about it.

In this view, the current focus on the exploitation of nature is consistent with our enthusiasm for individualism and competition that treats nature as robust. Thompson et al. (1990) raise the interesting perspective that increasing exposure to surprises, that show that nature is not robust, might lead to a shift to another "mode of seeing and being," with individualists shifting to hierarchy (emphasis on policy), egalitarianism (emphasis on solutions that emerge from interaction), or fatalism. In all, Douglas's work allows a direct link to be made between the "forms of social life" and ecological rationality.

CONCLUSIONS

Lubchenco's insight, that human beings have become a major force of nature, was linked to a strategic concern. As president of AAAS, she was worried that, since the cold war had ended, science had run out of things to do for society and would become a marginal concern. Luckily, the "eco-challenge" turns out to be a good rallying cry for a new social contract, with plenty of jobs and lots of funding for scientists.

We are intrigued by the implications of the eco-challenge. One is that the direction in which our blue planet is evolving is no longer a matter of self-regulation, self-organization, or of life creating the conditions for its own increasing complexity, as the Gaia hypothesis would have it. If we do indeed comprise a duality with our environment, if we are inescapably part of the complex web of life, to use Capra's (1996) fortuitous phrase, then human survival depends on our ability to maintain the ecological services that we have so far taken for granted.

The second is that, as a major force of nature, we must act with common purpose to manage the globe in a manner that is rational from an *ecological* point of view. The planet has become a soft system in that its direction emerges from the interaction of human projects. Those projects must now be guided by an ecological rationality.

The third implication is that an ecological rationality demands that we develop a "soft side to the earth," that is, develop the institutions, cognitions, norms, platforms for collective decision making, cosmovisions, and other social elements that allow us to *remain structurally coupled to context*.

These implications sketch a daunting outline of the challenge ahead. We have been deeply influenced by the work of Holling and colleagues who have linked the need for AM to the behavior of complex ecological systems. However, social learning has all too often been operationalized as "the need for scientists to tell us what is out there," to use Lubchenco's phrase. Our analysis makes clear that this would not be sufficient. What is required, rather, is an understanding of human interaction in dialogue with the environment.

We can illustrate this notion by reference to health promotion. Epidemiologists identify mortality and morbidity in a population. They also determine the major causes, including infection, hereditary factors, environmental factors, or behaviors, such as smoking, nutrition, exercise, or stress. Where behavior is the causal factor, as it is in the anthropogenic ecochallenge, dealing with that behavior is not just a question of explaining how the behavior affects health, e.g., showing what happens to lungs as a result of smoking. Dealing with behavior requires understanding behavior in interaction with its context. But, and this is a fundamental point, learning about human behavior cannot just be the dispassionate scientific activity of adding such learning to the body of scientific knowledge. The understanding must (1) form the basis of widespread self-reflection, on which voluntary, collective corrective action can be based and (2) be driven by informed emotional involvement and commitment to the preservation of life. Hence we need a theory of collective human behavior that people can buy into, a story that provides an alternative to an economic narrative that has palpably failed to safeguard the ecological conditions of human existence.

We have proposed the Santiago theory of cognition as the biological core of renewed action. It offers a dynamic model with great heuristic scope and a no-nonsense realism: to be effective is to maintain structural coupling with the domain of existence, the essential duality between organism and environment. Knowledge is effective action in the domain. Learning is adaptive to context and the flux of the continuous present. The theory allows us to formulate a definition of ecological rationality that reveals the poverty of economic theories of rationality and guides us toward institutional renewal. Finally, the Santiago theory allows us to model a collective human cognitive agent as a condition for a sustainable Earth.

A key question remains: *Is collective human cognitive agency possible?* Or are we doomed by default to destroy the conditions for our own survival? We have tried to answer this question by examining selected social science theories. The selection no doubt reflects the scope of our ignorance. We welcome the contribution of other disciplines in the development of the ideas presented in this chapter. However, on the basis of our explorations so far, we are led to some important conclusions. First, the Santiago theory of cognition can be seen as offering a unifying and widely sharable heuristic tool for thinking about ourselves. Second, there would appear to have been insufficient collaboration as yet in research into the possibilities of developing collective cognitive agency, at varying scales. Whether it is even theoretically achievable at a global scale remains unanswered. Third, our review offers glimpses of hope that people can dance to a different tune than strategic rationality. We saw that institutions can create conditions for serving the collective interest (see also Uphoff, 1992) but that there remains much to be explored concerning how to create these effectively in today's globalizing economies. Fourth, institutions clearly are more than the emergent property of the individual search for the satisfaction of self-interest (as methodological individualism would suggest). Indeed, as Mary Douglas points out, the search for the satisfaction of self-interest itself is institutionally determined. And, adds Eleanor Ostrom, people clearly are capable of managing their common pool resources in a sustainable manner through appropriate institutional designs. In sum, we believe that this chapter outlines an agenda for research and action of the utmost importance for our own future existence.

REFERENCES

Aaron, H., 1994. Public policy, values and consciousness, Distinguished Lecture on Economics and Government, *J. Econ. Perspect.*, 8(2), 3–21.

Arrow, K., 1994. Methodological individualism and social knowledge, Richard T. Ely Lecture, *AEA Pap. Proc.*,, 84(2), 1–9.

Arthur, B., 1994. Complexity in economic theory: inductive reasoning and bounded rationality, *AEA Pap. Proc.*, 84(2), 406–411.

Beck, U., 1994. The Reinvention of Politics: Towards a Theory of Reflexive Modernisation. Chapter 1 in Beck, U., A. Giddens, and S. Lash, Ed., *Reflexive Modernisation: Politics, Tradition and Aesthetics in the Modern Social Order,* Stanford University Press, Stanford, CA, 1–55.

Bos, A. H., 1974. *Oordeelsvorming in Groepen,* doctoral dissertation, Wageningen Agricultural University, Veenman, Wageningen.

Boulding, K., 1968. General systems theory: the skeleton of science, in *Modern Systems Research for the Behavioral Scientist*, Buckley, W., Ed., Aldine, Chicago, 3–11.

Callon, M. and Law, J., 1989. On the construction of socio-technical networks: content and context revisited, *Knowledge in Society: Studies in the Sociology of Science Past and Present*, 8, 57–83.

Capra, F., 1996. *The Web of Life, A New Synthesis of Mind and Matter*, Harper Collins, London.

Checkland, P., 1981. *Systems Thinking, Systems Practice*, John Wiley, Chicester, U.K.

Clark, A., 1997. *Being There: Putting Brain, Body and World Together Again*, The MIT Press, Cambridge, MA.

Damasio, A. R., 1999. *The Feeling of What Happens: Body and Emotion in the Making of Consciousness*, Harcourt Press, New York.

Dörner, D., 1996. *The Logic of Failure. Recognising and Avoiding Error in Complex Situations*, Addison-Wesley, Reading, MA (translated by Rita and Robert Kimber from the German: *Logik des Misslingens,* Rowolt Verlag, 1989).

Douglas, M., 1986. *How Institutions Think*, University of Syracuse Press, Syracuse, NY.

Douglas, M., 1996. *Thought Styles*, Thousand Oaks, London.

Durham Peters, J., 1999. *Speaking into the Air, A History of the Idea of Communication*, The University of Chicago Press, Chicago.

Fishbein, M. and Ajzen, I., 1975. *Beliefs, Attitude, Intention and Behaviour*, Addison-Wesley, Reading, MA.

Funtowicz, S. O. and Ravetz, J. R., 1993. Science for the post-normal age, *Futures,* 25(7), 739–755.

Funtowicz, S. O., Ravetz, J., and O'Connor, M., 1998. Challenges in the use of science for sustainable development, *Int. J. Sustainable Dev.,* 1(1), 99–108.

Giddens, A., 1984. *The Constitution of Society: Outline of the Theory of Structuration*, Policy Press, Oxford.

Gonzalez, R., 2000. *Platforms and Terraces, Using Participatory Methods and GIS in Linking Local Knowledge and Remote Sensing for Watershed Monitoring in Ifugao, Philippines*, doctoral dissertation, ITC, Enschede, and Wageningen University, Wageningen, The Netherlands.

Gunderson, L. H., Holling, C. S., and Light, S. S., Eds., 1995. *Barriers and Bridges to the Renewal of Ecosystems and Institutions*, Colombia Press University, New York.

Habermas, J., 1984. *The Theory of Communicative Action,* Vol. 1, *Reason and the Rationalisation of Society,* Beacon Press, Boston.

Habermas, J., 1987. *The Theory of Communicative Action,* Vol. 2, *Lifeworld and System, A Critique of Functionalist Reason*, Beacon Press, Boston.

Holling, C. S., 1995. What barriers? What bridges? in *Barriers and Bridges to the Renewal of Ecosystems and Institutions,* Gunderson, L. H., Holling, C. S., and Light, S. S., Eds., Columbia Press University, New York, 3–37.

Jiggins, J. and N. Röling, 1999. Interactive valuation: the social construction of the value of ecological services, *Int. J. Environ. Pollut.*, 5(4), 436–451.

Jiggins, J. and Röling, N., 2000. Adaptive management: potential and limitations for ecological governance, *Int. J. Agric. Resourc. Gov. Ecol.,* 1 (1), 28–42.

Jiggins, J., Hubert, B., and Collins, M., 2000. Globalisation and technology: the implications for learning processes in developed agriculture, in Learning Research Network, Cow up a Tree: Knowing and Learning for Change in Agriculture, Case Studies from Industrial Countries, INRA, Paris.

Keiter, R. B. and Boyce, M. S., 1991. *The Greater Yellowstone Ecosystem: Redefining America's Wilderness Heritage*, Yale University Press, New Haven, CT.

Kolb, D., 1984. *Experiential Learning: Experience as a Source of Learning and Development*, Prentice-Hall, Englewood Cliffs, NJ.

Kuhn, T. S., 1970. *The Structure of Scientific Revolutions*, 2nd ed., University of Chicago Press, Chicago.

Latané, B., 1997. Dynamic social impact: the social consequences of human interaction, in *The Message of Social Psychology, Perspectives on Mind in Society*, McGarty, C. and Haslam, S. A., Eds., Blackwell Publishers, Oxford, 200–220.

Light, S., Gunderson, L., and Holling, C. S., 1995. The Everglades: evolution of management in a turbulent ecosystem, in *Barriers and Bridges to the Renewal of Ecosystems and Institutions,* Gunderson, L. H., Holling, C. S., and Light, S. S., Eds., Columbia Press University, New York, 1-3–169.

Light, S. S., Carlson, E., Blann, K., Fagrelius, S., Barton, K., and Stenquist, B., 1998. Citizens, Science, Watershed Partnerships and Sustainability, The Case in Minnesota, Minnesota Department of Natural Resources, Surdna Foundation, and Science Museum of Minnesota, St. Paul.

Lubchenco, J., 1998. Entering the century of the environment: a new social contract for science, *Science,* 279, 491–496.

Maarleveld, M., 2000. *Social Learning in Dilemmas in Natural Resource Management, The Case of Sub-Terranean Drinking Water Resources in Gelderland, The Netherlands*, doctoral dissertation, Wageningen University, Wageningen.

Maarleveld, M. and Dangbegnon, C., 1999. Managing natural resources: a social learning perspective, *Agriculture Human Values,* 16, 267–280.

Maturana, H. R. and Varela, F. J., 1987 [rev. ed. 1992], *The Tree of Knowledge, the Biological Roots of Human Understanding*, Shambala Publications, Boston.

Merton, R., 1957. *Social Theory and Social Structure*, Free Press, Glencoe, NY.

New Scientist, 1999. 20 November, 4.

New Scientist, 2000. 22 January, 18.

North, D. C., 1990. *Institutions, Institutional Change and Economic Performance*, Cambridge University Press, New York.

Ostrom, E., 1990, 1991, 1992. *Governing the Commons, The Evolution of Institutions for Collective Action*, Cambridge University Press, New York.

Ostrom, E., 1998. Coping with tragedies of the commons, paper presented at 1998 Annual Meeting of the Association for Politics and Life Sciences, Boston, September 3–6.

Oversloot, H., 1998. De culturele theorie wellwillend belicht door een agnosticus, *Tijdschr. Beleid Pol. Maatschappij*, 5(4), 2–14.

Parson, E. A. and Clark, W. C., 1995. Sustainable development as social learning: theoretical perspectives and practical challenges for the design of a research program, in *Barriers and Bridges to the Renewal of Ecosystems and Institutions,* Gunderson, L. H., Holling, C. S., and Light, S. S., Columbia Press, New York, 428–461.

Platteau, J. P., 1996. The evolutionary theory of land rights as applied to sub-Saharan Africa: a critical assessment, *Dev. Change*, 27, 29–86.

Platteau, J. P., 1998. Distributional contingencies of dividing the commons, invited paper for Research School CERES Seminar "Acts of Man and Nature? Different Constructions of Natural and Resource Dynamics," Bergen, The Netherlands, October 22–24.

Ridley, M., 1995. *The Origins of Virtue*, Penguin Books, Harmondsworth, Middlesex, U.K.

Roe, E., 1999. Report to the Rockefeller Foundation, available at http://www.instantvision.com/rockefeller_report.

Röling, N., 1970. Adaptations in development: a conceptual guide for the study on non-innovative responses of peasant farmers, *Econ. Dev. Cult. Change*, 19(1), 71–85.

Röling, N., 1997. The soft side of land, socio-economic sustainability of land use systems, *ITC Journal*, Special Congress Issue on Geo-Information for Sustainable Land Management, 97, 3–4, 248–262.

Röling, N. and Maarleveld, M., 1999. Facing strategic narratives: an argument for interactive effectiveness, *Agric. Hum. Values,* 16, 295–308.

Romeijn, P., 1999. *Green Gold. On Variations of Truth in Plantation Forestry,* Treemail Publishers, Heelsum, The Netherlands.

Satz, D. and Ferejohn, J., 1994. Rational choice and social theory, *J. Philosophy*, 91(2), 71–88.

Searle, J., 1984. *Minds, Brains and Science*, The 1984 Reith Lectures, BBC, London.

Simon, H., 1969. *The Sciences of the Artificial*, MIT Press, Cambridge, MA.

Steins, N. A., 1999. *All Hands on Deck, An Interactive Perspective on Complex Common-Pool Resource Management Based on Case Studies in Coastal Waters of the Isle of Wight (UK), Connemara (Ireland) and the Dutch Wadden Sea,* doctoral dissertation, Wageningen University, Wageningen, The Netherlands.

Tamborini, R., 1997. Knowledge and economic behaviour, a constructivist approach, *J. Evol. Econ.,* 7, 49–72.

Thompson, M., Ellis, R., and Wildavsky, A., 1990. *Cultural Theory,* Westview, Boulder, CO.

Uphoff, N., 1992. *Learning from Gal Oya, Possibilities for Participatory Development and Post-Newtonian Social Science*, Cornell University Press, Ithaca, NY.

Von Bertalanffy, L., 1968. General systems theory: a critical review, in *Modern Systems Research for the Behavioral Scientist*, Buckley, W., Ed., Aldine, Chicago, 11–31.

9 On the Edge of Chaos — Crafting Adaptive Collaborative Management for Biological Diversity Conservation in a Pluralistic World

Jon Anderson

CONTENTS

INTRODUCTION

Adaptive management for the conservation of biological diversity requires understandings of, and abilities to influence, biophysical and socioeconomic processes and systems. These systems can be seen as structurally coupled, mixed, or unified. However, due to traditional separation of disciplines, they have often been treated

0-8493-0020-7/01/$0.00+$1.50

separately (Gunderson et al., 1995). Nevertheless, social scientists and ecologists working separately have come to almost parallel and similar understandings of some of the key elements and characteristics of both the biophysical and socioeconomic subsystems. These system characteristics include their inherent unpredictability, complexity, contingency, variation, and dynamics. "Complex adaptive systems" include such seemingly disparate systems as ant colonies, tropical forests, local communities, and national economies. The study of these systems also shows that they are adaptive — that they have the ability to undergo self-reorganization as the result of stress and that this reorganization and the emergent properties are not random (Waldrop, 1992).

Along with an emerging understanding of these systems, there is growing concern for the global and local "ecochallenges," such as global warming and biodiversity loss, and the livelihood challenges including millions of people who live in poverty (Roling and Wagemakers, 1998; Anderson, 1998). These challenges reflect a need to act urgently before certain thresholds of "no return" are reached where the environment is permanently compromised. Challenges include overexploitation of marine fisheries and fishery collapse, global warming, utilization of close to one half of the Earth's land and water resources, and large-scale extinction over the past several decades (Lubchenco, 1998).

Urgency is thus uneasily combined with complexity. There is mounting concern that classical models of management of natural resources and biological diversity may not be parts of the solution but parts of the problem. The characteristics of complex systems fit poorly with monolithic, static, and simplistic interventions and institutional arrangements.

This chapter first attempts to embed biological diversity in complex adaptive systems and to explore some elements of emerging understandings of the nature of complex adaptive systems. It then focuses on the pluralistic aspects of these systems and attempts to identify key elements for building robust and resilient management systems. Finally, the chapter looks at emerging approaches for undertaking adaptive management in pluralistic settings.

COMPLEX SYSTEMS

This section briefly describes some elements of the emerging understanding of both ecosystems and social systems. These understandings share commonalities that are then described as integrated complex adaptive systems.

Biological diversity, defined elsewhere in this book, is often described as having three main levels — genetic, species, and ecosystem or habitat diversity. At some levels biodiversity preservation and conservation outside of ecosystems, *ex situ*, is possible but often problematic. Much of the biodiversity conservation work requires *in situ* efforts at preserving habitats and species diversity at the ecosystems and landscape level. Biological diversity is embedded in ecosystems, directly or indirectly. Where it is threatened (and where humans can have impact) is at the intersection with the social system — those systems that are in proximity and impacted by humans. Biological diversity management requires the management of complex systems. Diversity is not only an emergent property of these systems but a constituent component. There is feedback between levels and types of diversity and ecosystem robustness. Diversity seems key to ecosystem function and adaptation, and ecosystems are key to biological diversity.

Ecosystems

Over the past decades the understanding of ecosystems has changed significantly. The vision of ecosystems was previously characterized by such concepts as steady state and equilibrium. Ecologists have always been aware of the importance of natural dynamics in ecosystems, but historically the focus has been on successional development of equilibrium communities (Pickett and White, 1985). In forestry, it was often assumed that development followed fairly predictable successional paths and that for certain fixed environmental conditions there was a single optimum climax state. More recent and emerging understandings emphasize the contingency of successional stages and the complexity and dynamic state of ecosystems.

History–contingency. For many systems, it appears that even given the same operating rules and principles as well as similar starting conditions, small differences in initial conditions can have dramatic effects on "final" outcomes. Some forested ecosystems, affected by an historical event such as a storm or logging, may in fact have their successional trajectory irreversibly changed. Ecosystems have many possible endings — there is more than one possible "climax" state (Holling, 1995) Complex adaptive systems have many possible endings — there is more than one possible sustainable land use for any particular piece of land (Anderson et al., 1999).

Dynamics. Forested systems are always undergoing stress and adapting. The field of patch dynamics encompasses the study of disturbance and its importance in ecological processes. Some ecologies remain "stable" only because of periodic, more or less unpredictable, disturbances. The ponderosa pine lands of the western United States are maintained through periodic burning and are, in a sense, a fire climax. Elimination of this burning changes the ecology and forest composition radically and over relatively short periods of time.

Complexity. There is an increasing understanding of the complexity of ecosystems. An intensive and extensive web of interactions exists between interdependent species and individuals. It is these interactions that give rise to potential for reorganization and adaptation — and to a certain robustness (Holling, 1995). However, there is a limited understanding and ability to understand these interactions and linkages.

Unpredictability. The complexity of interactions makes it very difficult if not impossible to predict how ecosystems will react to stress and disturbance. The introduction of the chestnut blight into North America totally eliminated one of the main constituents of the wooded ecosystem. However, instead of leading to a total collapse of the system, these ecosystems have recovered and the previous and actual systems are structurally and functionally remarkably similar. Projections done in the 1960s and 1970s of the total deforestation of Gambia turned out by the mid-1980s to have been greatly exaggerated. Surprise and change are constants. Gould (1996) stresses the unpredictability and contingency of any particular event in evolution.

Open systems. Another characteristic of these ecosystems is that they are essentially open. They are not closed systems that cycle through stages without external input. Systems are impacted by local disturbances such as storms and fires. But they are also affected and partly determined by distant outside factors such as global warming and El Niño events.

Variation. The variation in ecosystems is important. Gould (1996) points out the need to focus upon variation within entire systems, and not always upon abstract measures of average or central tendency. These more common measures can be a side product of expansion and contraction in the amount of variation in a system and "not a consequence of anything moving anywhere." This argument then throws doubt in the sense of linear progress, to a singular higher entity, at the same time that it values variation and particularity.

Diversity is both a contributing factor and an outcome of complexity. Frequent disturbances may be necessary for maximizing biological diversity. In those areas where disturbances have been eliminated, diversity tends to diminish. Mature woodlands tend to have a fewer number of species than highly disturbed and patchy systems. Increased diversity also tends to make systems more robust and better able to adapt to stress.

Much of the dialogue about sustainability is prefaced on an approach to the ecochallenge that promotes a shift from the present unsustainable scenario to a future sustainable state. Given what is known about ecosystems it may be futile to attempt to reach such a state. In fact "sustainable state" may be an oxymoron. Sustainability is perhaps not a "unique state" to be achieved "once and for all," but is a continuous adaptive process.

To reflect better the complexity of systems, some colleagues at CIRAD* have shifted their discourse from sustainability to viability to describe better the processes and range of potential "sustainabilities." Instead of the view of sustainability as a state, they emphasize a dynamic and transforming view of systems.

> [T]he notion of viability allows one to escape from … static and deterministic hypotheses. … It allows at each instant a plurality of potential solutions, without a particular optimum; these evolutions depend on the state of the system or the history of the system itself (Le Roy et al., 1996).**

Local Social Systems

Much of the literature on biological diversity conservation recognizes a tight coupling with the social system. Initially, people and their economic activities were conceived of as threats to biodiversity. Recently, more participatory approaches have been used to "enhance biodiversity conservation through approaches that attempt to address the needs, constraints and opportunities of local people" (Wells and Brandon, 1993). Much of this work has made fundamental assumptions about the nature of local communities — their skills, interests, motivations, and homogeneity.

---

* CIRAD is the *centre de cooperation internationale en recherche agronomique pour le development* — in Montpellier, France; colleagues are in the forestry program.

** Author's translation from French.

In a parallel sense to understandings of ecosystems, understandings of local communities are moving from equilibrium/consensus-based ideas to a dynamic and pluralistic understanding (Leach et al., 1997; Leach, 1999).

Key concepts of this emerging understanding are as follows:

History and dynamics. In the past there has been a tendency of outsiders to assume that local communities have no history (Wolf, 1982). However, at present there is a greater sensitivity to the historical itinerary of communities and how it affects and determines their present status and that of the environment in which they live. The history of communities is replete with contingent events such as the existence of charismatic leaders.

Dissensus and conflict. Much of the literature promoting the involvement or participation of local communities assumes that they are fairly homogenous and consensual. However, recent work has stressed the conflictual, differentiated, and pluralistic nature of communities. Leach et al. (1997) point out that, in fact, conflict rather than consensus characterizes many local communities that may be involved in biodiversity conservation. Brown (1998) has attacked the naive notions of participation and Enters and Anderson (1999) have questioned several key assumptions of decentralizing biodiversity conservation to local communities including those about the homogeneity of local communities.

Pluralism: Within any society or community, a variety of groups exist with different, autonomous, and sometimes conflicting interests and experiences. These differing views are resistant to reduction to a common perspective. Groups tend to be autonomous and work in parallel (Rescher, 1993). Essentially there is no central authority to control or manage the system even though in many cases a single entity attempts to impose its vision of the world on the others. Communities are differentiated and, even when there is cooperation, latent differentials and conflicts underlie this cooperation.

Consensus: Rescher's (1993) work on pluralism raises many questions about consensus as a concept allowing one to judge options in a postmodern world. Consensus can be valuable if achieved for the right reasons. Unfortunately, these reasons are difficult to assemble — consensus is difficult to achieve for any nontrivial questions of substance. In fact, in some cases the search for consensus can be harmful if it masks legitimate differences and disagreements and diminishes the power of checks and balances.

Learning: The understanding of how learning takes place and therefore how education needs to be undertaken is evolving. From a "banking approach" where knowledge is "deposited" in the learner's mind, there is a trend to establish more active roles and responsibilities for learners. Knowledge cannot be deposited, but the conditions for self-realization can be created. This throws into question the feasibility of imposing and transferring knowledge. Learning is contingent and internally driven. It requires a respect for different kinds of knowledge and an appreciation for multiple sources of information (Biggs, 1991). Le Roy et al. (1996) discuss also the "plural logic" of many local and traditional societies.

Open systems: In today's world there are no closed social systems. Globalization and liberalization have contributed to linking communities throughout the world and making them increasingly interdependent. Few communities are totally self-sufficient.

Calls for making natural resource management more effective by making it more appropriate and reflective of natural and social realities have existed for some time (Taylor and Soumare, 1986). More recently, there has been a questioning of the understanding of both the social and natural realities. Leach et al. (1997) question the "assumptions of homogenous, consensual 'communities' and the existence of stable, universally valued 'environments' and the potentially harmonious relationship between these." The integrative study of complex adaptive systems may help to generate ideas that could contribute to better management for biological diversity.

Understanding of Complex Adaptive Systems

Several groups, such as the Santa Fe Institute and the International Institute for Applied Systems Analysis, have been working to understand complex systems better. The term *complex adaptive systems* is used to describe many in the social and biological realm and can be applied to coupled or unified systems.

Complexity occurs as many individual agents interact with each other in various ways. The richness of these interactions allows the system as a whole to undergo spontaneous self-organization. Complex self-organizing systems are adaptive. They do not passively react to events.

Adaptation is the processing of feedback, the changing of internal models so that they correspond more closely with experience. Every complex adaptive system possesses a level of dynamism that makes it qualitatively different from complicated but static objects. As Waldrop (1992) describes it, the key elements of complex adaptive systems include:

- Many levels of organization — With agents at one level serving as the building blocks for agents at a higher level;
- Constant revising and rearranging of their building blocks as they gain experience — Learning evolution and adaptation are the "same";
- No practical way of finding the optimum — The most they can ever do is to change and improve themselves relative to what other agents are doing; complex adaptive systems are characterized by perpetual novelty;
- Rich webs of interactions among multiple agents.

Melanie Mitchell (in press) states that there are "many systems in nature in which a very large collection of relatively simple agents, operating with no central control and limited communication among themselves, collectively produce highly complex, co-ordinated and adaptive behaviours." They tend to be "a network of many agents acting in parallel" and the "control of a complex adaptive system tends to be highly dispersed — any coherent behavior arises from competition and co-operation among the agents themselves."

These latter characteristics highlight the pluralistic nature of these systems. Anderson et al. (1999) have attempted to deal with the issue of pluralism particularly in forestry and from the societal perspective. They have noted that pluralism can be defined as the existence within any society or system of a variety of groups or agents with different, autonomous, and sometimes conflicting interests, values, and perspectives. Further, these differing views cannot be reduced to a common perspective by the reference to an absolute standard. This work has also emphasized that groups or agents tend to be autonomous and work in parallel and that no central authority can control or manage the system.

The nature of systems as outlined here obviously has tremendous implications for how they are "managed." If the ecosystems and social systems themselves are truly dynamic and adaptive, then effective management must be similarly so. To be effective, management of biological diversity within ecosystems must reflect the nature of these systems. A management system that is static and fixed will fail over the long haul to be effective in managing a system that has adapted and evolved. Since these systems are in themselves dynamic, adaptive management must parallel this trait.

Given the complex and conflicted nature of local communities, the potential effectiveness of the new conventional wisdom on biodiversity conservation approaches that are based on the needs and desires of local communities can be called into question (Enters and Anderson, 1999). Contemporary rural communities have been assumed to be homogeneous and stable. Local tenurial, knowledge, and management systems are considered suitable for conservation and local people are uniquely interested and skilled in sustainable forest use and conservation. These assumptions need careful questioning. Methods are needed that more ably deal with conflict, plurality, and dissensus.

Although complex adaptive systems have many related characteristics that are difficult to separate because of their systemic linkages, the remainder of this chapter focuses mainly on the aspects relating to pluralism. This pertains to the presence of many autonomous agents and the impossibility of central control of complex systems.

PLURALISTIC ASPECTS OF SYSTEMS

This section examines the pluralistic aspects of complex adaptive systems in which biodiversity is embedded. The role of pluralism is somewhat paradoxical as it contributes to forces that can break down or inhibit cooperation and collaboration while also providing for essential elements of robustness and adaptability. Robustness is critical to self-assembling or self-repairing systems. Three key characteristics stem from the fact that many autonomous agents are in play: (1) redundancy, (2) self-correction, and (3) mutual learning. Pluralistic management systems recognize these characteristics of complex adaptive systems and attempt to bring a range of autonomous groups together in an interdependent way.

Giles Gunn's (2000) commentary on William James's (1907) brand of empiricism points out two aspects of this thought that are important in dealing with pluralism in biodiversity conservation. First is the "affirmation of a universe whose parts are not only multiple and diverse but always changing and often untidy." The second aspect relates to the democratic aspects of pluralism — "its public aim is not only to dissolve the hierarchical protocols and formalities that define topics and styles of typical disputation but also to enact the informality, directness and frankness of a dialogue between equals." In a sense, this is one of the keys to practical biological diversity management — that monistic, rational approaches built on vertical hierarchies of command and control must give way to conversations and negotiations between equals. Many can admit to the diversity and variety of experience and of perspective on biological diversity management. However, the tendency is to consider this variety as being somehow in error —the result of an inadequate understanding. There is one true understanding, and other views must be brought into line with this.

In terms of management organizations and systems, there is also often a recognition that a variety of organizations exist but the relation between them is often seen as a hierarchical or dependent one. In other words, a powerful organization like a forest service might recognize the existence of other actors but believe that these actors should and must work within a framework laid out by them. Agreements and contracts are signed for discrete activities and inputs, but other organizations are seen as proxies for the forest service. A more realistic view that has been seen is to accept autonomy in other groups and have conversations as equals.

It is useful to examine these three characteristics of pluralistic systems further: redundancy, self-correction (adaptation), and mutual learning.

Redundancy. Redundancy in a system means that a structure can have several functions or a function can have several structures. The existence of redundancy is sometimes viewed as inefficient since in a "perfectly tuned" system having this "duplication" increases costs. However, in many ways this "sloppiness" or untidiness is necessary for evolution, creativity, and adaptability. The importance of redundancy can be obscured by concerns about wasteful duplication. This calls into question facile critiques of duplication that have been popular in the development discourse.

Gould (1993) stresses redundancy and multiple use as the handmaidens of creativity (adaptability). "If each organ had only one function (performed with exquisite perfection), then evolution would generate no elaborate structures, and bacteria would rule the world. Complex creatures exist by virtue of slop, multiple use and redundancy." Many vital functions are performed by two or more organs, and one can change so long as the other continues to play the needed role. People can breathe through both their nose and mouth — and thank goodness, or everyone would be dead of colds. Evolution wins its required flexibility thanks to messiness, redundancy, and lack of perfect fit.

Pervasive redundancy makes adaptation and evolution possible. If it did not exist in complex systems, these systems would become brittle and tend toward extinction.

Self-correction and self-improvement. Effective management systems are aided by self-correcting mechanisms that control abuses and promote feedback. A system that provides institutional checks and balances is therefore at an advantage. It is crucial that systems have functions that are separated and autonomous. Systems with less than several autonomous groups often begin to act as monopolies, become rigid, and lose resiliency and credibility. The dismantling of Government Forestry Institutions in New Zealand in the early 1990s may be an example. While part of the impetus for the changes were economic and financial (downsizing government bureaucracies and cutting public costs), it is clear that the loss of forest service credibility was a major factor. "Although the rationale for the reforms was based on efficiency, the key drivers were the separation and clarity of functions." (Aldwell, 1998). In the reform process, functions were separated and new organizations formed. Crucial to this effort was separating production functions from conservation functions, limiting public sector involvement to "public goods" and market failure areas, separating funding from implementation and delivery, and devolving some authority and responsibilities to local and nongovernmental organizations (NGOs). While it may be too early to judge adequately the extent of these reforms, it does appear that transparency has improved and there is a greater scope for self-correction and increased credibility (Aldwell, 1998).

Mutual learning. Steele et al. (1999) state, "Only innovative human capacity building approaches, especially those that give attention to both individual and organisational learning, can address the complex challenges of sustainability and development. Stimulation of more learner-centred, interactive and critically reflective modes of learning is essential." Adaptive management clearly requires the processing of feedback and learning in many dimensions. Learning can be optimized through interactions between different knowledge systems. Learning is especially productive at the interface of two or more conflicting ideas. An example of this type of learning is the case of a conflict between the local community and Cornell University over procedures for the disposal of carcasses from the veterinary school. The school had announced a procedure to the local community whose immediate reaction was "not in my back yard." A new solution came about because of the conflict and the clash of the two perspectives. Neither group would have come up with it alone. (G. Thomas, personal communication.)

Pluralism promotes redundancy, self-correction, and learning — aspects that are key to any attempt to manage adaptively for biodiversity conservation. While one might promote these aspects individually, for example, by setting up systems for mutual learning, this would be a supply-driven approach that would be difficult to implement effectively and efficiently in the absence of an underlying commitment to and existence of institutional pluralism.

CRAFTING ADAPTIVE MANAGEMENT SYSTEMS IN PLURALIST SETTINGS

SUSTAINABILITY IN PLURALISTIC SYSTEMS — BUILDING ROBUST AND RESILIENT MANAGEMENT SYSTEMS

Conventional natural resource management systems reflecting "equilibrium centred, command-and-control strategies" (Gunderson et al., 1995) and institutionally based on the dominance of expert authorities and on narrow scientific knowledge paradigms, seem fundamentally ill-equipped to handle the sets of understandings and characteristics of complex adaptive systems. ACM is part of the search for management systems and approaches that coincide more closely with these characteristics and that are better able to adapt to and influence the underlying processes. There are many existing elements of response — coming from the fields of participation, institutional analysis and development, economics, and others. ACM is an evolving field that attempts to develop management systems that are sensitive to the underlying nature of the biophysical and socioeconomic situation and benefit from a range of existing tools and methods.

Margoluis and Salafsky (1999) have defined adaptive management as having three basic elements:

- Testing assumptions — Which is about systematically trying different interventions to achieve a desired outcome;
- Adaptation — Which is about systematically using the information obtained through monitoring to take action to improve interventions; and
- Learning — Which is about systematically analyzing and documenting process and results, and integrating lessons into institution-level decision making and sharing with broader communities.

A problem with much of the literature on adaptive management is that while it rightly puts the emphasis on learning and communication it inadequately describes the motivation — the why and how people should undertake such learning. This type of continuous learning with limited reliance on fixed frames of reference has high basic and transaction costs and is extremely difficult to promote. Hirst (1997) has pointed out that without models, one is condemned to constant learning — and people have amply shown that they are not very good at constant learning. Although flexible models may be necessary, groups need incentives and an environment in which learning and communication is subtly encouraged. One way to do this is to attempt to set up pluralistic institutional platforms to force dialogue.

Building Adaptive Collaborative Management

The pluralistic nature of governance needs for complex adaptive systems has implications for approaches to ACM of biological diversity. These pertain particularly to command-and-control approaches and consensus-based approaches.

APPROACHES, METHODS, AND TECHNIQUES

A series of approaches has been developed and tested that seem more relevant and appropriate to attempting ACM of complex systems. Several of these are noted in Daniels and Walker (1999) and Anderson (1999). They include appreciative inquiry, search conferencing/participative design workshops, constructive confrontation, collaborative learning, transactive planning, patrimonial mediation and subsidiarity, and communities of interest and open decision making. These approaches seem to share several characteristics. The integration of the following characteristics into attempts to manage biological diversity may help craft systems that are more responsive and effective.

Multistage process. The nature of complex adaptive systems indicates that approaches are needed that are interactive and iterative and recognize the need to pass through a variety of stages and phases. Clement (1999) describes the political process in pluralistic settings as a "never ending series of proposals and counterproposals, advances and withdrawals." Linear approaches and processes may be incompatible with the interactive and pluralistic nature of these systems.

Communication (facilitating collective action). A key to emerging patterns of cooperation and collective action is the ability to communicate. Approaches that have used participatory communication and communication for development techniques that promote "conversations and negotiations among equals" have usually fared better than those that use communication in a top-down manner to support a particular group or perspective.

Fullness of time (past and future). The contingent aspect of complex adaptive systems clearly argues for approaches that consider the past trajectory of these systems. In addition many approaches, such as patrimonial mediation (Babin et al., 1999), have shown that visioning futures and possible improvements aids in mobilization of groups in committing to improvements.

Experimenting and learning (facilitating interactions). Methods that promote the conservation of biological diversity must be able to adapt to adaptation, and to experiment and process feedback. Daniels and Walker's (1996) work on collaborative learning, which integrates soft system methodologies and lessons from conflict management, is one example. These approaches facilitate and encourage interactions between knowledge systems and different groups. There is an acceptance that the outcome of these interactions is not known *a priori* as is the case with some command-and-control approaches.

Leveling the playing field — power sharing (building pluralism). In most pluralistic settings some groups have more power than others in spite of the fact that they are fundamentally needed by the system. Many participatory approaches help build the capacities of marginalized or less powerful groups. This enables these groups to play a more active role in management and thus to improve its robustness. Proactive approaches to empowering different groups are needed if checks and balances are to function adequately.

Not consensus dependent. The concept of consensus is problematic in pluralistic settings. The rush to consensus can be unrealistic given the nature of communities and ecosystems. Many participatory methods seem to be consensus dependent and this can limit their effectiveness in pluralistic settings. Techniques that recognize and give voice to differences tend to be better suited.

Improvements to situations and processes — not final states. There has been a tendency to assume that one must move from an unsustainable state to a sustainable one. Some development literature as well as that on conflict, stresses on the other hand that steady states are not likely and that one should work for incremental improvements in present situations and processes instead of new states.

Coalition networks. Since control is dispersed in these types of systems, influence in managing them often depends on the formation of coalitions. Coalitions by their nature are temporary and often informal. They exist only as long as it benefits their members, sometimes only as long as it takes to make a single decision. They differ from partnerships in that they are issue based, not structurally based, and are openly temporary.

Site specificity (lessons from history). Brown (1998) states that "no one model of collaborative management can be offered which can be applied indiscriminately regardless of context." Gould (1996) also stresses the need to value variation over generalizations. Most successful approaches do not attempt to impose generalized and detailed models, but are careful to understand and work within the specificity of the site and ecosystem. Appreciative inquiry techniques, for example, seek to develop a detailed local understanding of social and ecological systems.

These approaches and understandings have implications both for policy and for institutional arrangements. Policy development will have to be more pluralistic, site specific, dynamic, and adaptable. This is true of the policies themselves — they should be viewed as temporary improvements and not final states. Policy will have to encourage the participation of a broad range of sometimes difficult to identify stakeholders. Some rural development strategies are doing this — for example, the PROAGRI* in Mozambique clearly attempts to be institutionally pluralistic. However, it is less clear whether this is the encouragement of diversity more than an encouragement of pluralism — the dependency relationships between the government and other actors remains vague and the motivations seem economic rather than the desire for a broad richness out of a web of autonomous agents. One way this can be done is to limit the power and the monopoly abilities of organizations, at least of those within the realm of control of the policy makers. Another approach would be to encourage recognition and support for a range of informal and nonclassical organizations including those in civil society.

The creation of institutional arrangements — that foster pluralism and dialogue and communication — is essential. Recently, much has been written about the creation of platforms or forums that allow diverse stakeholders to come together (Roling and Wagemakers, 1998). These platforms are significantly different from traditional hierarchical institutional relationships, which reinforce dependence.

CONCLUSIONS

Pluralism and adaptation are unavoidable. Ecosystems — in which biological diversity is embedded — are complex adaptive systems with strong elements of pluralism and with constant adaptive capabilities. Management systems that do not reflect these traits cannot expect to be effective over the long term.

For those involved in trying to "manage" or develop management systems for ecosystem management and biological diversity conservation, the lack of central control, the presence of many autonomous yet interdependent agents, the high dispersion of power and interests, may make the system seem to be on the edge of chaos. However, these characteristics may be the very ones that promote sustainability and viability.

The governance of these systems requires more than an acceptance of pluralism; their sustainability and adaptability require it. As Murray Gell-Mann is quoted as saying, "the world will have to be governed pluralistically or not at all" (Waldrop, 1992).

A continuing concern about these types of situations involving local communities and ecosystem management is that of equity. It is feared that in the absence of a "beneficent manager" the less powerful will be left out. There are several considerations

* PROAGRI is the Portuguese acronym for the 5-year National Program for Agricultural Development in Mozambique. It is the strategic instrument that is being used to transform the agricultural sector.

to be noted here. First, as part of the "web" weak actors are sometimes more powerful than other actors give them credit for — see Scott's *Weapons of the Weak* (1985). Second, beneficent managers often do not understand other actors and are often blind to their needs and their autonomy — thus inadvertently causing dependency and harm. Third, the simple fact of recognizing pluralism and accepting the lack of a central controlling power gives the weak more voice and visibility. Finally, actors are free to form coalitions and one set of actors can help build capacity in others to hold their own in the system. Pluralism gives the poor and less powerful an identity recognizing that they have something to say and offer that is on a par with others. It is a first step in empowerment and self-actualization. While pluralism is not without ideology it goes beyond tolerating differences and focuses on valuing and being enriched by them (Esack, 1997).

Although pluralism may offer improved chances for achieving equity, this goal is more a normative one than a realistic one. Equity is itself dynamic. Achieving equity would require overcoming strong existing biases, resource allocations, education, and social structural barriers.

The greatest need may be to come to a dynamic understanding of systems and learn better how to influence rather than to try and control them.

ACKNOWLEDGMENTS

The author is grateful to Norman Uphoff, Director of the Cornell International Institute for Food, Agriculture and Development (CIIFAD) and Lini Wollenberg, of the Center for International Forestry Research (CIFOR) for helpful comments on an early draft of this chapter.

REFERENCES

Aldwell, P., 1998. Restructuring of the forest research in New Zealand: review, impacts and prognosis: the case of the new Zealand Forest Research Institute Limited, in *Emerging Institutional Arrangements for Forestry Research*, Enters, T., Nair, C. T. S., and Kaosa-ard, A., Eds., FAO, Bangkok.

Anderson, J., 1998. Accommodating multiple interests in local tree and forest management: Some observations from a pluralistic perspective, paper presented at the 1998 international course on local level management of trees and forests for sustainable land use, September 23–October 5, International Agricultural Centre, Wageningen, The Netherlands.

Anderson, J., 1999. Four considerations for decentralised forest management: subsidiarity, empowerment, pluralism and social capital, in *Decentralisation and Devolution of Forest Management in Asia and the Pacific*, Enters, T., Durst, P., and Vistor, M., Eds., FAO, Bangkok.

Anderson, J., Crowder, L. V., and Clement, J., 1999. Accommodating conflicting interests in forestry — concepts emerging from pluralism, *Unasylva*, 49(194).

Biggs, S. D., 1991. A multiple source of innovation model of agricultural research and technology promotion, Overseas Development Institute Agriculture and Administration, Research and Extension Network, Paper 6, London.

Babin, D., Bertrand, A., Weber, J., and Antona, M., 1999. Patrimonial mediation and management subsidiarity: Managing pluralisms for sustainable forestry and rural development, in *Pluralism and Sustainable Forestry and Rural Development,* FAO, Rome.

Brown, D., 1998. The limits to participation in participatory forest management, paper presented at the 1998 international course on local level management of trees and forests for sustainable land use, September 23–October 5, 1998, International Agricultural Centre, Wageningen, The Netherlands.

Clement, J-C., 1999. The political and institutional aspects of pluralism in forestry, in *Pluralism and Sustainable Forestry and Rural Development,* FAO, Rome.

Daniels, S. and Walker, G., 1996. Collaborative learning: improving public deliberation in ecosystem-based management, *Environ. Impact Assessment Rev.,* 16, 71–102.

Daniels, S. and Walker, G., 1999. Rethinking public participation in natural resource management: concepts from pluralism and five emerging approaches, in *Pluralism and Sustainable Forestry and Rural Development,* FAO, Rome.

Esack, F., 1997. *Qur'an, Liberation and Pluralism*, Oneworld Publications, Oxford.

Enters, T. and Anderson, J., 1999. Rethinking the decentralisation and devolution of biodiversity conservation, *Unasylva,* 50(199).

Fairhead, J. and Leach, M., 1998. *Reframing Deforestation Global Analysis and Local Realities: Studies in West Africa*, Routledge, London.

Gould, S. J., 1993. *Eight Little Piggies Reflections in Natural History,* Penguin Books, London.

Gould, S. J., 1996. *Full House the Spread of Excellence from Plato to Darwin*, Three Rivers Press, New York.

Gunderson, L. H., Holling, C. S., and Light, S. S., Eds, 1995. *Barriers and Bridges to the Renewal of Ecosystems and Institutions*, Columbia University Press, New York.

Gunn, G., 2000. "Introduction" to *Pragmatism and Other Writings,* James, W., Penguin Classics, New York.

Hirst, P., 1997. *From Statism to Pluralism*, UCL Press Limited, London.

Holling, C. S., 1995. What barriers?, in *Barriers and Bridges to the Renewal of Ecosystems and Institutions,* Gunderson, L. H., Holling, C. S., and Light, S. S., Eds., 1.

James, W., 1907/2000. *Pragmatism and Other Writings*, Penguin Books, New York.

Leach, M., 1999. Plural perspectives and institutional dynamics: challenges for community forestry, paper presented at the International Agriculture Center Executive Seminar on "Decision-making in a natural resources management with a focus on adaptive management," 22–24 September, Wageningen, The Netherlands.

Leach, M., Mearns, R., and Scoones, I., Eds., 1997. Community based sustainable development: consensus or conflict? *Inst. Dev. Stud. Bull.* (Sussex), 28(4).

Lee, K., 1993. *Compass and Gyroscope. Integrating Science and Politics for the Environment,* Island Press, Washington, D.C.

Le Roy, E., Karsenty, A., and Bertrand, A., 1996. *La securisation fonciere en Afrique — pour une gestion viable des ressources renouvelables,* Editions Karthala, Paris.

Lubchenco, J., 1998. Entering the century of the environment: a new social contract for science, *Science,* 279, 491–496.

Margoluis, R. and Salafsky, N., 1999. Adaptive Management of Conservation and Development Projects: Transforming Theory into Practice, An Example Course Curriculum, Ref. 39, available at www.bsponline.org/publications/showhtml.php3?39.

Mitchell, M., in press. Analogy-making as a complex adaptive system, in *Design Principles for the Immune System and Other Distributed Autonomous Systems*, Segal, L. and Cohen, I., Eds., Oxford University Press, New York.

Perrings, C. A., Maler, K.-G., Folke, C., Holling, C. S., and Jansson, B.-O., 1995, *Biodiversity Conservation Problems and Policies*, Kluwer Academic, Amsterdam.

Pickett, S. T. A. and White, P. S., Eds., 1985. *The Ecology of Natural Disturbance and Patch Dynamics*, Academic Press, Harcourt Brace Jovanich, New York.

Rescher, N., 1993. *Pluralism: Against the Demand for Consensus*, Clarendon Press, Oxford.

Roling, N. G. and Wagemakers, M. A. E., Eds., 1998. *Facilitating Sustainable Agriculture: Participatory Learning and Adaptive Management in Times of Environmental Uncertainty*, Cambridge University Press, Cambridge.

Scott, J. C., 1985. *Weapons of the Weak: Everyday Forms of Peasant Resistance*, Yale University Press, New Haven, CT.

Steele, R., Nielsen, E., and Mbozi, E., 1999. Community learning and education in a pluralistic environment: implications for sustainable forestry, agriculture and rural development, in *Pluralism and Sustainable Forestry and Rural Development*, FAO, Rome.

Taylor II, G. F. and Soumare, M., 1986. *Strategies for Forest Development in the West African Sahel: An Overview,* Rural Africana, East Lansing, MI.

Waldrop, M. M., 1992. *Complexity the Emerging Science at the Edge of Order and Chaos,* Touchstone, New York.

Wells, M. P. and Brandon, K. E., 1993. The principles and practices of buffer zones and local participation in biodiversity conservation, *Ambio.,* 22, 157–162.

Wolf, E. R., 1982. *Europe and the People without History,* University of California Press, Berkeley.

10 Authority and Scale in Political Ecology: Some Cautions on Localism

Ronald J. Herring

CONTENTS

INTRODUCTION

Any consideration of conservation is framed by the dependence of natural systems on interaction with human systems. Conservation of biodiversity typically relies on public authority. But public authority at what scale of operation? How is public authority to protect natural systems constituted, realized on the ground at various levels (from village commons to nation states to global soft-law regimes), contested and undermined by pressures of livelihood and profit? Among these conunundra are those related to scale and authority. This chapter primarily treats the issue of scale, but the politics of nature* inescapably confronts authority; indeed, the two are

* Use of the term *nature* is risky — loaded with extraneous baggage of cliche and contestation, not to mention cultural loading — but purposeful in this chapter. Literature on ecological systems, politics, and policy too commonly runs the risk of validating reductivist views of nature — flattened to "natural resources," already incorporated into property systems, utilitarian norms, and use routines. That language naturalized what Karl Polanyi (1944/1957) called the thoroughly unnatural: commoditization of the biophysical world. Obviously there is no essentialist meaning of the word *nature*; what people construct as natural in any particular time reflects their experience, interests, "involvement" with the biophysical world (Lukacs, 1923, p. 324). William Cronon's (1991) salubrious and bucolic "second nature" was culturally constructed — from real terrain, myth, and memory — to ease the psychic and physical strains of urban satanic mills. Although Schama (1996) notes that environmental historians have "lamented the annexation of nature by culture" (1996, p. 12), in fact there is no alternative. No political ecology is complete without a phenomenology of nature. This is the greatest single lacuna in the understanding of authority in nature.

0-8493-0020-7/01/$0.00+$1.50

intertwined through considerations of legitimacy on the one hand — what scale of authority is acceptable to citizens on the ground? — and science on the other — how are claims to managerial authority in nature validated?

The question of scale is inescapable. Ecological systems impose imperatives of scale, or level, independent of alternative principles of social ordering; authority over and about nature first confronts scale. Societies draw administrative boundaries for reasons of convenience, cultural praxis, power, or simple historical residue. Natural systems seldom coincide with any of these products. Nature mocks administrative grids.

Nation states as a locus of authority are under attack. Localism as an argument for appropriate scale is currently popular with nongovernmental organizations (NGOs), academics, and some foundations and development agencies. Understanding that paramilitary command-and-control systems of aggressive states have often served neither conservation nor people, a discernible, and plausible, celebration of localism in scale and indigeneity in knowledge has taken the field. The irony of contemporary development thinking is that a premature celebration of the local (the product of what Pranab Bardhan, 1995 calls the "anarcho-communalists") — displacing responsibility from bad states to good communities — reinvents precisely the scale structure that has kept local oppression alive in many settings, including the southern United States (Herring, 1998).

Although the implied critique of control of nature by centralized states is experientially grounded and has obvious power, an emergent orthodoxy of localism runs unacknowledged risks. The concrete danger is that populist conceptualization of a fortuitous confluence of local knowledge, local practice, local institutions, and local communities is coming to premature closure around a celebration of the local. State denigration as a hangover from the liberalization *Zeitgeist* that constituted the "Washington consensus" on development has aided a counterreaction that displaces the question of authority by simply changing its locus — from bad states to good localities (often well endowed with "social capital"). This reversal of fortune of the state in the field of nature is ironic, for it threatens to obscure the core meaning of ecology itself — the interconnectedness of processes across levels.

This chapter is written in the belief that orthodoxies and fads regularly cycle through intellectual and policy communities, closing opportunities for progress. Any critique of localism confronts a powerful, in places hegemonic, vision of conditions for the preservation and restoration of nature: a robust triad of indigenous knowledge (ecological, institutional, and ethical), effective community, and decentralized political institutions. The critical countervailing narrative questions the romanticization of "local knowledge" on epistemological grounds, the assumption of "community" as social reality, the reality and stability of values rooted in local nature, and the dangers inherent in ceding public authority to local power.* Joining the strands of

* A schematic overview that does not do too much violence to the complexities of these positions is John Dryzek's (1997) book on environmental discourses; see also Sinha et al. (1997). A superb precis of the noble ecological other view is contained in Baviskar (1995, p. 44–47). Strands of these competing narratives will be discussed throughout the text. Arun Agrawal (1995) terms the local knowledge community the *neo-indigenistas*.

this conflict, the author suggests that the focus should be on authority as manifest in a dual sense: epistemological and political, each implicated in the terrain of the other.

PROBLEMS OF AUTHORITY IN NATURE: THE CLAIM OF THE STATE TO RIGHTFUL CONTROL

Preservation of nature rests on and evokes competing evaluative systems, grounded increasingly on authoritative claims about biodiversity — its value, fragility, conditions for persistence. Yet commoditized nature is in many ways the core of the legitimation practices of the state: the stuff of economic development, processed into use values, priced in markets. This is Karl Polanyi's (1944/1957) "great transformation" to market society — a long historical struggle, which in the contemporary world is accelerated by the victory of market economics as a science of growth. Yet authority in (especially) poor societies is buttressed by the claims of the specifically developmentalist state: the public good of economic growth legitimates state power. For the state to set aside swatches of "nature," or significant elements thereof, is to undermine this claim based on material progress. There are always opportunity costs. The knowledge disputes are profound: how does one value nature? Who has the authority to make the evaluation?

Answers to questions of authority in nature are too often avoided or assumed in the policy literature in the guise of being "practical." Nothing is more impractical than proceeding from a bad model. Shifting the conversation to "user groups" and "stakeholders" confronting problems of "natural resources" both sounds tidier than it is and obscures some of the most difficult problems in nature.

Authority, whether or not acknowledged, is at the bottom of much of the debate *about* nature as well as the conflict in and around natural systems. That conflict derives from two interdependent but distinct problems of authority: first, as legitimated power, in the ordinary language sense of political authority, and, second, in the epistemological sense, as in "scientific authority." Because states deploy science as the legitimation for their managerial regimes in nature, science too often becomes tainted with this instrumental use. The conceptual flattening of science to state claims of authoritative knowledge joins the reductivist flattening of a complex and robust nature to "natural resources," both buttressing the case for localism.

The notion that natural systems will be degraded by human use absent centralized political authority has long been dominant. To avert the "tragedy of the commons" is to constitute public authority: "Leviathan" in Hardin's original formulation, building on Hobbes's classic view of the state of nature in which everyone suffers for lack of authority. Significantly, Hardin's influential statement on human behavior and state power was published not in a journal of sociology, but in the periodical *Science*.* States from the ancient to the colonial to the contemporary have used the

* Hardin (1968); for a brief overview of the controversies provoked by the model, see Herring's piece for *Items* 1990. Elinor Ostrom's (1986) germinal statement stressed the flaws in the "tragedy model" and the potential for social learning.

presumed tragedy as legitimation for intervention, command, and control; these systems are now almost universally under attack as elitist and undemocratic (in Oregon as in Assam). Arguably, such systems of control by necessity reduce diversity of cultural and institutional forms and of the knowledge embedded therein. This reduction in diversity may be as dangerous in the social sphere as in the biological.* But is the obverse true? Does taking power from a central state enhance prospects for preserving natural system functions?**

State claims to managerial authority in nature rest overtly or covertly on knowledge claims legitimated by science. The tragedy of the commons is a classic case, but lacks the power of knowledge derived from the more established sciences; there are no Nobel prizes for sociology. Social science views on institutional desiderata are easier to dismiss as interested and ideological than are prescriptions from the natural sciences; yet all politically authoritative statements about public goods in nature and means of attaining them are rooted in ecological science.

Social science may be dismissed by the public at large; natural science has more authority (although clearly not without contests from below even in the most schooled societies). Yet the systematic study of natural science as a social and cultural phenomenon argues that scientific authority is, like political authority, constituted socially, contested, embedded in organizations, interests, and politics. It has its blind spots and reversals of conventional wisdom, its needs to claim more than is established, its resistances and conventions. Nowhere is this reality more evident than in ecological prescription, where the risk of not acting often precludes the level of certainty one would expect of, say, pharmaceutical effects.

In challenging the centralized Leviathan state in nature, critiques of "Western science," prominent in the ex-colonial world, join critiques of the institutional solutions premised on superior knowledge claims of specialized state agencies — managerial science. These critiques are reinforced by powerful currents emerging from intellectual cultures of the "West " even as the periphery of the world system complains of the hegemony of "Western science." Western science is counterposed to local knowledge. Yet, complicating the apparent dualism between local and external knowledge is the preference often *locally* manifest for the latter as superior, modern, efficacious. There is the common irony of outsider intellectuals in quest of local knowledge encountering among those who possess it a determination to appropriate modernity with all its technological and informational powers.***

These models of ecological knowledge may be ordered chronologically but more importantly recycle through space contemporaneously. Modern science dismissed local knowledge at the same time that consolidating nation-states prevailed over the

* James Scott's *Seeing like a State* (1998) argues the functional inevitability of rationalization, standardization, and simplification in bureaucratic state practice. On colonial arguments for village incompetence and the necessity of centralized control, see Guha (1990).

** Decay of the state, and of authority generally in Russia, for example, seems not to enhance freedom the way advocates of laissez-faire would argue, nor to improve substantive outcomes in public health or provision of social overhead capital or other common-good functions. Denigration of state power does not produce a reflexive solution in decentralization.

*** See, for example, Amita Baviskar (1995, p. 171–172); also, current work of Ann Gold and Bhoju Ram Gujar on education in Rajasthan, in process (personal communication).

agreed "idiocy of village life," as Marx had it. Included in the claim of village backwardness were widely held views of political incompetence, ignorance, parochialism.* Claims to superior knowledge both legitimated and disguised colonial interests (Guha, 1989b).

Resistance to the political effects — and errors — of centralizing modernization eventually undermined the hubris of this "dominating" or "colonizing" knowledge.** Contributing to this delegitimization was the persistent phenomenon of the state claiming science in its legitimation of control and criminalization of subsistence routines in nature (e.g., Peluso, 1992, on Java; Guha, 1989, on the Himalaya), such that "traditional" struggles of peasants against state and market took on an element of opposition to the hubris of "scientific" forestry.

The association of state and hubris (which became "science" in a transmogrification initiated by the state itself and perpetuated by critical theorists) facilitated a counterrevolution against the state, which points to community-based activities as solution to a multitude of social ills. Rather than sinks of ethnic hatreds, oppressive inequalities, and brutal practices, localities emerged as "communities" sociologically and as sources of lost institutional and technical knowledge epistemologically. In parallel, "social capital," lodged in the mundane associational life of face-to-face communities, trumps state power and undergirds governance. In its incarnation in the development assistance community and many foundation operatives' understanding, the model implies seamless communities, effortless knowledge, and frictionless institutions.

Much of the discourse on this subject either lacks history or assumes dichotomous histories of North and South. Simon Schama (1996) notes:

> [T]hough environmental history offers some of the most original and challenging history now being written, it inevitably tells the same dismal tale: of land taken, exploited, exhausted; of traditional cultures said to have lived in a relation of sacred reverence with the soil displaced by the reckless individualist, the capitalist aggressor. And while the mood of these histories is understandably penitential, they differ as to when the Western fall from grace took place.

Bacon and Descartes are popular choices.

> For some historians it was the Renaissance and the scientific revolutions of the sixteenth and seventeenth centuries that doomed the earth to be treated by the West as a machine that would never break, however hard it was used and abused (Schama 1996, p. 13).

---

* Jawaharlal Nehru said, "A village, normally speaking, is backward intellectually and culturally and no progress can be made from a backward environment." Dr. Ambedkar, a prime drafter of the constitution of newly independent India, said, "What is the village but a sink of localism, a den of ignorance, narrow-mindedness and communalism? ... I hold these village republics have been the ruination of India." In Mitra (1997, p. 3–4); Robert Wade's *Village Republics* (1988), on the economic conditions for collective action, rehearses the reasons for theoretical suspicion of local institutional capacity. Marguerite Robinson's 1988 book, subtitled *The Law of the Fishes,* demonstrates culturally grounded views of a Hobbesian world of local politics.

** Apffel-Marglin and Apffel-Marglin, 1996.

Colonialism is a popular choice. In nationalist historiography — such as Guha's (1989b) *The Unquiet Woods* — the fall is clearly associated with colonialism. The agent is forestry *science*, but in Guha's account science is merged with commercial interests which drive state land control. Guha writes (p. 195):

> While accurately pinpointing the inability of Western science to come to grips with the eco-crisis, the alternative proposed by this school implies a return to pre-industrial modes of living — a vision perhaps as elusive as Western science's claim to bring material prosperity for all.

The convergence here is diagnostic. "Science" personified as actor fails to deal with ecocrisis. Science earlier legitimated (and in the narrative stands for) state interests in criminalizing subsistence routines of peasants. Not only is science sullied by its incorporation into the hubris of paramilitary land managers, but it is burdened with *claims* "to bring material prosperity to all." With this view of science as foil, it is not surprising that local knowledge wins. Science as an agnostic method is replaced by a science implicated in instrumental state claims. This position then denies the emancipatory meaning of ecology — a meaning strategically deployed in biodiversity discourse. The science of ecology, for example, offers the grounds for new valuation of nature — beyond the reductivist "natural resources" forestry science sought to manage sustainably. If everything is connected to everything else, particular parts of nature cannot be readily reduced to their market value. Ecosystem functions require exploration and valuation. This understanding grounds a new politics, as global NGOs interested in preservation link with local activists against destructive state and capital. Nevertheless, oppositions of the kind Guha implies create additional grounds for celebrating the local: if global, or "Western," science is one of the problems, "local knowledge" is one of the solutions.

Sequestering nature in "protected areas" raises the knowledge and authority issue in starkest terms: how authoritative is the science that necessitates curtailment of use (whether grazing animals, thinning deadwood, or collecting grasses and other forest products), sets limits and thresholds and allocates dispensations among "multi-use" objectives? The model of large sequestered "wilderness" areas, pioneered in the United States, is both appropriated and rejected by less industrialized countries. In the periphery, integration of many people and their livelihoods is so bound up with nature that sequestering is difficult morally and politically. This problem in North America was solved by extermination and sequestration of peoples, not parks; comparable solutions today are hindered by NGOs protecting human rights and "cultural survival" and an international conscience premised on a fundamentally different moral order. Like polluting industrialization, confiscation and mobilization of natural resources were less problematic when the currently rich countries were doing it. A counterdialogue in the South argues from a different science against impractical set-asides of large areas for wilderness protection in favor of "miniparks" with corridors. Does anyone, local or supralocal, ever understand the ecology in sufficient detail and with sufficient certainty to make these decisions? The honest answer is that despite the deployment of knowledge claims by state agencies, local user groups, and NGOs, one seldom has the knowledge to answer these questions with confidence.

Struggles over sequestration of nature have led, in many parts of the world, to persistent challenges to state claims over territory, sometimes to a militarization of nature. The venerable link between insurgency and wild places is strengthened as state intrusion arouses local opposition. International soft law regimes such as the World Heritage Convention place an outer ring of authoritative certification of biological value in sequestered nature on a collision course with pressures of "development" and politics. Whether the *global* sequestration regime of the World Heritage Convention adds extra layers of assistance, expertise, and normative weight, or increases by its very distance nationalist and subnationalist backlashes against protected areas, is an open question.*

LOCAL KNOWLEDGE

Scientific tradition founded in positivism isolates and elevates a singular "real" knowledge built upon independently confirmable tests of theory-driven hypotheses. Such knowledge would be value-free and therefore in a different sphere from normative notions of the good, the desirable, the unacceptable. In recent decades a powerful discourse among scholars, activists, and NGO practitioners concerned with environmental degradation has argued for paying attention to alternative systems of grounded, contextualized knowledge of indigenous origins. Rejecting the impetus of outside "dominating" knowledge, this discourse suggests that locally generated knowledge offers appropriate solutions at the local level.**

A complex set of interrelated claims is woven through this discourse. Two strands among these are of particular interest. First, there is the claim that local knowledge, based on generations of experience, captures system dynamics (ecological wholes) in nuanced ways unavailable to the disaggregating analytical method presumed to characterize "Western" science. Local knowledge then takes into account variable geophysical particulars that are inaccessible to distant forest officers, planning commissions, or state operatives generally.*** In Jim Scott's (1998) recent work, "seeing like a state" necessitates flattening of local variance to produce the "legibility" on which bureaucratic control depends. Second, local knowledge is imbued with values (normative notions) from which it cannot be detached; claims for local knowledge, then, are both positivist and normative — although to practitioners the dichotomy itself is alien.

Criticism of this stance as obscuringly romantic or nostalgic comes from different directions. It is clear that ecological wholes are not so easily perceived from a very local view, no matter how intense the view; if amphibians are really falling

---

* The evidence on India is mixed. Certainly the Manas World Heritage Site, and to a lesser extent Kaziranga, has experienced resistance to the very idea of Delhi's — let alone Paris's — control of local terrain, fueling secessionist mobilization (see Herring and Bharucha, 1998).

** This discourse within environmental studies is, of course, part of a much broader postmodern and postcolonial questioning of the meta-narratives of Euroamerican thought such as linearity, progress, development, and the adequacy of representation. The literature is vast; for various histories relevant to authority in nature, see, for example, Apffel-Marglin and Marglin (1996); 1996; Arnold (1996); Banuri and Marglin (1993); Berman (1981); Glacken (1967); Merchant (1980).

*** However, see Robbins (1998) for an example of a village-born forest officer whose job and life entail conflict between the government story and his own contradictory local knowledge.

victim to increased ultraviolet radiation permitted by deterioration of the ozone layer, local diagnosis is not likely to be adequate, nor would the global Montreal Protocol be an imagined solution. Moreover, the evidence of local knowledge — extensive knowledge of categories and uses — typically is lacking in mechanisms of system dynamics. Local knowledge is weakest as an alternative ecology, strongest as a utilitarian inventory. Moreover, it is not clear that pro-nature values — or even the correlates of restraint and limited needs — either are dominant among "eco-system people"* or are surviving the seemingly inexorable erosion of particularistic values globally through commercialization and marketization. On the knowledge side, the movement seems to be the very pragmatic slow replacement of rules of thumb and folk wisdom with verifiable knowledge.

The presumed gulf between two modes of knowing nature is deeply rooted in European intellectual traditions, and has been thoroughly explored and historically contextualized.** Gellner succinctly characterizes local knowledge as "common-sense knowledge — socially embedded, unspecialized, unsystematized"; its opposite, with its capacity for exponential leaps, he calls "proper, abstract, socially liberated science" (p. 15). This affirmation of the properly abstract implies a freedom from the local. The distinction between socially embedded and socially liberated knowledge parallels the tradition of Karl Mannheim's social science: ideology reflects socially embedded interests and is thus the enemy of both liberation and truth (a connection carried on in the critical theory of Habermas).***

Among the most challenging of questions is the extent to which claims that "environmentally beneficial knowledge must be local knowledge" have a verifiable basis or place in positivist science. In other words, is indigenous knowledge "really right"? Alternatively, is it of another order, literally incommensurable?**** Can one tell? What are the criteria?

---

* In explaining her own voyage of disillusionment during fieldwork in western India, Amita Baviskar (1995, p. 173) concludes: "While reverence for nature is evident in the myths and many ceremonies which attempt to secure nature's co-operation, that ideology does not translate into a conservationist ethic or a set of ecologically sustainable practices." She is quick to point out the structural features that render both difficult, but nevertheless laments "the claims made about *adivasis* [tribals] by environmental activists keen to incorporate them into a discourse about environmental movements and sustainable development."

** The contrast is starkly posed in different terms by Vitebsky (1995): "How can one compare a kind of knowledge which is local with one which is global? If the latter is also universal or absolute, it should thereby negate the former by logical necessity so that there will be nothing further to discuss. If not, this already implies a recognition that the arena of operation of 'knowledge' is not just truth, but also appropriateness and applicability. It is through power relationships between knowledges that some of them can be turned into forms of ignorance" (1995, p. 183).

*** Various authors suggest varying chronologies and organize their descriptions according to varying motivating perspectives, including disciplinary and/or political orientations such as gender studies or postcolonial studies. On the former, see Merchant (1980); Shiva (1991); for the latter, see Alvares (1992); Bennagen and Lucas-Fernan (1996); Leach and Mearns (1996); Nandy (1990). Adorno and Horkheimer's (1991) vision of "instrumental reason" evokes much of what in more recent environmental knowledge discourse is labeled "decontextualized" or "epistemic." They pose an intimate connection between instrumental reason's domination of nature and its domination of subjects (Adorno and Horkheimer, 1991; Jay 1984).

**** A major area in which this has been extensively explored in many cultural and ecological settings is agriculture; see, for example, Clay (1988, Latin America); Lansing and Kremer (1993, Bali); Richards (1985, Africa). Appadurai's discussion of numbering and measuring in colonial India (1996) as well as his earlier work on agricultural estimations (1989) are relevant here.

Rather than denigrate "science" as a reified enemy of local knowledge, social scientists interested in conservation would do better — positively and normatively — to embrace the basic premises of ecology in their own work while incorporating appropriate science in their explanations. Ironically, much work in the social sciences misunderstands environmental phenomena for lack of attention to the natural sciences. Social scientists not surprisingly find social causes for phenomena that have natural explanations, sometimes in the form of blaming the peasants when nature is the culprit.*

Reconciliation of contending theories of knowledge should not be impossible. Certainly, discourses of generalizing science uncomprehending of indigenous systems of knowledge are likely to fail in the service of conservation. Positivist science has no answer to the question of value — the core issue in nature as a provider of "public goods" or center of an ethical system. To the extent the alternative epistemologies implied by the affirmation of local knowledge insist on the entwining of, for example, cosmology, morality, and technology, the knowledge systems are not in conflict but incomparable. Moreover, it is clear that many incarnations of local knowledge — such as those in the *Honeybee* network centered on Gupta's group in Ahmedabad — are fully compatible with inductive hypothesis testing, even if the full range of scientific theory is not rolled out in every case of innovation. But to clear the air for dialogue, one must not confuse science defined properly as a system of epistemology and method with "science" appropriated for reasons of power or legitimation of hubris. Although resting on the aura of ecological scientific authority, much *policy* represents guesses, interests, muddles of state agencies. The contradiction is therefore not essentially epistemological, but rather inherently political.

A second consequence of the rejection of the positivist distinction between value and fact is refocusing attention on the uses to which local knowledge has been put — the values entailed in its use. Consider the common observation that slaughter of protected endangered species cannot proceed without local knowledge (and often local collaboration; e.g., Menon, 1996). Local knowledge in the service of commercialized destruction of nature is still local knowledge. Indeed, as in the iconic case of Robin Hood, the inability of the state to apprehend those it defines as criminals in natural settings is hindered first and foremost by the lack of local knowledge in the arms of the state. Yet, the association of wilderness with freedom from the state seems more heroic in the case of Robin's merry men (Schama, 1996) than in its contemporary incarnation in the form of drug dealers, poachers, smugglers, right-wing bioregionalists, and militia survivalists.

Science will not resolve the problem of valuation. Of the multiple possible valuations of nature — for producing game, aesthetics, biodiversity, jobs, ecotourists, danger, risk, and the like — selection and weighting are parts of a political process. How does the representation of "tribal" or "indigenous" peoples as stewards of

---

* To take but one example, Blaikie and Brookfield (1987, p. 37–49) discuss erosion in mountain Nepal as "a crisis of the environment or a crisis of explanation" and convincingly demonstrate the latter. The problem was not bad farmers, or inadequate local knowledge, as reams of literature had diagnosed and government officers had tried to redress, but plate tectonics, over which there is no policy domain. The effects of El Niño in aggravating the current fires in Southeast Asia may prove to illustrate the same phenomenon.

nature pick and choose among disputed histories of ethnicity formation, settlement, and conquest? How do states reify local knowledge in collaboration with international NGOs eager for validation of their premises? How much of the creation of "paper parks" all over the world is a function of representations by governments of global environmental correctness? It is in answering these questions that one begins to understand the relationships among knowledge, authority, and conservation.

SCALE

In the current discourse about governance and nature, localism has increasingly been seen as the antithesis of and answer to state interventionism — already discredited by the (historically) recent liberalization *Zeitgeist*. One set of questions then concerns the locus of authority — and thus the unit of governance. Linkages across units and means of convening effective supra-local communities where none exists become central because of the very nature of ecological systems. There can be no purely local community in ecological terms.

Since at least the 1970s, academics, activists, administrators, and grassroots organizations sharing concerns for ecological recovery and social equity, have sought and often found solutions of appropriate scale and technology in community-based institutions.* Effective administration and demonstrable successes are attributed to small-scale participatory strategies, to projects based on local perceptions. Contributing to the elevation of community solutions were dramatic instances of local resistance to megadevelopment and its ecological damages, as well as critiques by intellectuals and activists of externally imposed and alien knowledge systems — resonating with centuries of intellectual tradition in the "North" as well as the contemporary "South."** Ideologies of empowerment for poor people and goals of social equity converged with heightened respect for local knowledge, conservation ethics, and administration by communities.

But "community" invites a more critical look. The theoretical critique may take the form of methodological individualism or structural stratification theory. From the view of methodological individualism, communities experience great difficulties forming themselves, for reasons well established in the theory of collective action (Wade, 1988). From the viewpoint of stratification theorists, communities may by well formed, but formed for purposes of pursuing interests of dominant fragments, whether defined by gender, caste, lineage, class, faction, or other particularizing social characteristic. In Rousseau's terms, the expression of will in such communities does not represent the "general" will, only the triumph of some particular will. The

---

* See, for example, the contributors to Poffenberger and McGean (1996). On theoretical reasons for participatory success, see Uphoff (1992). Note that Uphoff's great success story of new identity formation as "farmers" rather than as ethnic partisans, in the midst of a local and national civil war of great violence, was *not* based on community, but on practiced organic solidarity in face-to-face social relations.

** The motif of authoritative knowledge imposed from above (by those who know better) as necessary for the good of all emerges in Glacken's (1967) detailed tracing of nature and culture in Western thought. See also Sahlins (1994), Schulte (1994), Thompson (1975) on local resistance to such control in European history; see Grove (1995); Sivaramakrishnan (1995) for the colonial project in relation to nature in India; also Guha, (1989b); on contemporary India, Kothari et al. (1996). For Java, Peluso (1992).

empirical critique is that, comparatively, communities exhibit quite a range of environmental values and knowledge; scale is no guarantee of either process or substantive outcomes. The property rights/wise use movement in the United States uses terror against both officials of the state and environmentalists in defense of their claim ("right") to use public lands for private advantage (see Helvarg's *War against the Greens*). Devolution of authority to these (locally rooted) interests would be environmentally disastrous.

Localism has a strong history in the United States, where any form of central authority is deeply distrusted. A civil war was fought over issues of "states' rights," meaning in part the right to maintain slavery as a social system in the 19th century. In environmental protection, the "sagebrush rebellion" pits local organizations and the local state against national law, policy, and officials. One example is worth citing. The Humboldt–Toiyabe National Forest — the largest in the country — has been the site of violence and threat. When a district ranger's office was fire-bombed, and later his vehicle in front of his house as well, the Forest Service had to remove him from the area as it could not guarantee his safety. His offense to local vigilantes was to enforce the law. In 1999, when the Forest Service closed a federal road in the area for conservation reasons, a member of the Nevada State Legislature organized a mob, supported openly by the lieutenant governor of the state, to reopen the road forcibly (Ruch, 1999). Attacks and threats against the Bureau of Land Management and Forest Service employees and facilities doubled between 1995 and 1998. The Nevada forest supervisor resigned, citing not only threats, but also various forms of harassment and social boycott (Ruch, 1999). What the locals wanted in this case was a not uncommon or unreasonable desire: the right to convert nature into "natural resources" and convert natural resources into commodities for sale. Conservation attempts by the national government — which claims that public lands are managed for a larger public than that recognized by local timber and grazing interests — are undercut by local power. Their very locality is converted to a claim of proprietorship: what right can a government thousands of miles away, in Washington, D.C., or its agents, claim or exert on their local terrain?

The critique of localism from ecological science has to do with the scale of ecological systems — inevitably supralocal and therefore necessitating supralocal vision, monitoring, and institutions. The normative critique is that local claims to autonomy preclude spreading the benefits of use of nature to a larger population — justice, either intergenerational or cross-sectional, is not necessarily served if locality implies ownership. If locality A's traditional terrain (which may well have been taken from people now residing in locality B centuries back) contains gold, and locality B's turf only sagebrush and scrub thorn, is the windfall a legitimate proprietary claim of A based on territorial propinquity? How large is the public in "public interest" or "public lands"? Where would the authority to adjudicate reside?

Consistent with the anti-state mood of these times, celebration of localism is somewhat ironically paired with globalism. The "international system" is proposed as a more appropriate level of governing nature than the nation-state. At the international level, soft law promises governance coterminous with a planetary ecosystem — and "global community." The implications for local governance of nature are profound and often opaque to state officials who, claiming to represent nation-states, sign documents in large foreign cities. Commoditization of whole species — not just their genetic material — has raised an increasing threat to biodiversity — an industry worth at least $15 billion (U.S.) annually and infiltrated by internationally organized crime. The global response is to regulate markets in some species but not others, creating a globally rigged market for endangered species: the Convention on Trade in Endangered Species (CITES). This global soft law regime depends ultimately on local authority and local knowledge. Powers of *certification* mandated under treaty provisions institutionalize knowledge — and thus provide new niches for corruption.* Can the forest officer or custom official, in Canada or in India, distinguish among almost identical species of fish, reptiles, amphibians, tubers, orchids? Can governments in poor localities restrict livelihoods on grounds of the global value of biodiversity of species? Do people anywhere care about the minute differences between similar minnows that separate species (the problem of the snail darter)? This distinction is premised on the existence of a potential public good (ultimately derived from the value of biodiversity), which public authority pursues, but it is one poorly received, or actively contested, by people whose livelihoods are affected (e.g., Kothari, 1997; Herring and Bharucha, 1998).

One consequence of efforts to enforce such laws is violence between bearers of national and local public authority and hunters, traders, herders, smugglers. It is not clear that states are winning. Nevertheless, global soft law regimes proliferate; states sign on as representatives of their respective societies and then set about answering the countervailing claims to authority in their own periphery. At the extreme, conflicts over landscape autonomy feed redefinitions of community, turf, and rights to secession. Governance at the local level seems undermined by attempts of states to fulfill obligations accepted as global actors: the rise of bioregionalism, "ecological ethnicity" (Parajuli, 1996), sagebrush rebellions, wise-use movements.

Declining capacity for governance endangers broader societal values, as recognized most forcefully by the 1997 *World Development Report* from the World Bank. Departing from its economistic core, the Bank has broached the messy and expansive question of public authority as a *sine qua non* condition for "development." But that report discovers a dilemma in the scale of public authority: hegemonic Leviathan states are dysfunctional, yet "certain dangers are inherent in any strategy aimed at opening and decentralizing government … [including] the risk of gridlock or of capture by … interest groups." Conceivably, "the crisis of governance that now afflicts many centralized governments will simply be passed down to lower levels" (World Bank, 1997, p. 130). This groping for scale itself indicates both the general absence of authoritative organizational science — one simply does not know — as well as the reality that scale is no guarantee of substantive outcomes.

Advocates of decentralization and local control sometimes forget that in the United States it took a (highly centralized, elitist) Supreme Court and the (not very participatory) 101st Airborne Division to get black children into the locally controlled

* "Cultivated" specimen of "wild" endangered species are exempt from CITES. Authoritative discrimination between cultivated from wild is left in the hands, in India, of Divisional Forest Officers. In the ambiguity of both plants and animals raised for trade vs. those rounded up by village children for price, the local authority's decision may be influenced by a small gratuity. See Herring and Bharucha, 1998.

schools and colleges of the South in the 1950s. Writing on Java, Nancy Peluso (1992) excoriates the "custodial paramilitary" model of state resource control, which produces "secret wars and silent insurgency" inimical to conservation (1992, pp. 235, 236), and yet stops short of endorsing village power structures, which maintain exclusion and inequality.* This is the central dilemma.

Governance is a relational construct; it presupposes some workable relationship between various levels of state and society implying some form of legitimacy over power — i.e., authority. Effective governance is then relational in a cultural as well as structural sense. It must be approached in an interdisciplinary way if it is to be intelligible. When Robert Wade (1988) writes of the economic conditions for collective action at the local level (in the aptly titled *Village Republics*), he finds precisely what is missing from the institutional and economistic perspectives. Successful village republics tend to work on issues of clear public good absent redistributive conflicts; possibilities for expanding the pie create community. Nature seldom cooperates; evaluations contend rather than converge. Moreover, cross-cutting arenas — inevitable in natural systems — render governance problematic in both its structural and cultural dimensions.

The global–local problematic is then indissoluble, just as the bumper stickers say. Global governance of the natural commons presupposes local governance. Yet global governance depends on — works through, reinforces, legitimates — nation-states in the first instance, reinforcing the centralism legitimated by scientific forestry more than a century ago. Paramilitary command-and-control systems have not only failed but have fed the ethnoregionalism that ironically undermines stateness, while simultaneously low-intensity conflict defeats conservationist goals. Yet it is difficult to imagine the formation of global environmental regimes functioning without nation-states, given the structure of the international system.

Expansion of centralized state claims has historically reduced the institutional and cultural diversity of local arrangements. The pernicious effect of the state on local accommodations to natural systems may then be both structural and cultural; centralized states reduce both the political space within which local communities can work institutional solutions to perceived problems and the authority of existing institutions. Just as reduction of biodiversity precludes options, traditional state control of nature reduces the richness of institutional and cultural diversity from which governance can be constructed. Yet, if ecological systems impel ever higher levels of articulation, reconstitution of the nation-state system itself may be the necessary condition for global authority in nature.

CONCLUSIONS

Conjunctions of knowledge and authority often lie at the heart of political ecology — getting the prices, institutions, and interests right is not enough for a policy-useful,

* Walt Coward has responded to this critique (personal communication) with the observation that localism implies solutions at the local level, not solutions based on local communities. Fair enough, but the trade-offs are not so easily dismissed; as long as the locus of authority in Arkansas was Little Rock, attacking racial oppression necessitated assertion of national values over local power, at the local level.

theoretically robust, and conceptually dynamic political ecology. Authority does not obey the law of conservation of matter and energy; it is constantly being created and destroyed.

The rationale for government itself — the normative legitimation of power — is inadequate provision of public goods, or broadly conceived "market failure." Nature evokes persistently contested notions of the constituents of a public good and their relative ranking. Bruce Babbitt, as Secretary of the Interior of the United States, said: "We can't get re-authorization of the Endangered Species Act by emphasizing snakes." Absence of consensus on the very existence of public goods (snakes?), and their relative ranking, undermines authority by challenging the knowledge claims of those who nominally wield it.

Much of the contest animated by uses and abuses of nature is then over systems of authority in a dual sense. In overt resistances to power over "natural resources" — arson, evasion, poaching, felling, smuggling, assassinations, insurrection, secession — material interests and livelihoods are at stake, as well as control of political space. Normatively, regulation is locally perceived as criminalization of daily life by a distant state — "squatting" for habitation, theft for harvesting, "poaching" for providing calories, and so on* — not as pursuit of some public good. These overt conflicts reflect and illuminate a prior conflict over authority — the normative basis of determining what is a right, what a theft, what a concession, and which representation of the value of nature is correct. Phenomenologies of nature are necessary to make sense of interests.

In the dispute over the value of nature, science has borne a special burden. "Scientific management" of "natural resources" — the double redaction — was a colonial project in much of the world, enforced through domination. The conflicts are well documented, and outlived colonialism. But the conflating of an agnostic method of knowledge production and testing — i.e., science — with the instrumental "science" which legitimates state managerial imperatives has introduced an unfortunate rift into political ecology, much of it between local actors and nation states.

This chapter has argued that authority in the political sense — legitimated power — is in the sphere of nature uniquely contingent on authority in the discursive sense. That is, the power claim inherent in environmental "management" is typically a claim that some public good will result from following a set of rules placed in motion by public authority, rules that are an embodiment of scientific knowledge. This claim to authority is rooted in systematic assessment of "tipping points," "irreversible damage," and "threshold effects" in complex systems — systems too large and complex to yield to purely local knowledge or to purely local authority. Local knowledge and local public authority both fail to "scale up"; ecological dynamics are not respectful of parochial boundaries.

The antinomies of localism, local knowledge, and authority can be captured by looking at one example of a common pattern, an appropriate place to end. There is an almost universal understanding of the problem of local complicity in violation

* One of the finest consistent explications of this view over a long historical period and multiple dimensions is Nancy Peluso's *Rich Forests and Poor People: Resource Control and Resistance in Java* (1992, p. 19).

of international soft law regimes of both sequestered nature and protected species by organized criminal activity. International law is meaningless without local cooperation. Writing of the one-horned rhinoceros protected by international agreement under CITES and within a World Heritage Convention (WHC) Natural Site — Kaziranga — Vivek Menon (1996, p. 65) concludes:

> As most poachers come from far-off villages, they need a fringe village in which to wait and bide their time, as well as to return to after the poaching. Also, there is a need for a local person who has knowledge of routes, location of anti-poaching camps, patrol times and routes, etc. Further, if arms are to be stored for some time … again a local is ideally involved. This local could be a corrupt employee of the Forest Department or a villager; examples of both cases are known and not uncommon. It is stressed, however, that to malign the entire Forest Department or body of villagers local to rhinoceros reserves would be most inappropriate and undesirable. Indeed, it is only with the full co-operation of these groups that poachers' assistants can be identified from among them and extracted.

Menon's disaggregation is necessary and valuable. There is a difference between subsistence routines and slaughter; this distinction is locally recognized. CITES and WHC as global soft law depend fundamentally on the local state, which is inevitably embedded in local society. The global is then local, as is the state in any meaningful sense; the surveillance of local state and society is mutual. The robust implications of Menon's poignant example are obscured by demonizing either the state or international environmental regimes in favor of a premature celebration of the local.

ACKNOWLEDGMENTS

This chapter is a revision of ideas from a working group and, as such, benefits from the group and from textual commentary of members Barbara Lynch, Sheila Jasanoff, Norman Uphoff, and, especially, Ann Gold (particularly in the section on knowledge, where the prose is now beyond disentangling), none of whom bears any responsibility for this version. The author acknowledges as well useful commentaries from Manoj Srivastava, Arun Agrawal, Subir Sinha, John Richards, Norman Uphoff, and, especially, John Schelhas.

REFERENCES

Adorno, T. W. and Horkheimer, M., 1991. *Dialectic of Enlightenment,* Continuum, New York.

Agrawal, A., 1995. Dismantling the divide between indigenous and scientific knowledge, *Dev. Change*, 26, 413–439.

Agarwal, A. and Narain, S., 1992. *Toward a Green World*, Centre for Science and Environment, New Delhi, India.

Alvares, C., Ed., 1992. *Science, Development and Violence: The Twilight of Modernity,* Oxford University Press, New Delhi, India.

Apffel-Marglin, F. and Apffel-Marglin, S., Eds., 1996. *Decolonizing Knowledge: From Development to Dialogue,* Clarendon Press, Oxford.

Appadurai, A., 1989. Transformations in the culture of agriculture, in *Contemporary Indian Tradition*, Borden, C. M., Ed., Smithsonian Institution Press, Washington, D.C., 173-186.

Appadurai, A., 1996. *Modernity at Large: Cultural Dimensions of Globalization,* University of Minnesota Press, Minneapolis.

Arnold, D., 1996. *The Problem of Nature: Environment, Culture and European Expansion,* Blackwell Publishers, Oxford.

Bandyopadhyay, J. and Shiva, V., 1988. Political economy of ecology movements, *Econ. Polit. Wkly.,* June 11, 1223–1332.

Banuri, T. and Apffel-Marglin, F., Eds., 1993. Who will save the forests? in *Knowledge, Power and Environmental Destruction*, Zed Books, London.

Bardhan, P., 1995. Research on poverty and development: twenty years after redistribution with growth, *World Bank Research Observer*, p. 59–72.

Baviskar, A., 1995. *In the Belly of the River: Tribal Conflicts over Development in the Narmada Valley,* Oxford University Press, Delhi, India.

Bennagen, P. L. and Lucas-Fernan, M. L., Eds., 1996. *Consulting the Spirits, Working with Nature, Sharing with Others: Indigenous Resource Management in the Philippines*, Sentro Para sa Ganap na Pamayanan, Inc., Quezon City.

Berman, M., 1981. *The Reenchantment of the World*, Cornell University Press, Ithaca, NY.

Blaikie, P. and Brookfield, H., 1987. *Land Degradation and Society*, Methuen, London.

Clay, J. W., 1988. *Indigenous Peoples and Tropical Forests: Models of Land Use and Management from Latin America*, Cultural Survival, Cambridge.

Cronon, W., 1983. *Changes in the Land: Indians, Colonists and the Ecology of New England,* Hill and Wang, New York.

Cronon, W., 1991. *Nature's Metropolis,* W. W. Norton, New York.

Dryzek, J., 1997. *The Politics of the Earth*, Oxford University Press, New York.

Feeny, D., Birkes, F., McCay, B. J., and Acheson, J. M., 1990. The tragedy of the commons: twenty-two years later, *Human Ecology*, 18, 1.

Ferry, L., 1992. *The New Ecological Order,* The University of Chicago Press, Chicago.

Gadgil, M. and Guha, R., 1992. *This Fissured Land: An Ecological History of India,* Oxford University Press, Delhi.

Gadgil, M. and Guha, R., 1995. *Ecology and Equity*, Penguin Books, New York.

Gadgil, M., Joshi, N. V., and Patil, S., 1993. Power to the people: living close to nature, *Hindu Surv. Environ.,* 58–62.

Gellner, E., 1997. Knowledge of nature and society, in *Nature and Society in Historical Context.* Teich, M., Porter, R., and Gustafsson, B., Eds., Cambridge University Press, Cambridge, 9–17.

Glacken, C. J., 1967. *Traces on the Rhodian Shore: Nature and Culture in Western Thought from Ancient Times to the End of the Eighteenth Century*, University of California Press, Berkeley.

Gold, A. G., 1997. Wild pigs and kings: the past of nature and the nature of the past in a rajasthani Princedom, *Am. Anthropol.,* 99(1), 70–84.

Grove, R. H., 1995. *Green Imperialism*, Cambridge University Press, Cambridge.

Guha, R., 1985. Forestry and social protest in British Kumaun, c. 1893–1921, *Subaltern Stud.,* 4, 54–100.

Guha, R., 1989a. Radical American environmentalism and wilderness protection: a third world critique, *Environ. Ethics,* 11, 1.

Guha, R., 1989b. *The Unquiet Woods: Ecological Change and Peasant Resistance in the Himalaya*, Oxford University Press, Delhi, India.

Guha, R., 1990. An early environmental debate: the making of the 1878 Forest Act, *Indian Econ. Soc. Hist. Rev.,* 27, 1.

Guha, R., 1994. Switching on the green light, *Telegraph* (Calcutta), 25, 10–94.
Guha, S., 1995. Kings, commoners and the commons, paper presented at the conference on Science and Technology Studies, April 28–30, Cornell University, Ithaca, NY.
Hardin, G., 1968. The tragedy of the commons, *Science,* 162, 1243–1248.
Helvarg, D., 1994. *The War against the Greens*, Sierra Club Books, San Francisco.
Herring, R., 1990. Resurrecting the commons: collective action and ecology, *Items,* 44, 4.
Herring, R., 1998, Celebrating the local: Scale and orthodoxy in political ecology, in *Governance Issues in Conservation and Development,* Hyden, Goren, Ed., Conservation and Development Forum, Gainesville, FL.
Herring, R. and Bharucha, E., 2001. Capacity, will and governance: India's compliance with international accords, in *Engaging Countries: Strengthening Compliance with International Accords*, Brown Weiss, E. and Jacobson, H., Eds., MIT Press, Cambridge, MA.
Ingold, T., 1992. Culture and the perception of the environment, in *Bush Base: Forest Farm: Culture, Environment and Development*, Croll, E. and Parkin, D., Eds., Routledge, London, 39–56.
Jay, M., 1984. *Adorno*, Harvard University Press, Cambridge, MA.
Kothari, A., 1997. *Understanding Biodiversity: Life Sustainability and Equity*, Orient Longman, New Delhi, India.
Kothari, S. and Parajuli, P., 1993. No nature without social justice: a plea for cultural and ecological pluralism in India, in *Global Ecology: A New Arena of Political Conflict*, Sachs, W., Ed., Zed Books, London.
Kothari, A., Singh, N., and Suri, S., 1996. *People & Protected Areas*, Sage, New Delhi, India.
Lansing, J. S. and Kremer, J. N., 1993. Emergent properties of Balinese water temple networks: coadaptation on a rugged fitness landscape, *Am. Anthropol,,* 95, 97–114.
Leach, M. and Mearns, R., Eds., 1996. *The Lie of the Land: Challenging Received Wisdom on the African Environment*, The International African Institute, London.
Lukacs, G., 1923 (1971). *History and Class Consciousness*, Merlin, London.
Luke, T. W., 1997. *Ecocritique: Contesting the Politics of Nature, Economy, and Culture,* University of Minnesota Press, Minneapolis.
Menon, V., 1996. *Under Seige: Poaching and Protection of the Greater One-Horned Rhinoceroses in India*, Traffic, India: WWF-I, Delhi, India.
Merchant, C., 1980. *The Death of Nature: Women, Ecology and the Scientific Revolution*, Harper and Row, San Francisco.
Mitra, S., 1997. Making local government work: rural elites, Panchayati Raj and legitimacy in India, paper presented at the Conference on Against the Odds: Fifty Years of Democracy in India, Princeton University, Princeton, NJ.
Naben, G. P. and Trimble, S., 1994. *The Geography of Childhood: Why Children Need Wild Places*, Beacon, Boston.
Nandy, A., Ed., 1990. *Science, Hegemony and Violence: A Requiem for Modernity*, Oxford University Press, Delhi, India.
Ostrom, E., 1986. How inexorable is the "Tragedy of the Commons"? Distinguished Faculty Research Lecture, Indiana University, Bloomington, April 3.
Parajuli, P., 1996. Ecological ethnicity in the making: developmentalist hegemonies and emergent identities in India, *Identities*, 3(1–2), 15–59.
Peluso, N., 1992. *Rich Forests, Poor People: Resource Control and Resistance in Java*, University of California Press, Berkeley.
Poffenberger, M. and McGean, B., 1996. *Village Voices, Forest Choices,* Oxford University Press, Delhi, India.
Polanyi, K., 1944/1957. *The Great Transformation*, Beacon, Boston [originally published as *The Origin of Our Times*].

Raghunandan, D., 1987. Ecology and consciousness, *Econ. Polit. Wkly.*, 22(13), 545–549.

Richards, P., 1985. *Indigenous Agricultural Revolution: Ecology and Food Production in West Africa*, Westview Press, Boulder, CO.

Robbins, P., 1998. Paper forests: imagining and deploying exogenous ecologies in arid India, *Geoforum,* 29(1), 69–86.

Robinson, M. S., 1988. *Local Politics: The Law of the Fishes*, Oxford University Press, Delhi, India.

Ruch, J., 1999. Nature's guardians still face disrespect, *The New York Times,* Dec. 22, A29.

Sahlins, P., 1994. *Forest Rites: The War of the Demoiselles in Nineteenth-Century France,* Harvard University Press, London.

Schama, S., 1996. *Landscape and Memory,* Vantage, New York.

Schulte, R., 1994. *The Village in Court: Arson, Infanticide, and Poaching in the Court Records of Upper Bavaria, 1848–1910,* Cambridge University Press, Cambridge.

Scott, J. C. 1998. *Seeing Like a State: How Certain Schemes to Improve the Human Condition Have Failed*, Yale University Press, New Haven, CT.

Shiva, V., 1991. *The Violence of the Green Revolution: Third World Agriculture, Ecology and Politics,* Zed Books, London.

Sinha, S. and Herring, R., 1993. Common property, collective action and ecology, *Econ. Polit. Wkly,* xxviii(27–28), 1425–1433.

Sinha, S., Gururani, S., and Greenberg, B., 1997. The "New Traditionalist" discourse of Indian environmentalist, *J. Peasant Stud.,* 24(3), 65–99.

Sivaramakrishnan, K., 1995. Colonialism and forestry in India: imagining the past in present politics, *Comp. Stud. Soc. Hist.,* 37(1), 3–40.

Thompson, E. P., 1975. *Whigs and Hunters: The Origin of the Black Act*, Pantheon Books, New York.

Uphoff, N., 1992. *Learning from Gal Oya: Possibilities for a Participatory Development and Post-Newtonian Social Science,* Cornell University Press, Ithaca, NY.

Vitebsky, P., 1995. From cosmology to environmentalism: Shamanism as local knowledge in a global setting, in *Counterworks: Managing the Diversity of Knowledge,* Fardon, R., Ed., Routledge, London, 182–203.

Wade, R., 1988. *Village Republics*, Cambridge University Press, Cambridge.

Williams, R., 1980. Ideas of nature, in *Problems in Materialism and Culture*, Verso, London, 67–85.

World Bank, 1997. *World Development Report 1997: The State in a Changing World,* Oxford University Press, New York.

11 Tenure and Community Management of Protected Areas in The Philippines: Policy Change and Implementation Challenges

Maria Paz (Ipat) G. Luna

CONTENTS

0-8493-0020-7/01/$0.00+$1.50

INTRODUCTION

The Philippines holds the dubious distinction in biodiversity circles of being the hottest of the hot spots — hectare for hectare, it has the highest number of threatened endemic species of any country in the world. It is a strong candidate for an extinction spasm (Heaney and Regalado, 1998). In a policy breakthrough in 1992, Philippines Congress passed the National Integrated Protected Area Systems Act (NIPAS Act), which recognized the rights of indigenous peoples within protected areas and accorded rights to long term forest occupants. It was a breakthrough that needed an even stronger follow through.

This chapter analyzes the legal, political, and ecological backdrop within which these policy shifts have occurred. It first examines the policy environment and legal history that led to the crafting of the NIPAS Act. The two branches of policy change in question are discussed in turn. The first pertains to recognition of historical undocumented tenure and refers to indigenous peoples or those who have been in occupancy of lands and domains as far back as memory goes. The second relates to a concession granted by the state as an equity and human rights gesture. Here, the state grants rights because it recognizes both its incapacity to enforce a total ban on occupancy in parks where settlements continue to encroach, and it recognizes that settlements have formed inside national parks in the absence of sufficient enforcement of the laws. Most significantly, the policy shift reflects the tenet that granting of tenure to indigenous peoples and long-term migrants honors their stewardship of the land and provides incentives to protect it. It acknowledgeds that indigenous conservation norms exist, which had nearly disintegrated because of the failure of formal legal policy to reflect and respect them.

The first section elaborates the history of disenfranchisement suffered by indigenous peoples and migrant farmers within national parks prior to the NIPAS Act. The second section outlines the policy shifts that led to recognition of rights in laws and in rules and regulations. The third section considers interpretations of these policies by policy makers and influential sectors in the Philippines, and through a common property resources framework. The fourth section highlights issues and concerns that surfaced from experience in attempting to implement the new policy to and secure justice for indigenous peoples, and tenure for long-term migrants into protected areas. The chapter suggests that while conditions for collaborative management have been imbued with policy recognition and community-based management carries with it the premise of adaptability, a bureaucracy that is accustomed to performing a regulatory function must be convinced of its new role before there can be true adaptive collaborative management (ACM).

The Philippines has an active nongovernmental organization (NGO) community. In 1994, many of the largest and longest-standing NGOs in the country formed the NGOs for Integrated Protected Areas Incorporated (NIPA, Inc.) for the implementation of the NIPAS Act of 1992. The action involved nearly the entire spectrum of social development and environmental NGOs in the country coming together on a common agenda. In 1994, NIPA, Inc. and the Department of Natural Resources formed the Conservation of Priority Protected Areas Project (CPPAP) to support community-based protection and management planning of ten priority sites in the Philippines that are considered most critical and representative of Philippine ecosystems. Under the terms of the partnership, both NIPA, Inc. and the governmental body are accountable to the Global Environment Facility (GEF), which will fund the project through December 2001.

This narrative is written from the perspective of the author's direct experience with the CPPAP, first as legal counsel from 1995 to 1998, and then as program manager in 1998 and 1999. The author's role involved interpreting policy and addressing implementation issues to achieve the objectives of the NIPAS Act that would allow for genuine community-based management. During this time the author also provided legal assistance to several independent conservation initiatives that represented efforts by civil society to influence the implementation of the NIPAS Act.* The CPPAP itself is a useful experiment in ACM, with respect to the functioning of the partnership and the changes in project design that were deemed necessary during its course. The dynamics between the Department of Environment and Natural Resources and upland dwellers generates further insight into conditions needed for ACM.

THE POLICY ENVIRONMENT

POLITICAL HISTORY

The Philippines ranks sixth in the world with respect to the number of globally threatened species that it harbors (Mallari et al., 1999). Ahead of it in the list are such large countries as Brazil, China, and Indonesia. With 30 million ha and 77 million people, there is almost no patch of land in the country left untouched by human activity. While habitat degradation occurs at an alarming rate and affects 97% of all threatened species (Mallari et al., 1999) the remaining biodiversity is still characterized as one of the world's highest, with more endemicity than most other places. Many of these endemic species are also critically endangered. Despite the population density and the small size and fragmentation of remaining wildlife habitats, the ranking of the Philippines in biodiversity importance serves as testimony that harmony and coexistence between populations of people and populations of animals in the wild is possible to a degree. Even as the built environment and urbanization infringes on remaining natural habitats, villages and traditional communities still demonstrate customs and social norms of reverence for nature. Consider, for example, the practices of indigenous tribes such as the Dumagat, which show the strength of oral tradition in their hunting and gathering activities. The policy environment at this writing, however, gives one little reason to hope that this coexistence will last much longer.

* These initiatives included Haribon Foundation's Mt. Isarog National Park Conservation Project, Luntiang Alyansa ng Bundok Banahaw's activities on Mt. Banahaw, and the campaigns of Balik Kalikasan-Babilonia Wilner Foundation on Quezon National Park.

Since its "discovery" in 1521, the Philippines has had 368 years of Spanish colonial rule, 37 years of American occupation, 11 years of the Commonwealth Government under the United States, with 5 of these years being under the Japanese forces in World War II, 20 years as a new Republic of the Philippines until 1966, and 20 years under the Marcos dictatorship. The last 12 years from 1986 leading up to the celebration of the country's centennial in 1998 has provided the greatest opening for law and policy that reflects the aspirations of the people. It would be reasonable to presume based on this quick glance at the last 500 years of Philippine history that access to natural resources has been less than equitable and ideal. The legal history, however, will tell a different tale and if read apart from the history of implementation may seem to have dealt with native residents fairly. From the Laws of the Indies throughout the 17th and 18th centuries to the jurisprudence of the American occupation from 1898 through 1945, the law has consistently recognized prior vested rights of occupants to land.* The massive disenfranchisement, displacement of indigenous peoples, and nonrecognition of vested rights throughout the Spanish and American occupation spanning the 16th century all the way to 1945 were more the result of a failure in implementing the law than a reflection of official colonial policy. This is due mainly to the reliance of colonial governments on documentation, which indigenous peoples and poor farmers did not participate in. This made the law itself apparently just, while giving free rein to the colonizers to amass landholdings in what were presumed to be vacant lands for lack of documents showing otherwise.

Indigenous Peoples and Ancestral Domains

Native title refers to the rights to land held by people occupying it since before the colonization of the Philippines. Such rights exist even without the benefit of documentation to prove them. The Indigenous People's Rights Act of 1997 defined these lands as ancestral domains.** The legislation was enacted following persistent advocacy by groups who sought to have the legal and historical rights to land held by occupants for time immemorial clearly enshrined in statute. Many of the areas set aside as national parks have been declared as such without the benefit of prior survey or census. It is no coincidence that many of these areas are occupied and cared for by indigenous peoples and have been for generations. The declaration of these lands as national parks runs counter to a long line of colonial law and jurisprudence

* During the Spanish period (1521 to 1898), the Laws of the Indies, which were imposed by Spain on its colonies, recognized ownership of lands already occupied by the natives. The Royal Decree of 25 June 1880 emphasized that "all persons in possession of real property were to be considered owners provided they had occupied and possessed their claimed land in good faith since 1870."

It was also during this period that jurisprudence first recognized the existence of native title by the same terms. This strong legal basis for native title was not adhered to, however, in practice.

** Ancestral domains — Subject to Section 56 hereof, refer to all areas generally belonging to ICCs/IPs. It shall include ancestral land, forests, pasture, residential, agricultural, and other lands individually owned whether alienable and disposable or otherwise, hunting grounds, burial grounds, worship areas, bodies of water, mineral and other natural resources, and lands which may no longer be exclusively occupied by ICCs/IPs but from which they traditionally had access to for their subsistence and traditional activities.

that respected native title. Despite such legal recognition, the lack of documentation on these rights has allowed the rampant practice of disenfranchisement and evictions of indigenous peoples from their ancestral domains.

There is jurisprudence to support the argument that the inclusion of ancestral domains within national parks constitutes administrative error and is, therefore, void with respect to lands held by natives since time immemorial. In *Krivenko v. Register of Deeds* (79 Phil 461, 481, 1947), the constitutional classification of lands was deemed to be exclusive. Since there are only four constitutional classifications of public land in the constitution, namely, national park, mineral lands, forestlands, and agricultural lands, it follows that whatever was classified as national park was assumed to have been public lands subject to classification. This logic cannot hold in the face of a recognition that time immemorial occupation implied vested rights that could not be taken away by proclamation of an area as a national park. Every proclamation, therefore, that covers undocumented ancestral domains and makes it a national park would be void as the classification system cannot apply to private lands.

Migrant Farmers

Not all occupants of national parks have been there since time immemorial. There are those who entered into national parks in waves of upland migration during World War II and afterward, driven by failed agrarian reform policies and a general scarcity of land and livelihood in the lowlands. These migrations of farmers into national parks have always been treated as illegal and punishable under all national park laws that led up to the NIPAS Act.

Land distribution in the country has created a situation wherein almost four out of five farmers are landless or nearly so, and the poorest 40% of farmers receive only one seventh of all farming income (Myers, 1992). Although many claim that conversion of forestland to cropland sounds the death knell for much of the tropical forest zone, attempts to halt peasant deforestation require attention to the root causes, which are fundamentally structural (Vandermeer and Perfecto, 1995). Widespread shifting cultivation, called *kaingin*, is usually brought about by landless peasants following the logging roads into the forests and copying the systems practiced by the authorities in the larger estates. These upland migrants regard traditional swidden agriculture that allows for fallow periods as tribal and therefore backward (Hurst, 1990). Hurst concludes that this destructive practice is brought about by development policies advanced by the very people who blame the poor farmer for deforestation. Shifting cultivation, which shifts as easily from culprit to sustainable practice depending on the political interests of the opinion maker, has been found to be a method in which human practices is adapted particularly well to the local environment (Eckholm, 1976). Green Forum, one of the clusters belonging to NIPA, Inc., has advocated since the late 1980s for "the transfer of the rights to forest resources to upland communities" (Ofreneo, citing Green Forum, 1988).

Sustainable traditional practices such as shifting cultivation, however, have been criminalized and driven to unsustainable levels by the encroachment of settlers and commercial extraction license holders in areas that presumably are being kept fallow.

These encroachments have been relatively easy because there are no legal documents or any other evidence that such lands are occupied and part of a sustainable cycle of shifting cultivation. Green Forum and others believe that legal recognition of native title would lead to better protection of natural resources through the protection of traditional or long-term occupants.

By 1988, 2 years after the fall of the Marcos dictatorship, a World Bank study on Philippine forestry stated that regardless of the history of occupancy, forestland occupants have been legally considered squatters, and their numbers have been chronically understated in official statistics (World Bank, 1989). This study became the basis for conditions in subsequent loans to the country and resulted in pressure on the government to recognize tenure for upland occupants. It was the financial resources of the World Bank also that led to the draft legislation for a protected area system that recognizes ancestral land claims and accords tenure for long-term occupants (Leonen, 1998).

GAINS AND SETBACKS ON COMMUNITY MANAGEMENT POLICY

THE NIPAS ACT

Despite legal recognition and jurisprudence honoring rights to ancestral domain, the refusal of indigenous peoples to document their vested rights through registration with authorities has affected the issuance of documents that overturn prior public rights in protected lands. It became evident that lack of legal tenure for indigenous peoples and the criminalization of occupants in national parks were not alleviating the situation of continued forest degradation.

Strong community advocacy and NGO lobbying finally led to congressional action in 1992 when the NIPAS Act was passed. It became the first legislative enactment explicitly to bar the Department of Environment from evicting cultural communities from their present occupancy or resettling them to another area without their consent (Sec. 13, NIPAS Act). The NIPAS Act further defined tenured migrant communities as "communities within protected areas which have actually and continuously occupied such for five (5) years before the designation of the same as protected areas in accordance with this Act and are solely dependent therein for subsistence." This definition clearly accorded rights to occupants who have been in the national parks since 1987 to exercise certain private rights therein.

To this policy breakthrough, there was a range of reactions from civil society from celebration to skepticism. Many treated the enactment as an opportunity as it became the impetus for the CPPAP conservation project. This opportunity galvanized the NGO community in the Philippines into action that resulted in the passage of Resolution No. 1 of the Philippine Council for Sustainable Development urging the secretary of the Department of Environment and Natural Resources to communicate its wish for Philippine NGOs to implement the project. The request was given due respect and the coalition of NGOs that demanded it became NIPA, Inc. in 1994 and the coimplementor of the CPPAP which pilot tested the implementation of the NIPAS Act.

Community-Based Forest Management

In another dramatic policy shift, the then Philippine president declared in 1996 that the commercial concession system benefited only those with financial and political clout and declared the pursuit and promotion of a community forest management policy (Severino, 1998). Executive Order 263 (1996) established community-based forest management as the national strategy to achieve sustainable forestry and social justice. It provided the legal backdrop for the granting of community access to forestlands under long-term tenure instruments, provided communities employ environment-friendly, ecologically sustainable, and labor-intensive harvesting methods. This was carried out through DENR Administrative Order No. 29 (1996), which provides for the issuance of community-based forest management agreements in public forest, multiple-use zones, and buffer zones of protected areas.

Even further, the legislature recognized the ownership of lands and domains of indigenous peoples in the Indigenous Peoples' Rights Act of 1997 (RA 8371). This law recognizes private ownership of such lands and domains, accords indigenous peoples the right to have such lands and domains titled under the Public Land Act, and even defines the indigenous concept of ownership as being the material basis of the cultural integrity of indigenous peoples (Sec. 5, IPRA). Ancestral lands and domains are thus recognized under this law as private but community property and belonging to all generations. Such lands and domains therefore cannot be sold, disposed of, or destroyed.

Philippine Forestry Prospects

A scenario of Philippine forests projected to 2010 shows the country with a 6.6% forest cover. This condition, according to the map scenarios produced by the Environmental Science for Social Change (ESSC, 1999), will occur if community forest management does not become an accepted strategy and social forestry is allowed to become a basis for land speculation. On the other hand, 19% forest cover is projected for the same year if, among other factors, effective mechanisms are developed to address the issues of indigenous peoples, migration, and land speculation in protected areas. The series of legislative enactments and presidential pronouncements favoring the recognition of rights of indigenous peoples and long-term forest occupants would seem to signal a new dawn for Philippine forests and national parks.

Implementation of the policies that were put in place to enable the second ESSC scenario to materialize would constitute the next major hurdle. However, the very persons in the bureaucracy who were tasked to undertake the legislated changes were those promoting the concepts and practices that were prevalent prior to the new policies. Despite the legislative and executive fiat, the thinking of the major players in government remained entrenched in the old formulations. The context in which these policies needed to be implemented was less than ideal, and might even be said to be hostile. There remained hugely disparate access to natural resources among classes, accompanied by a general mistrust of the legal system by the poor. There is routine disregard of some laws, and corruption remains a pervasive social ill according to the World Bank progress report on anticorruption activities in the Philippines (Doronila, 1999).

These factors can be readily demonstrated. In the selection of sites for the CPPAP, for example, a national park that was identified for inclusion in the project was Mt. Iglit Baco in Mindoro, which hosts threatened and endemic species such as the Tamaraw (*Bubalus mindorensis*) as well as the dwarf water buffalo. The indigenous peoples in the area, the Mangyans, opted to oppose the implementation of the NIPAS Act in their area as they did not believe the Act would benefit them. Although the policies of the last 10 years may appear progressive and potentially effective, they cannot erase centuries of disenfranchisement of indigenous peoples. The ruling in *Rubi v. Provincial Board* in 1919 (39 Phil 660) that the Mangianes are citizens of a low degree of intelligence, and Filipinos who are a drag upon the progress of the State was [discriminatory but] formed part of Philippine jurisprudence early in the century. In 1980, Bontoc chieftain Macli-ing Dulag was killed after resisting all bribes and threats made by the Presidential Assistant on Tribal Minorities in 1980 (Myers, 1992) for him to approve the Chico River Dam Project. Because of such histories, the Mangyans were reluctant to embrace laws such as the NIPAS Act.

In 1998, actor and then vice president Joseph Estrada was elected president of the Philippines. Early in Estrada's administration, reversals had already been observed in the implementation of these laws. The appointment of Secretary Antonio Cerilles to the Department of Environment and Natural Resources caused a sensation when he declared that there should be no occupants in the forest zone. Consequently, the budget of the Community-Based Forest Management Office was diminished (LRC-KSK, 1999), and legal instruments for implementing rules and regulations for the certification of tenured migrants have not been developed as of this writing, although the Asian Development Bank supported a project to do so. This has resulted in the widespread lack of proof of tenured migrant status as provided for in the NIPAS Act.

LEGAL INTERPRETATION OF TENURE UNDER NIPAS

Legal Recognition of Ancestral Domains

Although the NIPAS Act explicitly recognized rights of indigenous peoples over ancestral domains, the nature and character of those rights were not specified. The IPRA strengthened the rights of indigenous peoples with the express statement that ancestral domains are owned by indigenous peoples under the indigenous concept of ownership. Questions still arise, however, whether the NIPAS Act or the IPRA would prevail in the determination of the right and authority to manage ancestral domains. The state retains authority to exercise police power over lands privately owned by its citizens. Some sectors may interpret the NIPAS Act to mean that a protected area status should cancel out the indigenous ownership recognized under IPRA. Even more specific are questions relating to possible conflicts between ancestral domain management plans as approved by councils of elders and the management plans to be prepared under the NIPAS Act. The rights of the secretary of the Department of Environment and Natural Resources under the NIPAS Act would also have to be tempered by the authority of indigenous peoples over their lands. These questions, although critical for the effective implementation of the law, have not been seriously posed due to the failure of the agency that was tasked over 2 years ago to process ancestral domain titles to begin operation. The National Commission on Indigenous Peoples (NCIP) has been paralyzed by leadership struggles, administrative cases against commissioners, lack of a sufficient budget, and a petition that was filed before the Philippines Supreme Court to declare the law unconstitutional.

These obstacles amount to a failure to implement the IPRA. Legally, however, such rights as are guaranteed therein, as well as rights recognized in *Cariño v. Insular Government*, continue to prevail as law. What becomes patent are attempts by certain sectors to set back the policy gains represented by the NIPAS Act and the IPRA.

Tenured Migrants

There remain no specific implementing rules for the granting of tenure instruments to migrants in protected areas. Two items of confusion were generated by provisions in the law that establish the conditions for earning tenure status, particularly (1) the duration of settlement in the area and (2) the level of livelihood dependence on resources from the area. In the general implementing rules for the NIPAS Act, which did not cover the procedure and granting of tenure instruments, the department attempted to deal with these issues theoretically. Department Administrative Order No. 25 (1992), Sec. 50 of these rules states:

> Any person who has actually occupied an area for five (5) years prior to its designation as part of a protected area in accordance with Section 5(a) of the Act and is solely dependent on that area for subsistence shall be considered a tenured migrant. As a tenured migrant he shall be eligible to become a steward of a portion of land within the multiple use management or buffer zone of the protected area, and from which he may derive subsistence. Provided, however, that those migrants who would not qualify for the category for tenure shall be resettled outside the protected area.

Section 5(a) refers to the designation of all previously proclaimed national parks as initial components of the protected area system, thereby placing the reference of the 5-year occupancy requirement at the passage of the Act and making May 30, 1987 the operative date of occupancy to be considered a tenured migrant.

A third difficulty arises from whether to define a qualifying migrant as a community or an individual occupant of a household. Although no policy pronouncements have yet been made on the issue, the census conducted by the department has established occupancy on a per-household basis. Operationally, therefore, if a community has been within the national park prior to May 30, 1987, each household in that community needs to pass the 5-year occupancy requirement to qualify to stay.

The occupancy requirement raises a fourth issue — that of actual and continuous possession relating to the specific portion of the land occupied. Certain categories of protected area are inconsistent with human occupation, such as natural parks, strict nature reserves, and resource reserves (DENR MC 35, 1993), as well as certain other zones as defined in DAO 25 (1992). However, categorization and zoning take place only after tenured migrants qualify as such. This activity has the potential of removing the rights assumed under the definition of tenured migrant and forcing the relocation, either within the protected area, if it is a zoning issue, or outside it, if it is a categorization that disallows human occupation. This removal of rights or relocation through the administrative act of categorization and zoning would be without due process of law. To avoid this, categorization and zoning must be undertaken with due consideration to tenured migrants and the lands they actually occupy. If the ecological imperative is to relocate them and the resulting zoning and categorization should conflict with these rights, due process of law must be observed including notice and hearing, compensation, and/or resettlement. Any resettlement into multiple-use zones or buffer zones must carry with it the rights of a tenured migrant as if there were actual occupation of the resettlement area.

As no final zoning has yet been undertaken and no rules released for tenured migrant certification and recognition, it is not possible to verify whether the DENR will carry out the categorization and zoning in the legally correct manner. It is possible that it may use a provision in the NIPAS Act that grants the secretary the power to control occupancy of suitable portions of the protected area and resettle outside of said area forest occupants therein, with the exception of the members of indigenous communities area (Sec. 10 [o], 1992).

In implementing the CPPAP, current drafts of the management plan do not include plans for resettlement. Should resettlement be necessary within the project as a result of ecological considerations, the World Bank guidelines for involuntary resettlement (OD 4.30, 1990) would need to be followed. The World Bank operational directive applies because the bank administers the CPPAP on behalf of the GEF.

A tricky issue with respect to the tenured migrant provision relates to seasonal farmers and nonresident farmers. Technically, the provision requires actual and continuous occupation for qualification as a tenured migrant. Whether this occupation relates only to having a residence or merely to farming within the area is not clear. Neither the NIPAS Act nor its implementing rules required residency within the protected area to qualify as tenured migrant. However, allowing nonresidents to qualify as tenured migrants would run counter to the terminology used as there is technically no "migration" into the protected area that would qualify the farmer as a migrant. On the other hand, failure to take such farmers into account would send a message that it is preferable to move into the protected area, as rights are accorded those who do, rather than to those who preferred not to move in but merely to benefit economically from the protected area land. This would need to be remedied in the final site specific legislation for establishing each protected area, based on the established facts in each candidate site and the impact of such seasonal and nonresident farmers on the biodiversity.

A cause for concern among protected area managers pertains to the rights of marriage partners and descendants among tenured migrants. During 4 years of implementing the CPPAP, expanded families have brought about observable threats to the protected area. They have been seen to cause additional clearings and new structures to live in. The issuance of a clear tenure instrument is one way these can be controlled. It is expected that the threat of cancellation due to violation of the terms of the agreement as reflected in the tenure instrument would be clear, thereby preventing violations. Any transfer of rights should be only to descendants and only of existing holdings and structures. Until the rules on tenure are passed, however, the very specter of expansion and additional in-migration sought to be avoided by Secretary Cerilles will actually take place.

A final issue with respect to tenured migrants relates to the rights held despite the absence of a certification proving the right. Under the current administration of Secretary Cerilles, many observers within the DENR and outside it have little hope that the rules on the tenure instrument will be approved. As such, it is necessary to pinpoint the moment at which the rights arise, whether such rights are proved by documentation or not. A rule in statutory construction counsels that the law should be construed so as to give it legal effect. The use of the word *tenured* when referring to persons in a statute can be construed by itself as a recognition of a right. The use of the same term in relation to a criterion implies that the right or tenure would automatically arise once the conditions are fulfilled. This leads to a conclusion that the rights of a tenured migrant arise by operation of law and are not contingent on the issuance of documentation thereof. Once threatened with resettlement or other coercions, the absence of a tenure instrument such as a community-based forest management agreement is not a justification and the right to tenure can still be exercised.

The most recent official word on the issuance of tenure instruments on protected areas is the Memorandum from the Director of the Protected Areas and Wildlife Bureau to the Secretary of DENR recommending changes to DAO 29 (1996) to accommodate the provisions of the NIPAS Act and allow the use of Community-Based Forest Management Agreements granted under these rules to serve as temporary tenure instrument. In the meantime, the DENR has yet to issue rules specific to tenure in protected areas. The problem with using DAO 29 in place of protected area–specific instruments is that DAO 29 awards rights under the authority only of the executive branch through rules and regulations, whereas tenured migrants derive their rights from a congressional act — the NIPAS Act itself. Therefore, by a mere recall of Executive Order 263 by the president or an amendment to DAO 29 by the secretary, the rights of the tenured migrant under a community-based forest management agreement can be voided.

Common Property Resources

Commons are resources that are subject to individual use, but not to individual possession, and benefit a number of users with individual use rights who can collectively exclude others who are not members of the collectivity (Blaikie and Brookfield, 1987). Commons are governed by rules that are arrived at by agreed-upon decision-making arrangements. Two kinds of rules are thus generated in commons: (1) the rules of the decision-making process itself and (2) the rules governing the use of property. Blaikie and Brookfield differentiate commons and private property resources but admit there are intersections between the two. The distinction may not be relevant where limitations to private property use constitute restriction sufficient to make the ownership question a percentage game regarding which party exercises authority or quantum of right sufficient to be designated the owner (Gray, 1995). Especially with respect to the ownership of public lands by the state and the corresponding right of an owner to exclude others from the resource, the percentage game applies. The exclusion or denial of the right of access to citizens whose welfare it is the duty of the state to secure must be based on something more tangible than the state's right to exclude its own citizens from resources it owns (Tanggol Kalikasan–Haribon, 1999). In the experience of Slovenia, it was advanced that legislation, theory, and legal practice should be based on the principle that ownership cannot hinder social justice (Supancic-Vicar, 1994). This is the case particularly when the agent that is tasked to uphold social justice is the owner of the resource.

In the sense that decision-making arrangements and resource-use rules govern both ancestral domains and community-based forest management areas of tenured migrants, both will qualify as commons. However, the IPRA has stated unequivocally that ancestral domains are private in character but commonly held. This kind of ownership should then constitute an exception to the distinction between private property resources and common property resources. For the purpose of rules on decision making and resource use, however, there would seem to be little difference so that both ancestral domain and tenured migrant areas may be treated as commons with the former being of a higher order perhaps than the latter.

The widely known theory about commons is the tragedy surrounding it when the rules on decision making start to break down and individual users begin to free-ride on the resource to get ahead (Hardin and Baden, 1977). Another influence on common property resources is the removal of responsibility of management from the community by the state. This happens when the state fails to recognize rights of the community to the resource or when it perceives that the local managers are not doing a good job in the management and conservation of the resource. State intervention is addressed for ancestral domains by the IPRA, which specifically states that resource use will be governed by the concept of self-determination by indigenous peoples. However, for tenured migrants, especially in the absence of clear tenure instruments reflecting their legally vested rights under the NIPAS Act, such uncertainty is likely to lead to the disastrous consequences observed in other areas where management authority has been moved from the local community to bureaucratic decree (Blaikie and Brookfield, 1987).

Only with state recognition that the management of ancestral domains and tenured migrant areas are ably managed through local systems of decision making and resource-use rules will there be sufficient impetus for recognition of community management regimes. Otherwise, the bureaucracy will be nominally and legally granting tenure and recognizing private ownership while continuing to reassert authority and state intervention. In so doing, it will undermine the local management system and consequently be detrimental to the resource.

COLLABORATIVE MANAGEMENT AND CONFLICTING INTERPRETATIONS: THE CPPAP EXPERIENCE

Collaborative management in the CPPAP is exercised through a partnership between NIPA, Inc. and the Department of Environment and Natural Resources. The dynamics between these implementers has been a continuous source of tension within the

CPPAP. Conflict between the partners reveals divergent perspectives regarding the legitimacy of local management of common property resources. It highlights a not unusual dilemma with progressive legislation, that implementation is vested in a bureaucracy that fundamentally resists the change. A danger persists that the gains in community-based forest management from the NIPAS Act and the IPRA will remain only paper victories. Additional examples of divergent views that could erode collaboration in protected area management are discussed below.

Motivational Factors of Collaborators

Protecting the forests as a motivation has met with little objection since the logging industry began to sunset. Following the Ormoc tragedy in 1990 wherein an estimated 4000 to 7000 people died or disappeared in a combination typhoon and sea surge, people's attention was swiftly brought to the state of the forests and alarm bells were rung by NGOs. How protection is to be achieved, however, is a subject of fierce debate. Secretary-designate for the Department of Environment and Natural Resources met with NGOs for the first time in June 1998 and enunciated his forest policy as one of fencing out forest occupants. While many NGOs were stunned by the pronouncement, which appeared to obliterate two decades of struggle for upland communities and their partners, others reserved judgment, giving the new appointee the benefit of the doubt. Many advocates laid down their swords believing that he would need to be briefed about these issues before he could make a correct judgment. Although a number of NGOs opposed the appointment of the secretary-designate at the Commission on Appointments, the commission nonetheless confirmed the appointment. A year and a half later the secretary is singing the same tune to the accompaniment of a mendicant's plaint — DENR does not have sufficient funds to monitor the forests and needs more money from congress. What is the likelihood that the department will obtain what is required to guard all designated forestland in the Philippines from its over 18 million residents, many of whom have legal rights to be there?

The secretary gave reason to some in the department to stick to their archaic guns on this issue. Despite Executive Order 263 making community-based forestry the primary forest policy of the Philippines, many in the bureaucracy dusted off their former prejudices and persisted in the belief that the forest occupants are the main problem.

The idea that forests need to be protected against the very people living inside them was allowed to hibernate in the decade of enlightened legislation and executive orders, but was reawakened with a secretary uninitiated in the long years of debate about forest occupants. There has been little serious effort among NGOs to convince lower-level DENR personnel to foster the intent of the new legislation, or at least not to hinder it. In effect, the current DENR bureaucracy appears intent on backtracking from the progressive policies of previous years and on resisting the implementation of provisions of the NIPAS Act on tenured migrants.

Key Result Areas as Incentives

A reliable way to ensure that policy is implemented by the DENR bureaucracy is to include it in the Key Result Areas (KRA) by which the performance of the implementor is measured. Although this ensures that certain things are accomplished, it does not provide for the quality of work. For example, Certificates of Ancestral Domain Claims (CADC) in Northern Sierra Madre proved to have been issued without consultation with the people whose names appear on the certificate. The certificates covered areas that are not the ancestral domains and hunting grounds of the indigenous peoples concerned, while failing to include areas that *are* part of the gathering grounds. The Dumagat tribe, to whom a CADC was granted, is so named because of the word *dagat* or sea from which their food is collected and speared. Although the coastal area should be part of their domain, the area recognized by the CADC is landlocked. The speed with which this CADC was issued was widely speculated to be due to the KRA targets that DENR personnel needed to meet for the implementation of Philippine Agenda 21. Although not all government officials undertake management by KRA to the detriment of the quality of work, there is a clear need for a general motivational force and reinvention of the bureaucracy and its belief system to keep these occurrences to a minimum.

A positive aspect of a KRA style of management is that it can be passed on from one administration to the next. Also, a KRA may be drawn directly from the vision, down to the mission and goals of the department. Because these are public documents, therefore, any citizen can question any particular KRA and challenge officials regarding whether or not the accomplishment is indeed in line with the vision.

A possibility for increasing motivation is to professionalize protected area work as separate from forestry, not by a board examination, but through an association of protected area professionals. It would be preferable for this to be a small group of dedicated professionals who continuously debate the issues and the gray areas, occasionally issuing papers and pushing the envelope on more creative and appropriate protected area work that adheres to principles of practical management, as well as social justice, human rights, and participatory democracy. Such activity was attempted by a small group of people within the CPPAP, some of whom participated in the 1997 biennial conference of the George Wright Society, an organization of protected area professionals within the United States. The Philippine interpretation of the model would be less academic and technical, oriented toward ensuring that protected area work would be used to correct historical wrongs, inequity, and social injustice. Unfortunately, the demands of project work within the CPPAP did not leave time for such a group to coalesce.

Implementation Delays and Burueaucratic Discretion

When asked to explain a decision or action, government bureaucrats often would say it was based on law or that it is the policy of the bureau or the department. They would imply that discretion was not exercised, nor were decisions or actions colored in implementing the NIPAS Act. The Act, however, has considerable directory provisions that are difficult to implement due to budget and time constraints of the department. For example, the law requires it to map nearly 200 initial components of the protected area system within a year from the passage of the Act. The absence of funds necessary for such an exercise, which includes consultations, surveys, and other preparatory work, make it impossible to achieve within the deadline in the law. Simply, the selection of priorities by the DENR makes each decision or action one of discretion. Its choices of which protected area management boards it will support, activate, or call for meetings, the determination of areas in which to work, and the thoroughness with which the steps are performed all are matters of discretion.

When it is accepted that law cannot be used as an excuse for a poor decision, then it can be determined whether such decision was reached after a careful balancing of factors that consider the public interest. As such, the many factors that affect the preservation of forests for the greatest number and in favor of those living in them are weighed and responsibility is accorded based on assignments of weight. Although this type of decision making is more challenging, it is responsive to social realities. It is also the best defense against bureaucrats who would use the law to advance their own self-interest through an action or decision.

The Danger of Supplanting Local Management Systems

Information, education, and communication, usually referred to as IEC, frequently is the mantra used for achieving and improving protected area management. Success for many is measured in terms of how successful the IEC has been. Frequently, heavy reliance on successful IEC comes from those who are skeptical of the abilities of communities to manage resources. For them, any success in protected area management depends on the success of their IEC in changing value systems and community norms to meet their own standards. The expectation in such instances is that the community resource norms are nonuseful or nonexistent and can be readily discarded. When this is the framework for IEC, any improvements in management of the protected area that result cannot be sustained. A corresponding belief system needs to accompany the information, and these do not change through IEC programs; they change over time along with changes in patterns of respect and trust in the people implementing the laws.

Belief in community authority and capacity for management corresponds with understanding that there are systems of rules in place within a community that has lived together and depended for a long time on a particular resource base. This system should be the basis for any IEC and must be used and enhanced rather than replaced.

CONCLUSIONS

The last decade of the millennium was one of enlightened policy changes in Philippine protected area management. The NIPAS Act, the IPRA, and the rules on community-based forest management all guaranteed rights and participation in management for indigenous peoples and long-term migrants into protected areas.

There remains, however, a failure by the implementing bureaucracy to understand the way communities traditionally managed resources. Instead of helping communities find ways to improve customary norms for management, many of the people who are tasked with implementing the new policy believe that they need to teach the community correct practice. It often appears as though the protected area manager is looking for a formula for successful management without examining the village itself. Instead of realizing the errors of past management, the bureaucracy is attempting to have the community manage the forest the way the bureaucracy did. This not only appears disempowering to local people, but any gains would be difficult to maintain, making management unsustainable in the long run. Evidence of this is apparent in the way political appointees in powerful positions within the bureaucracy have dictated the interpretation of policies and from the rigid, numerical success indicators that often are used in monitoring the implementation of the laws and policies.

To improve this situation, protected area managers need a set of guiding principles to apply to the various types of dilemmas that they encounter in their work. Such a framework for thinking could help turn the tide and establish mechanisms of protection. Genuine collaboration and adaptation to the cultural and management norms of local communities would need to be employed. These strategies should be anchored by an intimate knowledge of communities brought about by interaction in an atmosphere of respect and mutual gain. Without a belief in the viability of community power, a manager's role in mobilizing for protection in a presumably collaborative atmosphere likely will fail. Notions of community empowerment would continue to constitute lip service and would keep the community on guard as it learns whether or not to trust the sincerity of the partnership being offered by government.

Collaboration with government, on the other hand, teaches much about the constraints of working in a bureaucracy and the difficulties of realigning bureaucratic concepts with progressive new laws and rules. Success can be expected only if the collaboration is based on mutual respect and permeates to the bottom rungs of the hierarchy. This is needed to prevent political games in the higher rungs from undermining long-term efforts at collaboration and mutual learning.

A common framework of beliefs is required among collaborators that includes such basic precepts as the right of all living things to thrive, social justice, and local empowerment as the fundamental pillars. The process of arriving at and polishing such a framework may be best accomplished through an independent association of dedicated protected area professionals. Only through embracing such a framework in sincerity will adaptation lead to better outcomes instead of excuses, nonimplementation of progressive laws, and other costly divergence strategies.

REFERENCES

Blaikie, P. and Brookfield, H., 1987. *Land Degradation and Society*, Routledge, London.

Doronila, A., 1999. WB: corruption in RP deep-rooted, in *The Philippine Daily Inquirer*, November 7, p. 1.

Eckholm, E. P., 1976. *Losing Ground: Environmental Stress and World Food Prospects*, W. W. Norton, New York.

Environmental Science for Social Change (ESSC), 1999. Decline of the Philippine Forest, map and text, Bookmark, Inc., Makati, the Philippines.

Gray, K., 1995. The ambivalence of property, in *Earthscan Reader in Sustainable Development*, Earthscan Publication, London.

Hardin, G., 1977. The tragedy of the commons, in *Managing the Commons*, W. H. Freeman, San Francisco.

Hardin, G. and Baden, J., Eds., 1977. *Managing the Commons,* W.H. Freeman, New York.

Heaney, L. and Regalado, J. C., Jr., 1998. *Vanishing Treasures of the Philippine Rainforest 1998,* The Field Museum, Chicago.

Hurst, P., 1990. *Rainforest Politics: Ecological Destruction in South East Asia*, Zed Books, London.

Leonen, M. V. F., 1998. Human rights and indigenous peoples, in *Human Rights and Indigenous Peoples, A Public Forum*, Llenos, M. V., Ed., Quezon City, the Philippines. 1998.

LRC-KSK (Legal Rights Center–Kasama sa Kalikasan), 1999. Millennium millions: a look at budget allocations for critical natural resources, Occassional Paper, LRC-KSK, October 12, unpublished.

Mallari, N. A. D., Strattersfield, A. J., and Crosby, M. J., 1999. Birds and barometers, *Haring Ibon*, Nov. to Dec.

Myers, N., 1992. *The Primary Source: Tropical Forests and Our Future*, W. W. Norton, New York.

Ofreneo, R., 1988. Debt and environment: the Philippine experience, in *Asia's Environmental Crisis*, Howard, M. C., Ed., Westview Press, Boulder, CO.

Severino, H. G., 1998. Opposition and resistance to forest protection initiatives in the Philippines: the role of local stakeholders, Discussion Paper (DP 92), United Nations Research Institute for Social Development, Geneva, Switzerland, May.

Supancic-Vicar, M., 1994. The influence of the reprivatization of land on nature protection and protected areas, in *Protected Area Economics and Policy*, Munasinghe, M. and McNeely, J., Eds., IUCN and WB, Washington, D.C.

Tanggol Kalikasan–Haribon Foundation, 1999. Developments and issues in NIPAS Act implementation: six years hence, in *Primer on Protected Areas*, Haribon Foundation, Quezon City, the Philippines.

Vandermeer, J. and Perfecto, I., 1995. *Breakfast of Biodiversity: The Truth about Rain Forest Destruction*, Institute for Food and Development Policy, Oakland, CA.

World Bank, 1989. Philippines Forestry, Fisheries, and Agricultural Resource Management Study (FFARM Study), Sept. 1989, Washington, D.C.

World Bank, 1990. Operational Directive 4.30: Involuntary Resettlement, in The World Bank Operational Manual, June 1990, Washington, D.C.

Section III

Modeling Protected Area Human Activity Systems

12 Making Public Protected Areas Systems Effective: An Operational Framework

Andy White, Hans Gregersen, Allen Lundgren, and Glenn Smucker

CONTENTS

INTRODUCTION

Public protected areas have long been the cornerstone of biodiversity protection and many policy makers, both at national and international levels, are calling for the

0-8493-0020-7/01/$0.00+$1.50

establishment of additional areas (Kramer et al., 1997; Brandon et al., 1998). In April 1998, for example, the World Bank and the Worldwide Fund for Nature (WWF) jointly committed to establish 50 million ha of new protected areas and more effective management in another 50 million ha of existing protected areas before the year 2005. These calls for new protected areas coincide with the growing recognition that many of the existing protected areas are ineffective and that the currently accepted models for protection are under question and are perhaps flawed — if not in concept, then at least in practice (Brandon et al., 1998; Wells et al., 1998). A major review of protected area effectiveness commissioned by the World Bank and WWF in 1999 found that less that 25% of forest protected areas in surveyed countries were "well managed with good infrastructure," 25% were suffering from severe degradation, between 17 and 69% of protected areas had *no* management, and less than 1% could be considered secure in the long term (Dudley and Stolton, 1999).*

In addition, these studies recognize that poverty and population growth in many areas of the world will increase pressure on protected areas in the coming decades. These trends all point to the need to take a fresh, objective look at the conceptual foundations of the public protected area approach and the requisite conditions for its success.

The "protected area" has historically been the unit of analysis in the majority of studies, evaluations, assessment frameworks, and development targets. This chapter argues that protected areas cannot be fully understood in isolation; rather, many of the constraints faced by individual protected areas derive from the national protected area systems to which they belong. Individual protected areas are managed within national legal frameworks and according to national policies and regulations. For this reason, improving and sustaining effectiveness often, if not usually, requires reforming protected area systems.

The purpose of this chapter is to propose a simple yet comprehensive analytical framework for assessing national, public protected area systems. The framework doubles as an outline of the conditions for effective execution of the public protected area approach. While these may be considered as the "ideal" or "optimal" conditions, the authors recognize that seldom could all of them be fully achieved in practice. Rather, they represent a state that protected area planners should move toward for improved effectiveness.

The framework could be used to assess a particular protected area, but the focus of the text is on the protected area system. The framework draws on the existing literature on social assessment of protected areas and concepts proposed by the authors in other contexts and adapted for the protected area problem.** It is important here to reiterate that the chapter focuses on national and public protected areas and does not explicitly address the issues associated with public support for private protected areas — as are emerging in southern Brazil and southern Mexico, for

* The countries surveyed included China, Mexico, Brazil, Gabon, Indonesia, Papua New Guinea, Peru, Russia, Tanzania, Vietnam.

** See Geisler (1996), Cruz and Davis (1997), and Borrini-Feyerbend (1997) for presentations of social assessment methodologies for protected area management and case studies and Poole (1989) for a description of the particular challenge of developing partnerships among indigenous people, conservationists, and government planners.

example. Indeed, these innovations in protected area management are developing in response to some of the very challenges of establishing public systems addressed in this chapter.

As background, the chapter first describes the basic nature of public protected area systems. This section identifies key challenges for making systems effective and key imperatives for overcoming those challenges. These challenges and imperatives are understood to be inherent to the public protected area approach itself. Next, the chapter presents an operational framework for effective protected area systems. This framework entails: (1) providing an enabling institutional setup; (2) addressing critical operational issues; and (3) sustaining institutional performance. To illustrate use of the framework, the chapter then applies it to the case of Haiti's protected area system. This case is not intended to be representative of the execution of the protected area approach generally, although many of the issues raised in the case are common to many other protected area systems. The chapter concludes with some final thoughts on the task of making protected area systems effective and the feasibility of applying the public protected area approach.

BACKGROUND: THE NATURE OF PUBLIC PROTECTED AREA SYSTEMS

Public Property, Political Ventures, and Constituencies

National protected areas are set aside because they are deemed to have strategic, national value. They are national public properties set within a particular local context. Protected area systems, on the other hand, are national-level political systems organized to maintain a set of particular protected areas. As with all public properties, the maintenance of a protected area system and the management of protected areas are essentially political ventures and are inherently conflictive. Policy and practice are outcomes of political processes and negotiation between vested interests. In this sense, a protected area "system" does not pertain to the government, private, or civil sectors. Rather, these systems are political constructs generated by the dynamic interactions of the government, private, and civil sectors, at local, regional, national, and international scales. And thus, by definition, there is no fixed and permanent outcome — either in the particular management objectives or the organization of the system. Social tastes and preferences change over time, as do ecological systems, and for this reason protected area management is a dynamic and sustained process, and not a movement toward a predetermined, fixed state.

The only necessary condition for establishing and maintaining protected area systems is that the supportive *constituency* have sufficient power to convince the government decision makers to identify the areas and establish and support the minimal institutional framework. The constituency can include public, private, or civil actors, but in most cases contains strong civil society elements of both national and international character. Experience shows that the effective constituencies often combine public and private elements, acting in a constructive, collaborative manner. In this sense, the fundamental core of protected area management is the constituency, and the survival, power, and political effectiveness of that constituency.

Conflict and Collaboration

Of course, establishment of protected areas is not synonymous with effectiveness, and constituency is a necessary, but not sufficient condition for effective management. Given the public nature of protected areas, neither the government nor the broader constituency can act alone, and must skillfully negotiate and compromise with other stakeholders. Since protected area systems are national, political ventures and national public properties set within a local context, they have at least four sets of stakeholders: government and civil society actors at the local and national levels. Historically, many countries have maintained policies where national protected areas were managed by central, public agencies, and these areas were managed (at least in theory) for the national public good. Private and civil sectors of society, local populations, and even local governmental entities had little or no role in management. More recently, as the poor performance of centralized management is recognized and local and civic groups have gained political voice, these same groups have increasingly challenged conventional management systems and approaches.

Most conflicts between stakeholders surround issues of *authority* over management (i.e., who decides how the system and area is managed) and rights to the *benefits* from management (i.e., who benefits from the management). In many cases, there has been pressure to shift authority and benefits from *central* to *local* and from *public* to *private* spheres of action.* Of course, formally allocated authority is not effective unless the holder has the power to enforce its decisions. Similarly, formally allocated access to benefits (in a property-rights sense) is not effective unless the holder has the power to protect that access and the resources and knowledge actually to access the benefits.

These conflicts can only be effectively negotiated when the stakeholders agree on rules by which the authority and benefits are negotiated and distributed. And although government protected area managers often have the legal authority and the political power to act unilaterally, they are increasingly realizing the need to provide opportunities for civil society stakeholders to participate in management. Existing political systems in many countries now establish what has been called a "shared-power" context for public natural resources management — where power is shared, for either *de jure* or *de facto* reasons, between civil society and government stakeholders. So, except for the most recalcitrant of governmental agencies, *collaboration* has become a political imperative. Thus, the effectiveness of the system and the sustainability and productivity of protected areas depends not only on government law and enforcement of such, but also on the goodwill, participation, and collaboration by key stakeholders, including the government — at each level of negotiation.

Adaptation to Uncertainty

As described above, protected areas, like any area of nature used for various outputs, depend on thoughtful management to be maintained on a sustainable basis. However,

* See Smucker and White (1998) for a simple description of the range of institutional options for protected area management.

when it comes to humans and their different organizations "managing" nature, there is a great deal of uncertainty, a great deal that is unknown about the resource itself, and a great deal of politics regarding the use of national environmental assets. In addition, there is a great deal of uncertainty and ignorance with regard to spatial and temporal relationships and with regard to the effects of human use on the areas in question. Natural resource managers are increasingly aware of failures of past management actions to deal effectively with uncertainty and of the need to accept their relative ignorance.

This uncertainty derives not only from technical, or ecological, relationships, but from changes in social and political perspectives as well. As stated earlier, protected area systems reflect local, regional, national, and international social and political interests, and these interests change over time. Given these natural and social uncertainties and changes, the challenge for protected area managers is to establish mechanisms and adopt practices that recognize the dynamics and uncertainty of nature and of society. *Adaptive management* is a term indicating an explicit strategy of responding to technical uncertainty and managing through "learning by doing" (Holling, 1978; Walters, 1986; Bormann, 1994).*

Effective protected area management therefore involves a learning process. When a number of stakeholders are collaborating in management, the process of learning from doing often gets confusing and in some cases becomes ineffective. Explicit consideration should be given to the linkages between learning and adopting what is learned. This process of learning from experience should be orderly and conscious — following, so to speak, the same guidelines that any good research should follow and searching for objectiveness, reproducibility of results, and representativeness of samples used. At the same time, the process of transferring what is learned and adopting the results that make sense also should be a fairly conscious and in some cases structured process, or at least a process that involves a set of rules agreed upon by the collaborators in the management.

Three Challenges and Three Imperatives for Effectiveness

In sum, effective public protected area management is a daunting goal, largely because of three fundamental challenges: (1) they are public properties and the system is a product of a dynamic national polity; (2) they are the focus of perennial conflicts between disparate stakeholders at all levels of society; and (3) they seek to protect a resource about which there is great uncertainty and ignorance. These challenges can be overcome and protected area systems can be effective if (1) there are politically effective *constituencies*, (2) there is effective participation of, and *collaboration* between, stakeholders; and (3) if the system is operated in a sufficiently *adaptive* manner to take into the account the uncertainties involved. Unfortunately, this is no small task. The following proposed framework is intended to determine the status of existing systems and identify areas where they can be strengthened.

* For early examples in forestry, see Thompson (1968), Fight and Bell (1977), and Lundgren (1983).

AN OPERATIONAL FRAMEWORK FOR EFFECTIVE MANAGEMENT OF PUBLIC PROTECTED AREA SYSTEMS

This section suggests an operational framework for putting into practice the imperatives of building constituencies, creating collaborative mechanisms, and organizing management in such a way that it readily can adapt into effect. This framework is presented in synopsis form in Table 12.1 and is described in depth below. It entails (1) providing an enabling institutional setup; (2) addressing critical operational issues; and (3) sustaining institutional performance.

Providing an Enabling Institutional Environment

The overriding purpose of the institutional environment is to provide the legal and institutional space for stakeholders to fulfill their roles as adaptive, collaborative constituencies for protected area management. This enabling institutional environment includes three sets of institutions: (1) the legal and institutional framework of the protected area system itself (i.e., the mix of government and civil society organizations that govern protected areas); (2) the government structure charged with representing government interests in protected area management (i.e., the government stakeholders); and (3) the set of nongovernmental organizations (NGOs) that would represent civil and private sector stakeholders (i.e., the nongovernmental stakeholders). This enabling institutional setup is portrayed in Figure 12.1, along with the fundamental roles of government and civil society actors.

The minimum required responsibilities of national legal and policy framework would be to set the overall direction for protected area management, determine the roles and relationships among central, local, government, and civil society actors, and, more importantly, set the rules by which policy and operational decisions would be made. This legal and institutional framework would entail three levels of negotiation: the *macroinstitutional*, *policy making*, and *implementation/management* levels. These levels correspond to three types of rules governing the system: *constitutional*, *collective choice*, and *operational* (Ostrom, 1990).*

In the context of protected area systems, constitutional rules are those fixing the macrolegal and institutional setup and determining what entities are eligible to participate in the management of the protected area system, as well as the process for judging and punishing illegal activity. For example, constitutional rules determine the fundamental relationships between civil society and government, the array of government agencies, and the role of the judiciary. Collective choice rules are those determining how policy is made and how the entities engaged in protected area management are organized. Operational rules are those fixing the day-to-day operations of management. Conflict, negotiation, and collaboration take place at each level.

The minimum role of the government structure would be to monitor the protected area system, to ensure that management meets the legal and political requirements from a government perspective, and to voice proposed adaptations to government officials and policy makers. Government entities must also participate in the

* Of course, these three types of rules exist at any and every level and within each organization. For reasons of simplicity the authors have chosen to correlate the different rules with political level.

TABLE 12.1
An Operational Framework for Effective Management of Public Protected Area Systems

Providing an Enabling Institutional Environment	Addressing Critical Operational Issues	Sustaining Institutional Performance
1. The national governance environment, providing sufficient stability, justice and rule of law to permit government performance and engagement of civil society. 2. The legal framework of the protected area system itself. This would: • Set overall direction, identify which government agencies were involved and setting their respective responsibilities. • Enable and encourage active participation by stakeholders and civil society at all decision levels and arenas: legal, policy and operational. • Set the "rules of engagement" for interaction between stakeholders and enable them to build political constituencies, collaborate with each other and adopt an adaptable approach. 3. The government structure charged with representing government interests. Their minimum role would be to monitor the system and ensure it meets legal requirements. 4. The set of NGOs that would represent civil and private sector stakeholders. Their minimum role would be to monitor the system actively, hold the government accountable, and advocate for improvements.	1. A system for monitoring and evaluating both the status of the protected area and the behavior of all relevant stakeholders. • This would be set by the institutional mechanism (public/private) established for managing and administering areas, but must be transparent and perceived fair by all stakeholders. • This must regularly disseminate findings to all stakeholders and could entail different approaches for different types of protected areas. 2. An effective enforcement system; • The system should be established and agreed upon by key stakeholders in the management structure. • Self-policing rules and enforcement would be desirable, although some government oversight and backup would be required. 3. A formal mechanism to ensure active, direct contact between key stakeholders and encourage accurate communication of information between them. • This would encourage enough face-to-face communication to enable different stakeholders to learn the values and interests of others, and establish an evolutionary process of decision making (rather than creating situations where exit and "revolution" becomes the only option). • There is opportunity to take advantage of low cost, new ICTs (information, communication and technologies).	1. Increase knowledge in key areas of management uncertainty (including the relative values associated with different management strategies). • Again, make good use of ICTs and research related to protected areas. • Search out relevant research and make results known. 2. Generate the resources that are essential to carry out effective management. • Develop the means of tapping key sources of long-term funding to support protected areas. • Take full advantage of the private sector to access its management skills and other resources. 3. Develop the incentive or motivation for all stakeholders to participate actively in management. • Stakeholders benefiting should participate in paying the costs in amounts corresponding to their benefits, and those losing from the system should be compensated fairly. • Incentive schemes must be tailored to reflect circumstances particular to each system (market approaches will be preferable in some cases).

TABLE 12.1 (continued)
An Operational Framework for Effective Management of Public Protected Area Systems

Providing an Enabling Institutional Environment	Addressing Critical Operational Issues	Sustaining Institutional Performance
	4. A mechanism to coordinate interventions and intervenors. • This should be established in concert by the key stakeholders, not necessarily government. 5. A mechanism for sustaining financing the protected area system. 6. Mechanism to ensure that the local population has some vested interest in protection. • This would require establishment of practical incentive mechanisms on an institutional level.	4. Cultivate competent leadership by providing promising individuals with the knowledge, resources, and incentive to commit to protected area management.

coordination and planning of the protected area system. The extent to which government entities become involved in implementation varies, and different levels of intervention would be appropriate in different circumstances. Depending upon the level of capacity of the civil sector, and the will of the stakeholders, government intervention in management could be limited to holding the title to the protected areas, all other use and management rights seconded to civil and private actors.

The NGOs have a different minimum set of responsibilities, including acting as an independent monitor and advocate of the protected area system. This set of organizations both holds the government accountable and seeks to support it in a constructive way against forces damaging the system. The role of NGOs in implementation would vary according to national and local contexts, but it is critical that they engage at each level. It is clear from experience that strong civic participation at each level of negotiation can strengthen governmental performance. At the national level, organizations must act to lead and advocate for legal and constitutional reform strengthening the protected area system. At the policy-making level, civil society must constructively analyze, critique, and advocate for improved policies. At the implementation level, civil society must also participate in the governance of the particular protected area, at least in the form of comanagement committees and, at the most, in assuming complete responsibility for managing the protected area.

As has been stated previously, for the system to perform and function effectively, participation and collaboration is required at each level by representatives of each group of stakeholders. To achieve this collaboration, the setup must be perceived as fair and balanced. Domination of one level by either the government or civil society stakeholders (e.g., central government agency domination of policy agenda or civil society domination of the operational aspects of park management), then the system eventually becomes dysfunctional and ineffective. This is because both government and civil society groups have particular, legitimate, and unique functions to perform, and imbalance precludes effective fulfillment of that role. True partnership rests upon the precepts of trust, reciprocity, transparence, and fairness, and upon near equal standing in terms of power to effect the outcome. The absence of perceived fairness inevitably breeds antagonism and resentment, true and false accusations of abuse, and, eventually, all possible forms of resistance and subterfuge.

A number of variables would facilitate collaboration. Most importantly, the greater the degree of consensus between stakeholders concerning the existing institutional setup (i.e., the rules of the game) and the greater the degree of consensus regarding the level at which each conflict should be negotiated, the easier the collaboration. Collaboration would also be facilitated by positive interpersonal relationships among the stakeholders in management, an endowment of social capital (preexisting habits or institutions that facilitate trust and collaboration among stakeholders), and livelihood interdependence. Experience suggests that stakeholders whose lives are intertwined in dimensions other than the particular protected area issue would be more willing to negotiate and collaborate on that one particular thread of their relationships.

Additional facilitating factors would include a common vision for the protected area system, agreements that benefits and costs should be shared equitably, and, finally, recognition that an agreement to share equitably both benefits and costs involved in management does not necessarily imply sharing the same things — each

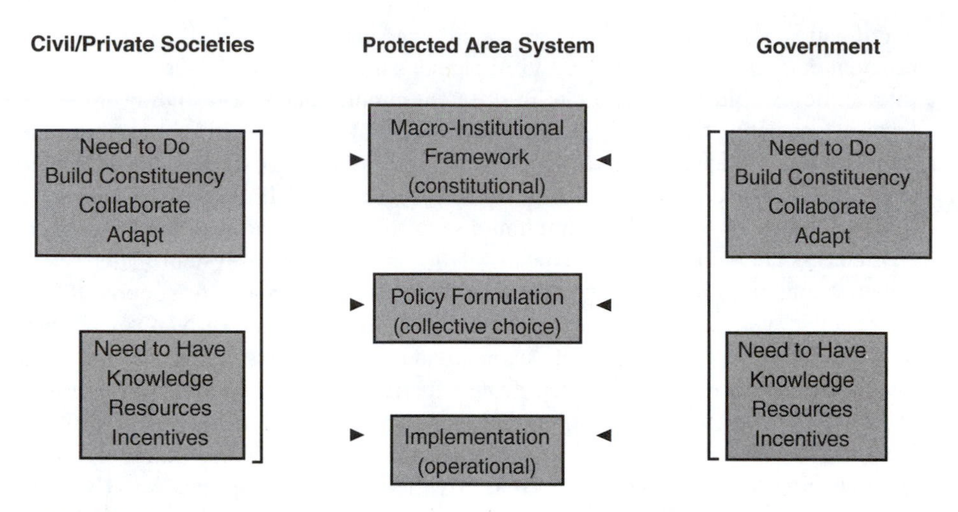

FIGURE 12.1 Effective protected area systems: operational framework.

stakeholder contributes to the system and benefits in its own particular fashion. Given the different values and interests of the stakeholders, these three factors would be very difficult to achieve, yet should remain as goals, for progress on each would go a long way toward facilitating effectiveness.

The notion that benefits and costs should be shared implies that through a process of consensus building the stakeholders can reach agreement on the various relative values involved, both on the cost and the benefit side. Thus, the roles of different collaborators may be quite different — one responsible for day-to-day supervision of use of the protected area; another responsible for the more conceptual and analytical task of assimilating and synthesizing the lessons from past management; and one responsible for raising funds for achieving the common management vision. The same differences may exist on the benefit side, some acquiring their main benefits from winter camping, others from spring and summer bird watching.

Given the permanent conditions of tension between stakeholders, uncertainty and lack of knowledge, the institutional setup would enable: flexibly mediating tensions between stakeholders; monitoring and sanctioning the behavior of the key actors; adapting to new conditions, preferences, and actors — and all in a manner perceived as fair by the relevant stakeholders.

Addressing Critical Operational Issues

Obviously, an enabling setup that is oriented to facilitating stakeholder collaboration is important, but not sufficient. The actors within this setup must effectively address critical operational issues. Operational issues to be addressed explicitly at the level of each protected area would include:

1. Establishing a transparent and credible system for monitoring and evaluating both the status of the protected area and the behavior of all relevant stakeholders;
2. Establishing and maintaining an effective enforcement system;

3. Establishing a system to make active, direct contact with key stakeholders and actively communicate accurate information to them;
4. Effectively coordinate interventions and intervenors;
5. Successfully sustaining credibility and financing for the protected area; and
6. Ensuring that the local population has some vested interest in protection.

In addition, there are a series of linkages that managers must consider in operationalizing protected area management, in the same way that such linkages must be considered in all integrated natural resources management (CGIAR/TAC 1996). The main ones of concern include the following.

Spatial linkages. As mentioned earlier, management at the site, plot, or field level must be linked explicitly to management considerations of the entire protected area and at the landscape or regional level. This need arises both on the technical front — an action in one area has technical implications for other areas, e.g., in a watershed management context; and on the benefit or need front — the value of an action that produces an environmental service depends on what happens at other sites within a landscape or watershed unit.

Temporal linkages. Intergenerational issues loom large in the management and use of natural resources. Justification for protection of many areas hinges on adding in expected benefits for future generations (although often protection can be justified on the basis of immediate future benefits alone). Also, given the lags that are often involved in nature and its changes, an explicit temporal perspective is needed.

Linkages between production and protection activities. All protected areas have costs associated with them, if nothing else than the cost of protection. It is essential that the benefits associated with protection are consciously linked to (and justified on the basis of) present and future production of goods and environmental services wanted by people. Agreement on the desired linkage must be achieved among the collaborators in protected area management; otherwise, the collaboration will become difficult to sustain.

In sum, effective management will require effectively dealing with critical operational issues. These include devising a system for monitoring and evaluating performance, coordinating action, and sustaining a flow of financial resources to cover management costs. Managers and stakeholders will also have to develop adequate responses to the spatial and temporal linkages between their area and external areas and actors. Among these linkages are those between protection and the production of goods and services demanded by key stakeholders, in particular, and the public, in general.

Sustaining Institutional Performance

The performance and sustainability of the protected area system would ultimately depend on whether the stakeholders fulfill their constructive roles on a sustainable basis. This in turn would depend on whether the stakeholders have the knowledge, resources, and incentives to contribute. In general, there is a perennial need to (1) increase knowledge in key areas of management uncertainty (including the relative values associated with different management strategies); (2) generate the resources that are essential to carry out effective management; and (3) develop the incentive or motivation for all collaborators to participate actively in the costs and the benefits associated with moving toward a common vision.

Increased knowledge depends on effective research, monitoring, and evaluation of management as it proceeds and on learning from others. Generation of resources depends on managing for outputs that are valued, and on managing in a way that is relevant to the wants of the stakeholders, whether they be local residents or the global society concerned with biodiversity protection, and carbon sequestration. Relevance leads to willingness to pay for it. Finally, incentive or motivation comes through a recognized sharing of benefits — a feeling among all stakeholders involved in the collaboration that they are being treated fairly and that benefits for them exceed the costs to them.

There is one more variable that is critical for sustained institutional performance, and that is *leadership*. Although leadership appears to have become somewhat passé and has disappeared from the discourse as researchers have gained appreciation for the collective sociological context, there is no social progress without leadership. For all their faults and foibles, charismatic leaders remain a critical force in conservation. Protected area stakeholders must consciously cultivate responsible and accountable leadership by providing leaders with the knowledge, resources, and incentive to contribute. This leadership is necessary to encourage and sustain constituency support and enthusiasm for protected area systems.

APPLYING THE FRAMEWORK: THE CASE OF HAITI'S PROTECTED AREA SYSTEM

BACKGROUND

Haiti is well known for being the poorest and most environmentally degraded country in the American Hemisphere. It is less well known that despite the degradation, poverty, and political turbulence, Haiti is also home to globally significant biodiversity — most of which is within the boundaries of Haiti's three national, natural protected areas, the Pic Macaya National Park, the La Visite National Park, and the Pine Forest National Forest. These three protected areas were established in the early 1980s and total approximately 35,000 ha. Each protected area is strategically located at the headwaters of major watersheds and two of the three are actively used by ecotourists.* As can be imagined in a country where poverty is pervasive, and governmental structures are weak, these protected areas face tremendous threats, including illegal logging and encroachment from preexisting settlements.

In 1996, the government launched a 5-year project aimed at strengthening the capacity of the country to manage its protected areas. This project was financed by

* This section draws heavily on Smucker and White's personal experiences in Haiti, where, since 1990, they have been active in assisting the reform of Haiti's protected area system. Between 1996 and early 2000, White managed the protected area project for the World Bank and Smucker was a leading advisor to the Ministry of Environment on the project. See Smucker and White (1998) for a more thorough review of the process of devolving protected area management in Haiti.

the World Bank and conceived as an Integrated Conservation and Development Project. It entails three components: institutional development (including technical training and technical assistance); protected area management (with funding for strengthening the capacity of the park and forest services and establishing management activities in the three protected areas); and buffer zone development (including agroforestry extension services to peasants via local NGOs and a demand-driven small development fund to finance capital improvements by local peasant organizations). Total project funding is $21.5 million (U.S.), with about 25% dedicated to the institutional development and protected area management components and about 50% dedicated to the buffer-zone development fund. The project has as key goals the reform of the legal and institutional framework governing protected area management, the devolution of management responsibility to local comanagement committees, and the improvement of household incomes in the buffer zones.

Applying the Framework

Providing an Enabling Institutional Environment

There is not, currently, a legal framework governing the protected areas. The Ministry of Agriculture administers both the national parks and the national forest. The two national parks were established by presidential decree and the Ministry of Agriculture has historically administered the national forest without specific guidance in the law. It is estimated that the government has provided less than U.S. $30,000/year since the 1980s for protected area management. The ministry has managed these areas from its central headquarters, with no field personnel in the two parks and limited and sporadic technical staff in the national forest. The ministry has not had a policy guiding management of these areas, but has governed them on an *ad hoc* basis with great variation of approach with each different minister. In 1996, the president established a Ministry of Environment by decree but this ministry does not yet have an organic law defining its responsibilities. Since establishment, the Ministry of Environment has lobbied for the transfer of protected area management authority to the Ministry of Environment. The Ministry of Agriculture has been reluctant to relinquish authority, and this has led to substantial debate and tension between the ministries.

Civil society in Haiti has historically been discouraged by the state, and is subsequently limited in organizational capacity and political voice. Peasant organizations, which are pervasive in rural Haiti, are growing in number, capacity, and accountability. Environmental NGOs, a recent phenomenon, are few, but are growing in number and capacity. Civil society, whether at the local or national level, has had no official role in protected area management. The local groups, rather, have expressed themselves via protest and subterfuge.

Since 1997, the government with World Bank support has undertaken a wide range of initiatives oriented to reforming the legal and institutional framework and provide an enabling environment for protected area management. A new legal framework has been drafted and reviewed by local and national-level stakeholders in the public, private, and civil sectors. This new law is awaiting the establishment of a new parliament for review and ratification. The project has also led the establishment of comanagement committees in each of the protected areas. These committees are composed of representatives of peasant organizations and government entities and now act as consultative bodies — reviewing annual plans and budgets and actively participating in planning, monitoring, and evaluation of protected area management. The new legislation would give them greater authority in comanagement with the government. The new law would also establish a new para-statal agency under the authority of the minister of environment to manage all protected areas — both the national forest and the national parks. The legislation would also establish authority for a national environmental trust fund whose function would be to generate funding for the protected areas on an adequate and sustained basis.

The project has also led to the establishment of a new, national-level, civil society lobbying group for protected area management — the Counseil d'Appui de Systeme Nationale des Aires Protegees (CASNAP). CASNAP has led independent evaluations of the project and protected area system, monitored the project, mediated disputes among the project, the government, and local representatives, and also actively lobbied for the reform of the legal framework. The project has also sought to strengthen local groups and national environmental NGOs both by encouraging and facilitating their participation in policy debates. Two of the three NGOs providing agroforestry services in the protected areas are national NGOs. In addition, the project has made concerted and unprecedented efforts to inform and educate elected officials, government technicians, and local people of the issues and challenges associated with protected area management.

In sum, from the perspective of the analytical framework, the government and its associated civil society partners are on the right track in improving the effectiveness of the Haitian protected area system. Despite unproductive political wrangling between the Ministries of Agriculture and Environment, and despite the government's stubborn resistance to devolving power to local people and jurisdictions, and despite the weak capacity of local organizations to assume that responsibility, the legal and institutional reforms necessary to enable collaborative and adaptable management are under way. Indeed, the commitment of numerous local people and organizations to improve management, despite the odds in Haiti, is often heroic. Unfortunately, these efforts are undermined by the sheer scale of the challenge of establishing protected areas amid such threats and the ongoing political turbulence. At this writing, the country is entering another phase of political stalemate, the legitimacy of the new parliament is contested, and donors are again threatening to cut off assistance. Nonetheless, despite these political issues, the project is strongly supported by the ministers of finance, agriculture, and environment and the government has requested a follow-on project. Their rationale for extending the project is to complete and consolidate the institutional reforms initiated by the project and to extend the buffer-zone development components to a much broader area of the country.

Addressing Operational Issues

With the assistance of the World Bank project, the Government is in the process of establishing protected area managers and technical teams in each protected area.

The managers and their teams are responsible for leading park improvements in coordination with local organizations and governments, monitoring and evaluation, and enforcing regulations in collaboration with local volunteers, the national police force, and the judicial system. This latter responsibility — enforcing regulations — is proving to be a major challenge. Illegal logging and encroachment remain a primary threat to the protected areas. Loggers defy the unarmed park guards, there are no police deployed in rural areas, and the judicial system remains incapable of effectively dealing with infractions.

To improve park management, the government is also in the process of devolving some operational responsibility to new advisory councils set up to comanage each protected area. The councils are composed of representatives of the diverse stakeholder groups, including peasant organizations, local government officials, local elected officials, and the head of the protected area itself. The councils work in partnership with the protected area manager and now engage in lobbying for local support of the park, monitoring and evaluating the performance of the park service, assisting in enforcing regulations, collecting and channeling local perspectives on park management, and mediating disputes between local people and the park service. The councils now receive a small operational budget from the government and each is in the process of acquiring legal status as an NGO. In the proposed legal reform these councils would become part of the formal park management system, with responsibility to review and approve the annual plans and budgets of each protected area.

The government is also engaged in addressing another key operational issue: establishing a sustained source of funds to finance the protected area system. With assistance from the Global Environmental Facility, the World Bank, and other donors, the government is in the process of establishing an independent trust fund that would finance the basic recurrent costs of managing the protected area system. This fund is being established as a NGO with minority government participation on the board of directors.

The government and its many partners are making progress on addressing these many operational issues, but the challenge is tremendous and the gains are tenuous. Making progress in establishing effective, collaborative management in Haiti requires making progress on several, more fundamental issues that have bedeviled Haiti for centuries: namely, establishing accountable, local representation; establishing effective, deconcentrated units of national government; and establishing effective local policing. And substantive progress on these fundamental issues is contingent upon national-level politics well beyond the control of the supporters of the national park system.

Sustaining Institutional Performance

According to the analytical framework, sustaining institutional performance requires ensuring that the key stakeholders have the knowledge, incentive, and resources to collaborate and adapt, as well as cultivating leadership for protected area management. In Haiti, the government and its civil society allies in improving the protected area system have largely undertaken this approach. Despite the pressures to allocate

World Bank resources to activities with short-term impacts, almost 25% of all project funds are dedicated to education and training. This includes establishing a technical college to train 60 new agroforestry and park technicians, running training programs for members of the protected area councils, the national lobbying group (CASNAP), national police, and elected officials on protected area management. This investment also strengthens the emerging social capital around the protected area system and new leadership.

To improve the incentives for collaboration, the government and allies have adopted different approaches for different stakeholders — including carrots to encourage pro-protected area behavior and coercion to discourage illegal activity. The category of positive incentives includes a U.S. $4,000,000 fund to finance rural projects that enhance local economic development requested by local organizations and governments, and an equal amount of funding to provide agricultural extension services in the buffer zones of the protected areas via NGOs. Both of these activities are pilot approaches to support rural development in Haiti. They have also been controversial, as both represent departures from the traditional approach of agricultural development led by the Ministry of Agriculture. The fund is the first demand-driven, rural investment fund managed by the government. It is also novel in having a decentralized decision-making system and operating with the specific objective of strengthening the capacity of peasant organizations. The fund is now functional and is working to improve its targeting and responsiveness, in particular to those who are now denied their illegal access to the parks. The agroforestry services are similarly improving their delivery of services to peasants in the buffer areas. The government now proposes extending these two approaches, the rural investment fund and the provision of agroforestry services to peasants via NGOs, to other areas of the country in a follow-on rural development program. To encourage the collaboration of other key stakeholders, such as the police, local government, and elected officials, the government now provides them voice in deciding issues pertaining to the protected areas and regularly invites them to participate in training events and conferences.

In terms of resources, the most critical in Haiti is financing. The trust fund to finance recurrent costs is the key element of government efforts to ensure sustained resources for protected area management. In terms of cultivating leadership, the government and allies have attempted to do so by investing in training, but especially by investing in new civil society organizations embodied in the new protected area framework. These organizations have given many local leaders their first opportunity to be directly involved in improving the protected area system, a major national issue. The many participatory events held, beginning with the design of the protected area project and including the many participatory evaluations and management events, have all given opportunity for new leaders to emerge and develop constituencies.

In sum, this case from Haiti illustrates some of the complicated and challenging issues faced by those who attempt to improve the effectiveness of protected area management. The story is still unfolding in Haiti, and it is too early to know the impact of the many initiatives. Steps have been taken within the context of each of the elements in the framework put forth earlier for effective protected area management; progress is evident. Although the country obviously has a long way to go in achieving effective protection, at least in terms of protected area management, there is movement in the right direction. That is all that can be asked of any society that has been through turmoil, disruption, and the depths of poverty experienced in Haiti.

CONCLUSIONS

National, public protected area systems are a critical element of biodiversity protection strategies globally. As argued in this chapter, full effectiveness of this approach is only possible if protected area management is part of the long-term objective of the governments and people involved. The success of individual parks is linked to the national systems to which they belong, and effective and sustainable protected area systems require that management continually adapt to the changing political and social contexts within which the areas exist. Successful adaptation requires constituencies that are politically effective and able to reach consensus on *what* areas need to be protected, *who* should and will protect the areas, and *how* the areas will, effectively, be protected. The process of effective protection involves collaboration among and participation of key stakeholders and adaptive management systems that involve learning and adapting as results of past management decisions emerge, rather than management focused on an initially defined end state.

Achieving these basic conditions is daunting in any context or country. Past experience with natural area management covers the range from utter failure to sustainable success, and indeed the latest research indicates that, on average, protection is far from effective in most countries. Given this great disparity between average performance and conditions for success, it is worth noting that the protected area approach derives primarily from the United States where it coevolved with democracy, increasing government transparency, accountability, and an ever-stronger civil society. Indeed, one could argue that the national park system in the United States has only approached full effectiveness in recent decades as government and civil society have matured.

Unfortunately, the protected area approach has been exported and imported to all corners of the globe to countries where these basic conditions for effectiveness frequently do not exist or are struggling to emerge — often without sufficient understanding of the requisite conditions for effectiveness. Trying to make the public protected area system work in these conditions is often like trying to fit a big square peg in a small round hole. It is at least a major challenge and perhaps completely impossible. The case of Haiti illustrates the many challenges of applying the protected area management approach in countries with weak democratic institutions and limited social capital. The challenge of applying this approach cannot be underestimated.

There is no doubt that public protection areas will and should remain a key approach in biodiversity protection, but the authors would argue that given what has been learned regarding the difficulty of making this approach work, conservation proponents should adopt a much more cautious and informed stance toward protected areas. First, conservationists must fully recognize the tremendous challenge of making this approach effective in many countries in the world, and, second, conservationists must dedicate more energy to identifying and promoting alternative and complementary approaches. Fortunately, new approaches are emerging, such as including conservation easements, private protection areas, tradable development rights, and conservation concessions, and showing promise in a number of different countries and contexts. Interestingly, many of these new approaches are market rather than government mechanisms, thus escaping some of the thorny issues and challenges associated with managing public land and government agencies.

It is clear that it is important to improve protected area systems wherever possible and to employ a new set of complementary approaches. The authors hope that the framework presented here will be useful in guiding efforts to strengthen protected area systems. But it is also clear that the larger challenge to achieving more effective biodiversity protection is mustering the leadership and political will to commit to conservation. Many countries have shown that such leadership can emerge and that such will can be mobilized if management is collaborative and adapts in response to changing social and political contexts.

ACKNOWLEDGMENTS

This chapter originated as a paper presented at the International Symposium entitled "Adaptive Collaborative Management of Protected Areas: Advancing the Potential" held at Cornell University, September 16–19, 1998. The authors thank Neil Bryon for his contributions and John Schelhas, Robin Mearns, Chuck Geisler, and Louise Buck for their constructive comments. The opinions in this document are those of the authors and not necessarily those of their respective institutions.

REFERENCES

Bormann, B. et al., 1994, Adaptive Ecosystem Management in the Pacific Northwest, General Technical Report PNW-GTR-341, USDA Forest Service, Pacific Northwest Research Station, Portland, OR.

Borrini-Feyerbend, G., 1997. *Beyond Fences: Seeking Social Sustainability in Conservation*, International Union for the Conservation of Nature (IUCN), Cambridge.

Brandon, K., Redford, K., and Sanderson, S., 1998. *Parks in Peril: People, Politics and Protected Areas*, The Nature Conservancy and Island Press, Washington, D.C.

CGIAR/TAC, 1996. Priorities and Strategies for Soil and Water Aspects of Natural Resources Management Research in the CGIAR, Document SDR/TAC:IAR/96/2.1, TAC Secretariat, Food and Agriculture Organization of the United Nations, Rome.

Cruz, M. C. and Davis, S. H., 1997. Social assessment in World Bank and GEF-funded biodiversity conservation projects: case studies from India, Ecuador, and Ghana, Environment Department Papers, Social Assessment Series, Paper 43, The World Bank. Washington, D.C.

Dudley, N. and Stolton, S., 1999. Threats to forest protected areas: A survey of 10 countries carried out in association with the World Commission on Protected Areas, Research Report from IUCN for the World Bank/WWF Alliance for Forest Conservation and Sustainable Use, The World Conservation Union (IUCN), Gland, Switzerland.

Fight, R. D. and Bell, E. F., 1977. Coping with Uncertainty — A Conceptual Approach for Timber Management Planning, General Technical Report PNW-59, USDA Forest Service, Pacific Northwest Forest and Range Experiment Station, Portland, OR.

Geisler, C. C., 1996. Adapting social impact assessment to protected area development, in The Social Challenge of Biodiversity Conservation, Davis, S. H., Ed., Working Paper 1, Global Environment Facility, Washington, D.C., 25–43.

Holling, C. S., Ed., 1978. *Adaptive Environmental Assessment and Management*, John Wiley & Sons, New York.

Kramer, R., Van Schaik, C., and Johnson, J., Eds., 1997. *Last Stand: Protected Areas and the Defense of Tropical Biodiversity*, Oxford University Press, Oxford.

Lundgren, A. L., 1983. Strategies for coping with uncertainty in forest resource planning, management, and use, in *Proceedings, New Forests for a Changing World, National Convention of the Society of American Foresters*, Society of American Foresters, October 16–20, Portland, OR, 574–578.

Ostrom, L., 1990. *Governing the Commons: The Evolution of Institutions for Collective Action*, Cambridge University Press, Cambridge.

Poole, P., 1989. Developing a partnership of indigenous peoples, conservationists and land use planners in Latin America, Policy, Planning and Research Working Papers, WPS 245. Latin America and the Caribbean Technical Department, The World Bank, Washington, D.C.

Smucker, G. and White, T. A., 1998. Devolution of protected-area management in Haiti: status and issues, paper presented at the 7th Common Property Conference of the International Association for the Study of Common Property in Vancouver, British Columbia, Canada, June 9–17, mimeo.

Thompson, E. F., 1968. The theory of decision under uncertainty and possible applications in forest management, *For. Sci.,* 14(2), 156–163.

Walters, C., 1986, *Adaptive Management of Renewable Resources,* Macmillan, New York.

Wells, M., Guggenheim, S., Khan, A., Wardojo, W., and Jepson, P., 1998 (June 19 draft). Investing in biodiversity: a review of Indonesia's Integrated Conservation and Development Projects, East Asia Region, The World Bank, Washington, D.C., mimeo.

13 Ecoregional Management in Southern Costa Rica: Finding a Role for Adaptive Collaborative Management

John Schelhas

CONTENTS

INTRODUCTION

Recent thinking in conservation biology stresses the importance of conservation at a regional scale that includes both protected areas and the lands that surround and connect them (Schelhas and Greenberg, 1996; Laurance and Bierregaard, 1997; Soulé and Terborgh, 1999). The need to develop governance systems that can incorporate the full diversity of landholders and interest groups that have decision-making power or an interest in regional land management is implicit in these

0-8493-0020-7/01/$0.00+$1.50

approaches (Cortner and Moote, 1999). This is the *collaborative* component of "adaptive collaborative management," which has received considerable attention in the literature in recent years (Sample, 1993; Western and Wright, 1994; Gunderson et al., 1995; Cortner and Moote, 1999). But it is also necessary to introduce state-of-the-art scientific knowledge into these collaborative processes to enable the *adaptive* management of regional ecosystems. Although "adaptive management" is a well established concept in ecology (Walters, 1986; 1997; Lee, 1993; Gunderson et al., 1995), there has been considerably less attention paid to it within the social sciences. Recent interest in collaborative approaches may have led some practitioners to believe that collaboration can substitute for social science. Yet, if larger landscapes in multiple ownerships are to be managed, it will be necessary to bring together an interdisciplinary understanding of both ecosystem processes at the regional landscape level and the human processes that increasingly are shaping these landscapes.

One approach to developing a science of human ecosystem interactions at the landscape level is formal modeling of relationships between social and economic factors and land cover patterns (Lee et al., 1992; Turner et al., 1996). Although formal modeling is an important tool, historical studies suggest that unpredictable exogenous variables such as changes in agricultural or wood product prices, or changes in laws and policies, can account for major changes in land use and cover (Turner et al., 1996). Since major social or economic reorientations and their impacts on land cover can often only be recognized in retrospect, a more diverse tool kit may be needed to capture new trends and ongoing processes for adaptive management. Conceptual modeling, which does not require quantitative data, can address this limitation and thereby complement (not replace) formal modeling. This chapter introduces the notion of conceptual modeling and makes a case for pluralistic modeling as a part of the adaptive collaborative management process. The chapter describes the situation related to the conservation of forests adjacent to La Amistad International Park on the Pacific slope of Costa Rica from the perspective of ecological anthropology, presents a conceptual model of land-use change processes for use in adaptive management, and discusses the potential for implementing a more collaborative approach to adaptive management at this site.

SITE DESCRIPTION

CONSERVATION SIGNIFICANCE

La Amistad International Park (PILA), which spans the border of Costa Rica and Panama, is one of the largest continuous forested areas and one of the largest protected areas in Central America (IUCN, 1992). The park was officially established in Costa Rica in 1982 (193,929 ha), and in Panama in 1988 (207,000 ha). In spite of its large size, the park has conservation limitations that stem from the fact that it comprises primarily higher elevation forests. Midelevation life zones are severely underrepresented in protected areas in Costa Rica, particularly on the Pacific slope (Powell et al., 1995/96; Guindon, 1996) because these are the areas best suited for growing Costa Rica's primary export commodity, coffee. The mid-elevational habitats adjacent to the Pacific side of PILA have been converted over the past 50 years from continuous forest to a mosaic of remnant forests, second-growth forests, forest plantations, coffee plantations, annual crops, and pastures (Schelhas et al., 1997). This conversion is of conservation interest not only because it threatens species endemic to the mid-elevation zone, but also because seasonal altitudinal migrations are common among Costa Rican bird, insect, and perhaps mammal species (Stiles, 1988; Loiselle and Blake, 1992; Guindon, 1996). If one part of a seasonal altitudinal habitat gradient is lost, it is likely that some of the species that use this gradient will not be able to survive. Thus, forest loss and change in the mid-elevation zone can have ecological repercussions across the full altitudinal gradient.

Prior to the 1990s, there was very little conservation biology research in southern Costa Rica. Consequently, little is known about biodiversity conservation issues in and around PILA. Recent research on birds, insects, trees, and mammals in this seminatural mosaic of forest and agricultural lands addresses this gap (Borgella, 1995; Daily and Erlich, 1995; Aldrich and Hamrick, 1998). Preliminary findings suggest that there are forest-dependent species inhabiting forest patches in the region and that ecological relationships among forest patches and the surrounding agricultural matrix are complex. However, no particular species or habitat of conservation concern has yet been identified on which to base the development of a landscape-level conservation plan involving forest corridors and/or stepping-stones. Major investment in any sort of formal biological corridor, along the lines of those being undertaken elsewhere in Costa Rica (e.g., La Selva, Talamanca, and Monteverde) would be unwise in the absence of a strong scientific rationale and a clear conservation objective. Nevertheless, it is reasonable to assume that the retention of forest patches and corridors on the private lands adjacent to PILA would have biodiversity conservation benefits and would also provide a wide range of other environmental and social benefits, including watershed protection, carbon sequestration, and provisions of forest products (Schelhas and Greenberg, 1996; Schelhas et al., 1997).

National-level maps of forest cover in Costa Rica generally show the area outside PILA as deforested. However, fine-scale mapping of forest patches from satellite images and air photographs (Figure 13.1) shows a network of forest patches interconnected by riparian forest corridors throughout the agricultural landscape adjacent to PILA (Wilson, 1998), that could serve as the foundation for regional, landscape-level forest conservation to complement and strengthen that in designated protected areas. Managing this landscape mosaic for biodiversity conservation, however, requires first understanding the social factors that are shaping it. This issue, and the application of an adaptive collaborative management approach, are the subject of this chapter.

Forest–People Interactions

Since 1992, the author has been conducting research in southern Costa Rica under two NSF-funded projects.* The first project is ecological and social research on forest patches. The second looks at the content and source of environmental values

* "Research Training Group, Ecological and Social Science Challenges of Conservation" (BIR-9113293293 and DBI-9602244), and "Policy, Norms, and Values in Forest Conservation: Protected Area Buffer Zone Management in Central America" (SBR-9613493).

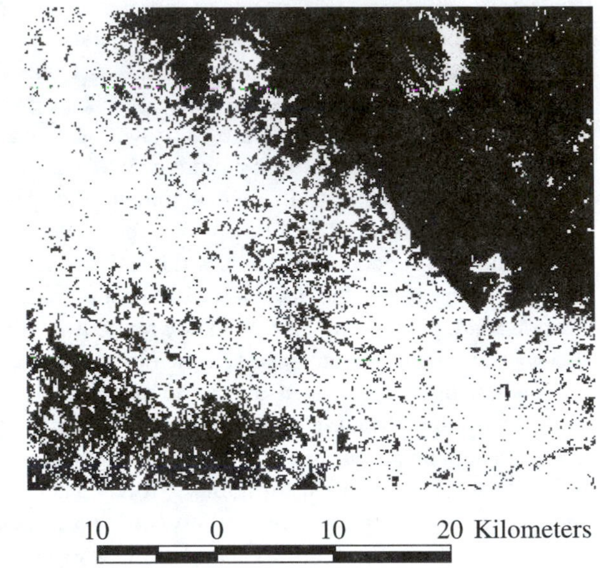

FIGURE 13.1 Fine-scale forest cover in southern Costa Rica (PILA boundary is the straight line to the right of the center of the figure). (After Wilson 1998.)

among rural landholders across individual, household, community, and national levels, as well as at related actual changes in amount and patterns of forest cover from satellite images.

The social research has generally focused on understanding farm household land-use choice behavior, including influences that are both economic and sociocultural. This research has found that farm-level land use is diversified to balance returns and risk; meet diverse needs for household products; meet sociocultural objectives; and provide environmental services (Schelhas, 1996a). Forests are a part of this diversified land-use mosaic for several reasons:

- They fill an economic niche in farming systems as a land use that accrues value over time while requiring low labor investments, once established. Thus, the research found forest more likely to occur when landholders are involved in highly remunerative off-farm labor, are older households, or otherwise have less labor available (Thacher et al., 1997). Forests may also be of significance to landholders as a complement to high-risk intensive agriculture (Schelhas, 1996b).
- Farm households often maintain forests as a source of materials such as timber, fuelwood, and vines for basketmaking (Schelhas, 1996a; Jantzi et al., 1998).
- One of the primary reasons given for farm forest conservation is for environmental services such as soil and water conservation (Thacher et al., 1997; Jantzi et al., 1999).

- Farmers give a number of "cultural" reasons for forest conservation, including as a legacy for their children, caring for God's creation, to maintain aesthetically pleasing environments, and to conserve wildlife and biodiversity (Thacher et al., 1997; Jantzi et al., 1999).
- It has become increasingly common in recent years to integrate a low and open canopy of nitrogen-fixing trees into coffee plantations, particularly those of small landholders, to minimize the use of agrochemicals (principally fertilizer by improving nutrient cycling) and thus minimize cash outlays, particularly when coffee prices are low (Schelhas, 1996a).

These are useful findings, but are limited by the fact that studies of individual and household behavior and values by themselves ultimately provide little insight into long-term land-use trends. The author argues here that developing a model of changing land-use patterns must go beyond individual decision making and biophysical models to also include attention to social structures and processes. The basis for such a model can be found in Rhoda Halperin's book, *Cultural Economies Past and Present*. Halperin (1994) notes that broader patterns and relationships are more difficult to analyze and observe than individual behavior, but that this does nothing to diminish their explanatory power.

Halperin (1994) discusses the use of "formal processual models," which are essentially ideal types that can be used as a standard of comparison. As such, they are very much like hypotheses for adaptive management. It is important to note that Halperin (1994, p. 252) emphasizes that these formal processual models are not commonsense understandings, which are culturally embedded and therefore often inaccurate. Rather, formal processual models develop a set of concepts that, when used analytically, illuminate the ways that economies function and how this differs cross-culturally by focusing on a set of economic processes in a generic model of the economy (Halperin, 1994, p. 51). Models are heuristic devices to help order data, and in formal processual models the units are complicated processes or sets of social relations (as opposed to formal atomistic models in which the units are individual actors) (Halperin 1994, p. 165). Formal processual models are perhaps best understood through examples, and one is presented here from research in Costa Rica.

PATTERNS AND RELATIONSHIPS IN SOUTHERN COSTA RICA

Land Speculation

On frontiers, land as a source of exchange value is generally at least as important as land as productive resource (Moran, 1988; Schelhas, 1996b). There is now general agreement that the rapid expansion of the frontier and concomitant widespread conversion of forests to pastures that took place in Central America from the 1960s through the late 1980s was driven by much more than the "Hamburger Connection" or any other productive use of land (Edelman, 1995). Equally important to cattle markets was a process of land speculation, abetted by government policies, in which land was cleared and kept clear of trees both as a part of a process of claiming public or absentee-owned lands for private benefit, and as a way of defending against these actions (Schelhas, 1996a).

What is perhaps most interesting here is that while the role of forest-to-pasture conversion in this process has diminished in recent years, the speculative process itself continues. What has changed is that now forests, rather than pastures, are the land use of choice for speculation. At first glance this may appear to be good, from a conservation point of view, since people are now conserving forests in the belief that they will add to increased land values more than would agricultural land uses. However, the long-term economic value and productivity of forests in Costa Rica is unclear, and the speculative retention and planting of forest may very well be on no firmer economic ground than was the cattle boom of the 1970s and 1980s. The hope and expectation that profits can be realized from forests and forested lands appear to be driven more by a combination of expatriate purchases of forested land and a boom in forest-associated ecotourism, both of which send a message that forests can be economically valuable, rather than on actual returns from forest-based enterprises. Like cattle pastures, forests will in the end provide attractive economic returns only in certain places under certain circumstances, if at all. Forest values can be expected to change as the trend toward speculation in forestlands matures.

National and Global Cultural Change

Complementing the change in the relative value of pasture and forests, which was described above, has been an overall cultural change in the way land is viewed in Costa Rica. Prior to the 1980s the prevailing land *myth* in Costa Rica was that of a nation of small, independent farmers claiming land by working it (Biesanz et al., 1982). Since the 1980s, this myth has been supplanted by a new myth of Costa Rica as an ecological paradise (Boza et al., 1995). Evidence of this can be found in the national media. As a Costa Rican friend points out, where before, as TV stations signed off for the night, images of *campesinos,* coffee harvests, and oxcarts flashed across the screen, now images of Costa Rica's national parks, plants, and wildlife are seen. This new myth was manifested in interviews with rural landholders, who express diverse values — for heritage, community, and aesthetic values as well as for products and services — related to forests and biodiversity and, perhaps more interesting, seem to be reinterpreting many of their farming practices in conservation terms. For example, they sometimes describe coffee and fruit tree planting as reforestation (Jantzi et al., 1999; Pfeffer et al., 1999). This suggests a complex and dialectical relationship between environmental values and behaviors that invites better understanding through further research and observation.

Global Economic Change

In addition to the influence of global cultural forces such as environmentalism, there are also influences from global market forces related to the liberalization of trade. Just as the process of forest-to-pasture conversion in the 1970s and 1980s was partially driven by international policies and market forces, so, too, are the more recent land use trends. The decline of the cattle market in Costa Rica was the result of reduction of government incentives because of international and domestic environmental pressures, as well as changing international markets and trade relations (Lehman, 1992; Müller, 1998; Abler et al., 1999). Similarly, the rise in economic importance of forests has several international sources. These include the elevation of tourism, much of it ecotourism, to the place of number one earner of foreign exchange in Costa Rica in the early 1990s (Pratt, 1999), when it overtook traditional commodity exports such as bananas and coffee.

An equally important economic trend has been the demand in developed countries for sustainably grown forest products. A number of sustainable forestry operations in Costa Rica have developed, and the U.S. and European markets for sustainably grown timber is expanding (Jenkins and Smith, 1999). A number of small farmers in southern Costa Rica are trying to tap into international demand for organic beans and organic and shade-grown coffee in an effort to garner premium prices for their crops by exploiting these niche markets. The potential of niche markets for sustainable or "green" forest products is increasing landholder interest in tree and forest conservation. Whether this continues will depend on the development of profitable and accessible markets for these products.

A third trend is the rise of plantation agricultural and *maquiladora* factories that produce everything from clothing to computers and related job growth in these sectors. Nontraditional exports are increasing more rapidly than traditional ones (Proyecto Estado de la Nación, 1996), and in 1998 the computer industry surpassed tourism as the number one earner of foreign exchange in Costa Rica (Pratt, 1999). As indicated above, there is evidence that involvement in well-paid* off-farm employment has a positive influence on forest retention on farms (Thacher et al., 1997). This is a hypothesis that should be examined over time, because, if free trade does what it has promised (bring about widespread economic growth and prosperity by stimulating nontraditional agricultural exports and industry), there may be a large-scale regeneration of forests similar to what occurred in the northeastern United States or Puerto Rico earlier (Williams, 1989; Franco et al., 1997; Koop and Tole, 1997). An alternative hypothesis is that, if trade liberalization fails to deliver benefits that exceed costs for the rural poor, there could be a return to shifting agriculture for subsistence production at the expense of forests.

Changing Forest Policies

Another important trend has been changing forest policies in Costa Rica. While many of the government policies that promoted unproductive deforestation in the past have changed (Watson et al., 1998), broader changes in the political process provide new cause for concern. Costa Rica has a tradition of alternating between its two major political parties in presidential elections. (Costa Rica elects a new president every 4 years, under a system that prohibits a president from running for

* Poorly paid off-farm employment may result in a simultaneous shortage of land and labor that can lead to intensive farming without the investments in resource management required for sustainability (Collins, 1987).

reelection.) Frequent changes of government, combined with a rise in technocratic policy making, have led to almost annual changes in forest policies. The result is that new policies are changed before they have time to filter into public consciousness, and the operative effect is that the only thing that landholders have confidence in is that forest and land-use polices *will* change. This uncertainty, in turn, creates a climate in which long-term forest land use and conservation are perceived as risky because of uncertainty about what future forestry practices will be permissible or favored. This suggests that too frequent policy adjustments may produce perverse results, and should serve as a cautionary note for adaptive management and any other process that frequently reevaluates and recommends changes in policies.

An additional issue is the recent dominance of an economic approach to environmental polices, emphasizing transfer payments by the government to individual landholders for conserving forests or reforestation. However sensible these payments may appear when the economic costs and benefits of forest conservation are analyzed (Kishor and Constantino, 1993), the author's interviews reveal hints that farmers participating in these programs may feel that they are caring for forests more for the government than for themselves, which *may* foreshadow an erosion of the local social and cultural mechanisms that promote forest conservation. This is exacerbated by the fact that the Costa Rica government, forced to reduce government expenditures on core services such as education and health care by austerity measures imposed by the international lending agencies, has proved unable to sustain reliably many of the incentive programs promoting forest conservation (Rohter, 1996; Escofet, 1998; Dulude, 2000). The result may be the worst of both worlds — a shift in attributed responsibility for forests from individuals to the government, combined with ineffective government forest conservation efforts. If the government is unable to sustain its financial incentives promoting forest conservation, it may be better off relying on social and cultural means rather than allowing forests to be eroded by short-term transfer payments.

CONCEPTUAL MODELS FOR ADAPTIVE MANAGEMENT

The above processes can be converted into hypotheses about the changing relationships between people and forests adjacent to one sector of La Amistad National Park, in Costa Rica. These hypotheses suggest important questions that should be asked in research and observation of forest–people relationships in Costa Rica. A set of hypotheses would include:

1. Speculative land markets drive land-use choice in Costa Rica as much as productive land-use value. Much of the recent interest in forestlands is speculative, and therefore may not be sustained.
2. Costa Rica is undergoing a cultural shift in national identity from that of an "agrarian democracy" to one of an "ecological paradise." The results of this are a mixed amalgam of increased valuing of trees and forests and reinterpreting existing land-use practices in favorable ecological terms.

3. The Costa Rican economy is shifting under trade liberalization to greater orientation toward export production (both industrial and nontraditional agricultural products). To the extent that export industries provide large numbers of well-paying jobs (by Costa Rican standards), Costa Rican forests will recover substantially, particularly on lands that are marginal for agriculture.
4. Frequent changes in forest policy may create uncertainty about future returns from forests that discourages forest management regardless of the substance of the policies.
5. Economic mechanisms to promote forest conservation may undermine sociocultural mechanisms.

These hypotheses that can help form a conceptual model for adaptive management and orient long-term research to produce the cumulative learning that will provide the basis for natural resource management in the future. Such an approach would differ substantially from the fragmented and *ad hoc* way that forest policy and management have been approached in the past.

Although this chapter has emphasized "formal prosessual models," the larger point is that there is a need for pluralism in the development of models for adaptive management. It is unlikely that one will be able to develop "super-models" that rigorously include all the different scientific approaches — quantitative and qualitative — that can inform forest management and policy. However, the example above found that models of process have important nodes of articulation with individual choice models (e.g., off-farm employment and changing forest cover) and one can expect there to be important and instructive linkages between other models. A more realistic goal than a single model may be a disciplinary pluralism that promotes the development of many different conceptual and formal models, rather than the development of a single model, and allows managers and scientists together to sort out the lessons that these models provide. For example, in the Costa Rican case, the conceptual model interfaces well with economic and policy models (Lutz and Daly, 1991; Kishor and Constantino, 1993; Abler et al., 1999) and would be complemented by ecological models as well.

Finally, while it is suggested here that multiple scientific models have a very important role to play in adaptive management, it is equally important to include models held and developed by residents of the region — including farmers, land managers, and business people (i.e., "folk models"). Thus it is argued that the use of participatory processes is important, and that making models through participatory processes is complementary to scientific models. This leads to the collaborative component.

ADDING A COLLABORATIVE COMPONENT

There are few, if any, examples of collaborative natural resource management from southern Costa Rica, and the author does not have much to report regarding the use of collaborative approaches in the region adjacent to PILA in southern Costa Rica beyond his involvement in several Participatory Rural Appraisals. Perhaps the lack of collaborative approaches is more logical than it seems. There appear to be several reasons collaborative approaches to forest management in southern Costa Rica are problematic, and understanding these is perhaps the first step to beginning to think about how a collaborative component might be added to the adaptive management component whose development was outlined above.

Uneven Concentration of Power

One issue is the fact that Costa Rica has been historically characterized by a concentration of power at the two extremes — the individual/farm household level and the national government level. It is also true that national government agencies dealing with natural resource management on private lands have been historically weak. Community, provincial, and regional organizations and institutions in Costa Rica are relatively undeveloped, even in contrast to other Central American countries. Although Costa Rica is a long-standing democracy, the lack of intermediate organizations and institutions has in general made it difficult to develop the nested hierarchies of institutions that are needed to support farmer organizations and community-based conservation and sustainable development at the watershed and ecoregional levels (Ostrom, 1990; Uphoff, 1993; Pritchard et al., 1998). Coto Brus, where the author's group is working, is particularly disadvantaged in this regard because it is a relatively isolated region with a near total lack of conservation or development projects and a very low level of services from government ministries or NGOs.

Governance Changes and Economic Changes

Costa Rica has been influenced by the recent worldwide trend toward devolution of government power, and there is an ongoing effort to transfer national government power and responsibilities down to the municipal level. Perhaps not surprisingly, as in many other places, the central government is transferring the responsibilities but not the money and other resources to local governments. Local governments must raise the money through the implementation of a new property tax. Increasing property taxes are generally considered to be a detriment to forest management (GAO, 1978; Coughlin, 1980; Greene, 1994). It is not clear that adequate allowances are being, or can be, made in the new tax codes to provide incentives to maintain land in forests as property taxes are instituted. This will depend on how forest conservation is prioritized at the community and municipal levels relative to development or meeting municipal financial needs. Thus, it is possible that strengthening local institutions in combination with the institution of property taxes could have a deleterious effect on forestland uses and conservation. Even if a mechanism such as easements were regarded as a viable option in this area, there is a very serious question of who would manage and enforce the easements (see Gustanski and Squires, 1999). Paradoxically, the apparent road to collaboration via devolution is paved with pitfalls.

Uneven Distribution of Costs and Benefits of Forest Conservation

The way forest values are traded off with other values is fundamental to the amount of support forest programs have from local residents and decision makers (Satterfield and Gregory, 1998). Focus group interviews in one community in Coto Brus suggest that awareness of environmental problems is widespread, but people not directly involved in conservation committees may not rank them as highly in importance as other concerns (Schelhas, 1996c). This may be because many of the costs and benefits of forest conservation accrue across different levels of the global to local continuum. For example, conservation of Costa Rica's biodiversity is perhaps of greatest interest at the national and global levels. This is *not* to say there is no local interest in biodiversity conservation — focus group and interview results have indicated that there is — but at the local level it is more likely to conflict with human livelihoods and development aspirations. In another example, soil and water conservation, which research indicates is the biggest motivations for forest conservation in southern Costa Rica, splits costs and benefits between upstream and downstream landholders and communities. Only by developing a full set of nested institutions can these conflicts between levels and places be addressed.

But the question of the likelihood of this happening remains. Most of the success stories in the literature are examples where economic dependence on forest products — often nontimber forest products — is strong. This is not the case in Costa Rica. Economic returns from forests through ecotourism, small- and large-scale timber cutting, and nontimber forest products are low and received by relatively few people. Although there is considerable evidence that rural people value forests, including their local forests, forest values in southern Costa Rica are grounded in less tangible ecosystem services values, particularly watershed values. Interestingly, many of the communities, for example, Siete Colinas and Alpha (Jantzi et al., 1999; O'Connor, 1998), in which forests are currently being protected in the belief that this will maintain local water supplies are soon going to be tied into an expanded aqueduct system that is bringing water from high in the mountains. It remains to be seen whether dispersed and intangible forest values can be significant enough to promote the development of and participation in institutions for forest conservation.

The above discussion of collaborative issues suggests additional hypotheses for adaptive collaborative management. These include:

1. Nested-level hierarchies of governance are necessary for successful ecoregional forest management.
2. Devolution of forest decision making authority in Costa Rica, in combination with the implementation of property taxes, is both increasing and decreasing incentives for forest management. The results, in terms of changes in forest cover, will depend on the way that differential assessments for forestlands are implemented and on the economic value of forests themselves.
3. Strong local forest conservation mechanisms are unlikely to develop in an agrarian economy in the absence of a significant economic value for forest products.

CONCLUSIONS

The adaptive component of managing land use adjacent to protected areas must be pluralistic in including different types of scientific models that focus on patterns and relationships, as well as on individual behaviors. This chapter has used an example based on Halperin's notion of formal processual models for understanding forest and tree use and conservation on privately held farms in southern Costa Rica — a model that is distinctly nonquantitative — to illustrate this point. Both the human–forest interactions and the forest cover–biodiversity relationships are highly complex at the regional level. Conceptual modeling to specify important relationships can provide some propositions for adaptive management and policy making, and also identify important research questions that can help fill in critical gaps in the knowledge for long-term adaptive and collaborative management of regional human-occupied ecosystems.

The prospects for the development of *comprehensive* formal models that cross disciplines appear dim. A better approach may be to use a suite of *partial* models systematically — formal quantitative, conceptual, and folk — and seek linkages among them. These can be used to develop the hypotheses on which to base adaptive management and research. Collaborative processes are fundamental to this, yet there are significant obstacles to implementing them in southern Costa Rica. By making these obstacles explicit, the collaborative and adaptive approaches can be merged to inform and improve efforts to develop the art and science of adaptive collaborative management in southern Costa Rica and elsewhere.

ACKNOWLEDGMENTS

The author thanks Charles Geisler and Jeff Langholz for their thoughtful comments on an earlier draft of this chapter.

REFERENCES

Abler, D. G., Rodriguez, A. G., and Shortle, J. S., 1999. Trade liberalization and the environment in Costa Rica, *Environ. Dev. Econ.,* 4(3), 357–373.

Aldrich, P. R. and Hamrick, J. L., 1998. Reproductive dominance of pasture trees in a fragmented tropical forest mosaic, *Science,* 281(5373), 103–105.

Biesanz, R., Zubris Biesanz, K., and Hiltunen Biesanz, M., 1982. *The Costa Ricans*, Prentice-Hall, Englewood Cliffs, NJ.

Borgella, R., 1995. Population Size, Survivorship, and Movement Rates of Resident Birds in Costa Rican Forest Fragments, M.S. thesis, Cornell University, Ithaca, NY.

Boza, M. A., Jukofsky, D., and Wille, C., 1995. Costa Rica is a laboratory, not ecotopia, *Conserv. Biol.,* 9(3), 684–685.

Collins, J., 1987. Labor scarcity and ecological change, in *Lands at Risk in the Third World: Local-Level Perspectives*, Little, P. D., Horowitz, M. M., and Nyerges, A. E., Eds., Westview Press, Boulder, CO, 19–37.

Cortner, H. J. and Moote, M. A., 1999. *The Politics of Ecosystem Management,* Island Press, Washington, D.C.

Coughlin, R. E., 1980. The economic impact: differential assessment and the conversion of land to urban uses, in *Property Tax Preference for Agricultural Land*, Roberts, N. A. and Brown, H. J., Eds., Allanheld, Osmun, Montclair, NJ, 43–64.

Daily, G. C. and Ehrlich, P. R., 1995. Preservation of biodiversity in small rainforest patches: rapid evaluations using butterfly trapping, *Biodiversity Conserv.,* 4, 35–55.

Dulude, J., 2000. Pioneer program in peril, *Tico Times,* February 4, VI(5), available online at http://ticotimes.co.cr

Edelman, M., 1995. Rethinking the Hamburger thesis: deforestation and the crisis of Central America's beef exports, in *The Social Causes of Environmental Destruction in Latin America*, Painter, M. and Durham, W. H., Eds., University of Michigan Press, Ann Arbor, 25–62.

Escofet, G., 1998. Suspensions of forest incentives blasted, *Tico Times,* 42(1483), 1, 8.

Franco, P. A., Weaver, P. L., and Eggen-McIntosh, S., 1997. Forest Resources of Puerto Rico, 1990, Resource Bulletin SRS-22, Southern Research Station, USDA Forest Service, Asheville, NC.

GAO, 1978. Effects of Tax Policies in Land Use, CED 78-97, General Accounting Office, Washington, D.C.

Greene, J. L., 1994. State tax systems and their effects on nonindustrial private forest owners, in *Proceedings of the 1994 Society of American Foresters/Canadian Institute of Forestry Convention,* Society of American Foresters, Bethesda, MD.

Guindon, C. F., 1996. The importance of forest fragments in the maintenance of regional biodiversity in Costa Rica, in *Forest Patches in Tropical Landscapes*, Schelhas, J. and Greenberg, R., Eds., Island Press, Washington, D.C., 168–186.

Gunderson, L. H., Holling, C. S., and Light, S. S., 1995. *Barriers and Bridges to the Renewal of Ecosystems and Institutions*, Columbia University Press, New York.

Gustanski, J.A. and Squires, R. H., 1999. *Protecting the Land: Conservation Easements Past, Present, and Future*, Island Press, Washington, D.C.

Halperin, R. H., 1994. *Cultural Economies Past and Present*, University of Texas Press, Austin.

IUCN, 1992. *Protected Areas of the World: A Review of National Systems,* Vol. 4: *Neararctic and Neotropic,* IUCN, Gland, Switzerland.

Jantzi, T., Schelhas, J., and Lassoie, J. P., 1999. Environmental values and forest patch conservation in a rural Costa Rican community, *Agric. Hum. Values,* 16, 29–39.

Jenkins, M. B. and Smith, E. T., 1999. *The Business of Sustainable Forestry: Strategies for an Industry in Transition*, Island Press, Washington, D.C.

Kishor, N. K. and Constantino, L. F., 1993. *Forest Management and Competing Land Uses: An Economic Analysis for Costa Rica*, The World Bank, Washington, D.C.

Koop, G. and Tole, L., 1997. Measuring differential forest outcomes: a tale of two countries, *World Dev.,* 25(12), 2043–2056.

Laurance, W. F. and Bierregaard, R. O., 1997. *Tropical Forest Remnants: Ecology, Management, and Conservation of Fragmented Communities,* University of Chicago Press, Chicago.

Lee, K. N., 1993. *Compass and Gyroscope: Integrating Science and Politics for the Environment*, Island Press, Washington, D.C.

Lee, R. G., Flamm, R., Turner, M. G., Bledsoe, C., Chandler, P., DeFerrari, C., Gottfried, R., Naiman, R. J., Schumaker, N., and Wear, D., 1992. Integrating sustainable development and environmental vitality: a landscape ecology approach, in *Watershed Management: Balancing Sustainability and Environmental Change*, Naiman, R. J., Ed., Springer-Verlag, New York, 499–521.

Lehman, M. P., 1992. Deforestation and changing land use patterns in Costa Rica, in *Changing Tropical Forests: Historical Perspectives on Today's Challenges in Central and South America*, Steen, H. K. and Tucker, R. P., Eds., Forest History Society, Durham, NC, 58–76.

Loiselle, B. A. and Blake, J. G., 1992. Population variation in a tropical bird community: implications for conservation, *BioScience,* 42(11), 838–845.

Lutz, E. and Daly, H., 1991. Incentives, regulations, and sustainable land use in Costa Rica, *Environ. Resourc. Econ.,* 1, 179–194.

Moran, E., 1988. Social reproduction in the agricultural frontier, in *Production and Autonomy*, Bennet, J. W. and Bowen, J. R., Eds., University Press of America, Lanham, 199–212.

Müller, E., 1998. Land use policy and secondary forest management in the northern zone of Costa Rica, in *Ecology and Management of Tropical Secondary Forest: Science, People and Policy*, Guariguata, M. R. and Finegan, B., Eds., CATIE, Turrialba, Costa Rica, 11–18.

O'Connor, K. A., 1998. Riparian Land Use and Riparian Land Use Change in a Rural Community in Costa Rica, M.S. thesis, Cornell University, Ithaca, NY.

Ostrom, E., 1990. *Governing the Commons: The Evolution of Institutions for Collective Action,* Cambridge University Press, Cambridge.

Pfeffer, M., Schelhas, J., and Day, L., 1999. Forest Conservation, Value Conflict, and Interest Formation in a Honduran National Park, unpublished manuscript.

Powell, G. V. N., Bjork, R. D., Rodriguez, S. M., and Barborak, J., 1995/96. Life zones at risk: gap analysis in Costa Rica, *Wild Earth,* 5(4), 46–51.

Pratt, C., 1999. Intel ousts tourism as top earner, *Tico Times,* 43(1521), 1, 8.

Pritchard, L., Jr., Colding, J., Berkes, F., Svedin, U., and Folke, C., 1998. The Problem of Fit between Ecosystems and Institutions, IHDP Working Paper No. 2, International Human Dimensions Programme on Global Environmental Change, Bonn, Germany.

Proyecto Estado de la Nación, 1996. *Estado de la Nación en Desarrollo Humano Sostemible*, Lara Segura, San José, Costa Rica.

Rohter, L., 1996. Costa Rica chafes at new austerity, *New York Times,* September 10.

Sample, V. A., 1993. Building Partnerships for Ecosystem Management on Forest and Range Lands in Mixed Ownership, Forest Policy Center, Washington, D.C.

Satterfield, T. and Gregory, R., 1998. Reconciling environmental values and pragmatic choices, *Soc. Nat. Resourc.,* 11, 629–647.

Schelhas, J., 1996a. Land use choice and forest patches in Costa Rica, in *Forest Patches in Tropical Landscapes*, Schelhas, J. and Greenberg, R., Eds., Island Press, Washington, D.C., 258–284.

Schelhas, J., 1996b. Land use choice and change: intensification and diversification in the lowland tropics of Costa Rica, *Hum. Organ.,* 55(3), 298–306.

Schelhas, J., Ed., 1996c. An Interdisciplinary Analysis of Sustainable Landscape Management in Siete Colinas, Coto Brus, Costa Rica, Department of Natural Resources, Cornell University, Ithaca, NY.

Schelhas, J. and Greenberg, R., 1996. *Forest Patches in Tropical Landscapes,* Island Press, Washington, D.C.

Schelhas, J., Jantzi, T., Kleppner, C., O'Connor, K., and Thacher, T., 1997. Meeting farmers' needs through forest stewardship, *J. For.,* 95(2), 33–38.

Soulé, M. E. and Terborgh, J., 1999. *Continental Conservation: Scientific Foundations of Regional Reserve Networks*, Island Press, Washington, D.C.

Stiles, F. G., 1988. Altitudinal movements of birds on the Caribbean slope of Costa Rica: implications for conservation, in *Tropical Rainforests: Diversity and Conservation*, Almeda, F. and Pringle, C. M., Eds., California Academy of Sciences and Pacific Division, American Association for the Advancement of Science, San Francisco, 243–258.

Thacher, T., Lee, D. R., and Schelhas, J. W., 1997. Farmer participation in reforestation incentive programs in Costa Rica, *Agrofor. Syst.,* 35, 269–289.

Turner, M. G., Wear, D. N., and Flamm, R. O., 1996. Land ownership and land cover change in the Southern Appalachian Highlands and the Olympic Peninsula, *Ecol. Appl.,* 6(4), 1150–1172.

Uphoff, N., 1993. Grassroots organizations and NGOs in rural development: opportunities with diminishing states and expanding markets, *World Dev.,* 21(4), 607–622.

Walters, C., 1986. *Adaptive Management of Renewable Resources,* Macmillan, New York.

Walters, C. J., 1997. Challenges in adaptive management of riparian and coastal ecosystems, *Conserv. Ecol.,* 1(2), 1, available at http://www.consecol.org/vol1/iss2/art1.

Watson, V., Cervantes, S., Castro, C., Mora, L., Solis, M., Porras, I. T., and Cornejo, B., 1998. *Making Space for Better Forestry: Costa Rica Country Case Study,* IIED, London.

Western, D. and Wright, R. M., 1994. *Natural Connections: Perspectives in Community-Based Conservation,* Island Press, Washington, D.C.

Williams, M., 1989. *Americans and their Forests: A Historical Geography,* Cambridge University Press, Cambridge.

Wilson, C. L., 1998. A Comparative Analysis of Tropical Forest Fragmentation Patterns Using Aerial Photograph Interpretation and Satellite Image Processing Techniques, M.S. thesis, Cornell University, Ithaca, NY.

14 Population Dynamics, Migration, and the Future of the Calakmul Biosphere Reserve

Jenny A. Ericson, Mark S. Freudenberger, and Eckart Boege

CONTENTS

0-8493-0020-7/01/$0.00+$1.50

INTRODUCTION

The international conservation community continues to focus on the environmental consequences of human settlements in and around protected areas. Particularly in less-developed regions of the world, protected areas are often little more than polygons on a map; and administrative capacity is insufficient to control both the impact of colonists seeking land and the traditional use of resources by local populations (McNeely and Ness, 1996). Conservationists are beginning to confront the reality that hard-earned conservation gains are threatened by an influx of people into national parks, biosphere reserves, extractive reserves, and other types of protected areas (Cruz, 1996; Barton et al., 1997; Caudill, 1997; Dompka and Allcott, 1998; Engelman, 1998). For example, in the rain forests of the Central African Republic, miners extract diamonds from the rivers and streams of the Dzanga-Sanga Special Dense Forest Reserve located in the buffer zones (Freudenberger et al., 1996). In the Annapurna Conservation Zone of Nepal, employment generated by ecotourism draws in migrant workers whose dependency on the extraction of firewood and other forest products increases pressure on surrounding resources (Bajracharya et al., 1997). The rich literature on settlement along roads constructed into Amazonian forests paints a grim picture of the difficulties encountered in mitigating the impacts of in-migration (Hecht and Cockburn, 1989; Rudel and Horowitz, 1993; Pichon, 1993). And, as in southeastern Mexico, landless rural populations in Central American countries are encouraged to settle around national parks, which constitute some of the last remaining agricultural frontiers (Bilsborrow and DeLargy, 1991; Pasos et al., 1995).

The history of the human species is certainly one of migration. Demographers warn the conservation community that "of all population problems, those of migration appear the most intractable. … [M]igration streams are very difficult to start when people are not willing to move, and very difficult to stop when they do wish to move" (Ness with Golay, 1997, p. 119). With increasing frequency, conservationists are recognizing that landscapes harboring rich pockets of biodiversity will soon be radically transformed due to unprecedented levels of in-migration and that actions must be taken now to mitigate the ecological impacts of this phenomenon.

This chapter is a description of population dynamics and in-migration around the Calakmul Biosphere Reserve (CBR) located in the southern Yucatán Peninsula of Mexico. It presents results from the initial phases of an applied research program developed with the following objectives:

1. Generate dialogue between stakeholders (e.g., community members, regional authorities, and national and international governmental and non-governmental institutions, or NGOs) about the impact of population growth or distribution on biodiversity conservation around the reserve;
2. Develop an effective, low-cost population monitoring system; and
3. Establish a culturally and politically appropriate participatory land-use planning process that takes account of complex local and regional human population dynamics.

The program is a collaborative effort between Pronatura Península de Yucatán, A.C. (PPY), World Wildlife Fund (WWF), and the University of Michigan Population–Environment Fellows Program.

Information gathered primarily through the use of participatory and other qualitative research methods illustrates the complex mosaic of causes and environmental consequences of in-migration around the reserve. The applied research shows that the future of the reserve is compromised by both a steady influx of migrants and rapid natural population growth rates in the *ejido** communities. Moderating the environmental consequences of in-migration involves the creation of effective community land-use planning practices coupled with supportive national and regional economic and social policies. However, numerous deep structural challenges confront institutions committed to the promotion of conservation and just economic and social development.

THE CALAKMUL BIOSPHERE RESERVE AND CONSERVATION INITIATIVES

Covering 723,185 ha, the CBR is the largest tract of protected tropical forest in Mexico and an important site for biodiversity conservation. Established in 1989 by presidential decree, the reserve is located in the state of Campeche on the Yucatán Peninsula. It is an important element in a larger system of protected areas that form an ecological corridor of over 2 million ha stretching between central Yucatán and the Belizian forests (Figure 14.1). The main objective of the CBR is the long-term maintenance of biodiversity. Accepted into the UNESCO network of biosphere reserves in 1993, the CBR is divided into *core* and *buffer* zones. Although, to date, no management plan has been approved, the general understanding is that ecologically sustainable production activities are allowed within the buffer zone, while no human activity is permitted within the core zone (Galindo-Leal, 1996). Conflict arises from the fact that the borders of the core zone cut across the territory of preexisting *ejido* communities and privately held properties (Figure 14.2). Agriculture, forestry, cattle ranching, and subsistence-level hunting are practiced on *ejido* and privately held lands that overlap with the reserve. The buffer zone of the southern division of the reserve is composed of forest extension lands belonging to *ejido* communities north of the reserve.

* An *ejido* is a land grant administered by a group of individuals called *ejidatarios* who hold the usufruct rights to their *ejido* accorded to them by the Mexican federal government.

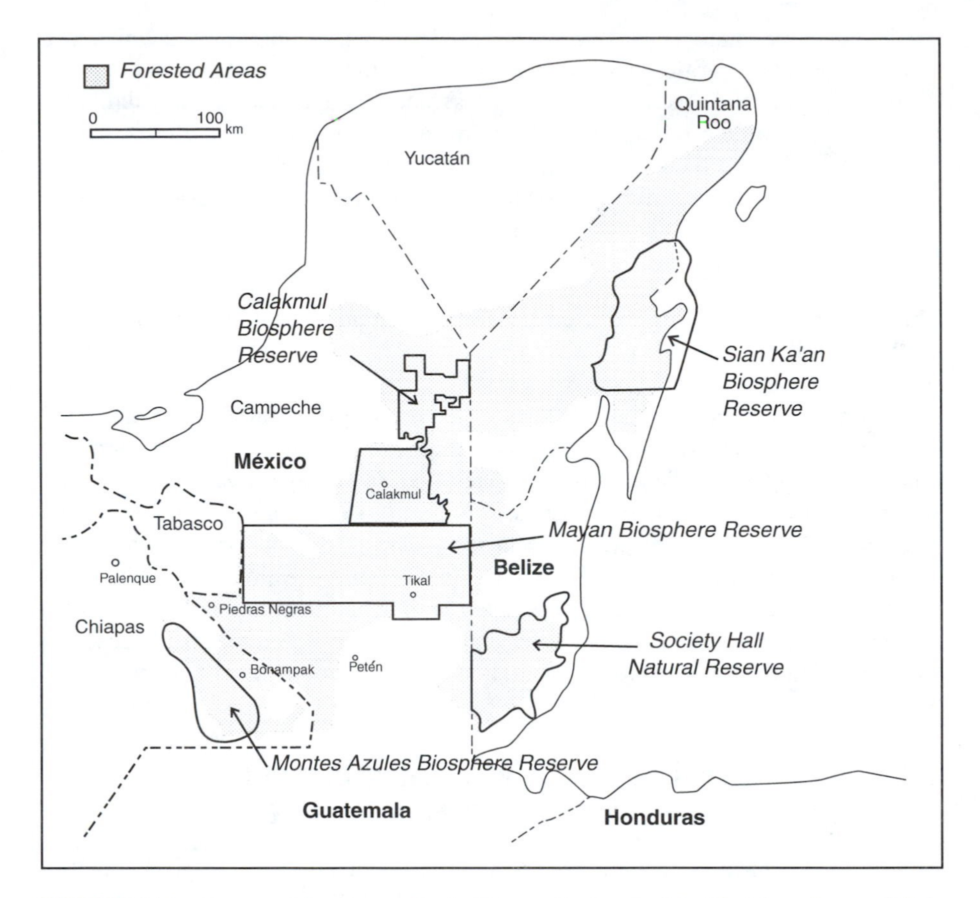

FIGURE 14.1 Protected forest areas in the Yucatan Peninsula. Modified from Boege, 1995.

The CBR is a patchwork of mature disturbed forest, secondary growth vegetation of less than 25 years, and savanna-type floodplains. The present state of the forest both within the reserve and in the *ejido* communities that surround it is a result of timber extraction, forest clearing for agriculture, and cattle ranching. The most abundant tall trees include chicozapote (*Manilkara zapota*) and ramón (*Brosimum alicastrum*) (Miranda and Hernandez-Xolocotzi, 1985; Rzedowski and de Rzedowski, 1989). Prominent commercial species are mahogany (*Swietenia macrophylla*) and Spanish cedar (*Cedrela odorata*). Biological inventories indicate that 18 endemic plant species are found in the larger Peten ecosystem.

Despite transformations in the landscape, the CBR is home to charismatic threatened and endangered mammals such as jaguars (*Panthera onca goldmani*), howler monkeys (*Alouatta pigra*), spider monkeys (*Ateles geoffroyi*), and tapir (*Tapirus bairdii*). Of the bird species sighted in the reserve, 30% breed in the United States and Canada and use these forests as their wintering grounds (Berlanga and Wood, 1997). Some of these neotropical migrants, such as the hooded warbler (*Wilsonia citrina*) and the swainson's warbler (*Limnothlypis swainsonii*), are threatened or endangered species (NOM, 1994).

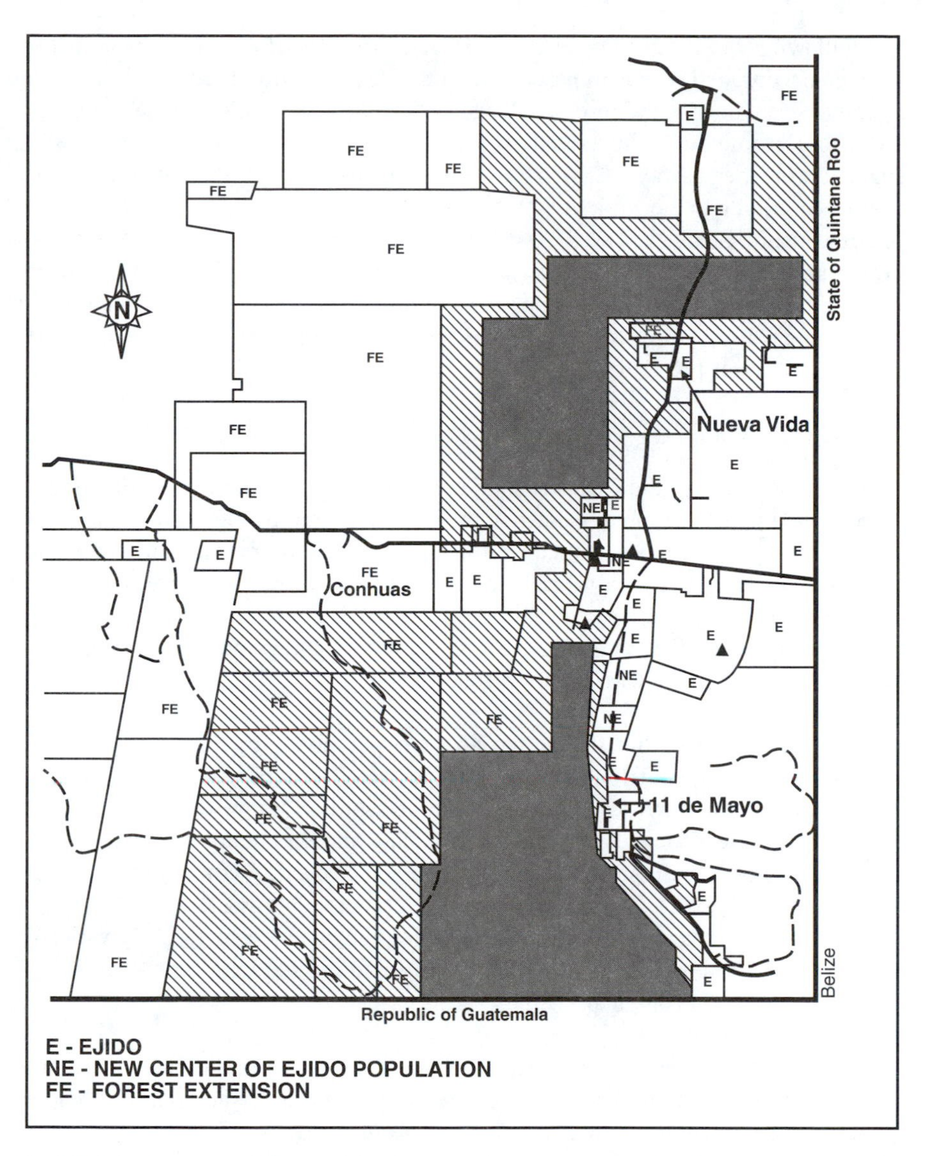

FIGURE 14.2 Map of CBR and land ownership patterns. Modified from Ericson, J., 1996.

Inconsistency and variability characterize the climatic conditions of the region. Historically, variation in precipitation levels on the Yucatán Peninsula has affected agricultural production and caused famine particularly when dry years succeed one another (Farriss, 1984; Murphy, 1990). Powerful hurricanes can cause severe flooding and crop devastation.

Shallow lowland basins called *bajos* are scattered throughout the reserve. Lined with heavy clay soils, *bajos* retain water during the wet season and quickly dry up when precipitation levels decrease during the dry season.

Following the establishment of the CBR in the late 1980s, the government of Mexico encouraged NGOs to participate in the design and implementation of management strategies for the reserve (Boege, 1995). PPY, a Mexican NGO founded in 1988, has promoted an integrated conservation and development approach in the region during the past 5 years. WWF, the Nature Conservancy, and other international donors contribute to the work of PPY by providing project assistance funds, training, and technical assistance. The PPY strategy is to involve the people who live in the communities surrounding the reserve in environmental education and sustainable development activities. This strategy is based on the theory that, if people living in and around natural areas receive economic and social benefits from these areas, their convictions about preserving these areas will be strengthened.

Research Program on Population

The administration of the CBR has long noted the importance of understanding and monitoring the population dynamics of the rural communities located in and around the reserve. Concern exists about the probable impact of migrant populations settling in this frontier agricultural zone and the need to formulate appropriate policy responses. The applied research program and its aforementioned objectives were developed through discussions between the reserve management and collaborating parties. They are being carried out in four phases:

1. Assessment of the forces that affect population growth at the regional level and identification of the critical areas around the reserve where population increases appear to place pressure on the reserve;
2. Literature review of available census data and secondary sources;
3. Diagnostic case studies carried out in selected *ejido* communities using the Participatory Rural Appraisal (PRA) methodology;
4. Use of research results to launch a regionwide debate through various public fora on the threats posed by migration to the long-term economic, social, and ecological viability of the reserve.

Stakeholder debate will generate the foundation for the formulation of appropriate policy responses to migration and shape the form of culturally and politically appropriate land-use planning in the communities around the reserve.

Research Methodology: PRA and Land-Use Planning Team

To assess the complex interrelationships between population growth and migration, tenure regimes and land-use practices, the PRA techniques are employed as the primary tool in performing the diagnostic case studies. An NGO associated with the University of Yucatán, Grupo DIP, trained the principal researchers and a team of ejido members in use of the methodology. PRA facilitates local people's participation in the collection, sharing, and analysis of a broad spectrum of data with a reduced amount of researcher-imposed bias (Chambers, 1983; Bruce, 1989; Freudenberger, K.S., 1994). It strives to deepen already existing knowledge and generate specific

Ejido Name	Inhabitants		C.G.R.*	Tdouble**	Land Area	Pop. Density		Ethnic	Legal Date
	1990	1995	(per year)	(years)	(hectares)	1990	1995		
Conhuas	250	398	9%	7.45	59,840	0.42	0.67	mixed	1931
11 de Mayo	80	253	23%	3.01	4,116	1.94	6.15	mixed	1995
Nueva Vida	72	163	16%	4.24	2,500	2.88	6.52	mixed	1984

* Crude Growth Rate
** Doubling Time in Years at *Current* Rate
Notes: All calculations based on census information acquired from the Mexican Instituto Nacional de Estadistica Geografia e Informatica (INEGI),1990 and 1995. Legal date refers to the date the ejido was officially recognized by the Mexican government although the settlement may be older.

FIGURE 14.3 Demographic characteristics of diagnostic case study *Ejidos*.

hypotheses, perhaps with recommendations for intervention (Freudenberger, K.S., 1994). The concept of *triangulation* is used to minimize bias and dependency on any one tool, information source, or perception based on the gender, ethnic group, or intellectual background of the researcher (Anderson and Rietbergen-McCracken, 1994).

Of the 114 communities located in and around the CBR, 3 were selected for participation in the case studies: Conhuas, 11 de Mayo, and Nueva Vida. The selection of these *ejidos* was based on a matrix of criteria, including population growth rates, ethnic composition, and geographic relation to the reserve. All experienced rapid population growth rates from 1990 to 1995, and, as is typical of the region, population densities in these *ejidos* are low (Figure 14.3). Four main themes were selected for these case study *ejidos* (Figure 14.4). Each community has a unique mix of indigenous* and nonindigenous groups from southeastern, northern, and central Mexico. Conhuas is located along an interstate road and extends into the buffer zone, 11 de Mayo borders the core zone in the remote southeast near the border of Guatemala, and Nueva Vida is just north of the region's most urbanized area and does not border the reserve (Figure 14.2). Conhuas has had minimal exposure to conservation and development projects, whereas the other two are actively involved with Pronatura Península de Yucatán.

Fieldwork carried out by a six-member team trained in PRA averaged 14 to 21 days of residency in each community. Team composition is diverse in terms of gender, age, ethnic origin, and educational background. Four members of the team live in the region and arrived there 15 to 20 years ago as migrants. Three of them are *ejidatarios* who play active roles in the social, economic, and political development of their communities. A woman social anthropologist of Mayan origins and an American woman with a graduate degree in resource management complemented the team.

Research results were presented to the *ejido* assembly, the community's governing body, at the completion of each case study for its approval prior to distribution

* In this context *indigenous* refers to ethnicity. It is not meant to imply that these people are native to the Calakmul region. Very little is known about the inhabitants of the region prior to the arrival of chicle and timber concessionaires in the early 1900s.

Theme		Objectives
1	Migration and changes in population over time.	History of the community and how it was founded; how migrants learned about the location of the community; migrant's reasons for arriving; where they came from; how long they tend to stay; why those that choose to leave the community do so.
2	General and reproductive health among community members.	Type of health services available in the community; health conditions and illnesses commonly occurring among community members; how illnesses and health problems are treated; problems encountered among pregnant women; the level and understanding regarding reproductive health; the woman's role in the community.
3	Land use and exploitation of natural resources.	Use and type of forest on community land; animal and soil resources; methods of natural resource exploitation; economic importance of production activities; resource relations between *ejido* communities; casual factors leading to changing land use patterns.
4	Planning mechanisms and discussions about the future at both family and community levels.	How community members intend to manage land distribution once their sons are grown and in need of land for crop cultivation; current internal community laws with respect to new migrants and land distribution; how recent migrants are treated; the decision-making process regarding location and amount of land available to recent migrants.

FIGURE 14.4 Themes and objectives of diagnostic case studies. Source: Maas Rodriguez, R. and Ericson, J., 1998.

to outside groups. The document containing the results was presented as an educational tool for use by the community to share and discuss its own history and future plans among its members, with other communities, and with interested NGOs and governmental organizations. It is illustrated with photographs of the community, maps, diagrams, calendars, and matrices produced by community and team members during the fieldwork. The text was carefully written and discussed by the team to ensure that the document can be easily understood by local audiences.* The final document is intended to represent the voice of the community and reflect the complexity of rural reality.

POPULATION DYNAMICS IN THE CALAKMUL BIOSPHERE RESERVE

The forests of the CBR shelter numerous ruins of the Preclassic and Classic Mayan civilization, which gives them cultural as well as ecological value. Archaeologists

* This has been one of the most challenging parts of the work due to the difference in educational levels and cultural perspectives among team members. Local resident team members become the mentors of the nonlocal team members because their perception of rural realities is closer to that of community members.

suggest that this area was once one of the largest and most powerful urban centers in the region, possibly the principal rival of neighboring Tikal in Guatemala (Folan, cited in Garrett, 1989). Archaeological discoveries indicate that the Classic Mayans used intensive agricultural practices and elaborate hydrological works to support substantial population densities (Adams, 1977; Turner, 1978; Thomas, 1981). Relic terraces, drainage systems, and related stone works indicating raised fields, and field demarcation have been detected over an area of approximately 10,000 km^2 in the Rio Bec zone. Drought and the disappearance of natural water sources in the region may have contributed to the drop in population levels of the Classic Maya and led to the shift of the ancient power center toward the northern Yucatán Peninsula (Dominguez and Folan, 1995).

Since the decline of the Classic Maya civilization (circa A.D. 950), the southern lowland region is thought to have been sparsely populated. Evidence exists that scattered communities of the Chontal Maya existed in the area during the Terminal and Post-Classic periods (Sharer, 1994); but it is believed that the Cehache, originally associated with Rio Bec, constituted the majority of the population (Antochiw, 1997).

For centuries during and after the conquest, the region's forests, considered uninhabitable by the Spanish, provided a safe place of refuge for Mayan resistance (Jones, 1989). In the 1800s logwood trees (*Haematoxylon campechianum*) found in swampy parts of the forest were felled and exported to Britain for their dye. At the turn of the century, rapidly rising international demand for chewing gum brought chicle tappers to the region to extract resin from the chicozapote trees (*M. zapota*). The chicle industry encouraged conservation of standing chicozapote trees and, in so doing, sought to protect mature forest areas containing these trees. In the late 1940s petroleum-based chicle substitutes were developed and global demand for the natural variety declined, causing a depression in the extractive tree crop economy. Also in the 1940s lumber camps, like the town of Zoh Laguna, were built to support the extraction of precious hardwoods such as mahogany (*S. macrophylla*) and Spanish cedar (*C. odorata*). Until the 1970s, the interior was largely the domain of lumber companies holding private concessions. Population levels were low and government investment in public infrastructure was minimal.

To the north of the reserve, evidence found in Maya communities seems to indicate that they may have been continuously populated since ancient times. Based on the size of Franciscan friaries and churches built in the early 1600s, it appears that there was a fairly large population in the area at the time of first contact by missionaries. A recent surface survey conducted in the *ejido* Pich revealed ceramic shards and the remains of buildings stylistically dated to the Late or Terminal Classic (A.D. 850 to 1000) (Faust, 1998). Other communities to the north were established in the late 1800s by Mayans fleeing the War of the Castes in the state of Yucatán. During the 1940s some of these northern communities were granted extensions to their lands for increased access to forest products such as chicle and precious hardwoods. Nine such extensions comprise the buffer zone of the southern division of the reserve (see Figure 14.2).

Much of this region was colonized prior to the formation of the reserve, as a result of government policy encouraging the settlement of areas considered to be

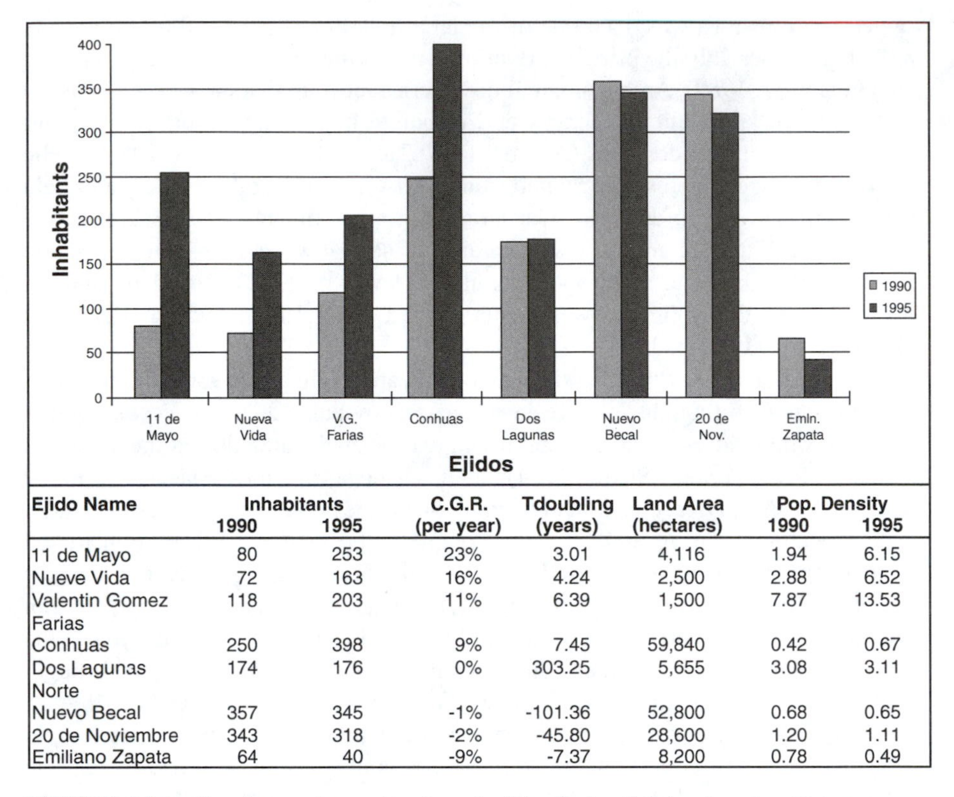

Ejido Name	Inhabitants 1990	Inhabitants 1995	C.G.R. (per year)	Tdoubling (years)	Land Area (hectares)	Pop. Density 1990	Pop. Density 1995
11 de Mayo	80	253	23%	3.01	4,116	1.94	6.15
Nueve Vida	72	163	16%	4.24	2,500	2.88	6.52
Valentin Gomez Farias	118	203	11%	6.39	1,500	7.87	13.53
Conhuas	250	398	9%	7.45	59,840	0.42	0.67
Dos Lagunas Norte	174	176	0%	303.25	5,655	3.08	3.11
Nuevo Becal	357	345	-1%	-101.36	52,800	0.68	0.65
20 de Noviembre	343	318	-2%	-45.80	28,600	1.20	1.11
Emiliano Zapata	64	40	-9%	-7.37	8,200	0.78	0.49

FIGURE 14.5 Population change in selected *ejidos* in the Calakmul region. Data sources: number of inhabitants — INEGI census 1990 and 1995. From Ericson, J., 1998.

underutilized. The lands on the eastern side of the CBR and the strip stretching between the northern and southern divisions of the reserve have become rapidly changing frontier zones during the last 20 years, most of them inhabited primarily by pioneer settlements dependent on a largely subsistence-level economy (Boege, 1995; Ericson, 1996). As is typical of frontier zones, a great deal of fluctuation and instability characterizes population dynamics in the communities. Figure 14.5 shows trends of both positive and negative growth in six *ejidos* between 1990 and 1995. Reduced out-migration seems to be the result of the arrival of government-relief programs, especially during times of drought and flooding, and the improvement of infrastructure, such as road construction and water catchment structures.

Ejido status was acquired by most of these communities during the 1980s, although many were settled as much as a decade earlier. Figure 14.6 shows that in 1995 *ejido* population averaged around 215 inhabitants with a regional population of 24,295 inhabitants in the 114 ejido communities of the Calakmul municipality. Numbers of *ejidatarios* vary from 25 to 200/*ejido*. Population density for the municipality, which includes the biosphere reserve itself, is approximately 2.5 persons/km^2. While land distribution within the *ejidos* is determined by internal regulations, parcel allotment to individual *ejidatarios* generally varies from 20 to 100 ha.

Community Name	Inhabitants 1980	1990	1995	C.G.R.* (per year)	Tdoubling** (years)
State of Campeche	*420,553*	*535,185*	*642,516*	*4%*	*18.96*
11 de Mayo	n.d.	80	253	23%	3.01
16 de Septiembre (L. Alvarado)	57	125	71	-11%	-6.13
20 de Noviembre	211	343	318	-2%	-45.80
A. Obregon, Gen. (Zoh Laguna)	791	1,098	985	-2%	-31.91
Aguas Amargas (San Isidro)	n.d.	48	68	7%	9.95
Aguas Turbias	n.d.	n.d.	15	n.d.	n.d.
Alacranes, Los	77	156	158	0%	272.06
Alianza Productora	n.d.	91	102	2%	30.37
Altamira de Zinaparo	175	1,016	1,139	2%	30.33
Amapola, La	n.d.	n.d.	24	n.d.	n.d.
Angeles, Los	n.d.	311	390	5%	15.31
Arroyo de Cuba	n.d.	n.d.	128	n.d.	n.d.
Arroyo Negro	44	131	182	7%	10.54
Becan	n.d.	100	164	10%	7.01
Bel Ha	n.d.	71	95	6%	11.90
Bella Union de Veracruz (Los Chinos)	n.d.	68	72	1%	60.63
Benito Juarez Garcia No.3 (Lic.)	n.d.	177	240	6%	11.38
Blaisillo	n.d.	124	98	-5%	-14.73
Bonanza	n.d.	17	20	3%	21.33
Cana Brava	n.d.	81	102	5%	15.03
Carlos A. Madrazo	22	26	36	7%	10.65
Carlos Sansores Perez (La Paz)	n.d.	37	123	24%	2.89
Carmen II	n.d.	229	290	5%	14.68
Centauro del Norte	n.d.	56	179	23%	2.98
Centenario	481	792	760	-1%	-84.03
Central Chiclera Villahermosa	n.d.	7	12	11%	6.43
Cerro de las Flores	n.d.	n.d.	74	n.d.	n.d.
Chan Laguna	210	503	539	1%	50.14
Chichonal	n.d.	56	79	7%	10.07
Concepcion	65	180	189	1%	71.03
Constitucion	500	726	898	4%	16.30
Cristobal Colon	111	278	337	4%	18.01
Dos Lagunas	n.d.	n.d.	176	n.d.	n.d.
Dos Lagunas	n.d.	174	228	5%	12.82
Dos Naciones	n.d.	106	190	12%	5.94
E. Eugenio Castellot I	35	44	66	8%	8.55
E. Eugenio Castellot II (El Carrizal)	n.d.	120	210	11%	6.19
Emiliano Zapata	158	64	40	-9%	-7.37
Felipe Angeles	n.d.	220	192	-3%	-25.46
Felipe Angeles II	n.d.	51	63	4%	16.40

FIGURE 14.6 Demographic statistics of communities around CBR. Data Source: Instituto Nacional de Estadistica Geografia e Informatica (INEGI) 1980, 1990, and 1995.

THE MOSAIC OF HUMAN PATTERNS OF LAND OCCUPATION

Anthropological studies identify the laborers in the chicle and timber industries as the first two of three distinct waves of migrants to enter the region in the 1900s (Boege and Murguia, 1990). Because of the seasonal nature of the work, these

Community Name	Inhabitants 1980	1990	1995	C.G.R.* (per year)	Tdoubling** (years)
Flor de Chiapas	n.d.	72	218	22%	3.13
Guadalupe, La	n.d.	217	298	6%	10.93
Guillermo Prieto	n.d.	102	94	-2%	-42.43
Gustavo Diaz Ordaz (San. Antn. Soda)	259	362	435	4%	18.87
Heriberto Jara	n.d.	131	195	8%	8.71
Hermenegildo Galeana	n.d.	137	98	-7%	-10.35
Innominado	n.d.	n.d.	8	n.d.	n.d.
Jobal , El	n.d.	99	100	0%	344.84
Jose Lopez Portillio No. 1 (Lic.)	n.d.	127	289	16%	4.21
Jose Morelos YP. (Civ alito)	102	162	253	9%	7.77
Josefa O. de Dominguez (Icaiche)	n.d.	107	156	8%	9.19
Justo Sierra Mendez	109	94	104	2%	34.28
Kiche Las Pailas	n.d.	185	284	9%	8.09
Km 120 (San Jose)	5	75	140	12%	5.55
Laguna Grande	n.d.	556	550	0%	-319.42
Lazaro Cardenas II (Ojo de Agua)	n.d.	239	328	6%	10.95
Ley de Fomento Agrop. (La Misteriosa)	n.d.	78	123	9%	7.61
Lopez Mateos (Lic. Adolfo)	340	583	454	-5%	-13.86
Lucha , La	n.d.	180	232	5%	13.66
Lucha , La	n.d.	28	105	26%	2.62
Manantial	n.d.	271	319	3%	21.25
Mancolona. La (Union 20 de Junio)	n.d.	191	270	n.d.	n.d.
Manual Castillo Brito	226	255	380	8%	8.69
Manuel Crecencio R.	68	189	271	7%	9.62
Maravillas, Las	171	123	97	-5%	-14.59
Mirador, El	n.d.	n.d.	21	n.d.	n.d.
Narciso Mendoza	n.d.	301	273	-2%	-35.50
Ninos Heroes	n.d.	184	209	3%	27.20
Nueva Vida	n.d.	72	163	16%	4.24
Nuevo Becal (El 19)	262	357	345	-1%	-101.36
Nuevo Campanario	n.d.	189	254	6%	11.72
Nuevo Conhuas	250	250	398	9%	7.45
Nuevo Paraiso	9	126	115	-2%	-37.94
Nuevo Progreso	n.d.	46	35	-5%	-12.68
Nuevo San Jose	n.d.	22	208	45%	1.54
Nuevo Veracruz	n.d.	97	184	13%	5.41
Pablo Garcia	166	545	611	2%	30.32
Paraguas	n.d.	n.d.	26	n.d.	n.d.
Pioneros del Rio Xnoha	n.d.	n.d.	234	n.d.	n.d.
Placeres	n.d.	17	10	-11%	-6.53
Placeres	n.d.	17	10	-11%	-6.53
Plan de Ayala (5 de Mayo)	n.d.	163	250	9%	8.10

FIGURE 14.6 (Continued.)

migrants were largely transient and dependent on fluctuating employment opportunities tied to variable international market conditions. The third wave of migrants are *ejidal* colonists who began arriving in the region in the 1970s and continue to arrive today. These colonists have been *pushed* from their places of origin by lack of land, lack of employment, displacement by commercial agriculture, ecological

Community Name	Inhabitants 1980	1990	1995	C.G.R.* (per year)	Tdoubling** (years)
Plan de San Luis	n.d.	50	50	0%	no growth
Pollos, Los	n.d.	n.d.	15	n.d.	n.d.
Porvenir, El	n.d.	20	32	9%	7.37
Porvenir, El	n.d.	362	34	-47%	-1.47
Puebla de Morelia	77	67	104	9%	7.88
Refugio, El	n.d.	71	107	8%	8.45
Ricardo Flores Magon (Lag. Cooxli)	n.d.	151	175	3%	23.50
Ricardo Payro Jene, Ing. (Polo Norte)	285	428	594	7%	10.57
San Antonio	3	20	18	-2%	-32.89
San Dimas (Alianza II)	n.d.	n.d.	18	n.d.	n.d.
San Miguel	n.d.	82	78	-1%	-69.30
Santa Lucia	298	240	245	0%	168.08
Santo Domingo	n.d.	n.d.	14	n.d.	n.d.
Silencio, El	n.d.	n.d.	6	n.d.	n.d.
Silvituc	386	639	739	3%	23.84
Solidaridad	n.d.	n.d.	137	n.d.	n.d.
Tambores de Emiliano Zapata, Los	n.d.	134	159	3%	20.26
Tepeyac	n.d.	17	12	-7%	-9.95
Tesoro, El	n.d.	191	288	8%	8.44
Tomas Aznar (La Moza)	151	167	176	1%	66.03
Tombola, La	n.d.	n.d.	25	n.d.	n.d.
Tres Reyes	n.d.	n.d.	68	n.d.	n.d.
Unidad y Trabajo	86	61	83	6%	11.25
Union , La (Dos Arroyos)	n.d.	n.d.	108	n.d.	n.d.
Valentin Gomez Farias	10	118	203	11%	6.39
Veintidos de Abril	n.d.	12	40	24%	2.88
Veintiuno de Mayo (Lechugal)	n.d.	130	163	5%	15.32
Victoria, La	n.d.	136	120	-3%	-27.69
Virgencita de la Candelaria, La	n.d.	272	325	4%	19.47
Xbonil	319	443	490	2%	34.37
Xpuhil	339	865	1,213	7%	10.25
Yazuchil	n.d.	n.d.	4	n.d.	n.d.
TOTAL:	**6,858**	**19,331**	**24,295**	**4%**	**19.42**

* Crude Growth Rate
** Doubling Time in Years at *Current* Rate

NOTES: Crude Growth Rate and doubling time based on 5-year period (1990-1995). Total CGR adjusted for comparison i.e., excluding communities for which there is no data in 1990. Between 1980 and 1990 there were **60** new communities founded. Between 1990 and 1995 **18** new communities were founded. In 1997 there were approximately **57** *ranchos* in Calakmul. Negative doubling times indicate the number of years necessary for the population to decrease to half its present size. Most of the communities listed are *ejidos*.

FIGURE 14.6 (Continued.)

catastrophe, and social unrest occurring in other parts of Mexico, including the state of Chiapas in recent years. They are subject to the *pull* of available land and the opportunity to establish new lives in a relatively unpopulated and still peaceful area. A fourth wave of in-migration, mostly government and service-industry workers, can be anticipated with the recent creation of the municipality of Calakmul, the strengthening of rural infrastructures, and the development of tourism in the region.

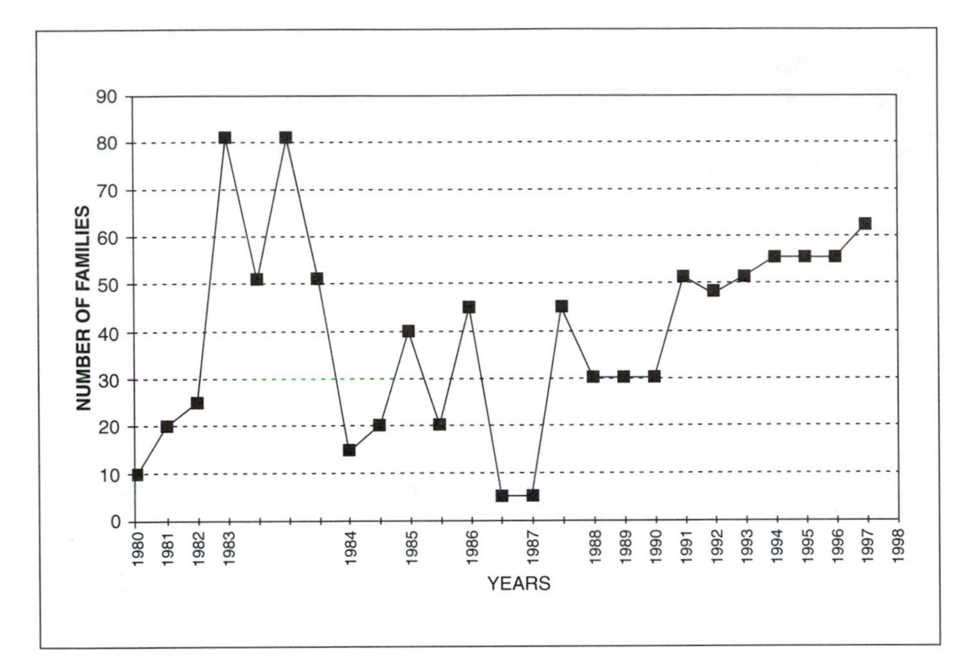

FIGURE 14.7 Evolution in numbers of families in *ejido* 11 de Mayo. Adapted from Montejo Alvarez, A. et al., 1997a.

Colonization Adjacent to the Calakmul Biosphere Reserve

Ejidos in the frontier zone contain an important percentage of mature primary and young secondary forests with some possessing as much as 50,000 ha of forested land. Rapid population growth rates in these *ejidos* indicate critical areas where land use may be subject to greater change. Of primary concern is the continuing in-migration to *ejidos* contiguous with the reserve. The *ejido* 11 de Mayo is located along the southeastern border of the reserve not far from the border with Guatemala. The first 11 years of the history of the community, shown in Figure 14.7, are characterized by a series of extreme changes in population level. Establishment of the *ejido* was prompted by agrarian reform policies in the early 1980s. Subsequent migrants are friends and family of original settlers.* In recent years, the number of Chol and Tzeltal families has increased as a result of the social and political unrest in the neighboring state of Chiapas. Most members of these families speak Spanish, some are bilingual, and others, mainly adult women, speak only the language of their ethnic group.

During the 5-year period between 1990 and 1995, the population of 11 de Mayo grew at a rate of 23% according to government census data. Case study research shows that most of the growth in this period took place during 1990, just after road

* In the *ejido* Conhuas it was found that many new families arrived during the past few years because of the availability of work in government-sponsored archaeological site excavations. This type of migration tends to be less permanent and lasts the duration of the job, which may be anywhere from 1 to 5 years.

access was improved, and that population growth slowed significantly after 1991. The limit of the *ejido* of 55 *ejidatarios*, established by internal law and registered in the agrarian reform offices, was reached in 1991 and there was no more available land for distribution to newcomers. Each *ejidatario* has been allotted 60 ha plus a half-hectare plot in the "urban" zone of the *ejido*. Population growth slowed but the level did not diminish during this period in part due to improved infrastructure provided by the government, including solar powered electricity, a primary school, and a water catchment system. Population density was calculated at 6.15 people/ha in 1995.

As illustrated in Figure 14.8, an oral history of 11 de Mayo according to one of its founding members, the mid-1980s brought fear and internal conflict. Division within the community over whether to apply for *ejido* status or small-property-owner status resulted in violence, and population levels fell to their lowest at this time. The *ejido* was not legally recognized by the agrarian reform offices until 1994. For the first 14 years of its existence, the community operated with an informal governing body, which allocated parcels of land to older residents and withheld it from newcomers. Declines in the number of people living in the *ejido* during the early years also resulted from climatic conditions. The site is located on the border of the reserve in a lowland *bajo*, which often floods during the rainy season and where water is retained in pools during much of the dry season. Severe flooding has on more than one occasion caused families to leave the community.

A particularly important *ejido* in the frontier zone is Xpujil. This community is quickly becoming urbanized as it has become the seat of the new municipality of Calakmul and is a major transport hub located at the crossroads of two main roads running through and alongside the reserve. In Xpujil, it is evident that the number of individuals with *ejidal* rights to the land, *ejidatarios*, has stabilized, whereas the number of *pobladores*, individuals without *ejidal* rights to the land, continues to increase. *Pobladores* are opening shops, providing services and looking for work in the administration of the new municipality, the biosphere reserve, and the tourism sector.

Natural Population Growth

Although migration is the primary force behind the rapid rate of increase in population regionwide, high rates of natural increase are beginning to show their effects. A recent study based on samples from four communities shows 5.8 children/family (Ericson, 1996). Global fertility rates based on government census data range between 3.9 and 5.2 live births/woman (CONAPO, 1996).* Women in their childbearing years, between age 15 and 49, make up 21% of the population (Loudiyi, 1994). As in many rural areas of developing countries, 51% of the population is under the age of 15 years with only 2% of the population 65 years or older (WWF, 1994). Confronted by a situation of high fertility and rapid rates of natural growth,

* These figures represent fertility rates calculated for the municipalities of Hopelchen and Champoton which, prior to the establishment of the Calakmul municipality in 1997, included the communities around the biosphere reserve. At this time, fertility rates based on government data for the Calakmul area are not available.

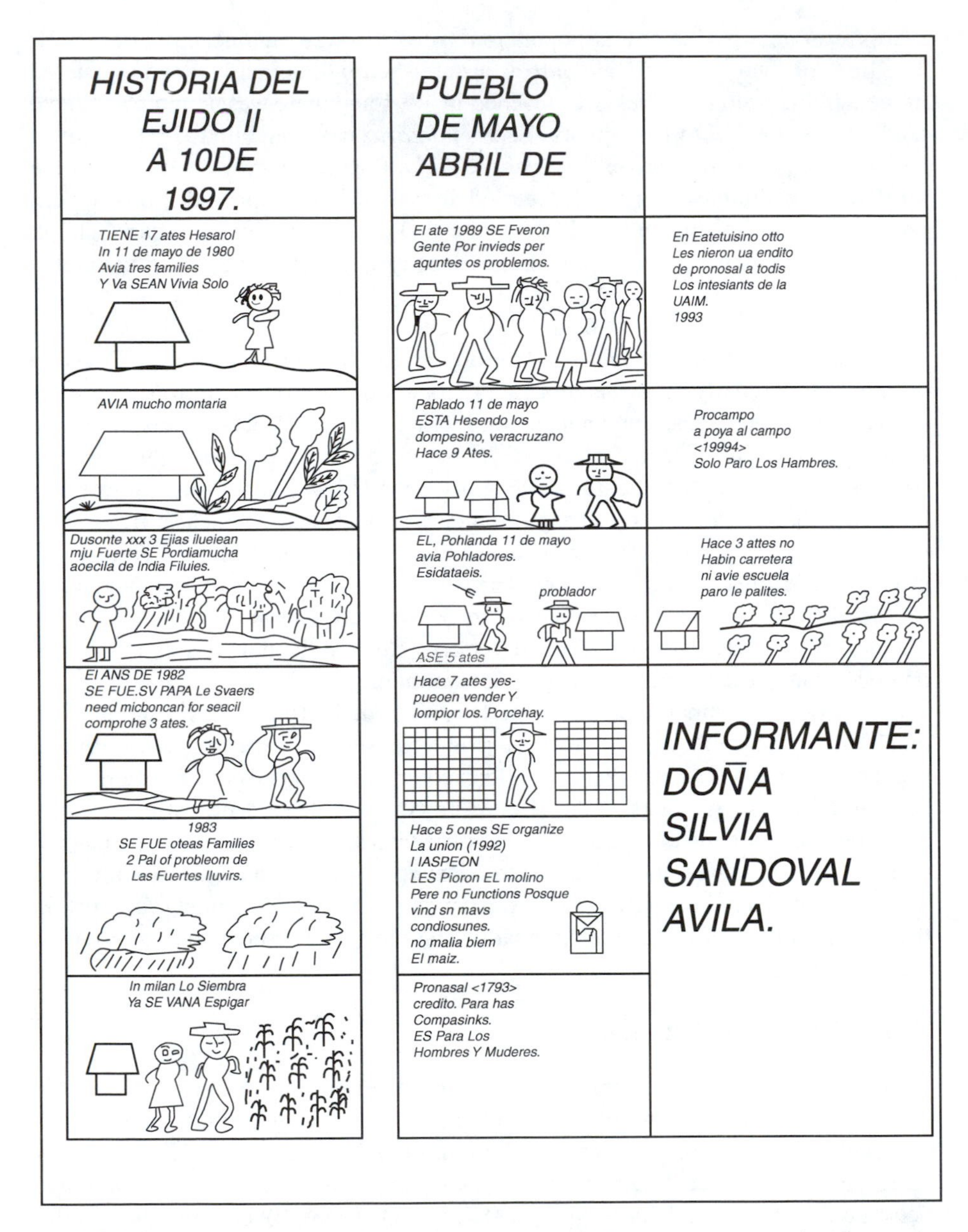

FIGURE 14.8 Oral history of the community of 11 de Mayo. Source: Montejo Alvarez, A. et al., 1997a.

PPY is incorporating an educational reproductive health program into the population component of its work around the CBR. The program is designed to increase family access to existing reproductive health services offered by government health services.

A sample of 17 women from the *ejido* 11 de Mayo shows that the average number of pregnancies among Chol, Tzeltal, and Spanish-speaking women is calculated at 5.2. When this same group of women was asked how many children each had desired at the time of marriage, many said they had never thought about it while others reported a desire for an average of 2.8 children. This large discrepancy between children desired at the time of marriage and number of pregnancies may reflect male rather than female preferences in use or nonuse of family-planning methods.*

Inappropriate Agricultural Technologies

Along with aspirations for a better future, colonists often bring with them land-use practices that may be unsuitable to the ecological conditions of their new home. Although most of the farmers in the economically marginalized communities of the region practice subsistence and small-scale agriculture, migrants from central and northern Mexico have a propensity to employ mechanized agriculture and agrochemicals for cash crop cultivation. These farming technologies contribute to reduction in the forest cover as well as declines in soil fertility and structure of this tropical ecosystem. In contrast, colonists already familiar with the tropical forest of the Yucatán Peninsula, such as the indigenous Maya, use the swidden system of agriculture that allows the forest to regenerate. Regeneration of secondary growth during the fallow period is encouraged by the Mayan populations because various forest products have use and exchange values (Murphy, 1990). In both systems, the use of fire to clear brush and crop residue before the onset of the rainy season can be sorely destructive. Unless watched with an experienced eye of farmers familiar with the ecosystem, fires burn quickly out of control.

Cattle ranching is another land-use practice that can be destructive. The cattle hoofs tend to compact the fragile limestone soils, while the tight webs formed by the roots of forage grasses prohibit the regrowth of forest vegetation. In the communities participating in this study, informants spoke of plans to convert agricultural lands to pasture (Figure 14.9). From a conservationist viewpoint, the aspiration of colonialists to raise livestock such as goats and cattle is viewed with considerable alarm. Yet, local communities view cattle as a symbol of wealth. Wealth-ranking exercises performed with informants in these communities put cattle owners among those with the fewest financial difficulties. The conversion of forest to pasturelands may result in a profound and perhaps irreversible land-use transformations within the *ejido* communities. The future of the CBR is at stake if this forest conversion in surrounding *ejidos* becomes more extensive. With rapidly growing rural populations occurring around the reserve, it is likely that political pressure will increase to convert the forest resources of the reserve into pastures and farmed lands.

The trends toward mechanized agriculture and cattle ranching are currently kept in check by financial and ecological constraints. Both activities require a financial investment greater than that available to the majority of the region's colonists. Although very few farmers can afford to purchase a tractor, many of them pay to

* Measuring pregnancy intentions has been shown to be problematic, but some studies indicate that fertility intentions are closely related to whether or not a woman has another child (Tan and Tey, 1994).

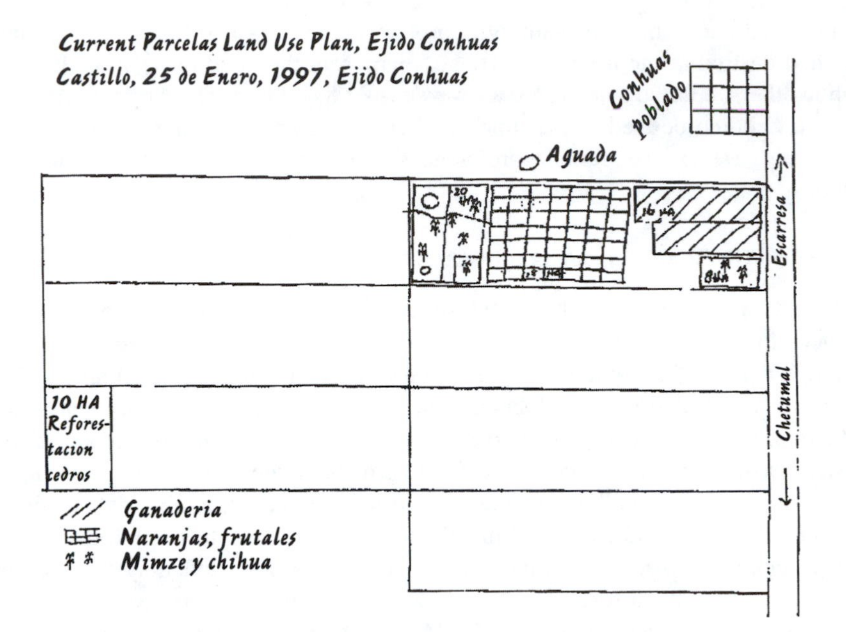

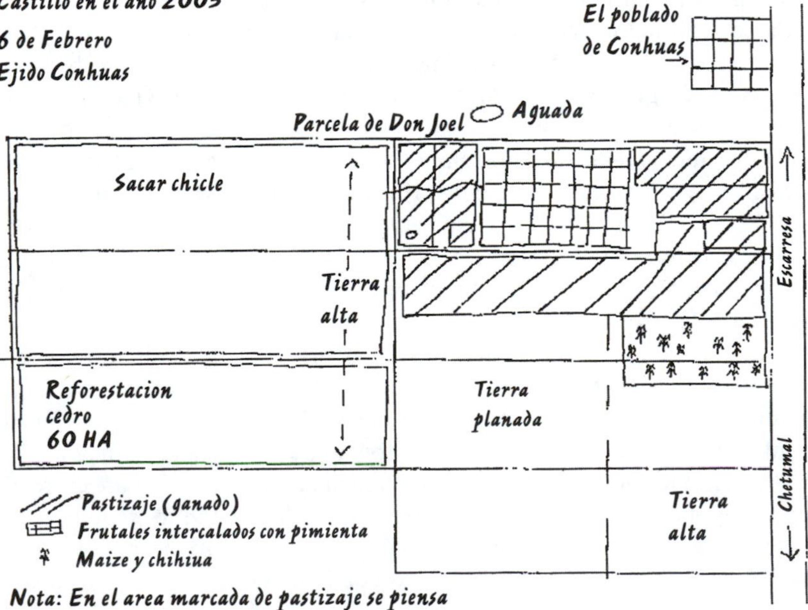

FIGURE 14.9 One family's present land use (a) and future (year 2005) land-use plan: *ejido* Conhuas. Source: Montejo Alvarez, A. et al., 1997b.

have 1 or 2 ha plowed each year by a regional *campesino* organization. In some communities, poor soil conditions and a shortage of permanent water sources prohibit both these activities. Conservation programs in the communities focus on stabilizing land use and curbing the expansion of agricultural and pasturelands by improving incomes from forestry-based enterprises such as sustained-yield timber harvesting, chicle extraction, apiculture, and agroforestry. However, it is uncertain that increased income generation from forest product extraction will reduce the amount of land in agriculture without effective land-use planning. Anecdotal experience in the region suggests farmers are investing incomes from commercial forest product extraction into land clearing of prime forestlands and agricultural inputs detrimental to the environment (Stedman-Edwards, 1997).

Resource Conflicts and the Calakmul Biosphere Reserve

The establishment of the CBR has caused conflicts between local communities and the reserve management. Although various studies were carried out in the region during the 1980s when the creation of an archaeoecological park was under consideration,* the reserve boundaries were drawn with insufficient consideration for the existence of *ejido* and private lands. The boundaries of the reserve as currently delineated are problematic because of previously existing land tenure arrangements and current resource use patterns. As a result, a number of *ejido* communities and small landholdings are either within or straddle the border of the core area in both the northern and southern divisions of the reserve.** Considerable unease exists within the communities regarding their tenuous tenurial position.

According to national forestry law, forestry studies and a management plan are required to receive permission from the federal government for felling authorized volumes of specified forest tree species in designated locations. Enforcement of this law has been inconsistent, which causes some residents to express frustration and anger at the prohibition of unauthorized timber exploitation on *ejido* lands and the complete prohibition within the core zone of the reserve. Many ignore the restrictions and continue cutting timber. This attitude is more evident in communities in which alternative income-generating projects have not yet been introduced. Many local residents believe that conservation of standing forest limits livelihood strategies necessary for survival.

No formal mechanism yet exists by which surrounding communities can share in the funds collected from tourists entering the CBR to view wildlife and archaeological

---

* Ecological, archaeological, legal, and political studies were carried out in the early to mid-1980s by the Centro de Investigaciones Históricas y Sociales of the Universidad Autónoma de Campeche under the direction of Dr. William Folan in collaboration with the Secretaria de Desarrollo Urbano y Ecología and the Instituto Nacional de Antropología e Historia. The WWF in collaboration with PPY and the Centro de Estudios para la Conservación de los Recursos Naturales, A.C. also performed studies during the latter part of the 1980s (see Vasquez Sánchez, 1991).

** In 1993 the state government succeeded in relocating one of the communities located within the core zone of the reserve. The community was persuaded to move because they were offered another site with improved access to communication and services. In addition, the government constructed housing for each family in the new site and provided solar power for each house.

sites. However, a tourist guide service run by a handful of local residents trained through a collaborative effort between NGO and governmental institutions exists in Xpujil. And, with help from a regional campesino organization, two *ejidos* with excavated archaeological sites in their locality have organized tourist services.

Land Markets and Land Sales

Land markets are emerging as a result of changes in the federal constitution and major land tenure reforms in the early 1990s. Land can now be bought and sold within *ejidos*. During the past couple of years more than 100 ha of land in Xpujil have been sold. Buyers are generally investors from outside the region. Much of this land has been purchased for construction of shops and hotels in anticipation of a growing demand for tourist facilities. Land has also been purchased for construction of an airstrip in Xpujil.

Concern exists about the market sales of forest extension lands located within the buffer zone of the reserve and along its western border. These extension lands belong to *ejido* communities located to the north of the reserve. Large distances exist between the extension lands and the communities to which they belong, indicating that these lands are probably not in use and *ejidatarios* may choose to sell them. In 1996, the forest extension lands of the *ejido* Dzibalchen, which border the northern division of the CBR, were expropriated by the federal government for the relocation of indigenous people from Chiapas. After Chiapas decided not to relocate its people to Campeche, the land was received by the Campeche state government. The future use of this land is now uncertain. In the case of Dzibalchen, a damage payment significantly lower than the commercial value of the land was paid to *ejidatarios*.

Infrastructural Development

Recent infrastructural developments may stimulate increased in-migration in the coming years. Scarcity of water resources in the region has in the past often encouraged the out-migration of large numbers of people. A large-scale water project currently underway will carry water 85 km from a shallow lagoon, Laguna de Alvarado, near the border with Guatemala to service 22 communities in the frontier zone. The aqueduct is intended for use during the dry season as a complement to already existing water catchment systems in the *ejidos*. Increased water security will decrease out-migration and may attract more migrants. However, based on population projections calculated on current growth rates, it is estimated that this new system will only be able to service the needs of these communities until the year 2005 (CNA, 1996).

The CBR is crossed by two major roads. An east–west interstate highway runs through the narrow midsection separating the northern and southern core areas. And, there is a paved road running north-south along the eastern side of the reserve through the upper core area and down to the *ejidos* located near the Guatemala border. The road will facilitate market and services access to the *ejidos* located along it and inevitably affect population growth rates in these communities. This road, as well as an airstrip under construction in Xpujil, will also encourage greater tourist access to the natural and archaeological sites within and around the reserve. The Mexican government has proposed that this road be extended to Tikal, Guatemala.

CHALLENGES AND OPPORTUNITIES FOR MITIGATING MIGRATION IMPACTS

FEDERAL GOVERNMENT LAND-USE POLICIES

The CBR was created by the Mexican federal government for the protection of the biological and cultural wealth of southeastern Campeche state. While some ministries and departments of the government support the policy to establish and protect the reserve, others promote policies that may undermine conservation goals. Federal governmental bodies may not yet recognize the ecological and social consequences of contradictory policies because the region is remote and relatively understudied. The contradictions are clearly present regarding policies to expand roads into the Calakmul region and to subsidize agricultural production.

As noted above, road construction through core areas of the reserve opens up the region for agricultural and social development. In-migration and its associated environmental consequences are stimulated by the growth of a rapidly improving road infrastructure. Federal government policy explicitly encourages the social and economic development of this frontier area. The Mexican press reports that officials in the federal offices of the ministry of social development (SEDESOL) consider this region to be among the 91 priority regions of the country due to its development potential (*Diario de Yucatan*, November 7, 1996). Unless effective land-use planning occurs, it is difficult to envisage how the municipality of Calakmul can be transformed into a development zone without undermining the rich natural resources found within and around the biosphere reserve.

During the presidential administration of Salinas de Gortari, a countrywide agricultural program called PROCAMPO was initiated to cushion the impact of the removal of trade barriers and price subsidies by improving incentives for basic food crop cultivation.* Scheduled to continue for a period of 15 years, the program provides cash subsidies directly to farmers on a per-hectare basis for planting crops such as corn, beans, and rice, regardless of yields produced. These politically popular subsidies provide much-needed economic assistance in the region. However, not only does this program encourage the reduction of forest cover by supporting agriculture, but research in other parts of Mexico has shown that PROCAMPO financially benefits livestock producers and the commercial sector more than it benefits the agricultural sector (Yuñez-Naude et al., 1995). Officially, only areas that had been planted for 3 years (1989 to 1992) prior to the initiation of the program are eligible for the subsidy; however, it is likely that farmers have also claimed the subsidy for newly cleared lands, especially in tropical areas like the Calakmul area where soil conditions for agriculture rapidly deteriorate.

* National Farm Modernization Program (Programa Nacional de Modernización del Campo).

State of Campeche Land-Use Policies

In January 1997, the region around the CBR was given municipal status to facilitate a greater governmental presence in the region.* According to state government officials, the creation of the Calakmul municipality is designed to facilitate the economic development of the region around the administrative capital of Xpujil. The fledgling municipality of Calakmul is being promoted as an ecological municipality because of the presence of the biosphere reserve within its borders. Thus, the concept of an ecological municipality exists as somehow distinct from the typical municipality. However, there is still no clarity about what distinguishes an *ecological* municipality from a standard municipality. Although the state of Campeche plans to designate a protected area within each of the state's municipalities, some observers suggest that this is simply an administrative mechanism for attracting additional international donor funding to the state for environmental programs.

Contradictions exist between the reserve, destined for the protection of the environment, and the newly established municipality designated as a priority area for development. Because the objectives of conservation and development are often incompatible, land-use planning and the creation of strategies for sustainable development in the region will be a topic of continual tension in the struggle to balance interests among stakeholders. Unfortunately, local-level institutional structures most likely to be involved in planning initiatives at this time (i.e., the administration of the new municipality, the administration of the CBR, and the various *campesino* councils, such as the Consejo Regional Agrosilvopecuario de Xpujil) lack experience in this field.

Although implications are not yet clear, the creation of the ecological municipality will probably stimulate another wave of migration into the region. Government and service-oriented workers are erecting homes and offices in Xpujil, the seat of the new municipality. Greater development in the infrastructure of the region is inevitable, with meeting the needs of the region's water requirements as, no doubt, the largest hurdle to be overcome.

Community Land-Use Planning Practices

The applied research conducted for this study in selected *ejidos* around the CBR suggests that effective land-use planning is intimately linked to a sense by residents that their rights to land and other natural resources are secure. According to Mexican agrarian reform law, residents are divided into *ejidatarios*, those with rights to the land and most often the founders of the *ejido*, and *pobladores*, those without rights to the land and most often the recent in-migrants.** In some communities, a considerable gulf exists between *ejidatarios* and *pobladores* because of differences that exist between the two groups in terms of access to social services from government

* The eastern border of the Calakmul municipality is being disputed by the neighboring state of Quintana Roo due to a historical conflict over the exact location of the border marker, the *punto Put*, lost to the jungle many years ago.

** The term *poblador* is commonly used in southeastern Campeche to refer to those without rights to the land. The term *avecindados* is used in agrarian law to signify the same.

agencies, decision making power, and natural resources. *Pobladores* may not invest labor and financial resources in sustainable agricultural technologies if they do not have secure access to land. In contrast, *ejidatarios* have much greater authority and institutional incentives to manage resources. Registered with the agrarian reform offices, *ejidatarios* are responsible for formulation and enforcement of internal *ejido* laws and they must participate in *ejido* management through an assembly. *Ejidatarios* are usually men, but women can also be *ejidatarias* under special circumstances specified in agrarian reform law.

In *ejidos* such as 11 de Mayo and Nueva Vida, *pobladores* still constitute a minority of the population. They are newcomers hoping to acquire *ejidatario* status after a stated period of time, or sons of *ejidatarios* expecting to become *ejidatarios* themselves. To be admitted to the *ejido*, an applicant must present a letter of recommendation from the authorities of his or her previous *ejido* or village and pay the required fee. Although they often have to share in communal work, contribute to *ejido* funds, and participate in assemblies, *pobladores* have neither a voice nor a vote in the ejido. Depending on the size of the *ejido*, *pobladores* are sometimes provided with the same amount of land allotted to *ejidatarios*, albeit located in a less desirable part of the *ejido* farther from the settlement area and covered with tall trees of the primary forest — an undesirable situation because considerable investments of labor must be allocated for forest clearing.* This land can be taken away at any time if the holder's behavior fails to satisfy the assembly of *ejidatarios*. When not provided with a *parcela*,** *pobladores* may be lent land on an informal basis by an ejidatario. Without land of their own, *pobladores* are an important labor pool for clearing fields and harvesting the crops of ejidatarios. However, the supply of wage labor is scarce in most of the communities of the region. *Pobladores* often prefer to leave their family to travel to other communities, cities, or, in the case of some young people, to the United States in search of employment.

Conflict, corruption, and violence are recognized problems in *ejido* life (Ronfeldt, 1973; Haenn, 1997). Mischievous acts are sometimes used to discourage *pobladores* from becoming too comfortable within an *ejido*. This is due to the fear that *pobladores* will gain excessive authority within the community. In Conhuas, for example, the footpath to the borrowed land of a particularly outspoken *poblador* was blocked repeatedly with fallen logs. He had been promised *ejidatario* status within 1 year of his arrival in the *ejido*, but almost 2 years passed before he received recognition by the assembly. Tensions are growing in the Conhuas community because the number of *pobladores* almost equals the number of *ejidatarios*. Although the *ejido* assembly plans to recognize all current *pobladores* as *ejidatarios* before closing the door to new arrivals, the *pobladores* endure a constant stream of criminal acts performed by a clandestine group of young people from some of the founding families of the *ejido*. While the community intends to close the door to newcomers,

---

* More labor is required to convert a hectare of primary-growth forest to agricultural land or pasture than is required to convert a hectare covered with secondary-growth vegetation. Primary forest of tall trees is thus less desirable for the cultivation of basic crops, such as maize.

** A *parcela* is the worksite assigned to an *ejidatario* or *poblador* by the *ejido* assembly.

this is easier said than done because it is very difficult to refuse settlement of newly arrived family members.

Informal, and to some extent formal, land-use planning already occurs in the *ejidos*. Land allocation decisions are made by the *edjido* assembly. Within the *parcelas*, households determine land uses largely based on the evaluation of environmental parameters such as soil type. But, land-use planning at the community and household level is sharply constrained by the legitimate tendency of individuals to seek out short-term economic benefits for the household. In poverty-stricken communities, this short-term calculation is certainly understandable. Yet, planning for conservation of biodiversity at the *ejido* level requires a much longer term vision about how land-use arrangements should be shaped to guarantee the maintenance of habitats and ecological processes. To the economically and politically marginalized peoples living in and around protected areas such as the CBR, it is difficult to derive a long-term vision of how to conserve resources in perpetuity when a family is confronted by immediate pressures to produce sufficient food and monetary benefits for household survival.

The pressures between short-term survival strategies and long-term conservation objectives play themselves out in the illegal trade for tropical hardwoods. Illegal extraction of precious and other marketable hardwoods occurs from individual *parcelas* and communally managed forest areas on *ejido* lands. Illegal timber cutting also occurs within the core area of the reserve and along both sides of the border shared with Guatemala. Both the *ejidos* and the reserve management staff lack an effective enforcement system to control this extraction. Enforcement structures are difficult to build when demand for precious hardwoods is high. Some of the timber is sold to carpentry workshops in the communities north of the reserve (Faust, personal communication, 1997), but most of it is purchased by distributors from states as far away as Veracruz and Jalisco.

Household Livelihood Strategies

Throughout the case study research conducted in the three *ejidos*, the rural people interviewed stressed the importance of improving the conditions of life for community members. Women seek to meet immediate household needs such as firewood, clothing for the children, roofing materials, and wire fencing for protecting domestic livestock, as shown in Figure 14.10. Men voice the need to improve crop yields primarily through mechanized agriculture. They seek to obtain credits from the government, and take advantage of services afforded by NGOs working in the area. Both men and women agree that water, electricity, schools, health services, and roads are priorities.

The three *ejido* communities participating in this study declare a strong preference and expectation of government assistance. This expectation is fueled by the history of job creation programs introduced just after the creation of the biosphere reserve in the early 1990s with the support of the government program Solidaridad and the national program for reforestation (PRONARE). Farmers largely depend on government subsidies from programs like PROCAMPO and on food aid provided by the federal agency for family development (DIF). Solidaridad also provides

PLANNING FOR THE FUTURE OF THE EJIDO 11 DE MAYO BY TZELTAL-SPEAKING WOMEN			
	HOW IS THE EJIDO PRESENTLY? (What do they have and what is missing?)	HOW TO RESOLVE OR ATTAIN WHAT THEY WANT IN THE FUTURE	HOW WOULD THEY LIKE THEIR EJIDO IN THE FUTURE, YEAR 2005 (what would they like to have?)
	They think the ejido has 4,116 ha. They think there are 500 inhabitants.		**They think it will have 4,115 ha. They think it will have 1,000 in habitants.**
SERVICES Problems/ Solutions	• Not enough water • They have to go far to look for water and fuelwood • School teachers do not always come to class every day of the week	• They will ask the authorities for everything they want	• Each house has a water tank • Piping/drinking water/another fenced water catchment basin so that people can drink the water • Cars to go to the water hole to wash clothes and carry fuelwood • Conasupo (government-owned store run by the community that provides basic necessities) • Electricity
COMMUNICATIONS Problems/ Solutions	• The road is bad; no cars can pass through when it rains	• They will ask the authorities for everything they want	
POPULATION Problems/ Solutions	• Sheep and other animals trespass into their backyards	• Work with Pronatura; ask them for animals • Get an operation to have less children	• Wire fencing for the backyards of houses • Sewing machines for the women • Clothes for the children • Stuffed animals • Tin roofs for the houses
FORESTS Problems/ Solutions	• Not everyone has a plot of land • Small harvests because of lack of rain • Sheep and other animals trespass into their plots	• Mechanize the land so that their children can have places to work/mechanize the land to save the forests • Buy gas stoves to save fuelwood	• Not everyone has a plot of land • There will be less forests and fuelwood

FIGURE 14.10 Planning priorities of Tzeltal-speaking women in the *ejido* 11 de Mayo. Adapted from Montejo Alvarez, A. et al., 1997a.

financial assistance to families with small children to encourage primary school attendance. This dependency on state subsidies tends to undermine community self-determination.

Interviews with women in these communities indicate that family reproductive decisions are most often made by men. Successful families are often those with many children who become *ejidatarios* in their own right and combine forces with other family members in productive activities, or become professional people such as teachers, or migrate to the United States and send remittances back home. Thus, the life projects of many rural women revolve around or are constrained by reproductive decisions made by their husbands and are based on hopes of increasing the family living standard once the children are old enough to join the labor force.

IMPLICATIONS FOR COMMUNITY-BASED RESOURCE CONSERVATION PROGRAMS

Consideration 1: Opportunities and Dangers of Participatory Research

PRA is a *participatory* research method, designed to empower informants as well as collect data. Working in a participatory manner with *campesinos* creates an opportunity for them to tell their own story by giving them tools to communicate their concerns to outside groups. It has great potential for use in multilevel community and regional planning as it creates opportunities to build mutual understanding among different parties.

However, the effectiveness of a participatory methodology must be considered within the context of power structures that characterize the society at large (Boesveld and Postel-Coster, 1991). Because *campesinos* are a relatively nonempowered group within Mexican society, the risk exists that data collected may be used to the detriment of informants in the design of top-down management plans. This risk is related to the fact that unempowered groups are generally unfamiliar with information management and unaware of how it can be manipulated by decision makers. Organizations basing their work on participatory field research methodologies must always ask themselves at what point does divulged information damage the interests of informants. Free and open exchange of all information can break the bonds of trust so carefully constructed during the research process.

Consideration 2: Diverging Perspectives between Conservation Objectives and "Life Projects" of Local Communities

Participatory research may not always provide results expected by the conservation organization. Conservation organizations must be willing to accept and work with the perspectives of rural people that may indeed differ from those of the conservationists. This study shows that around the CBR considerable efforts must be undertaken to narrow the gap that currently exists between local community concerns and the objectives of conservation organizations. This study suggests that community members become interested in ecologically sustainable activities, such as beekeeping or intensified agriculture, if these activities are economically viable. The marginalized people of this region seek financial advancement, and the package of inputs so far presented by conservation organizations may not be generating economic benefits that are sufficient to meet their expectations. In the harsh physical and social environment of the Calakmul area, communities around the reserve tend to see conservation agendas as a force that undermines survival strategies. This seems to indicate that environment and development programs may need to lobby for policy changes that increase the cost-effectiveness of sustainable agricultural technologies.

Consideration 3: Challenges for Building Population Monitoring Systems

As population dynamics are bound to continue to change rapidly over the coming years in the communities around the CBR, decision makers have requested the development of a multilevel population-monitoring system that includes the participation of local residents. Conservationists seek a system that accurately measures trends at both the microlevel of the *ejido* and the macrolevel of the region. Decision makers are interested to know where and why people settle the way they do and to understand better the determinants of fertility. This request harkens back to the ethical problems accompanying participatory research stated earlier. People divulging information about where they live may result in government restrictions and even forced removal from their lands. Alternatively, local people may surmise that such information might bring benefits and so distort the information accordingly. Local people need to be involved in this monitoring, but, unless they can trust that the information will not undermine survival strategies, they may not be truthful.

Consideration 4: Community Land-Use Planning and Conservation Planning

Little cultural homogeneity exists in the *ejidos* surrounding the CBR. Within each community, divisions can be uncovered along ethnic, kinship, or religious lines and conflicts are an ever-present reality. Inheritance customs with respect to the land vary between ethnic groups. In addition, because these communities were only recently established, they have limited experience undertaking collective action. Few rural institutions are present in the communities, and political parties seem to encourage internal divisions in the communities though few seem to receive benefits from political participation. Working with the rural communities around the CBR is thus a major challenge for conservation and development organizations.

Land-use planning at the local level entails the formulation and enforcement of rules on how land is to be used in the present and future. Certainly, within the current *ejido* system of governance, local communities have the freedom and legal protection to enforce internal rule-making. When it is in their interest to do so, *ejidos* can be effective units for land-use planning.

However, community-level planning alone in the *ejidos* in the region may not be strong enough to withstand the pressures exerted from paternalistic political and administrative structures, land tenure insecurity, the limited internal skills for community-level resource planning, and major development and conservation programs. Community-based planning alone cannot stem biodiversity loss linked to the growth of population densities that in themselves are associated with external factors such as the construction of new roads and water systems in the region. These pervasive forces overwhelm the ability of local communities to control the settlement of in-migrants and land uses within the *ejido*. While conservationists tend to treat increasing growing populations as a threat to the ecological viability of the reserve, many local residents see this in quite a different light. From their perspective, population growth contributes to the taming of the hostile and difficult physical environment.

Conservation NGOs such as WWF and PPY should begin to take a larger role in promoting environmental education programs that build on the concerns of communities to assure household survival. Programs should continue that propose economically sound sustainable agricultural practices linked to effective community-level land-use planning. Rural communities request access to family planning, but often for cultural and economic reasons lack access to reproductive health services. Although it is somewhat easier to focus on providing technical extension services to rural communities, conservation and development organizations should begin to look at more complex institutional and policy questions.

Consideration 5: Building Conditions for Dialogue on Population Dynamics and Conservation

The applied research program launched as part of the PPY population initiative indicates that fostering public debate and dialogue on population dynamics is a critically important tool for building consensus on future courses of action. But first, conservation organizations need to listen carefully to local residents, enter into discourse with them, and attempt to view conservation and development challenges from their perspective. Evidence provided by the diagnostic case studies is sobering in that it indicates the extent of work that must be done to assure that the environmental agenda is included in community-level planning activities. Following these case studies, an opportunity now exists for organizations like PPY to begin to launch dialogue and discussion on these issues and to learn to struggle together with local residents to come up with new solutions to old problems. This is the essence of environmental education. The PRA research process of this study experimented with tools to generate dialogue to see what type of responses are recommended by rural communities. Numerous insights have been generated, although it takes considerable foresight to link the current lack of community-level environmental awareness with conservation objectives. For example, conservation of wildlife is viewed only as a peripheral part of the worldview of *campesinos*, but it might become more important to them if direct benefits can be derived from a tourism industry interested in conserving both forests and wildlife.

Consideration 6: Meeting Unmet Needs for Public Health and Reproductive Health Care

A concerted effort was made throughout the diagnostic case studies to link the study with one carried out by the PPY reproductive health team. This latter study sought to increase understanding of factors that contribute to natural population growth rates in the *ejidos* and to integrate reproductive health issues into community-level discussions. From this research, it seems clear that the provision of reproductive and family health-care assistance can help build trust and confidence with community members because improvements in health tend to offer rapid returns to the household.

This approach is based on the hypothesis that local people will be more open to initiatives that offer long-term gains, such as conservation, once immediate community concerns have been addressed. In addition, resolving conflicts over health-care provision may be a step toward resolving a wide array of impediments to planning for economic development and sustainable uses of natural resources. The PPY reproductive health project attempts to respond to an expressed need for reproductive health services through an educational process based within the communities. As government-sponsored reproductive health services already exist in the communities but are generally not promoted, the project also reinforces local capacity to take advantage of these services.

CONCLUSIONS

The future of the CBR is at stake. Rapid population growth, primarily attributed to a dramatic rate of in-migration, threatens the long-term viability of the reserve unless effective land-use planning approaches and practices are put in place. At this time in the Calakmul region, there are few structures for empowering local communities to determine their future destinies. Caught in a web of dependency, local communities have neither the incentives nor the institutional support to plan for a just and sustainable future. Government at the federal and state level plays vital roles in building options and incentives for economic development and sustainable land uses. At the same time, NGOs face great challenges to encourage the creation of institutional structures that increase community participation in decision making and land-use planning, and ultimately lead to sustainable use of natural resources.

ACKNOWLEDGMENTS

The authors acknowledge the helpful editorial assistance of Lars Bromley on the initial version of this publication. The work was published originally as Occasional Paper No. 1, by the Program on Population and Sustainable Development (PSD) of the American Association for the Advancement of Science (AAAS) Summer 1999.

REFERENCES

Adams, R. E. W., 1977. Rio Bec archeology, in *The Origins of Maya Civilization*, Adams, R.E.W., Ed., University of New Mexico Press, Albuquerque, 85–99.

Anderson, S. and Rietbergen-McCracken, J., 1994. *El diagnostico participativo: un manual aplicado de tecnicas*, El Grupo DIP, Merida, Yucatan.

Antochiw, M., 1997. La cartografia y los Cehaches, in *Gobierno del Estado Libre y Soberano de Campeche*, Calakmul: Volver al Sur, Campeche, Mexico, 23–32.

Bajracharya, S. B., Gurung, L. P., and Manandhar, A., Population Dynamics and Resource Conservation Initiative in Annapurna Conservation Area Project, Nepal, WWF, Nepal, March.

Barton, T., Borrini-Feyerabend, G., de Sherbinin, A., and Warren, P., 1997, *Our People, Our Resources*, The World Conservation Union (IUCN), Gland.

Berlanga, M. and Wood, P., 1997. Las aves de la Reserva de la Biosfera de Calakmul, Campeche, Mexico, Internal report, Pronatura, Península de Yucatán, A.C., Mérida, Yucatán.

Bilsborrow, R. D. and DeLargy, P., 1991. Land use, migration, and natural resource deterioration: the experience of Guatemala and the Sudan, *Popul. Dev. Rev.,* Suppl. 16.

Boege, E., 1995. The Calakmul biosphere reserve, Mexico, working paper No. 13, paper presented at international conference, UNESCO South-South Cooperation Programme, Paris.

Boege, E. and Murguia, R., 1990. Diagnostico de las actividades humanas que se realizan en la reserva de la biosfera de Calakmul, estado de Campeche, Internal report, Pronatura, Península de Yucatán, A.C., Mérida, Yucatán.

Boesveld, M. and Postel-Coster, E., 1991. Planning with women for wise use of the environment. Research and practical issues, in *Landscape and Urban Planning*, Elsevier Science, Amsterdam, 20, 141–150.

Bruce, J., 1989. Rapid appraisal of tree and land tenure, *Food Agric. Organ.,* Community Forestry Note 5.

Caudill, D., 1997. *Lessons from the Field: Integration of Population and Environment*, World Neighbors, Oklahoma City.

Chambers, R., 1983. *Rural Development: Putting the Last First*, John Wiley & Sons, New York.

CNA (Comision Nacional del Agua), 1996. Internal report: Abastecimiento de agua potable en la region de Xpujil, municipios de Hopelchen y Champoton, Campeche, Gerencia estatal de Campeche, Campeche.

(CONAPO) Consejo Nacional de Poblacion, 1996. Situacion demografica del estado de Campeche, Campeche.

Cruz, M. C. J., 1996. Management options for biodiversity protection and population, in *Human Population, Biodiversity and Protected Areas: Science and Policy Issues*, Dompka, V., Ed., American Association for the Advancement of Science, Washington, D.C., 71–89.

Diario de Yucatan, November 7, 1996. Xpujil, region prioritaria para el Gobierno Federal, dice el titular de Desarrollo Social, Merida, Yucatan.

Dominguez, M. D. R., and Folan, W. J., 1995. Calakmul, Mexico: aguadas, bajos, precipitacion y asentamiento en el Peten Campechano, paper presented at IX simposio de investigaciones arqueologicas en Guatemala, Museo Nacional de Arqueologia y Etnologia, July, Guatemala.

Dompka, V. and Allcott, P., 1998. Do numbers matter: population impacts on environmental projects, American Association for the Advancement of Science and World Wildlife Fund, Godalming, U.K.

Engelman, R., 1998. *Plan and Conserve: A Source Book on Linking Population and Environmental Services in Communities*, Population Action International, Washington, D.C.

Ericson, J., 1996. Conservation and Development on the Border of the Calakmul Biosphere Reserve, Masters thesis, Humboldt State University, Arcata, CA.

Ericson, J., 1997. Regional assessment: Calakmul population-environment initiative, Internal report, World Wildlife Fund Mexico program, Mexico City.

Ericson, J., 1998. Population Dynamics in the Ejidos around the Calakmul Biosphere Reserve: Summary, Internal document, World Wildlife Fund, Washington, D.C.

Farriss, N., 1984. *Maya Society under Colonial Rule: The Collective Enterprise of Survival,* Princeton University Press, Princeton, NJ.

Faust, B., 1998. *Mexican Rural Development and the Plumed Serpent*, Bergin and Garvey, Westport, CT.

Freudenberger, K. S., 1994. Tree and land tenure: rapid appraisal tools, Food and Agriculture Organization, Community Forestry Manual 4.

Freudenberger, M., Mogba, Z., Zana, H., and Missosso, M., 1996. Migrations humaines et leurs impacts sur la conservation des ressources naturelles dans la Réserve de Dzanga-Sangha: Etude de as de l'economie de diamant à Ndélengué, République Centrafricaine, WWF, Social Science and Economics Program.

Galindo-Leal, C., 1996. La biosfera de Calakmul y el desarrollo sustentable, in Voz Comun: un espacio de integracion social en Campeche, 30, 20–21.

Garrett, W. E., 1989. La ruta Maya, in *Natl. Geogr. Mag.*, 176(4), 424–479.

Haenn, N., 1997. Creating communities through acceptance/rejection of migrants in frontier Campeche, paper presented at meeting of Latin American Studies Association, April 17–19, Guadalajara, Mexico.

Hecht, S. and Cockburn, A., 1989. *The Fate of the Forest: Developers, Destroyers and Defenders of the Amazon*, Verso, New York.

Jones, G., 1989. *Maya Resistance to Spanish Rule: Time and History on a Colonial Frontier*, University of New Mexico Press, Albuquerque.

Loudiyi, D., 1994. Calakmul demographic report, Internal report, World Wildlife Fund, Washington, D.C.

Maas Rodriguez, R. and Ericson, J., 1998. Memoria: Metodo y Tecnicas de la "Iniciativa de Poblacion y Medio Ambiente para Calakmul," Internal document: Pronatura Peninsula de Yucatan, A. C. Merida, Yucatan.

McNeely, J. A. and Ness, G., 1996. People, parks, and biodiversity: issues in population–environment dynamics, in *Human Population, Biodiversity and Protected Areas: Science and Policy Issues,* Dompka, V., Ed., American Association for the Advancement of Science, Washington, D.C., 19–70.

Miranda, F. and Xolocotzi, H. E., 1985. Los tipos de vegetación de México y su clasificación, in *Xolocotzia,* Vol. I, Universidad Autonoma Chapingo, México.

Montejo Alvarez, A. et al., 1997a. Estudio Rural Participativo, Poblacion y Medio Ambiente, Ejido, 11 de Mayo, Calakmul, Campeche, internal document, Pronatura Peninsula de Yucatan, A. C. Merida, Yucatan.

Montejo Alvarez, A. et al., 1997b. Estudio Rural Participativo, Poblacion y Medio Ambiente, Ejido: Conhuas, Calakmul, Campeche, internal document, Pronatura Peninsula de Yucatan, A. C. Merida, Yucatan.

Murphy, J., 1990. Indigenous forest use and development in the "Maya zone" of Quintana Roo, Mexico, Major paper, York University, Ontario, Canada.

Ness, G., with Golay, M. V., 1997. *Population Strategies for National Sustainable Development: A Guide to Assist National Policy Makers in Linking Population and Environment in Strategies for Sustainable Development,* Earthscan Publications, London.

NOM (Norma Official Mexicana) NOM-059-ECOL-1994.

Pasos, R., Girot, P., Laforge, M., Torrealba, P., and Kaimowitz, D., 1995. El ultimo despale, la frontera agricola centroamericana, Fundacion para el Desarrollo Economico y Social de Centroamerica (FUNDESCA), Panama.

Pichon, F., 1993. Agricultural Settlement, Land Use and Deforestation in the Ecuadorian Amazon Frontier: A Micro-level Analysis of Colonist Land-Allocation Behavior, Ph.D. dissertation, Department of City and Regional Planning, University of North Carolina, Chapel Hill.

Ronfeldt, D., 1973. *Atencingo: The Politics of Agrarian Struggle in a Mexican Ejido,* Stanford University Press, Stanford, CA.

Rudel, T. and Horowitz, B., 1993. *Tropical Deforestation: Small Farmers and Land Clearing in the Ecuadorian Amazon*, Columbia University Press, New York.

Rzedowski, J. and de Rzedowski, C. G., 1989. Transisthmic Mexico: Campeche, Chiapas, Quintana Roo, Tabasco and Yucatán, in *Floristic Inventory of Tropical Countries: The Status of Plant Systematics, Collections, and Vegetation, plus Recommendations for the Future*, Campbell, D. G. and Hammond, H. D., Eds., The New York Botanical Garden, New York.

Sharer, R. J., 1994. *The Ancient Maya*, Stanford University Press, Stanford, CA.

Stedman, E., 1997. Socioeconomic root causes of biodiversity loss: the case of Calakmul, Mexico, World Wildlife Fund Mexico Program, Mexico City.

Tan, P. C. and Tey, N. P., 1994. Do fertility intentions predict subsequent behavior? Evidence from Tenninsular, Malaysia, in *Studies in Family Planning,* 259(4): 222–231.

Thomas, P. M., Jr., 1981. Prehistoric Maya Settlement Patterns at Becan, Campeche, Mexico, Middle American Research Institute, Publ. 45, Tulane University, New Orleans.

Turner, II, B. L., 1978. Ancient agricultural land use in the central Maya lowlands, in *Pre-Hispanic Maya Agriculture*, Harrison, P. D. and Turner II, B. L., Eds., University of New Mexico Press, Albuquerque.

Vasquez Sánchez, 1991. Documento de trabajo para la elaboracion del plan de manejo de la reserva de biosfera Calakmul, Campeche, Centro de Estudios para la Conservacion de los Recursos Naturales, A.C. (ECOSFERA), Merida, Yucatan.

WWF, 1994. Summary of demographics in Calakmul, Internal report, World Wildlife Fund, Washington, D.C.

Yuñez-Naude, A., Taylor, J. E., and Rodriguez-Gonzalez, M. D. R., 1995. Impactos de las reformas economicas en una poblacion ejidal: una propuesta de analisis cuantitativo, presented at El Taller del Proyecto de Investigacion sobre la Reforma Ejidal, Centro de Estudios Mexico-Estados Unidos, University of California, San Diego, August 25–26.

15 Toward Social Criteria and Indicators for Protected Areas: One Cut on Adaptive Comanagement

Carol J. Pierce Colfer
with Ravi Prabhu, Eva (Lini) Wollenberg, Cynthia McDougall, David Edmunds, and Godwin Kowero

CONTENTS

INTRODUCTION

The Center for International Forestry Research (CIFOR) has developed a rigorously tested process for establishing criteria and indicators (C&I) of sustainable forest management. The concern of the organization with alleviating poverty and improving human welfare within an environmentally friendly context has generated particular interest in social criteria for evaluating the conditions in forests. Both the process and the product of the social C&I initiative appear to offer strategic insight, approaches, and tools for operationalizing the concept of adaptive comanagement in and around protected areas.

This chapter, after a brief overview of recent literature related to assessing and monitoring conditions in protected areas, examines some demonstrated and potential roles of social C&I in adaptive comanagement of protected areas. We then turn to

0-8493-0020-7/01/$0.00+$1.50

an examination of the social C&I CIFOR has developed within the context of industrial timber exploitation, as "templates" or "base sets" for possible use in conservation areas. We are interested in how they may be used in monitoring human and environmental conditions, as well as the efficacy of agreements or policies that may ultimately form the basis of collaborative management decisions. Next we identify ways in which we expect to advance understanding of the potential and effectiveness of adaptive comanagement (ACM) approaches in managing protected area forests through the development and application of social C&I of sustainable forest management. Finally we suggest practical ways to use these C&I, adapted both to protected area concerns and to local conditions, as part of the process of creating the conditions (or criteria) specified as important for sustainability in the real world.

THE DEVELOPMENT OF SOCIAL CRITERIA AND INDICATORS

Although the idea of social indicators is not new, in recent years there has been a resurgence of interest, as part of a global concern to develop C&I for sustainable forest management. In addition to the research and fieldwork reported in this book, within the field of conservation, a number of monitoring and assessment tools have been identified — only a sprinkling of which are mentioned here.

International Union for Conservation of Nature and Natural Resources (IUCN) has been involved in a number of efforts to improve monitoring and evaluation relating to human well-being. Borrini-Feyerabend, with Buchan (1997), has published a two-volume work aimed at improving our ability to achieve "social sustainability in conservation." Although this work is organized into key questions, rather than indicators per se, it deals with many of the issues we discuss below, indicating "warning flags," where pertinent.

Poffenberger (1996) reports case studies from India, Nepal, British Columbia, Panama, and Ghana, and draws conclusions about important issues in sustainability. In common with the authors, he has identified as important such issues as participatory planning with local communities, building on existing community management, devolving implementation to local communities or user groups, and policy support for local management. He also shares with the authors methodological opinions in his recommendation of techniques like mapping of traditional forest territories, use of traditional forest knowledge, and social fencing based on community management agreements. Jackson (1997), who reports some of the tools and methods that have been used in southern Africa to evaluate collaborative management of natural resources, emphasizes participation issues. Prescott-Allen's (1995) concept of a "barometer of sustainability" is particularly appealing, because of its simplicity. It, like the CIFOR approach discussed below, includes the idea that "human wellbeing is dependent on the wellbeing of the ecosystem" (IUCN, 1995a draft). In this system, indicators are considered context specific and are selected by the users (IUCN, 1995b).

The relationship between human well-being and the well-being of the ecosystem requires some qualification. Human well-being may be improved in the short run by ecosystem degradation, and there may be healthy ecosystems in protected areas where human well-being has been adversely affected (e.g., by excluding people). However, over the long haul, we would argue for a close link between the two. As Fisher (personal communication, 1998) has put it, in the long term, a healthy ecosystem's well-being is a *necessary but not sufficient* condition for human well-being.

Much of the growing literature on community-based management is also pertinent. Stevens (1997) uses C&I to assess the sustainability of a Turkish forest village ecosystem. He explicitly recognizes a link between "the state of the natural resource base, and the social and financial indicators that depend upon them" (p. 30). Work by the Asia Forestry Network has included relevant assessment methods (e.g., Poffenberger et al., 1992a, b), as have their case studies (e.g., Poffenberger and McGean, 1993a, b; 1994; Josayma et al., 1996; Poffenberger, 1998). Their methods books are written in clear language, available to many users. Their case studies provide readers with important details about local experience in a number of forested contexts in India, Thailand, Vietnam, the Philippines, and Hawaii. The books and manuals are easily accessible through the Asia Forestry Network.

Similarly, the *Rural Development Forestry Study Guides* provide useful and comprehensive insights (Carter, 1996; Hobley, 1996), again with both methodological and case study materials of great utility. Carter's book includes case material from Nigeria, Ecuador, Mexico, Ghana, Nepal, Kalimantan, and Uganda; Hobley's focuses on India and Nepal. A recent review by the Ford Foundation (1998) also stresses a number of the issues CIFOR has found to be important. Thomson and Schoonmaker Freudenberger's (1997) field manual on crafting institutional arrangements, along with other contributions from the FAO Community Forestry Series, will be particularly useful in our own future work. Resolve (1994) and Chandrasekharan (1996) are two excellent examples focused on conflict resolution in natural resource issues. Burford de Oliveira (CIFOR) and her colleagues have tested C&I in four community-managed forest contexts (two in Indonesia, and one each in Cameroon and Brazil; personal communication).

The research we report here began in 1994, when CIFOR in Bogor, Indonesia, began a research project entitled "Assessing Sustainable Forest Management: Testing Criteria and Indicators."* From the beginning, the importance of social issues was recognized. If a general set of social C&I could be found, the set could be used to attract attention to the plight of people residing in, and dependent on, the world's disappearing forests.

Within the project, there was agreement that research should be field based, collaborative, and iterative. At that time, existing sets of social C&I had all been developed as desk studies. We have felt that there needed to be a continual oscillation between desk work and fieldwork in developing useful C&I. The stress on collaboration derived from our conviction that there is a sizable value dimension to any

---

* This project has been led by Dr. Ravi Prabhu, and initially funded by GTZ. Over the years, the project has also received funding from the European Union, USAID, Ford Foundation, MacArthur Foundation, African Timber Organization, IDRC, JICA, and the Swiss Government, as well as CIFOR itself. Initially prompted by an interest in timber certification, the focus quickly shifted to sustainability in general.

discussion of sustainability and that it was therefore important to involve people with a variety of experiences and perspectives. Access to these various experiences and perspectives is critical in the attempt to clarify the concept of sustainability (important both ethically, to respect different values, and practically, because when values are not respected, conflict arises).

Finally, our commitment to an iterative approach represents recognition that we are groping in new ground and that no one actually knows how to manage forests sustainably or how to enhance human well-being within particular decision making contexts. Given this fact, a process of conceptualization, field trials, evaluation of results, revisions, and more field trials seemed the most practical approach. This iterative emphasis remains critical as we move from C&I development into use in the world's vastly differing human and environmental contexts. Between 1994 and 1996, interdisciplinary field tests of existing C&I sets for sustainable forest management (SFM) were conducted in six countries (Germany, Indonesia, Côte d'Ivoire, Brazil, Austria, and Cameroon). One activity was the development, early in 1995, of a conceptual framework for dealing with social issues (Colfer et al., 1995) based on social science literature, field experience, and perusal of existing sets of C&I. The C&I sets selected for testing were considered the best available at the time (Hahn-Schilling et al., 1994), and included those of the United States-based Rainforest Alliance, the Soil Association (United Kingdom), GTW from Germany, Lembaga Ekolabel Indonesia, and the Dutch DDB (Prabhu et al., 1996). Those of ATO (African Timber Organization) were tested in the final six-team test in Cameroon (Prabhu et al., 1998a).*

The results of these tests were analyzed in a number of ways, including a search for commonalties across sites (Prabhu et al., 1996; 1998a). The agreement on a number of issues pertaining to human well-being (which we considered an integral part of SFM) was particularly surprising, considering the heterogeneity of the social science expertise included.**

One conclusion that emerged from these first activities (subsequently called "Phase I") was a recognition that one set of C&I would never be globally applicable. The CIFOR teams have made progress in identifying a subset of criterion-level issues that applied in all the areas we tested; and we will be surprised if there are important, location-specific changes at the higher levels of the C&I hierarchy (principles, criteria). However, we are convinced that the indicator and verifier sets that CIFOR

* This 1996 Cameroon test differed from the previous ones in including six teams, comprising three rather than five experts, testing C&I in the same location (East of Kribi in the Wijma timber concession area). This test, besides examining the relevance of the C&I for Cameroon conditions, was meant to test the replicability of the method. A series of additional tests (with CIFOR collaboration) began in 1998 in West and Central Africa, under African Timber Organization (ATO) auspices, to test and adapt the ATO set to conditions in the respective countries. A test was conducted in Gabon in April, and another is under way in the Central African Republic (late 1998).

** The Indonesian team included an anthropologist (Laksono); the Brazil team, a German sociologist (Jan Kressin); in Côte D'Ivoire, a Dutch social psychologist (Heleen van Haaften) and an Ivoirean rural sociologist (Ahui Anvo). In the Austrian test, there were an Austrian forest farmer/communications specialist (Franz Rest) and an anthropologist (Friedld Grunberg); and in Cameroon, there were two economists (Alain Karsenty and Martine Antona) and four anthropologists (Francis Nkoumbele, John Mope Simo, Bertin Tchikangwa, and Jolanda van den Berg).

has developed will have to be tailored and adapted to individual contexts. Making these sets of C&I practical implements will, in our view, also require input from local managers (both formal and informal) — to ensure that desired conditions can be agreed upon and monitored by relevant stakeholders.* C&I must be approached as flexible devices, adaptable to varying local conditions. They cannot be applied mechanistically.

Another important conclusion was unanimous dissatisfaction with the methods available for assessing these important human issues quickly, inexpensively, and reliably. In response to this dissatisfaction, and driven by the conviction that success in this endeavor could have potentially important consequences for forests and forest people, we planned Phase II.

Phase II of the project began in 1996 and drew to a conclusion in early 1999. The social component of Phase II addressed two broad issues: (1) improvements in assessment methods and (2) increased understanding of the links between SFM and three issues widely agreed to be important to human well-being:

1. Security of intergenerational access to resources (including tenure, use rights, sharing of forest benefits);
2. Rights and means to manage cooperatively and equitably (participation, voice); and
3. Identification of relevant stakeholders ("forest actors").**

Our first step was to assemble and pretest a series of methods we considered useful in assessing the above issues in timber concession areas.*** This process, which has continued to inform the development of social C&I, began with a literature survey, the development of eight methods pretested in and around Danau Sentarum Wildlife Reserve in West Kalimantan in 1996 (Colfer and Wadley, 1996; Colfer et al., 1996a; 1997a, b; Dennis et al., 1998), then evaluation and revisions into a "methods binder" consisting of 12 methods. These methods were tested in two locations in East Kalimantan and a number of locations in Cameroon. Draft methods manuals, including the CIFOR "Best Bets" for social C&I (see below), were completed in early 1998 (recently finalized in Colfer et al., 1999a, b, c), based on test results from Sardjono et al. (1997), Tiani et al. (1997), and Mt. Cameroon (1997). Additional methods tests were completed in Cameroon by Diaw et al. (1998) and

* Fisher (personal communication, 1998) suggests developing a set of "participative procedures" to complement the C&I template. The ACM approach described later in this chapter is one planned attempt to address this issue.

** See Colfer (1995) for a preliminary attempt to differentiate "forest actors" — or those people outside managers need to take into consideration — from the wider category of "stakeholders." This issue is dealt with in more detail in Colfer et al. (1999a, b).

*** Use of the term, *issues*, rather than *principles* or *criteria*, derives from our dissatisfaction with the hierarchical framework generally accepted by those working on C&I (principles, criteria, indicators, and norms or verifiers, in a hierarchy of decreasing abstraction) — including ourselves. Although Lammerts van Bueren and Blom (1997) have made an excellent effort to standardize global usage of the terms, some of the problems with this framework are probably more fundamental than definitional. We use the hierarchical framework while continuing to seek a more congenial approach that can better capture the variation in importance and/or fluidity of levels of C&I in different locales.

Tchikangwa et al. (1998) in early 1998, and more recently in two locations in Brazil (by Porro and Porro, 1998) along with another partial test in Long Loreh, East Kalimantan (McDougall, 1998).

Current work focuses on analysis of the results of the methods tests, which we hope will shed additional light on the relationships among SFM and access to resources, on the one hand, and SFM and management rights and responsibilities, on the other. The former refers to the economic and livelihood basis of people's way of life and the latter to the norms and decision-making structures that affect use of local resources. We also hope that our methods tests will provide useful information in clarifying the continuing question of stakeholder identification, or "who counts" in SFM. Finally, based on observations in the course of testing, we are working on gender and other issues of stakeholder representation.

RELEVANCE FOR CONSERVATION AREAS

Although our work began with a focus on "forest management units," implicitly defined as forested areas managed for timber extraction, we found a more complex reality in the field. There were, of course, people living in and around those areas who had managed and continued to manage those forests in a multipurpose fashion for products like wildlife, foods, medicines, fibers, and a variety of services such as soil enrichment, maintenance of water quality, shade. It was also not unusual for both artisanal and industrial timber extraction to be under way near, and sometimes in, protected areas. Several of the research sites selected for social science methods testing in Indonesia and Cameroon were adjacent to or included protected areas. Each of these locations represents a mixture of conservation, timber exploitation, and more than one form of community-based forest management (Table 15.1).

TABLE 15.1
Test Sites with Conservation Implications

Conservation Areas	Villages	Main Ethnic Groups
Danau Sentarum Wildlife Reserve (West Kalimantan, Indonesia)	Ng. Kedebu', Danau Seluang, Kelayang, Bemban, Wong Garai[a]	Melayu (fishers) Iban (swiddeners)
Kayan Mentarang National Park (East Kalimantan, Indonesia)	Paking, Long Loreh	Abai, Lun Daye, Merap, Lepo' Ke (swiddeners); Punan (hunter/gath)
Mt. Cameroon–Limbe (southwestern Cameroon)	Mbongo	Balondo (indigenous) Ibibio, Ibo (Nigerian)
Dja Reserve (central Cameroon)	Messok and Mbaya Sembe, Pohempoum, Bareko	Baka (pygmies) Bareko, Nzime, Kako, Njem, Badjoue

[a] These are pseudonyms used to protect people's privacy. The research in West Kalimantan also included Reed Wadley and Emily Harwell; Kayan Mentarang work was led by Cynthia McDougall and Mustofa Agung Sardjono; the Mt. Cameroon team was coordinated by Mary Ann Brocklesby; and the Dja Reserve work was undertaken under Bertin Tchikangwa's leadership.

Given these common overlays of management systems, we have concluded that there are many areas in which the social C&I developed for timber concession areas apply to protected areas as well. Following is a reproduction of the most recent "Best Bets" for social C&I.

1. Forest management maintains or enhances fair intergenerational access to resources and economic benefits
   - 1.1 Local management is effective in controlling maintenance of and access to the resource.*
     - 1.1.1 Ownership and use rights to resources (inter- and intragenerational) are clear and respect preexisting claims.
     - 1.1.2 Rules and norms of resource use are monitored and enforced.
     - 1.1.3 Means of conflict resolution function without violence.
     - 1.1.4 Access to forest resources is perceived locally to be fair.
     - 1.1.5 Local people feel secure about access to resources.
   - 1.2 Forest actors have a reasonable share in the economic benefits derived from forest use.
     - 1.2.1 Mechanisms for sharing benefits are seen as fair by local communities.
     - 1.2.2 Opportunities exist for local and forest-dependent people to receive employment and training from forest companies.
     - 1.2.3 Wages and other benefits conform to national and/or International Labor Organization (ILO) standards.
     - 1.2.4 Damages are compensated in a fair manner.
   - 1.3 People link their and their children's future with management of forest resources.
     - 1.3.1 People invest in their surroundings (e.g., time, effort, money).
     - 1.3.2 Outmigration levels are low.**
     - 1.3.3 People recognize the need to balance numbers of people with natural resource use.
     - 1.3.4 Children are educated (formally and informally) about natural resource management.
     - 1.3.5 Destruction of natural resources by local communities is rare.
     - 1.3.6 People maintain spiritual links to the land.***
2. Concerned stakeholders have acknowledged rights and means to manage forests cooperatively and equitably.

---

* This criterion is obviously very closely connected with criteria addressed from ecological and formal "forest management" perspectives. This assumes additional attention in other parts of the overall assessment.

** Indicators 1.3.2 and 1.3.3 contain a potential contradiction. Low levels of outmigration (I1.3.2) indicate that people link their and their children's future to maintaining the forest; yet recognizing the need to balance numbers of people with natural resource use (I1.3.3) may lead them to favor outmigration. This contradiction would likely occur when conditions are deteriorating.

*** In and around protected areas, it may be necessary to add a condition that local people continue to have access to the land for such spiritual purposes, recognizing the frequency with which local people have been excluded from their homelands.

2.1 Effective mechanisms exist for two-way communication related to forest management among stakeholders.

2.1.1 >50% of timber company personnel and forestry officials speak one or more local language, or >50% of local women speak the national language.

2.1.2 Local stakeholders meet with satisfactory frequency, representation of local diversity, and quality of interaction.

2.1.3 The contributions of all stakeholders are mutually respected and valued at a generally satisfactory level.

2.2 Local stakeholders have detailed, reciprocal knowledge pertaining to forest resource use (including user groups and gender roles), as well as forest management plans prior to implementation.

2.2.1 Plans/maps exist showing integration of uses by different stakeholders.

2.2.2 Updated plans, baseline studies, and maps are widely available, outlining logging details like cutting areas and road construction, with timing.

2.2.3 Baseline studies of local human systems are available and consulted.

2.2.4 Management staff recognize the legitimate interests and rights of other stakeholders.

2.2.5 Management of Non Timber Forest Products (NTFP) reflects the interests and rights of local stakeholders.

2.3 Agreement exists on rights and responsibilities of relevant stakeholders.

2.3.1 Level of conflict is acceptable to stakeholders.

3. The health of forest actors, cultures, and the forest is acceptable to all stakeholders.

3.1 There is a recognizable balance between human activities and environmental conditions.

3.1.1 Environmental conditions affected by human uses are stable or improving.

3.1.2 In-migration and/or natural population increase are in harmony with maintaining the forest.

3.2 The relationship between forest management and human health is recognized.

3.2.1 Forest managers cooperate with public health authorities regarding illnesses related to forest management.*

3.2.2 Nutritional status is adequate among local populations (e.g., children's growth conforms to international standards of height for weight; infant and <5 year mortality levels are low).

3.2.3 Forestry employers follow ILO working and safety conditions and take responsibility for the forest-related health risks of workers.

* In areas bordering protected areas, threats to human safety and security from wild animals may be a serious health problem relating to forest management.

3.3 The relationship between forest maintenance and human culture is acknowledged as important.
   3.3.1 Forest managers can explain links between relevant human cultures and the local forest.
   3.3.2 Forest management plans reflect care in handling human cultural issues.
   3.3.3 There is no significant increase in signs of cultural disintegration.

Although no field-testing has been done to differentiate C&I that are relevant for protected areas vis-à-vis industrial timber operations, we can take the first step in this chapter, and examine possible differences conceptually. Below we examine the "Best Bets" critically, with protected area considerations in mind.

Much of principle 1,* "Forest management maintains or enhances fair intergenerational access to resources and economic benefits," would appear to apply as much to a protected area as to a timber concession. The most notable exceptions are under 1.2, "Forest actors have a reasonable share in the economic benefits derived from forest use." 1.2.2, for example, might better read: "Opportunities exist for local and forest-dependent people to receive employment and training from protected area management."** 1.2.4 ("Damages are compensated in a fair manner") is likely to involve different levels of damage and compensation in a protected area than in a timber concession. Timber extraction regularly entails road building through community orchards or sacred groves, soil compaction in potential agricultural sites, damage to forest canopy, loss of valued fruit trees, wood preservation chemicals polluting streams, etc., most of which are rare in protected areas (although roads and tourists may well cause problems). Eviction, common for protected areas, represents one serious kind of damage that is less common in timber concession areas. Crop loss and the threat of attack from wild animals can be other serious issues.

Turning to criterion 1.3, "People link their and their children's future with management of forest resources," 1.3.2 ("Outmigration levels are low") might be less appropriate in a protected areas, if there is an interest in decreasing or stabilizing the resident population over time. In efforts to comanage Danau Sentarum Wildlife Reserve, for example, Dudley and Colfer discussed several ideas for reducing the population within the reserve with local people in search of a mechanism that would not disadvantage or coerce them while reducing extraction levels in fisheries.

Much of the second principle, "Concerned stakeholders have an acknowledged right and means to manage forests cooperatively and equitably," also seems to apply

---

* We call this level "principles," but the identification of level of abstraction remains somewhat arbitrary. In other versions, we have written about one social principle ("maintaining or enhancing human well-being") with the next level down, as "criteria." In this version the previous "criteria" are presented as "principles." As noted above, we have reservations about the organization of C&I into a hierarchical framework. Although we are working on a network approach to these issues (Prabhu et al., 1998b), we have not yet progressed far enough to replace the hierarchical framework in common use.

** It is appropriate here to acknowledge the very significant involvement of other scientists in the development of these ideas. The collaborators and other CIFOR scientists who have worked with us through this whole iterative process are too numerous to list, but we do appreciate and acknowledge their input and roles.

to protected areas, with minor changes. Indicator 2.1.1, for example, on the ability of timber company personnel to speak local languages, would simply apply to protected area managers rather than to timber company personnel. The plans/maps, mentioned in indicator 2.2.1, showing different uses, were considered important in timber concession areas because timber companies are usually required to make maps defining their timber-harvesting activities. Although protected area managers may use maps differently, the kinds of maps specified in the indicator would serve to formalize a cooperative planning process between local people and protected area managers, both of whom may in fact be managing the same area with different goals and regulations. A number of protected areas are now actively involved in mapmaking with local people (e.g., Danau Sentarum Wildlife Reserve and the Kayan Mentarang National Park, both in Kalimantan). Indicator 2.2.2, specifying detailed, reciprocal knowledge of forest uses and plans prior to implementation, would refer to conservation-related activities rather than the logging details of cutting areas, road building, and timing. Such conservation activities might include plans to build a field center, participation in an international volunteer program, efforts to control human activity in "buffer zones," or strengthening of efforts to protect a particular species.

Where the first two "Best Bets" principles above have undergone extensive field testing and revision, the third one, "The health of forest actors, cultures, and the forest is acceptable to all stakeholders," has not. This principle did not receive the same unanimity of selection in Phase I tests as did the first two, but the issues were widely agreed to be important. There do not appear to be any of these C&I that would not be relevant, as written, in protected area management.

It seems probable that many of the C&I identified as important in forests managed for timber are also important in forests managed for conservation. Important next steps will include

1. Critical examination of these C&I by conservation specialists/experts in the social impacts of protected areas, in search of gaps or issues that pertain exclusively to protected areas and have therefore not been addressed in the CIFOR tests;
2. Participatory field tests focused on conservation areas (including critical examination of the C&I by local people living in or closely adjacent to such areas); and
3. Adaptation of (and tools for adapting) the resulting conservation area C&I, in the field for specific locales.

In sum, the C&I that were originally designed for forests that are managed for timber and other extractive use *appear* also to apply to protected areas, with a few minor changes (Table 15.2). Inevitably, additions or modifications will be necessary when we try to use them in specific protected areas, but that is, in fact, also true in forests managed for timber production. The CIFOR tests have identified a core of common issues, a "template," if one will, but there are always site-specific differences, C&I that are uniquely important in different locales. In all tests, team members have felt the need for additions, differences in emphasis, and/or deletions. We consider this a good sign, a recognition of the existing global variety in human and

TABLE 15.2
Comparison of Probable Differences between Social C&I in Areas Managed for Timber and for Conservation

In some cases, reference to "timber company personnel" needs to be changed to "conservation area managers."
Share in forest benefits and damages owing to local people are likely to differ in substance and quantity, although the issues remain important.
The possible need to stabilize or reduce population within protected areas may require changes in the issue linking children's futures to forest management.
The importance of maps, as communication devices pertaining to sharing of information among stakeholders, is important in both contexts, but the other uses of the maps (and thus their appearance and contents) may differ.

natural resource conditions, and also of the evolving nature of the human view of sustainability. Our effort has not been to develop *the* correct set, but rather to make progress toward an evolving definition of sustainability, on the ground. Prabhu and his team (personal communication) have been developing a tool (CIMAT) that will help some users adapt the CIFOR template, based on their own unique site conditions.

PARTNERSHIPS, ACTION RESEARCH, AND ACM*

With CIFOR Phase II drawing to a close, we have of course devoted considerable thought to what the logical "next steps" might be. We are well along in the process of developing a "toolbox" to help others who want to assess SFM in their locales. We are satisfied that we have contributed in a significant way to the ongoing process of defining sustainability in a way that can be assessed and that the C&I template will be of use.

But C&I can only begin to fulfill their potential, to contribute to real sustainability, when the conditions they specify apply in the field. Continuing in our iterative, collaborative, and field-based orientation, we now plan to work to implement the C&I that we have identified, in collaboration with formal and informal managers, in the field. This will require a greater emphasis on the process-oriented aspects of management and the empowerment of local people in those processes.

That we have encountered protected areas in our supposed focus on areas managed for commercial timber extraction should be no surprise. Rarely are areas managed for one purpose or by one user group in the tidy manner characteristic of the typical plan or report depicted by Ministries of Forestry and/or Environment in the world's capital cities. Instead, we find overlays of management, with different purposes and different managers, in the same areas. We have found, as discussed above, that many of the human issues that are important in areas managed for

* It is appropriate here to acknowledge the very significant involvement of other scientists in the development of these ideas. The collaborators and other CIFOR scientists who have worked with us through this whole iterative process are too numerous to list, but we do appreciate and acknowledge their input and roles.

commercial timber exploitation are also relevant in protected areas. Indeed, these findings are among the factors that have impelled us to try the multistakeholder approach described below.

We would like to develop, with both formal and informal local managers, in several locations, a C&I-based monitoring arrangement. This would reflect a move from our current, somewhat removed, semiacademic approach to an action research mode we are calling "Adaptive Comanagement" (ACM). This approach, as we have conceived it, represents a merging of global experience in comanagement and adaptive management.

A considerable amount of research has been done on comanagement. Claridge (1997, p. 19), for example, defines comanagement as "the active participation in management of a resource by the community of all individuals and groups having some connection with, or interest in, that resource." He further defines comanagement regimes to include:

- Sharing of authority and responsibility for resource management according to arrangements which are understood and agreed by all parties;
- Social, cultural and economic objectives are an integral part of the management strategy; and
- Sustainable resource management is a major objective.

We suspect that comanagement, on the ground, will involve the active participation of individuals and groups who, through a painstaking process of alliance building and networking, come to share certain goals for forest management. We focus on comanagement with the goals of improving human well-being — particularly that of forest residents — and of sustainable use of forest services and resources. We see empowerment and "making space" for local people as part of a viable strategy in pursuit of these goals, particularly vis-à-vis the policy process.

"*Adaptive* Comanagement," as we use the phrase also builds on the ideas of people like Hilborn and Walters (1992), Holling (1978), Lee (1993), Shindler et al. (1996), Stankey and Shindler (1997), and Stankey and Clark (1998), although we add a stronger emphasis on collaboration among the relevant stakeholders than appears to have been the case in the U.S. experiments they describe (see Prabhu et al., 1998b). Mayers and Kotey (1996) apply some of these ideas in an African context, focusing more on institutions, as we also hope to do.

We explicitly define ACM, in our context, as including:

- A conscious learning process in management;
- An equitable integration or involvement of all relevant stakeholders in the management process; and
- Decisions reflecting recognition of the dynamism and complexity of human and natural systems that touch on management.

The identification of new and/or effective institutional arrangements and policies that affect forest management can contribute, we believe, to the well-being of people and tropical forests. We hope to discover:

- Means (e.g., policies, institutions) of improving center–periphery links and the effects of such improvements on human well-being and sustainable forest management;
- The effects of ACM on human well-being and sustainability in particular forests;
- Specific mechanisms whereby constructive interaction among stakeholders can improve forest management and human well-being.

The scenario we think holds the greatest likelihood of success in addressing the current problems in the human–forest interface is one that adequately takes into account the interests of all relevant stakeholders and builds on the managerial capacities of local communities. It will also have to identify policies and institutions that promote human well-being, enhance the political power of generally disempowered forest residents, and sustainable forest management by communities, in most cases, in coordination with surrounding forest managers. Bringing about such a situation requires identification of stakeholders, including appropriate representatives, and improved mechanisms for ensuring their voice in decisions about local forests. It also requires working with local stakeholders and with government representatives creatively to establish partnerships and to solve problems in a variety of contexts.

Our efforts are based on several important conclusions:

- That all relevant stakeholders have a part to play and a contribution to make in sustainable forest management, environmental protection, and in enhancing human well-being.
- That without some consensus, or at least a working agreement, among the significant majority on how and for what purposes the forest is managed, resource degradation and rural poverty will be accelerated.
- That forests and human systems are complex, that no single solution is likely to be generally successful, and that iterative improvements in management are required.

Ultimately, the distribution of power among all stakeholders and the impacts of the exercise of that power on human well-being and on the forests themselves are the foci of our attention.

RESEARCH APPROACH

The approach we are taking builds on previous CIFOR (and other) research, and melds the ideas of ACM with our ongoing devolution research.*

CIFOR Experience Leading to the Development of ACM

The CIFOR interdisciplinary teams have conducted significant research on stakeholder identification (including attention to subgroups like gender, age, wealth,

* By devolution, we mean a purposeful process designed to transfer varying amounts of formal authority from "higher," "centralized" levels of government to "lower," more "peripheral," local-level governmental and NGOs. We are particularly interested in the bundle of policies that together serve this function.

ethnicity); techniques for assessing livelihoods in protected areas and the linkage between conservation and economic incentives; methods for NTFP conservation and development; sustainability monitoring tools like criteria and indicators and methods for adapting them to local conditions; a typology of community forest management; methods for assessing biophysical conditions in forests (whether used for timber, plantations, conservation, or community management); methods for assessing key social issues such as security of intergenerational access to resources ("tenure") and rights and means to manage cooperatively and equitably ("participation"); progress on decision support tools (like MCA, FLORES, and CIMAT); and research on the links among local contexts, institutions, and national policies.

Such experience provides an excellent conceptual and methodological pool from which to draw in the process of developing ACM in the field. We have begun with relatively inexpensive, global literature surveys and reviews; newly initiated case studies in selected sites will give us focused and directly comparable information;* and participatory action research on ACM in a small number of sites will allow us to put our findings into practice with local communities. Our previous work, combined with the work described here, will help us develop alternative management strategies and mechanisms to make or strengthen institutional linkages among stakeholders, both "vertically" and "horizontally."

Our previous research has led us to conclude that to reach the desirable scenario outlined above, in which local communities, protected area managers, and other stakeholders collaborate in truly sustainable forest management, we will need specifically to:

- Identify or develop policy instruments that support SFM, including both biophysical and social elements.
- Identify or develop local institutional arrangements that support socially and environmentally SFM.
- Improve methods for planning and monitoring conditions of sustainable forest management, including human well-being and ecological integrity.
- Experiment with the implementation of existing C&I in forests managed for various purposes, by stakeholders and by researchers in cooperation with relevant stakeholders.
- Identify, improve our understanding of, and further develop mechanisms for local empowerment in stakeholder negotiations, including ways communities can influence policy.
- Seek out practical means of improving participation of women and less represented/powerful stakeholders in the research and other stakeholder processes.

Our iterative and adaptive emphasis will require ongoing monitoring and evaluation of how our own efforts progress.

---

* This group, comprising primarily Madhu Sarin, Neera Singh, Antonio Contreras, David Edmunds, Louise Buck, and Lini Wollenberg, recently summed up their primary interest in these case studies as "creating space for community action in forest management" (Wollenberg, informal presentation 20/9/98).

We have begun with a set of literature reviews, relevant to ACM, including issues like accommodating multiple stakeholder interests, future scenarios, and people's strategies for influencing policy. Supportive research includes policy and historical surveys in Asia, Africa, and Latin America* that examine two-way influences between local events/actors and policy. These reviews will provide a systematic overview of what has been done, what has worked, and what has not worked.

Our case study research includes, but is not limited to examining the following:

- Field-level impact of devolution policies and institutions;
- Factors that affect impacts of devolution policies;** and
- Strategies and methods that allow people in "the periphery" to affect policy positively, including particularly ACM strategies.

In some areas, the case studies will serve as a source of information about devolution mechanisms and experience with devolution policies; other cases will serve as a prelude to more action-oriented ACM work.

The literature surveys and case studies set the stage for testing these approaches in paired research environments, one of which is exposed to participatory action research. In both research settings, forest and human conditions are monitored; but in one area, we work with the various local stakeholders to agree on and implement a C&I-based (self-)monitoring arrangement and to improve integration of all relevant stakeholders.*** We expect to begin in Asia (Indonesia, the Philippines, and Nepal), moving on to Central Africa and the Miombo Woodlands of southern Africa, and concluding with Latin America.

The comparative participatory action research approach will require direct input from social and biophysical scientists in adapting, monitoring, and implementing the C&I, together with local stakeholders.**** Scientists with both social and biophysical expertise will conduct the case studies, with backup support from CIFOR core staff. This is crucial for linking institutional issues, like stakeholder differences, local management, power, and policies, with variations in forest health on the ground.

A first step in the process will include techniques for identifying stakeholders, user groups, and underrepresented members of communities (Colfer, 1995; Mt. Cameroon team, 1997; Guizol, 1997; Wollenberg, 1997), "future scenarios" in micropolicy planning (as developed by Buck and Wollenberg, 1997),***** techniques for dealing

* The work, organized from Bogor by Lini Wollenberg and David Edmunds, involves a number of collaborators. In China, the work is coordinated by Liu Dachang; in India by Neera Singh and Madhu Sarin; in Madagascar, by Louise Buck and Lalaina Rakotoson; in the Philippines, by Antonio Contreras; and in Bolivia and Central America, by David Kaimowitz. Liu Dachang, Madhu Sarin, Neera Singh, and Antonio Contreras are including both historical reviews and 12 to 15 case studies in each country.

** This includes such factors as forest type, population density, cultural and political ecology, policy design and implementation, and the role of extrasectoral actors and institutions.

*** We acknowledge that finding such paired cases may be difficult or impossible, and are prepared to alter our research design if necessary.

**** Richard Ford's emphasis (see Chapter 20) on a significant period of "engagement," or trust-building, before progress can be made in joint action is well taken, and will inform our planning on site.

***** Manuel Ruiz-Perez was an early proponent of future scenarios work at CIFOR. His work has moved toward "scenario-based models," focusing on NTFPs, but may still prove useful in our future work.

with conflict (e.g., Resolve, 1994; Chandrasekharan, 1996; see also Chapter 5, by Fisher), and resource management agreements (Buck and Wollenberg, 1997; Nguinguiri, in press; see also Chapter 21, by Cowles et al.), among others to be developed. Our literature surveys and an ongoing study on stakeholder representation in several sites will contribute additional pertinent techniques. These techniques represent initial ideas about how both to link local stakeholders together in a horizontal fashion and to link the local up through the political chain to the national and back.

Our research will traverse a variety of forest types and locations in Asia, Africa, and Latin America. From the standpoint of our overall research, we see our ACM sites lending structure to a multidimensional space traversed by three main continua: *wet* to *dry*, *natural* to *planted* (with secondary forests falling between), and production of *goods* to production of *services* (conservation). By examining these kinds of differences, we hope to identify and/or develop more appropriate and tailored management and monitoring tools. From the standpoint of protected areas, we anticipate merging CIFOR case study research in Kayan Mentarang (East Kalimantan), in Danau Sentarum Wildlife Reserve (West Kalimantan), and in Madagascar, with our C&I work in production forests to form a fruitful marriage. We also expect to build on ongoing CIFOR C&I research on local forest management systems that typically coexist with other kinds of forest management. In all our research locations, we expect to include local communities with multiple uses of forests and multiple stakeholders' involvement in management for those uses.

CONCLUSIONS

In the broadest terms, the outputs from this project will include C&I-based monitoring arrangements to be used in ACM, mechanisms for effective and beneficial devolution, and decision support systems — all of which have potential application in protected areas. These represent our ideas of the most direct tools needed to address the goals we identified at the beginning relating to center–periphery links among stakeholders, ACM as a strategy, and mechanisms for constructive interaction among stakeholders. Having spent 4 years developing C&I for sustainable forest management, we are now in a position to find out how difficult it is to "practice what we preach." Anyone working on protected areas knows there is no time to lose.

REFERENCES

Borrini-Feyerabend, G. and Buchan, D., 1997. *Beyond Fences: Seeking Social Sustainability in Conservation,* IUCN, Gland, Switzerland.

Buck, L. and Wollenberg, E., 1997. Decision-making among diverse interests: the use of future scenarios in local forest management policy: a proposed methodology, paper presented at the International Seminar on Community Forestry at a Crossroads: Reflections and Future Directions in the Development of Community Forestry. Bangkok, Thailand, 17–19 July.

Carter, J., 1996. Recent approaches to participatory forest resource assessment, *Rural Development Forestry Study Guide 2*, Rural Development Forestry Network, ODI, London.

Chandrasekharan, D., 1996. Proceedings, electronic conference on "Addressing Natural Resource Conflicts through Community Forestry" (January–May), Community Forestry Unit, Forests, Trees and People Programme, Forestry Department, FAO, Rome, Italy.

Claridge, G., 1997. What is successful comanagement? in *Community Involvement in Wetland Management: Lessons from the Field*, Claridge, G. and O'Callaghan, B., Eds., Wetlands International, Kuala Lumpur, Malaysia, 19–21.

Colfer, C. J. P., 1995. Who counts in sustainable forest management? CIFOR Working Paper 7, CIFOR, Bogor, Indonesia.

Colfer, C. J. P. and Wadley, R. L., 1996. Assessing participation in forest management: workable methods and unworkable assumptions, CIFOR Working Paper 12, CIFOR, Bogor, Indonesia.

Colfer, C. J. P., with Prabhu, R. and Wollenberg, E., 1995. Principles, criteria and indicators: applying Ockham's razor to the people-forestry link, CIFOR Working Paper 8, CIFOR, Bogor, Indonesia.

Colfer, C. J. P., Wadley, R. L., Woelfel, J., and Harwell, E., 1996. Assessing people's perceptions of forests in Danau Sentarum Wildlife Reserve, CIFOR Working Paper 13, CIFOR, Bogor, Indonesia.

Colfer, C. J. P., Wadley, R. L., Harwell, E., and Prabhu, R., 1997a. Inter-generational access to resources: developing criteria and indicators, CIFOR Working Paper 18, CIFOR, Bogor, Indonesia.

Colfer, C. J. P., Wadley, R. L., Woelfel, J., and Harwell, E., 1997b. From heartwood to bark in Indonesia: gender and sustainable forest management, *Women Nat. Resourc.*, 18(4), 7–14.

Colfer, C. J. P., Brocklesby, M. A., Diaw, C., Etuge, P., Günter, M., Harwell, E., McDougall, C., Porro, N. M., Porro, R., Prabhu, R., Salim, A., Sardjono, M. A., Tiani, A. M., Tchikangwa, B., Wadley, R. L., Woelfel, J., and Wollenberg, E., 1999a. The BAG (basic assessment guide) for human well-being, Criteria and Indicators Toolbox Series, 5, CIFOR, Bogor, Indonesia.

Colfer, C. J. P., Brocklesby, M. A., Diaw, C., Etuge, P., Günter, M., Harwell, E., McDougall, C., Porro, N. M., Porro, R., Prabhu, R., Salim, A., Sardjono, M. A., Tiani, A. M., Tchikangwa, B., Wadley, R. L., Woelfel, J., and Wollenberg, E., 1999b. The grab bag: supplementary methods for assessing human well-being, Criteria and Indicators Toolbox Series, 6, CIFOR, Bogor, Indonesia.

Colfer, C. J. P., Salim, A., and McDougall, C., 1999c. The scoring and analysis guide for assessing human well-being, Criteria and Indicators Toolbox Series 7, CIFOR, Bogor, Indonesia.

DDB, 1996, *Evaluating Sustainable Forest Management* (compiled by Stortenbeker), DDB The Netherlands.

Dennis, R., Puntodewo, A., and Colfer, C. J. P., 1998. Fishermen, farmers, forest change and fire, *GIS Asia Pacific,* February/March, 26–30.

Diaw, C., Oyono, R., Sangkwa, F., Bidja, C., Efoua, S., and Nguiebouri, J., 1998. Social science methods for assessing criteria and indicators of sustainable forest management: a report of the tests conducted in Cameroon Humid Forest Benchmark and in the Lobe and Ntem River Basins, CIFOR internal report.

Ford Foundation, 1998. *Forestry for Sustainable Rural Development: A Review of Ford Foundation-Supported Community Forestry Programs in Asia,* The Foundation, New York.

Guizol, P., 1997, Informal Presentation of MAS at CIFOR, March, Bogor, Indonesia.

Hahn-Schilling, B., Heuveldop, J., and Palmer, J., 1994. *A Comparative Study of Evaluation Systems for Sustainable Forest Management (Including Principles, Criteria, and Indicators),* Arbeitsbericht, Bundesforschungs-anstalt fur Forst- und Holzwirtschaft, Hamburg.

Hilborn, R. and Walters, C., 1992. *Designing Adaptive Management Policies. Quantitative Fisheries Stock Assessment: Choice, Dynamics and Uncertainty,* Chapman & Hill, New York, 487–514.

Hobley, M., 1996. Participatory forestry: the process of change in India and Nepal, *Rural Development Forestry Study Guide 3*, Rural Development Forestry Network, ODI, London.

Holling, C. S., 1978. *Adaptive Environmental Assessment and Management,* John Wiley & Sons, New York.

ITW, 1994. *Assessment of Sustainable Tropical Forest Management* (compiled by Heuveldop, J.), Kommissionsverlag Max Wiedebusch, Hamburg, Germany.

IUCN,1995a. Monitoring and Assessment of Local Strategies For Sustainability: A Guide for Fieldworkers Carrying Out Monitoring and Assessment at Community Level (draft), IUCN, Gland, Switzerland.

IUCN, 1995b. Assessing progress toward sustainability, IUCN International Assessment Team Strategies for Sustainability Programme, Gland, Switzerland.

Jackson, B., 1997. A workshop on tools and methods for monitoring and evaluating collaborative management of natural resources in Southern Africa, IUCN Workshop Report, Gland, Switzerland.

Josayma, C., with Burgett, J., Case, L., Leialoha, J., Ross, L., Stormont, B., and Ziegler, M., 1996. Facilitating collaborative planning in Hawaii's natural area reserves, Asia Forest Network, Report 8, Berkeley, CA.

Lammerts van Bueren, E. and Blum, E., 1997, *Hierarchical Framework for the Formulation of Sustainable Forest Management Standards: Principles, Criteria and Indicators*, Tropenbos, Leiden, The Netherlands.

Lee, K. N., 1993. *Compass and Gyroscope: Integrating Science and Politics for the Environment,* Island Press, Washington, D.C.

Lembaga Ekolabel Indonesia, 1997. From the Web page of Lembaga Ekolabel Indonesia at http://www.iscom.com/~ekolabel/buku1.html (2 October 1997).

Mayers, J. and Kotey, E. N. A., 1996. Local institutions and adaptive forest management in Ghana, IIED Forestry and Land Use Series 7, IIED/ODA, London.

McDougall, C., 1998. Draft methods evaluation, internal document, CIFOR, Bogor, Indonesia.

Mt. Cameroon Project, 1997. Cameroonian test of social science methods for assessing criteria and indicators for sustainable forest management, Internal report, CIFOR, Bogor, Indonesia.

Nguinguiri, J.-C., 1999. Les approches participatives dans la gestion des ecosystemes forestiers d'Afrique Centrale, CIFOR/CORAF/FORAFRI Occasional Paper 22, CIFOR, Bogor, Indonesia.

Poffenberger, M., Ed., 1996. *Communities and Forest Management: With Recommendations to the Intergovernmental Panel on Forests*, IUCN–The World Conservation Union, Cambridge, U.K.

Poffenberger, M., 1998. Stewards of Vietnam's upland forests, Research Network Report 10, Center for Southeast Asian Studies, University of California, Berkeley, CA.

Poffenberger, M. and McGean, B., Eds., 1993a. Community allies: forest comanagement in Thailand, Research Network Report 2, Center for Southeast Asian Studies, University of California, Berkeley.

Poffenberger, M. and McGean, B., Eds., 1993b. Upland Philippine communities: guardians of the final forest frontiers, Research Network Report 4, Center for Southeast Asian Studies, University of California, Berkeley.

Poffenberger, M. and McGean, B., Eds., 1994. Policy dialogue on natural forest regeneration and community management, in *Proceedings, Asia Sustainable Forest Management Network*, Report 5, East West Center, Honolulu.

Poffenberger, M., McGean, B., Ravindranath, N. H., and Gadgil, M., 1992a. *Field Methods Manual,* Vol. I, Society for Promotion of Wastelands Development, New Delhi, India.

Poffenberger, M., McGean, B., Khare, A., and Campbell, J., 1992b. *Field Methods Manual,* Vol. II, Society for Promotion of Wastelands Development, New Delhi, India.

Porro, R. and Porro, N. M., 1998. Methods for assessing social science criteria and indicators for the sustainable management of forests: Brazil test, CIFOR Report, Bogor, Indonesia (October).

Prabhu, R., Colfer, C. J. P., Venkateswarlu, P., Tan, L. C., Soekmadi, R., and Wollenberg, E., 1996. Testing criteria and indicators for the sustainable management of forests: Phase I final report, CIFOR Special Publication, Bogor, Indonesia.

Prabhu, R., Maynard, W., Eba'a Atyi, R., Colfer, C. J. P., Shepherd, G., Venkateswarlu, P., and Tiayon, F., 1998a. Testing and developing criteria and indicators for sustainable forest management in Cameroon: the Kribi Test, final report, CIFOR Special Publication, Bogor, Indonesia.

Prabhu, R., Ruitenbeek, R. J., Boyle, T. J. B., and Colfer, C. J. P., 1998b. Between voodoo science and adaptive management: the role and research needs for indicators of sustainable forest management, paper presented at the IUFRO Conference, 24–28 August, Melbourne, Australia.

Prescott-Allen, R., 1995. Towards a barometer of sustainability for Zimbabwe, Draft report to IUCN (July), Gland, Switzerland.

Rainforest Alliance, 1993. *Smart Wood Certification Program*, Annex 6, The Alliance, La Jolla, CA.

Resolve, 1994. The role of alternative conflict management in community forestry, Working paper, Community Forestry Unit, Forests, Trees and People Programme, Forestry Department, Rome.

Sardjono, M. A., with Rositah, E., Wijaya, A., and Angie, E. M., 1997. A test of social science assessment methods concerning indicators and criteria for sustainable forest management in East-Kalimantan, internal document, CIFOR, Bogor, Indonesia.

Shindler, B., Steel, B., and List, P., 1996. Public judgments of adaptive management: a response from forest communities, *J. For.,* 94, 4–12.

Soil Association, 1994. *Responsible Forestry Standards for the United Kingdom,* The Association, Bristol, U.K.

Stankey, G. H. and Clark, R. N., 1998. Adaptive management areas: roles and opportunities for the PNW Research Station, Review draft. Pacific North West Research Station, Portland, OR.

Stankey, G. H. and Shindler, B., 1997. Adaptive management areas: achieving the promise, avoiding the peril, General Technical Report PNW-GTR-394, U.S. Department of Agriculture, Forest Science, Pacific Northwest Research Station, Portland, OR, 21 pp.

Stevens, P., 1997. *Measuring the Sustainability of Forest Village Ecosystems — Concepts and Methodologies: A Turkish Example*, CSIRO Forestry and Forest Products Technical Report 103.

Tchikangwa, N. B., with Sikoua, S., Metomo, M., and Adjudo, M. F., 1998. Test des mèthodes en sciences sociales de vérification des critères et indicateurs d'aménagement durable des forêts: périphérie est de la Réserve du Dja (Sud-Cameroun), internal document, CIFOR, Bogor, Indonesia.

Thomson, J. T. and Schoonmaker Freudenberger, K., 1997. *Crafting Institutional Arrangements for Community Forestry*, FAO, Rome, Italy.

Tiani, A. M., with Mvogo, B. E., Oyono, A., and Kenmegne, D. N., 1997. A Test of Social Science Assessment Methods (near M'balmayo, Cameroon), internal document, CIFOR, Bogor, Indonesia.

Wollenberg, E., 1997. Sampling stakeholders, in *The Grab Bag: Supplementary Methods for Assessing Human Well-Being,* Colfer, C. J. P., Brocklesby, M. A., Diaw, C., Etuge, P., Günter, M., Harwell, E., McDougall, C., Porro, N. M., Porro, R., Prabhu, R., Salim, A., Sardjono, M. A., Tiani, A. M., Tchikangwa, B., Wadley, R. L., Woelfel, J., and Wollenberg, E., Criteria and Indicators Toolbox Series 6, CIFOR, Bogor, Indonesia.

16 Overview of a Systematic Approach to Designing, Managing, and Monitoring Conservation and Development Projects

Nick Salafsky and Richard Margoluis

CONTENTS

0-8493-0020-7/01/$0.00+$1.50

INTRODUCTION

One of the premises behind the concept of adaptive collaborative management (ACM) is that it enables people involved in conservation and development projects to develop better monitoring systems. There is a growing movement to improve the monitoring of conservation and development projects. Monitoring can be defined as "the periodic collection and evaluation of data relative to stated project goals, objectives, and activities." Many people often also refer to this process as monitoring and evaluation (M&E).

Monitoring can potentially serve two important functions within a project:

1. *Adaptive Management* — Helping communities and project implementers systematically collect, analyze, and use the information they need to mange their local resources more effectively.
2. *Impact Assessment* — Enabling project teams and donors to learn from projects and draw more-generalized lessons regarding effective conservation strategies.

Despite the near universal agreement on the importance of monitoring, few community-based conservation and development projects have had much success in developing and implementing monitoring systems (Baron, 1998; Johnson, 1999). In many cases the question of monitoring causes a major "disconnect" between donors and groups implementing projects, as outlined in Figure 16.1. Donors demand that project teams design and implement monitoring systems. The teams typically agree in principle, but in practice either do not implement monitoring systems or implement systems that collect but do not use data.

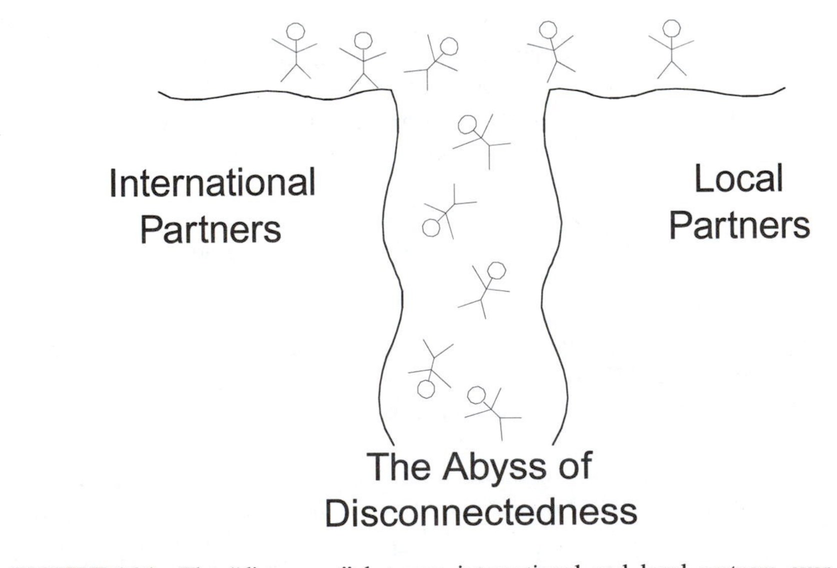

FIGURE 16.1 The "disconnect" between international and local partners over project monitoring. © 1999 WWF.

At least five main constraints keep project teams from developing and implementing monitoring systems and using the data from them.

Lack of Time and Money. Most field-based project teams face enormous time and financial pressures. On any given day, the team members must juggle a host of different tasks, such as developing and implementing complex program activities, maintaining working relationships with and among factions of local community stakeholders (who often have been feuding for generations), managing difficult logistical problems in communicating and getting supplies, dealing with complex staffing problems, and complying with requests from donors. Even if the team wants to monitor, it often ends up as a marginal activity that slips as more immediate crises demand action.

Perceived Lack of Qualified Staff. Project staff traditionally see monitoring as the domain of scientists. Many practitioners believe that rigorous monitoring work requires a team of Ph.D. scientists with white laboratory coats and elaborate equipment.

Little Connection between Project Interventions and Monitoring. Senior members of the implementing group who live in the country's capital city, or in places like Washington, D.C., are in many cases responsible for designing projects. The project team in the field thus often has little or no idea about the conceptual design of the project — what the goals and objectives of the project are and how the interventions are designed to achieve them. As a result, it is often difficult for the project staff to determine what they need to monitor to assess project success. Furthermore, project staff often treat monitoring as a separate set of activities instead of integrating it into the overall project plan.

Difficulty in Determining What Specific Data Need to Be Collected. Even if project teams can decide what information they need, they often have difficulty selecting the appropriate methods to use. In particular, since most "monitoring" staff come from specific disciplinary backgrounds, they tend to apply the methods particular to that discipline with little or no regard to whether they are necessary or appropriate.

Difficulty in Analyzing and Using Data. Despite these constraints, many projects succeed in collecting data. Most of the time, however, the data lie idle and the project never analyzes or uses them. This problem occurs because either the project collected the wrong data or because teams do not have the experience and expertise to perform the analyses.

These constraints are real, but project teams must try to overcome them. Over the past few years, the authors have worked with practitioners to develop "Measures of Success: A Systematic Approach to Designing, Managing, and Monitoring Conservation and Development Projects" (Margoluis and Salafsky, 1998). The hope is that this approach can help bridge the "disconnect" and lead to more successful projects, as depicted in Figure 16.2.

This chapter presents one method for working with community members and project managers to develop effective monitoring systems. The objectives are to (1) describe the evolution of the Measures of Success approach to monitoring in the context of the project cycle, and (2) provide an overview of the steps in the Measures of Success approach to monitoring. The authors hope also to illustrate that monitoring cannot be a stand-alone activity, but must be integrated into project design and management.

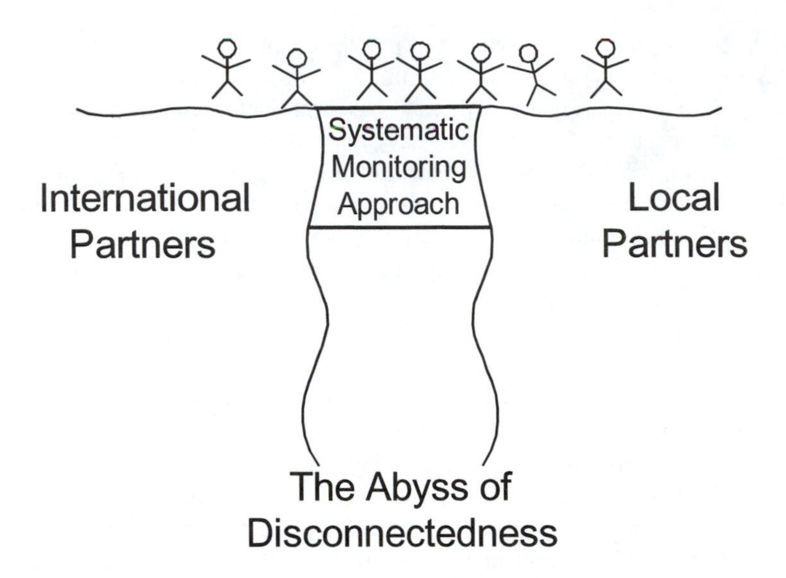

FIGURE 16.2 Bridging the gap with a systematic approach to designing, managing, and monitoring projects. © 1999 WWF.

THE EVOLUTION OF THE MEASURES OF SUCCESS APPROACH TO MONITORING

As illustrated in Figure 16.3, there are a number of sources for this approach. From a theoretical perspective, the approach draws on techniques developed by business, development, and scientific research. From a practical perspective, this approach draws on the authors' field experience working with conservation and development projects. The authors have developed and field-tested this approach in conjunction with their colleagues from many different Biodiversity Support Program (BSP) projects in Latin America, Africa, and Asia.

One of the most important of these sources was the efforts of the BSP Biodiversity Conservation Network (BCN). To illustrate this approach, its evolution in the context of the BCN program is briefly discussed. First, however, a brief overview of the BCN program is presented (see also BCN, 1999a, b; Salafsky et al., 1999; Salafsky and Wollenberg, 2000).

THE BCN EXAMPLE

BCN was established to fulfill two goals (BCN, 1999a):

1. Support enterprise-oriented approaches to biodiversity conservation at a number of sites across the Asia/Pacific region; and
2. Evaluate the effectiveness of these enterprise-oriented approaches to community-based conservation of biodiversity and provide lessons and results to BCN clients.

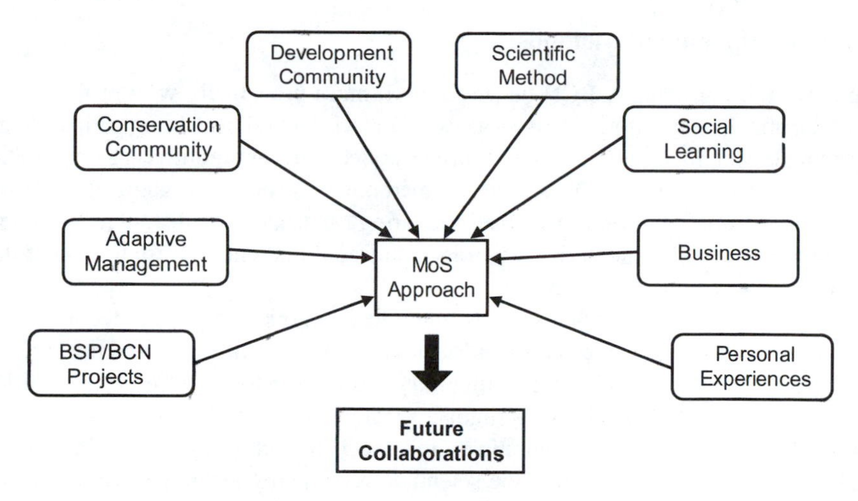

FIGURE 16.3 Sources behind the Measures of Success systematic approach. © 1999 WWF.

To achieve these goals, BCN brought together organizations in Asia, the Pacific, and the United States in active partnerships with local and indigenous communities. The network provided grants for projects that encourage the development of enterprises that depend on sustained conservation of local biodiversity.

The BCN core hypothesis was that, if enterprise-oriented approaches to community-based conservation are going to be effective, they must (1) have a direct link to biodiversity, (2) generate benefits, and (3) involve a community of stakeholders. In effect, the hypothesis was that if local communities receive sufficient benefits from an enterprise that depends on biodiversity, then they will act to counter internal and external threats to that biodiversity.

The History of BCN Efforts to Assist its Partners with Monitoring

BCN went through a number of different phases in crafting ways to help its partner organizations develop and implement monitoring plans.

Monitoring in the Initial Project Design

From its earliest stages, BCN understood the important role monitoring would play in fulfilling both of its goals — documenting project success and testing its core hypothesis. Initially, however, BCN thought that good-quality monitoring would result if project partners developed detailed biological, social, and enterprise monitoring plans in their project proposals.

The project teams did indeed develop lengthy plans on paper. Ultimately, on average, BCN projects allocated over 30% of their budgets to monitoring activities, a percentage far in excess of most conservation and development projects (BCN, 1995a). Over the first 2 years of the program, however, it became clear that many of the projects were running into the constraints outlined in the first section of this chapter and thus having difficulties implementing their monitoring plans.

Matrices of Different Methods

To solve these problems, BCN began to work more proactively with project teams on their monitoring efforts. This work was initially aimed at helping project teams determine which methods they could use to collect relevant monitoring information in a cost-effective fashion. This focus on methods was roughly organized according to academic disciplines and involved preparing "matrices" of different biological, social, and enterprise methods that project staff could potentially use to collect data about the BCN-funded projects.

Within each set of methods, BCN attempted to rank comparable techniques in terms of the trade-off between cost and accuracy of results, trying to find the methods that would be most suitable for community-based monitoring efforts. In addition, BCN also assembled panels of distinguished scientists to get their input on how best to select techniques that communities and local project teams could implement. Interestingly, however, although the scientists knew many techniques for collecting data, for the most part they were at a loss to explain how to do low-cost, community-based monitoring.

After distributing the matrices of different methods to the project teams, it soon became apparent that these were not sufficient to solve the problems the teams were having with monitoring. Instead, BCN began to realize that the project teams were having difficulty determining what information the project needed — the step that comes *before* selecting methods.

Comprehensive Guidelines of Monitoring Questions

To help its partners determine what information to collect, BCN began developing lists of potential monitoring questions the projects could ask about the biological, social, and enterprise components of their efforts. These lists of questions initially focused on the monitoring methods matrices, but soon expanded into a comprehensive listing of almost every conceivable question relevant to a BCN-type project (BCN, 1995b).

The idea behind these comprehensive guidelines was not to suggest that each project try to answer all of the questions, but rather to provide a resource guide that the group could use to determine what specific questions it needed to ask. When BCN sent this massive list of questions to project teams, however, it generally overwhelmed people and left them more confused than ever over what questions they needed to address in their specific project.

Common Sets of Questions

In an attempt to give partners more guidance in selecting specific questions, BCN next drew on experiences of all its partners to select the most pertinent questions in each of the three disciplinary areas. To this end, BCN convened a workshop in May 1995 among its South Asian grantees to review the comprehensive list of questions in each discipline and boil it down to a "common" or "minimum" set of critical information needs that all the projects could address (BCN, 1995c).

BCN made some progress toward this goal at the workshop, but most groups still were having difficulty in coming up with specific questions that they needed to address at their sites. There was a growing realization that, ultimately, conservation needs to be site specific and that project teams need to design monitoring not as a supplemental package organized by various academic disciplines, but instead as an integral part of the project design process.

Site-Specific Monitoring Plans in the Context of the Project Cycle

One useful technique is to view a project as going through a series of steps in a cycle, as outlined in Figure 16.4. To this point, BCN basically had been starting the process of helping groups develop their monitoring efforts with Step C, assuming that the projects had already progressed through the previous steps. BCN realized, however, that it needed to help project teams complete the earlier steps before planning monitoring efforts. Furthermore, project teams needed to design monitoring to meet the specific needs of each project site.

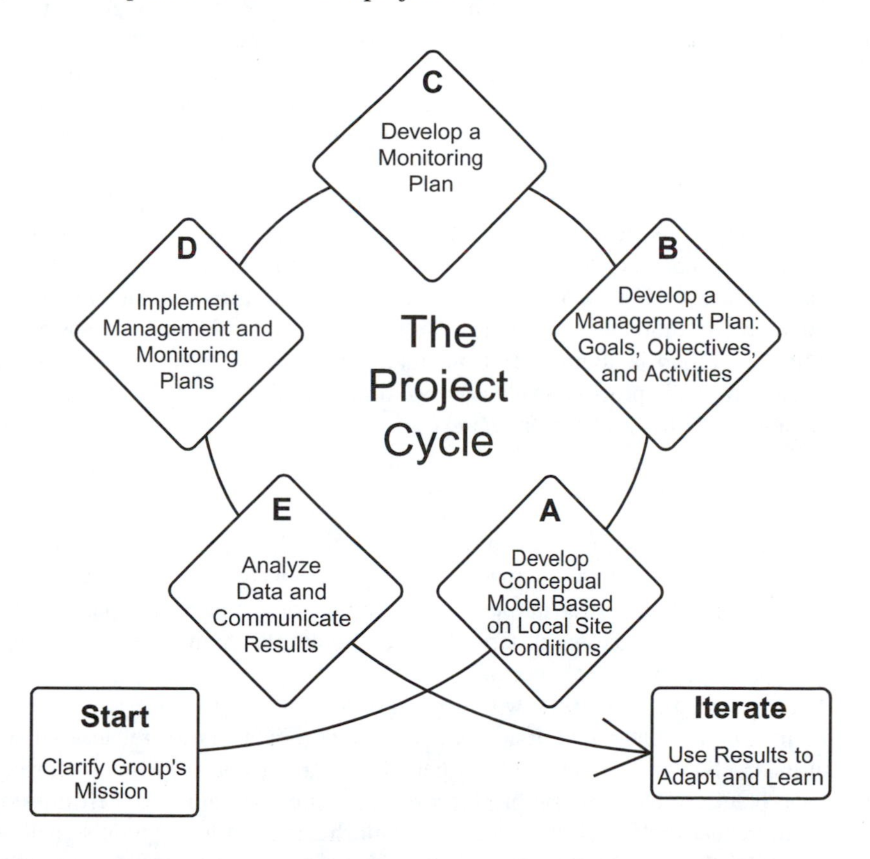

FIGURE 16.4 Monitoring in the context of the project cycle. (From Margoluis, R. and Salafsky, N., *Measures of Success: Designing, Managing, and Monitoring Conservation and Development Projects,* Island Press, Washington, D.C., 1998. © 1999 WWF.) With permission.

BCN convened two more workshops in September 1995 for its southeast Asian and Pacific grantees in which the network presented to grantees a new approach to performing monitoring in the context of the project cycle (Margoluis and Salafsky, 1998). Participants quickly recognized that this approach had the potential to solve many of the earlier problems. The work of BSP with partners in other parts of the world confirmed these findings. BCN has refined and adapted the approach since that initial workshop, and it continues to evolve even today.

STEPS IN THE BSP/BCN APPROACH TO MONITORING

This section presents the BSP approach to monitoring in some detail. Full details of the methodology can be found in Margoluis and Salafsky (1988). Steps in the procedure are illustrated using an example from the BCN-funded Pacific Heritage Foundation (PHF) project (see BCN, 1999a, for details about this project).

The overall approach is based on the project cycle shown in Figure 16.4. In addition to the starting and ending boxes, the diagram contains five diamonds, each of which represents a different step in the overall cycle. These steps generally need to occur in the order as represented by the letters A through E. The steps themselves, however, are part of an iterative process that involves going through the cycle numerous times.

The process is presented from the perspective of the group implementing the project. A *project* is defined as any set of actions undertaken by any group of managers, researchers, or local stakeholders interested in achieving certain defined goals and objectives. For example, a project could be steps that community members take to revive traditional resource-harvesting customs. Furthermore, whether the implementing group is composed of outsiders or members of the community, an important part of the process involves consulting with the local stakeholders at the project site in all stages of the project cycle.

START: CLARIFY YOUR GROUP'S MISSION

Before setting out to design a new project, one must have a clear understanding of the group's *mission*. A mission statement provides a vision for the future of the group — its long-term desired purpose, its strategies for achieving this purpose, and the values that will guide its work. Groups generally develop their mission statements through a strategic planning process.

If the group plans to work with other groups on the new project, it is also important to understand their missions and how the group's mission relates to theirs. As outlined in Figure 16.5, it is unlikely that any two groups participating in a project will have precisely the same set of purposes, strategies, or values in their mission. These differences make it all the more important that each group explicitly spell out its mission so that it is possible to see where overlap exists (the shaded areas) and where the differences are (the unshaded areas). Without a clear sense of what a group wants to accomplish and an understanding of what its partners are trying to

FIGURE 16.5 Overlapping missions of project groups. © 1999 WWF.

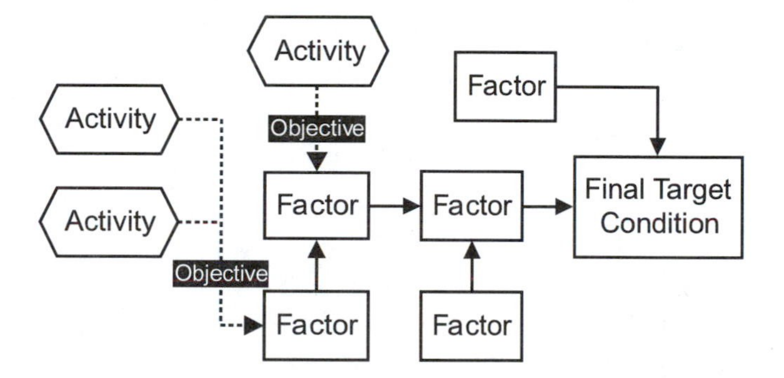

FIGURE 16.6 Diagram of a "generic" conceptual model. © 1999 WWF.

do, it will be difficult to design, manage, and monitor effective projects. Specific steps in this part of the process include:

1. Define the group's mission.
2. Find common ground with the project partners.

Design a Conceptual Model Based on Local Site Conditions

A *conceptual model* is the foundation of all project design, management, and monitoring activities. As illustrated in Figure 16.6, a conceptual model is basically a diagram of a set of relationships between certain factors that are believed to impact or lead to a *target condition.* In conservation and development projects the target condition is generally related to biodiversity. As illustrated in Figure 16.7 from the PHF project, the model is first built using existing information to present a picture of the project area prior to the start of the project.

In particular, the model should illustrate the key *direct* and *indirect threats* to the target condition. In this example, major direct threats include logging and mining

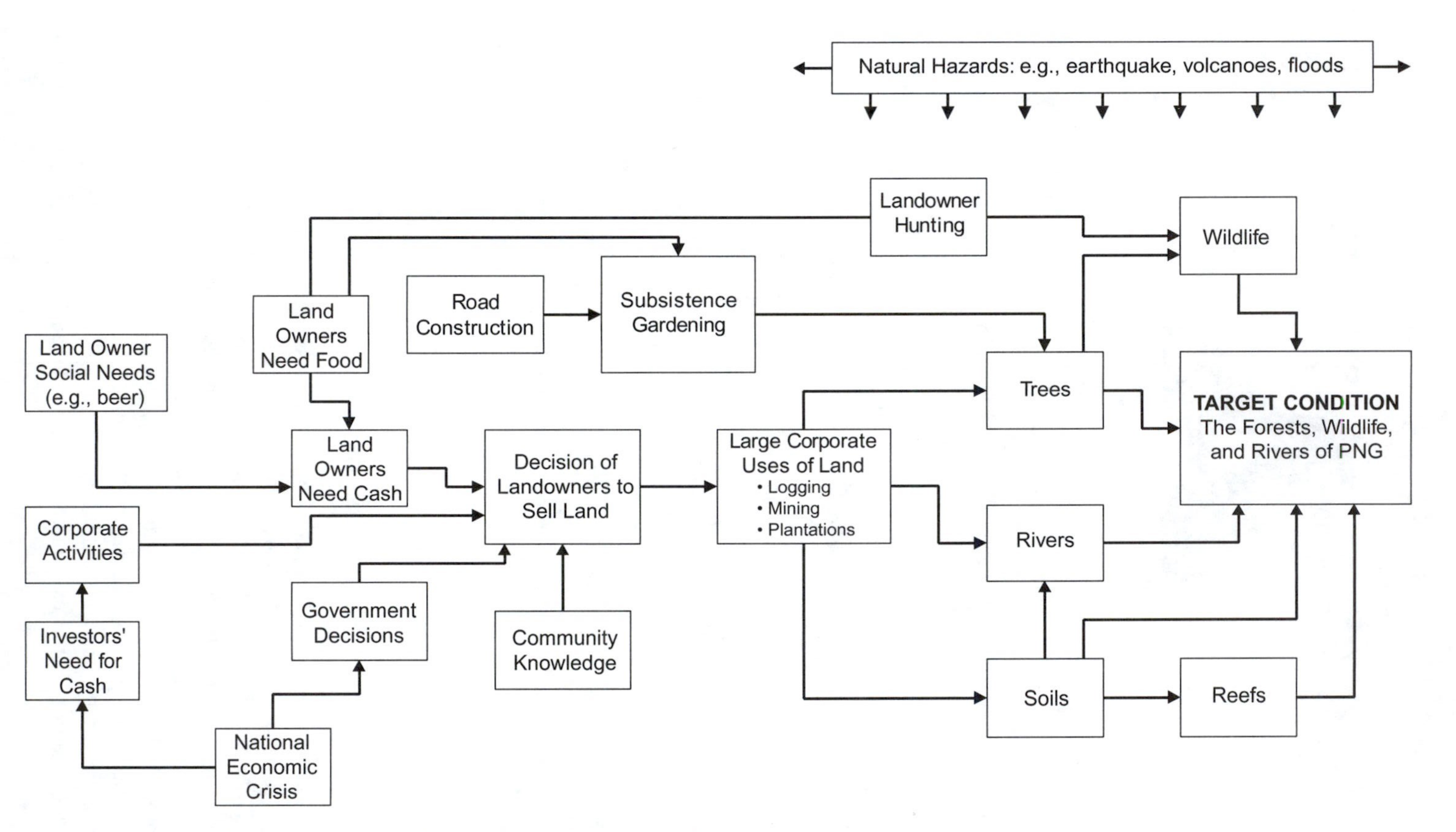

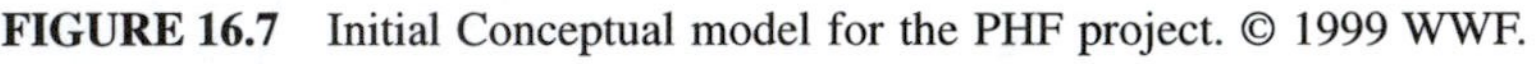
FIGURE 16.7 Initial Conceptual model for the PHF project. © 1999 WWF.

operations conducted by large corporations, expansion of subsistence agriculture gardens, and hunting. Project staff next present the model to local communities, revise it according to their input, and then use the model to identify and rank the key threats to biodiversity that the project will address. Specific steps in this part of the process include:

1. Review and compile existing information about the project site.
2. Develop an initial conceptual model of the project site.
3. Assess local site conditions to refine and improve the model.
4. Identify and rank threats at the project site.

Develop a Management Plan

A *management plan* describes the explicit *goals*, *objectives*, and *activities* designed to address the threats identified in the conceptual model. Goals, which are derived from the target condition of the project are broad statements of the desired state toward which the project is directed. Objectives are more specific statements of the desired outcomes or accomplishments of the project. Activities are specific actions undertaken by project participants designed to reach each of the project objectives, which in turn should lead to realization of the project goal. All activities should be linked to specific objectives that target critical threat factors identified in the conceptual model. These linked chains of activities and factors are the assumptions of the project. Once the management plan has been completed, the activities and objectives can be added to the project conceptual model.

Figure 16.8 illustrates part of a management plan for the PHF project. Figure 16.9 shows a *project conceptual model* for the project and depicts the expected impact of the management plan. Specific steps in this part of the process include:

1. Develop a goal for the project.
2. Develop objectives for the project.
3. Develop activities for the project.

Goal: Conserve the forests, wildlife, and rivers of PNG

Objective 1: Within 6 months from the start of each of 5 small-scale logging projects in the Bainings area, income increases by 200 Kina per week.

Activities for Objective 1:
1. Enterprise loans
2. Enterprise training
3. Marketing assistance

Objective 2: 80% of clan chiefs in the project site know about the importance of biodiversity after 1 year.

Activities for Objective 2:
1. Hold awareness workshops
2. Take chiefs to logging sites

FIGURE 16.8 Excerpt from the management plan for the PHF project. © 1999 WWF.

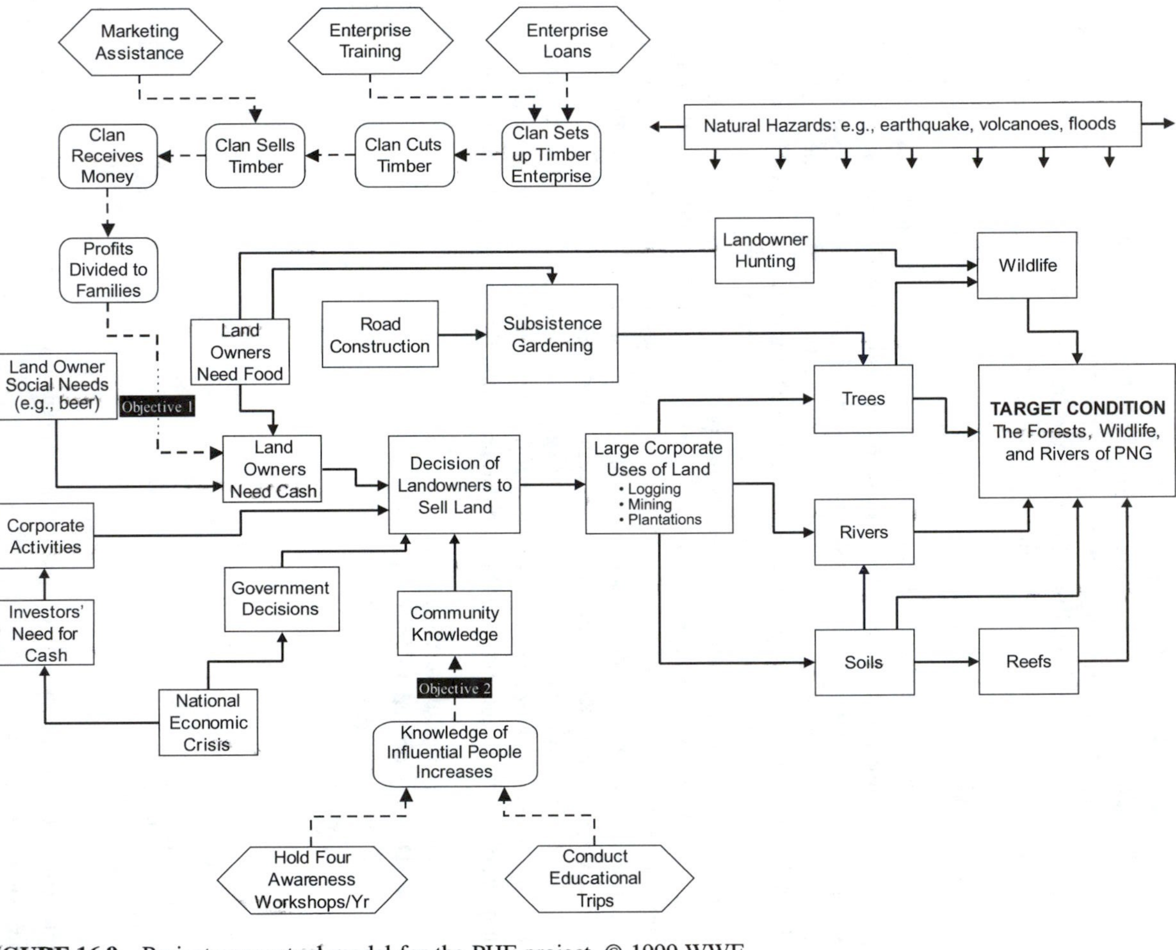

FIGURE 16.9 Project conceptual model for the PHF project. © 1999 WWF.

Goal, Objective, or Additional Information:					
Objective 1: Within 6 months from the start of each of 5 small-scale logging projects in the Bainings area, income increases by 200 Kina per week.					
WHAT	**HOW**	**WHEN**	**WHO**	**WHERE**	**COMMENT**
Household income	Inspect project records	Every 6 months	Enterprise managers	Enterprise offices	Look at income from sawmills
Kg rice consumed per month	Household survey	Every 6 months	Project social scientist	5 project villages	Proxy indicator for wealth

FIGURE 16.10 Excerpt from the monitoring plan for the PHF project. © 1999 WWF.

Develop a Monitoring Plan

A *monitoring plan* describes how one will assess the success of the project interventions. If the interventions of the project are not monitored, then there will be no way to know whether its goal and objectives have been achieved or what is needed to improve the project. The plan starts by identifying the internal and external audiences, what information they need, what monitoring strategies will be employed to obtain the data to meet each of these needs, and the specific indicators that will be measured. The remainder of the plan lists how, when, by whom, and where data for these indicators will be collected.

The key here is to be as specific as possible in writing down what and how data will be collected. An excerpt from the PHF monitoring plan is shown in Figure 16.10. Specific steps in this part of the process include:

1. Determine audiences, information needs, monitoring strategies, and indicators (*why* and *what*).
2. Select methods and determine tasks necessary to collect data (*how*).
3. Determine when, by whom, and where data will be collected (*when*, *who*, and *where*).
4. Develop a monitoring plan for project activities.

Implement Management and Monitoring Plan

The project conceptual model, management plan, and monitoring plan taken together comprise a complete *project plan*. This step involves implementing this project plan.

There is little that can be said about this step in a general context — it basically involves putting into action the work done in the previous steps. Unless the plan is implemented, there will be no hope of achieving the goals and objectives of the project. Specific steps in this part of the process include:

1. Implement the management plan.
2. Implement the monitoring plan.

Analyze Data and Communicate Results

Once data have been obtained, they must be analyzed and the results communicated to the internal and external audiences. The challenge here is to take the data that have been collected and turn them into useful *information* that can be made available to the project partners, other stakeholders in and around the project site, and outside audiences. Specific steps in this part of the process include:

1. Analyze data.
2. Communicate results to the internal and external audiences.

Iteration: Use Results to Adapt and Learn

Iteration means to repeat a process or sequence of steps that brings one successively closer to a desired result. It is the key step in a*daptive management,* where the work invested in monitoring can pay off by helping incorporate the information that has been obtained to improve the project and move forward. In this step, one first completes the process of testing assumptions and adapts the project plan based on the monitoring results. One then should also document and share the knowledge gained with others, so that they can improve their conservation efforts. Specific steps in this part of the process include:

1. Put the assumptions to the test.
2. Adapt the project based on the monitoring results.
3. Use results to refine the project and knowledge of conservation techniques.

CONCLUSIONS

The start of this chapter stated that monitoring has two primary functions:

1. *Adaptive Management* — Helping communities and project implementers systematically collect, analyze, and use the information they need to mange their local resources more effectively.
2. *Impact Assessment* — Enabling project teams and donors to learn from projects and to draw more-generalized lessons regarding effective conservation strategies.

The Measures of Success systematic approach to designing, managing, and monitoring conservation and development projects provides a useful framework for this complex and important process. The hope is that this overview of the BSP approach demonstrates not only that monitoring is essential to project success, but that it must be integrated into project design and management.

ACKNOWLEDGMENTS

The authors thank the many people who have contributed to the development of the Measures of Success approach to designing, managing, and monitoring conservation and development projects including, in particular, the project partners who worked with them to develop and refine these ideas. This work was supported by the U.S.

Agency for International Development (USAID) under the terms of Grant DHR-5554-A-00-8044-00. A version of this chapter was previously published in Saterson et al. (1995).

REFERENCES

Note that most of the BCN and BSP reports are available online at www.BSPonline.org or at www.BCNet.org.

Baron, N., 1998. Keeping Watch: Experiences from the Field in Community-Based Monitoring, *Lessons from the Field,* Issue 1, Biodiversity Support Program, Washington, D.C.

BCN, 1995a. *Annual Report,* Biodiversity Support Program, Washington, D.C.

BCN, 1995b. *Guidelines for Monitoring and Evaluation of BCN-Funded Projects,* Biodiversity Support Program, Washington, D.C.

BCN, 1995c. *Final Report on the BCN Monitoring Workshop, 22–26 May 1995, Bangalore India,* Biodiversity Support Program, Washington, D.C.

BCN, 1995d. *Final Report on the BCN Monitoring Workshop, Los Baños, Philippines,* Biodiversity Support Program, Washington, D.C.

BCN, 1999a. *Evaluating Linkages between Business, the Environment, and Local Communities: Final Stories from the Field,* Biodiversity Support Program, Washington, D.C.

BCN, 1999b. *Patterns in Conservation: Evaluating Linkages between Business, Communities and the Environment,* Biodiversity Support Program, Washington, D.C.

Johnson, A., 1999. Measuring our Success: one team's experience in monitoring the Crater Mountain Wildlife Management Area Project in Papua New Guinea, *Lessons from the Field,* Issue BCN-3, Biodiversity Support Program, Washington, D.C.

Margoluis, R. and Salafsky, N., 1998. *Measures of Success: Designing, Managing, and Monitoring Conservation and Development Projects,* Island Press, Washington, D.C.

Salafsky, N. and Margoluis, R., 1999. *Greater Than the Sum of Their Parts: Designing Conservation Programs to Maximize Impact and Learning,* Biodiversity Support Program, Washington, D.C.

Salafsky, N. and Wollenberg, L., 2000. Linking livelihoods and conservation: a conceptual framework for assessing the integration of human needs and biodiversity, *World Dev.,* 28, 1421–1438.

Salafsky, N., Cordes, B., Parks, J., and Hochman, C., 1999. *Evaluating Linkages between Business, the Environment, and Local Communities: Final Analytical Results from the Biodiversity Conservation Network,* Biodiversity Support Program, Washington, D.C.

Saterson, K., Margoluis, R., and Salafsky, N., Eds., 1999. *Measuring Conservation Impact: An Interdisciplinary Approach to Project Monitoring and Evaluation,* Biodiversity Support Program, WWF, Washington, D.C.

17 Anticipating Change: Scenarios as a Tool for Increasing Adaptivity in Multistakeholder Settings*

Eva (Lini) Wollenberg, David Edmunds, and Louise E. Buck

CONTENTS

INTRODUCTION

In contrast to past work on scientific adaptive management (Holling, 1978; Walters, 1986), a new paradigm of adaptive management is emerging that takes into account multiple interests in landscape management (Taylor et al., 1997; Johnson, 1999; Maarleveld and Dangbégnon, 1999). The new adaptive management seeks to be responsive to local demands, to engage in field-based problem-solving, and to facilitate collaboration among multiple stakeholders (McLain and Lee, 1996; Lessard, 1998). The term *adaptive comanagement* (ACM) is used in this chapter to indicate the bottom-up orientation and stakeholder focus of this new approach.

* Reprinted from *Landscape and Urban Planning,* 47, Wollenberg, E., Edmunds, D., and Buck, L., Using scenarios to make decisions about the future: Anticipatory learning for the adaptive co-management of community forests, pp. 65-77, Copyright 2000, with permission from Elsevier Science.

0-8493-0020-7/01/$0.00+$1.50

Although ACM seeks to achieve adaptation through iterative social learning among stakeholders, the focus of learning has been in practice on the monitoring of past actions (Prabhu et al., 1996; Margoluis and Salafsky, 1998; Datta, 1999). Yet anticipating and exchanging perspectives about the future can be an equally important source of learning. This chapter shows how anticipatory scenario techniques can be used as practical tools of ACM to enable forest stakeholders not only to respond to change, but also to be prepared to adapt to it. The range of general scenario methods at present in use are described, with focus on methods involving multiple scenarios and systems analysis as those with the potential for most improving anticipatory learning.

Scenario-based approaches are examined in the context of community-based forest management. Community-based forest management refers to common pool forests where the people living near them have significant rights and responsibilities for management. Given the trend to devolve forests to local authorities in many countries, the authors propose that there will be an increasing need for new methods of learning and adaptation to forces for change and interdependencies with relevant stakeholders. Community forest managers are often at the lower rungs of local power structures and therefore not fully informed of the policy, investment, and development plans created around them that affect their forests. The authors suggest that a more proactive stance to learning about potential changes should thus be an essential part of any adaptive community forest management process.

The chapter discussed first what scenarios are and describes several types of scenario approaches. The methods used to conduct multiple, systems-based scenarios are reviewed, including their potential application to community forest management. The authors then identify the features that make scenario analysis well suited for use within an ACM framework and conclude by identifying key principles for application of the scenario method to assist readers in devising their own approaches.

WHAT ARE SCENARIOS?

Although the term *scenario* is associated with several distinct approaches for gaining information about the future (Millet, 1988; Fischhoff, 1988; Sapio, 1995) and its meaning has shifted with different historical contexts (van de Klundert, 1995), the scenario *method* refers to a general category of techniques associated with creative visioning. Figure 17.1 shows how creative visioning techniques differ from other general approaches to thinking about the future, such as projecting and forecasting, assessment of potential impacts, and exchange (Deshler, 1987). Unlike projections, scenarios do not indicate what the future will look like. Scenarios instead stimulate creative ways of thinking that help stakeholders break out of established patterns of assessing situations and planning actions, so that they can better adapt to the future. They are most appropriate under conditions where complexity and uncertainty are high (Schoemaker, 1993), as is generally the case in tropical forests where communities are found. If these systems were more predictable, linear techniques of extrapolation would be sufficient for future planning. Where uncertainty exists, especially where the interests and plans of multiple stakeholders are not fully known, creative processes for anticipating change such as scenarios are useful.

Creative visioning is an approach intended "to challenge existing mental barriers to make use of creative intuition and construct visions or plans for a desirable or preferred future" (Deshler, 1987, p. 87). It is a response to the "human tendency to be bound by what we already know" (Deshler, 1987, p. 87). Visioning is used to discover interconnections between events, especially macroevents on microenvironments. Techniques include imaging, scenarios, and futures history writing.

Projection and forecasting are techniques that produce relatively precise quantitative predictions. This approach requires historical precedents, regularities of cause and effect, data availability, and short time periods. Some consider forecasts appropriate for single variables: prices, population, etc. (Blythe and Young, 1994), rather than complex phenomena. The method of arriving at the answer may be complex and is usually not transparent to decision makers. Methods include Delphi techniques, trend extrapolation, computer modeling, and cross-impact analysis.

Assessment of potential hazards is an approach for identifying the possible impacts of a new policy or practice. The method requires prior determination of criteria and indicators for assessment. Common techniques include environmental and social impact assessments.

Exchange and dialogue methods aim to release people from socially imposed and unexamined expectations. The method enables people to understand other group's plans and visions and stimulate dialogue. Techniques include discussions of literature, self-assessment, games, and simulations.

FIGURE 17.1 Four approaches to gaining information about the future. (Adapted from Deshler, 1987.)

Scenarios are thus stories of what might be. Community forest managers can use them to evaluate what to do now based on different possible futures. The options for the future reflect usually a combination of the extrapolation of current stable trends and a range of possible introduced changes, such as policies and natural events.

Three types of related scenario methods are commonly used. These are (1) a vision of the desired, ideal future; (2) comparison of the desired future with the present; and (3) comparison of assessing multiple scenarios. Common to all three techniques is the aim of developing new images of the future. In selecting among these techniques, however, users have to consider their purpose in applying scenario methods. To choose whether to produce a single desired future or a range of possible futures, the user needs to assess the degree to which information is needed about a normative "desired state" (e.g., to empower forest communities to imagine achieving their goals, or as a team-building exercise to create a shared vision for a set of stakeholders) or about a range of possible states (e.g., to develop contingency plans, assess risks, or determine trade-offs among different desired end points). To decide whether or not to link the present and the future, the user needs to assess whether simply producing a vision is sufficient (e.g., to build awareness or communicate to another group) or whether a more-detailed understanding of a sequential process is necessary (e.g., for planning change).

The user also needs to decide on the level of detail and data collection necessary. Any of the three scenario exercises can be implemented with relatively simple, low-cost participatory rapid appraisal (PRA) methods. Alternatively, they can be implemented with more-detailed sampling, data collection, modeling, and analysis. PRA-based techniques have focused mostly on the use of illustrations created through group processes graphically to show a vision or present conditions. These techniques have been used as empowerment, awareness, and planning tools. They include "possible futures" (Slocum and Klaver, 1995), "story with a gap" (Narayan and Srinivasan, 1994), and "guided imagery" (Borrini-Feyerabend, 1997) exercises. An analysis of the available resources, constraints, and forces for change ("force field analysis") can also be performed to discuss how the group plans to attain their ideal vision (Narayan and Srinivasan, 1994).

If the user is interested in knowing why a particular scenario may occur, they can employ more involved techniques for analyzing the uncertainties, drivers of change, and causal relationships associated with each alternative (see below). These techniques are used as one means for generating a range of scenarios that reflect different types of drivers of change and causal pathways. The analysis can encourage critical thinking about risks and systems relationships. It has often been carried out in conjunction with in-depth data collection and probability-based modeling.

As a tool for anticipation, people can use any of the three types of scenario methods to adapt their current mental model to changing circumstances. To the extent multiple stakeholders develop new shared mental models, the scenarios can be said to achieve social learning. Adaptation of mental models is the key aim of using scenarios. During times of rapid change or complexity, existing mental models include assumptions that are no longer valid or habits of observation that prevent seeing new relationships (Wack, 1985b). Scenarios introduce hypothetical possibilities that spur people's imagination and enable them to adjust their mental habits. They enable stakeholders to overcome cognitive biases to (1) undervalue that which is hard to remember or imagine, (2) better remember and give more weight to recent events, (3) underestimate uncertainties, (4) deny evidence that does not support one's views, (5) overestimate their ability to influence events beyond their control, (6) be overconfident about their own judgments, and (7) overestimate the probability of desirable events (Becker, 1983; Schoemaker, 1993; Bunn and Salo, 1993). Pierre Wack (1985b), one of the main developers of the Royal Dutch Shell Corporation scenario approach, calls this adjustment of mental models the "gentle art of reperceiving" (p. 147).

The new mental model derives its power of explanation by taking a systems view. For community forest systems, for example, macrolevel and environmental forces can be given special attention in scenario construction as sources of risk and drivers of change. Scenarios, therefore, can encourage an understanding of the outside world and how people's inside world (the household, the landscape, a local organization) interacts with it (Wack, 1985b). This sort of analysis is crucial for effective community-level decision makers operating in the context of larger social and environmental landscapes with many stakeholders. Furthermore, to the extent scenarios encourage a systems view of more than one future, they open up the possibilities for yet more creative thought and critical understanding through comparison of those futures. The authors propose that the potential for anticipatory learning and adaptation is therefore higher to the extent multiple scenarios are explored and the relationships among events, resources, and actors involved in these scenarios are well understood.

> The point … is not so much to have one scenario that "gets it right," as to have a set of scenarios that illuminate the major forces driving the system, their interrelationships and the critical uncertainties. The users can then sharpen their focus on key environmental systems aided by new concepts and a richer language system through which they exchange ideas and data (Wack, 1985b, p. 146).

Given the greater potential learning involved in the use of multiple scenarios and systems analysis, a review of the elements of this scenario-based approach follows.

ELEMENTS OF A SCENARIO-BASED APPROACH

Although informal scenario-based methods have existed for centuries, systems-based methods using multiple scenarios are attributed to the developers of the atom bomb around 1942, who needed to understand very unpredictable systems (Schoemaker, 1993, p. 194). The technique become more popularly known with its development by the Rand Corporation in the 1960s (Kahn, 1965) and by SRI and Royal Dutch/Shell Corporation in the early 1970s. Scenario methods have since been adapted to scores of applications, including land-use planning (Foran and Wardel, 1995; Yin et al., 1995). Using the term *scenario* loosely, Van de Klundert (1995) suggests that the application of scenarios has evolved in ways that reflect the historical context of planning. Scenarios in the 1960s emphasized prediction based on existing stable trends, whereas those in the 1970s and 1980s accentuated coping with uncertainty. Scenarios in the "stakeholder" 1980s and 1990s have emphasized public discussion and shared decision making.

A number of sources provide excellent overviews of scenario approaches (Becker, 1983, Wack, 1985a, b) and information about how to construct scenarios (Becker, 1983; Huss and Honton, 1987; Deshler, 1987; Schoemaker, 1993; Bunn and Salo, 1994; Fahey and Randall, 1998; Bossel, 1998). These sources are drawn on to review the common elements of scenario construction and identify those elements suitable for community forests. Techniques related to the qualitative scenario method (Huss and Honton, 1987) are emphasized, in recognition of the limited technical resources available in most community forest management settings.

The four elements common to scenario analysis are as follows:

1. Definition of the purpose of the scenarios;
2. Information about a system's structure and major drivers of change;
3. Generation of the scenarios;
4. Implications of the scenarios and use by decision makers.

The changes in policy environments, markets, and alliances among interest groups that community forest managers have faced in the last two decades, and the multiscale nature of these phenomena, indicate an urgent need for scenario-type planning suitable for community forests. Although the scenario methods literature is replete with examples of applications about forests (Foran and Wardel, 1995), there is unfortunately little available on the methods appropriate for community forestry management.

Four traits distinguish community forest applications. First, attention to negotiation about preferences and aggregation of different views is especially important at several levels: (1) within the community for common pool forests, (2) with other groups that comanage or use the forest outside of the community, and (3) with the people using or responsible for the agricultural lands, waterways, or other land uses that affect the forest or are influenced by it. Information from these other interest groups must be included in the construction and evaluation of the scenarios. Because many community forests involve people disadvantaged by their ethnic or class background, however, care must be taken that stakeholders' power relationships do not bias who has a say in the scenario exercises. Scenarios need to be able to integrate planning about the uses and impacts associated with different interest groups for a given landscape, but this need not mean that all groups participate equally in every stage of scenario construction and analysis. Second, differences in sophistication among stakeholders in community forests requires designing understandable, transparent methods for each participating stakeholder group, including villagers who may not be able to read (Stewart and Scott, 1995). Third, creativity may be required to encourage villagers to express their ideas about the future, where culture and environmental conditions support a belief in fate and unwillingness to talk of what might be. Fourth, if the method is to be replicable, costs need to be minimal in terms of specialists, transaction costs of involving stakeholders, time and the collection of information.

Below are summarized the methods involved for each element and their application are discussed in the context of community forests.

PURPOSE

Scenarios are more effective tools for learning to the extent their purpose is situated within a clear decision making context. The context should be defined in terms of the issue requiring a decision and include the relevant time frame, location, and actors associated with the issue. The issue might be concern about a potential disturbance to the community's harvesting plans, unexpected Non Timber Forest Products (NTFP) market opportunities, impacts of the community's forest on the larger watershed, or the implications of a new national forest protection strategy. Normally, some set of stakeholders has already identified an issue that would benefit from scenario analysis. Additional relevant stakeholders may be called to assist in defining the decision context from their point of view.

The purpose should be clear whether the scenarios are to be applied to identify or assess decision options. Stakeholders may use scenarios to identify feasible options in light of possible major changes, such as shifts in economic conditions or population movements (Kahane, 1992). Or, they may use the same scenarios to test the viability of an existing practice such as a policy giving tenure security to customary lands against the backdrop of the hypothesized changes. The choice depends on the problem at hand. In practice, the two purposes are often combined through iterative scenario generation.

Choosing which decision makers should be represented in a scenario exercise for community forest management requires attention to the roles of different groups in management (forest owner, user, beneficiary, regulator, sponsor, competitor, or neighbor), their positions on the decision issue, and their role in society at large (Colfer, 1995; Borrini-Feyerabend, 1997; Farrington, 1996). The views of these different groups become "anchor points" that can have a significant impact on framing subsequent discussions and decisions (Bazerman and Neale, 1992); hence, care is needed in the selection of who participates and how they represent different interests. Creativity is required to enable people with different social status or access to power to meet and exchange ideas (Edmunds and Wollenberg, in press; Anderson et al., 1999). Local villagers may wish to work with a third party such as a nongovernmental organization (NGO), although this raises questions about whose views are really being expressed, that of the villagers or of the NGO.

Scenario methods are themselves adaptable, and have used various forms of stakeholder input to inform the scenario process and help make it relevant to users. Many examples come from methods for land-use planning scenarios. Stakeholders can be a source of information about the criteria with which to evaluate scenarios (Stewart and Scott, 1995). They can screen or assign preferences to scenarios and their impacts (Van Huylenbroeck and Coppens, 1995). Interest groups may even identify the risks, goals (Yin et al., 1995), or policies that define the scenario themes (Stewart and Scott, 1995). Given sufficient technical support, they can work with scenarios interactively, by providing the specifications, for example, for Geographic Information System (GIS) and decision support systems (Veldcamp and Fresco, 1997; Malafant and Fordham, 1997). Importantly for ACM, scenarios can also be used to develop shared perceptions of different possible futures and create platforms for joint learning and negotiation (Stewart and Scott, 1995).

STRUCTURE AND DRIVERS OF THE SYSTEM

The second element common to scenario methods is the collection of information about the forces shaping the system. These include:

1. The structure of resources, actors, institutions, and events and the relations among them;
2. Identification of slowly changing, predictable trends (such as amount of forest area, internal population growth, and road infrastructure; whether these parameters are slow-changing needs to be determined on a site-by-site basis);
3. Identification of uncertainties and potential major drivers of change (such as the opening of a new market for forest products, the introduction of a new harvesting technology, a new policy supporting customary forestland ownership, or rural-to-urban migration).

The intent here is to provide enough information to community-level decision makers and other relevant stakeholders to allow them to construct plausible, distinct scenarios, not to achieve a comprehensive understanding of how each hypothetical future works. Indeed, one of the functions of scenario analysis is to simplify complexity about the future. Structural elements of the system and slow-changing phenomena are singled out for their relative predictability. For community forest systems, a minimum set of factors might include identification of forest uses, users, relations among users, rules about forest use, and relationship of the forest to local households' economic needs, to agriculture or livestock, and to water quality. The dynamics of the system originate from locally relevant uncertainties and slow-changing phenomena. Uncertainties and drivers of change form the nucleus around which each of a set of multiple scenarios is then constructed. Uncertainties may revolve around anticipated drivers of change, not only those that have been strong influences in the past. In community forest systems, key uncertainties often include natural calamities (flood, winds), land conversion, market fluctuations, the policy environment, and actions of competing users of the forest.

The basis for each trend and uncertainty (and its assumed impacts) can be discussed carefully among stakeholders to identify the arguments for and against the likely occurrence of these phenomena (Schoemaker, 1991). This step is crucial if scenarios are to play a useful role in making forest management both more adaptive and more collaborative. It helps to build a partially shared, negotiated perception or working agreement among stakeholders of the values and assumptions underlying the construction of the scenarios. At the same time, it highlights potential alliances and areas of conflict among community-level decision makers, and between them and other stakeholders. This step can also result in the exchange of substantial new knowledge between community and outsiders about, for example, legal trends or environmental degradation. Each of these functions is critical to adaptive and collaborative management.

Identifying trends and uncertainties constitutes the first important part of the learning process. Once agreement about these trends has been reached, the relationships among trends can be mapped (Figure 17.2).

GENERATING THE SCENARIOS

Scenarios are generated based on an understanding of the system. The selection of the scenario themes may be based on any one of a combination of underlying logics, including cases demonstrating the implications of key uncertainties, desirable and undesirable cases, or likely and unlikely cases. The array of scenarios to be compared should be directly linked to the decision issue and purpose of analysis.

For the purpose of adaptive management, it is assumed that the logic of greatest interest is exploring uncertainties and forces for change. This logic requires identifying two to three plausible values for the uncertainties or change agents. The scenarios are then constructed using these values or realistic combinations of them. Scenario analysis specifies "uncertainty across, rather than within scenarios ... [the scenarios thereby] *bound* the uncertainty range" (Schoemaker, 1993, p. 196). Scenarios help stakeholders cope with uncertainty, not by eliminating it, but rather by framing it and understanding the range of associated implications.

Scenarios could be based on different sources of risk or levels of risk, or a comparison of desirable and undesirable situations from which risks can be extrapolated. If the purpose is to explore unexpected risks, the scenarios could be set up

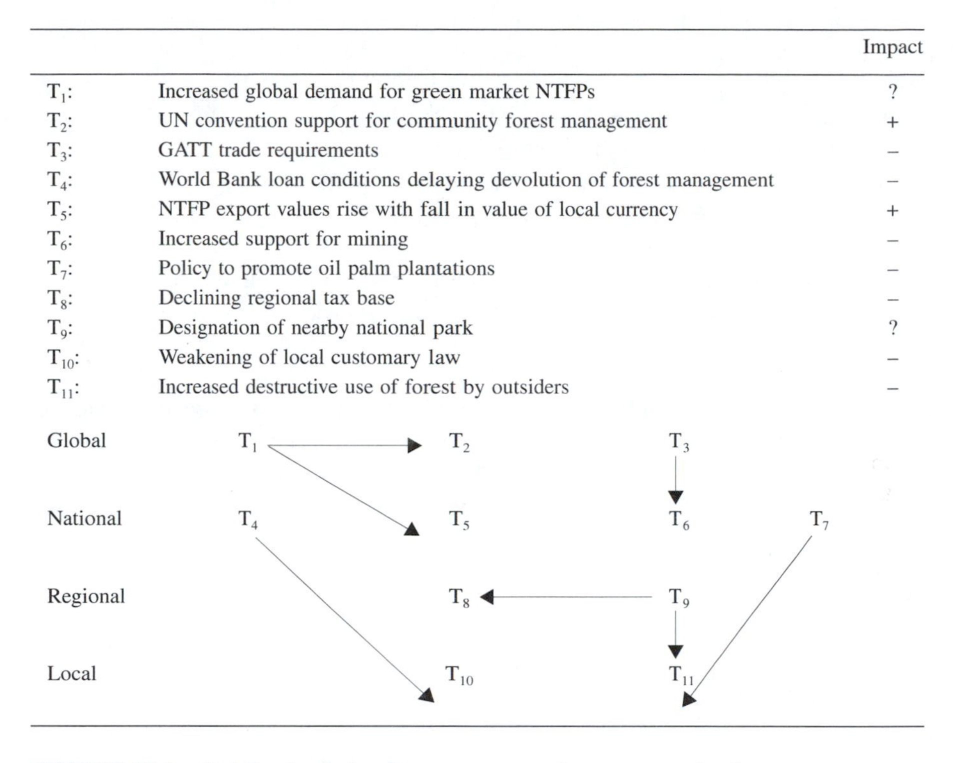

		Impact
T_1:	Increased global demand for green market NTFPs	?
T_2:	UN convention support for community forest management	+
T_3:	GATT trade requirements	–
T_4:	World Bank loan conditions delaying devolution of forest management	–
T_5:	NTFP export values rise with fall in value of local currency	+
T_6:	Increased support for mining	–
T_7:	Policy to promote oil palm plantations	–
T_8:	Declining regional tax base	–
T_9:	Designation of nearby national park	?
T_{10}:	Weakening of local customary law	–
T_{11}:	Increased destructive use of forest by outsiders	–

FIGURE 17.2 Multilevel relationships among trends on community forest management. (Adapted from Shoemaker, 1991, p. 553.)

to explore the opportunities across drivers of change, i.e., (1) possible changes in markets and pricing, (2) possible tenure policy changes, or (3) competition with external agents for forest benefits. For each driver of change, it may be desirable to specify a further set of scenarios showing a range of possible values. These ranges would be selected based on assumptions or principles about what it is that is important to compare, especially in terms of risk (Huss and Honton, 1987). As an example, community members might feel it important to compare scenarios showing the influence of increased transportation availability, a new taxation scheme, or logging by a neighboring concession. Alternatively, they could construct scenarios showing different transport options, a pair of scenarios comparing the convergence of favorable and unfavorable trends in taxation, or a range of scenarios showing the possible impacts of different logging plans.

Stakeholders should select scenario themes to challenge their thinking in ways that lead them to "a-ha" experiences or new insights (Wack, 1985b; Blythe and Young, 1994). Although scenarios can be used to overcome common biases about the way one thinks about the future (see above), the selection of scenario themes can suffer equally from these biases if the people conducting the exercise are not aware of ways of overcoming them. Wack (1985a) suggests that any first iteration of scenarios is unlikely to lead to insights because of the tendency to examine only obvious uncertainties. Techniques for stimulating creativity and overcoming biases include:

1. Using extreme outcomes, not just predictable ones;
2. Creating disruptions to historic trends;
3. Selecting scenarios that are distinct, not ones that reflect a gradient such as high, medium, and low values, or a positive and negative scenario;
4. Including undesirable scenarios;
5. Starting the construction of the scenario from an imagined future, rather than from extrapolation of current trends (Wack, 1985a; Schoemaker, 1991; 1993; Bunn and Salo, 1993).

As the intent of the scenarios is not to predict the future but to improve abilities to adapt to it, such extreme and non-continuous elements should not be considered "unrealistic."

Cultural attitudes may make overcoming these biases difficult in community forests situations. For example, people are often reluctant to predict or even talk about the future in concrete terms, especially the rural poor. People accustomed to a lack of control over their lives may prefer to acknowledge the power of or defer to fate, luck, or God's will rather than to make predictions. There may be a need to develop a willingness among the audience to face uncertainty and to understand the forces driving it (Wack, 1985b). Extra effort is usually required to develop undesirable scenarios because of tendency to deny evidence that contradicts people's hopes (Bunn and Salo, 1993). Similarly, to the extent people express their ideas in front of more powerful stakeholders, they are likely to feel less free to be creative, try to give the answer they hope the others want to hear, or avoid putting their ideas on the line for comment by others. Despite this reluctance, villagers managing trees and forests obviously think about the future, and often engage in planning for forest use, sometimes in cooperation with historical antagonists. The issue is therefore how to elicit this information.

The possibility of selecting participants in such a way as to facilitate dialogue has already been mentioned, but this may come at the cost of a more complete and realistic scenario discussion. Another possibility for eliciting people's visions of the future may require getting people to talk about the future in present terms. Villagers could be asked to indicate what they would like to see remain the same about their current lives and local forest management and what they would like to see changed.

Assuming these barriers can be overcome, the scenarios need to obey certain rules to be useful. They should be internally consistent; coherent, plausible; feasible, i.e., based on real forest resources, natural processes, logic, and ethics; linked to the present and understandable by the scenario user (Blythe and Young, 1994; Bossel, 1998). These requirements result in some trade-offs with creativity, but are necessary to ensure the learning is relevant to the real world. Users are more likely to comprehend and remember the relationships and causalities in scenarios to the extent information is conveyed in a storylike narrative and each story is given a label (Schoemaker, 1993; Bunn and Salo, 1993). They should be approximately the same length and involve the same amount of detail and comprehensiveness to avoid biases in their comparison (Bunn and Salo, 1993).

The recommended number of scenarios to use varies in the literature. Most authors suggest comparisons of three to nine scenarios (Wack, 1985b; Deshler, 1987; Stewart and Scott, 1995). The number of scenarios, of course, depends on the purpose of the analysis. One scenario may be sufficient for simple exercises where the intent is to facilitate group learning. More scenarios are necessary where a decision of consequence must be tested for its robustness against a large number of uncertainties (Bunn and Salo, 1993). Evidence from cognitive research indicates that people are also only capable of comparing a maximum of five to nine scenarios at one time (Stewart and Scott, 1995).

Wack (1985b) suggests no more than four scenarios. He recommends an ideal number of three, with one showing the surprise free world, and two showing critical uncertainties. He explains that the use of only two scenarios creates a tendency for one to be the pessimistic and one the optimistic view. People make judgments by taking a metaphorical average of the two scenarios. Where three scenarios are used, their themes should be selected to reflect different uncertainties. If the themes are only different values of the same uncertainty, people tend to select the middle one as the most desirable scenario.

Although people can only compare a limited number of scenarios at one time, large numbers of scenarios may be used during the course of a scenario exercise. A scenario exercise is usually repeated iteratively, with each iteration generating new scenarios. Stewart and Scott (1995) suggest conducting a first iteration of coarse scenarios that address the widest possible range of options. These first scenarios are used to identify a smaller subset of scenarios that are constructed at a finer level of resolution. Scenarios can also be nested. Nesting has the added advantage of addressing different scales (Wack, 1985b). For community forests, scenarios could be nested to include user group scenarios, larger forest-level scenarios, regional economy scenarios, and finally country-level and international scenarios. Both iterative and nested scenarios facilitate learning by community-level decision makers with limited knowledge of or experience with other stakeholders operating at larger scales.

The form of the scenario and its presentation should be designed with the different stakeholders' capacities and preferences in mind. The presentation of the scenario need not be written or on paper. Tan-Kim-Yong (1992), for example, found that three-dimensional models of local landscapes facilitated lively exchange of stakeholders' views about land-use planning. The use of simple materials for some audiences should be balanced against the need to keep all the stakeholders involved and stimulated. The degree to which the method is transparent and understandable to all the stakeholders will further aid their ability to work with the scenarios and learn together from them (Blythe and Young, 1994).

GIS and maps can be used to represent scenarios in ways that make them more tangible and "present" (Malafant and Fordham, 1997). Community-based management interventions commonly involve GIS and the generation of maps. These tools have proved popular and useful for strengthening local management. The skills for mapping and maps are increasingly widespread among NGOs and forest communities.

GIS-generated scenarios have the advantage of being interactive and more readily manipulated to show different scenarios. Care should be taken, however, to avoid negative impacts on group dynamics based on different levels of familiarity with or access to such technologies.

IMPLICATIONS OF THE SCENARIOS AND USE BY DECISION MAKERS

The final element is the discussion and analysis of the implications of each scenario for making decisions. Although scenarios can benefit community-level decision makers simply by bringing stakeholders together and facilitating the exchange of information, they are most useful to the extent they influence each stakeholders' thinking and actions to enable coordination and improved management. The scenario must "come alive in 'inner space,' the manager's microcosm where choices are played out and judgment exercised" (Wack, 1985b). Wack suggests that the biggest challenge in scenario analysis is successfully reaching the decision makers, not the construction of the scenarios itself. This requires that true learning occurs, i.e., the scenarios are clearly understood and internalized among decision makers (Wack, 1985b, p. 142).

As mentioned, the stories associated with each scenario can be used as platforms for the stakeholders to articulate their views. Schoemaker (1993) observes that scenarios work well because of their cognitive appeal as stories and metaphors. This appeal may be used to facilitate interactions among stakeholders, including eliciting a range of views of management and enabling negotiation using the scenario as a basis for discussion (Van Huylenbroeck and Coppens, 1995; Stewart and Scott, 1995). They are perhaps also more appropriate than technically focused discussions and formal negotiations where community representatives are involved, as they can reduce the differences in rhetorical skill among stakeholders. A well-told story can, however, also generate expectations that the scenario is more probable than warranted.

The analysis of a first round of scenarios commonly leads to the identification of new forces for change and new themes for scenario development. Scenario development may require several cycles before stakeholders feel that they have explored sufficient possibilities.

OPTIONS IN SCENARIO CONSTRUCTION

Although these are the basic elements of multiple scenario exercises based on systems analysis, the range of variation in scenario-related methods is broad. Several typologies provide guidance to key differences among methods (Ducott and Lubben, 1980; Huss and Honton, 1987; Bunn and Salo, 1993). The differences related to the purposes of scenarios were discussed earlier. Differences related to the methods for generating and analyzing scenarios are summarized in Table 17.1 and discussed below.

Although most reviews of scenario methods distinguish between quantitative and qualitative methods, the boundaries between these two approaches have become increasingly blurred by techniques that make use of both kinds of methods and

TABLE 17.1
Dimensions of Variation in Scenario Exercises

Dimension along Which Scenarios Vary	Range of Extremes	
Methods of construction and analysis	Quantitative, "hard," formal models Statistical forecasting Trend-impact analysis Cross-impact analysis	Qualitative, "soft" Visioning Intuitive logic
Source of information	Rational, scientific observation	Judgment and intuition of decision makers
Role of stakeholders	Passive objects of analysis	Active participants in construction and evaluation
Use of forecasting or predictive models	High	Low
Selection of scenario themes — explicitness of values	Normative, e.g., scenarios reflect the desired and "good" or the undesired and "bad"	Descriptive, not based on social preferences
Comprehensiveness, complexity, and detail of scenarios	High	Low
Degree to which the scenarios reflect current conditions	Reflect the unexpected, hypothetical, and extreme	Extrapolated from current trends
Length of scenario path	Short: "snapshot" at one point in time	Long: story of events linked to present
Starting point of pathway	Future, uses backward inference, deductive	Present, uses future inference, inductive

[a] The two columns do not represent coherent pairs. One could find a quantitative and descriptive scenario, for example.

Source: Adapted from Ducot and Lubben (1980), Bunn and Salo (1993).

information (Bunn and Salo, 1993). Nevertheless, some methods are recognized as more quantitative such as *cross-impact analysis*, which incorporates the probabilities of outcomes affecting other outcomes into the scenarios (Duval, 1975; Harrell, 1978), and *trend impact analysis,* which uses forecasting to quantify the impacts of trends (Huss and Honton, 1987). Qualitative techniques include the *intuitive logics* approach, which relies largely on interviews and interactions with decision makers (Huss and Honton, 1987). Bunn and Salo (1993) critique the quantitative methods for treating decision makers as passive entities, and therefore being less relevant and less engaging a tool for changing users' thinking.

For the purposes of supporting ACM in community forests, it is clear that stakeholders must be treated as actors and involved in the scenario process. Their perceptions and knowledge are key to creating and interpreting the scenarios. For transparency purposes, as well, there is likely to be a need for relatively qualitative techniques. Forecasting techniques may be relevant for highly predictable phenomena, but the cost and human resources for such activities may not be available. The role of values in the selection of scenario themes and degree of comprehensiveness and detail in the content of the scenarios will be site specific. Whether the analysis begins in the future and is prospective or begins in the present and is projective (van de Klundert, 1995) will also depend on case-by-case needs. To the extent the scenarios are intended to identify visions and alternatives, the stakeholders will need to give less attention to the present and current trends. To the extent there is a need to integrate learning about the past with learning about the future, it will need to focus on the present and forward trends. Cultural preferences for learning styles may — as mentioned — influence how the approach is adapted.

LESSONS FOR SCENARIO ANALYSIS IN COMMUNITY FOREST MANAGEMENT

Several generalizations may be drawn about the application of scenario methods to community forest landscapes.

First, scenario analysis provides opportunities for important *forward*-looking learning for the ACM of community forest landscapes. The authors suggest that the long-term and dynamic nature of interactions among local people's livelihoods, sustainability objectives, and the biophysical conditions of community forests make prediction and simple feedback loop-type learning problematic. The interdependence of these relationships among stakeholders further complicates learning. More-open-ended, forward-looking methods are needed that address complexity and risk, particularly methods that can provide community-level decision makers with information on multilevel social and environmental processes. The four common elements of scenario methods (purpose, structure and drivers, scenario generation, and use) that make this kind of learning possible can be applied to most community forest settings.

Second, to the extent community forest systems involve many and competing interests, especially across groups with vastly different influence and power, scenario methods will need to give special attention to accommodating differences among these groups. It may not be desirable or cost-effective to work with all stakeholders. Communication differences and the possibility for unfair decision making are likely to increase where powerful players like timber companies are matched with weak ones like a nomadic group of hunter/gatherers (Edmunds, 1999; Anderson et al., 1999). In these cases, parallel rather than joint scenario processes are warranted. The scenarios can serve as a platform for debate among relatively cooperative stakeholders and be used to communicate interests in a common language among more antagonistic stakeholders. Scenarios may help to highlight interdependencies among interest groups and thereby also foster cooperation. Scenario creation could also be used selectively with community stakeholders to empower them, with the understanding that a subsequent stage of analysis of existing scenarios, decision making, facilitation, and negotiation would engage other relevant stakeholders. Costs will increase proportionally with duplicate processes. It is therefore necessary to understand fully the players needed to participate in a decision and develop a strategy of using joint scenarios, parallel scenarios, or a less-intensive alternative with each group.

Third, the information necessary to build scenarios in community forests is often lacking. Information either is not available (e.g., the biological characteristics of many nontimber forest products) or is held in places or among people that rarely exchange their knowledge. At most sites, an extra investment is likely to be required just to collect the required data. Some stakeholders may not wish to share their information with others (e.g., the location of a valued resource, plans for illegal harvesting), which again suggests the need for parallel rather than joint scenario processes where negotiation occurs after the scenarios have been created. One advantage of the scenario method is that it can help prioritize information needs so that data can be collected more efficiently. As a story, the scenario can also be used to exchange information effectively and create a shared understanding among stakeholders. These are the features of the scenario method (rapid knowledge acquisition, exchange and share understanding) that contribute most directly to adaptiveness (McLain and Lee, 1996).

Fourth, to be transparent, usable by community members, and replicable, the principles of simplicity and tangibility need to be applied to every step of the scenario exercise. The decision issue is best grounded using a map, a story, pictures, photographs, a three-dimensional model, or an existing document such as a management plan with which the community is very familiar. Qualitative methods are likely to be more user-friendly than quantitative ones. The number of working scenarios is best kept to a minimum of two or three. To cope with multiple actors and scales, scenarios will most likely need to be nested. A third party may be useful not only to facilitate the scenario creation and analysis, but also to help refine them for presentation to other audiences. Such refining is already common in participatory mapping, where NGOs help communities produce maps similar to those used by government and therefore communicates persuasively to groups accustomed to using these maps. Lessons can be drawn from this experience about methods for aggregating and communicating views among different groups, as well as the limits to the role of the third-party facilitator.

CONCLUSIONS

Scenario methods differ from other tools for ACM by providing a framework for anticipating the future. They should prove useful in community forestry by encouraging analysis of processes and actors operating at landscape and larger scales. They can broaden perspectives about how the forest might change in unexpected ways and serve as a platform for reaching agreement among different stakeholders. Scenarios involving multiple stakeholders can speed up the process of information exchange and enhance adaptivity by expanding the availability and flow of information for decision making, particularly from sources outside the community. These impacts can enhance the preparedness of the stakeholders involved in community-based forest management in coping with change.

The feasibility and impact of using scenario methods effectively in practice remains to be tested. To facilitate the process of using and testing scenarios for anticipatory learning in community forest settings, the authors have prepared a guide for users (Wollenberg et al., 2000). The authors anticipate that a body of empirical evidence will emerge as they gain experience with their research partners in various protected forest settings throughout the world in combining the elements of scenario construction and analysis to foster social learning.*

The review of methods for the construction of scenarios indicates the broad scope of possibilities for using scenarios and their relevance to ACM of community forests. Although it may be appealing to consider scenario methods as generally applicable to most resource management settings, there are special conditions of community forestry that must be considered to adapt the method successfully. As these conditions are not unique to community forest landscapes, the authors trust that some aspects of the approach they develop should be more generally applicable to other settings where complex stakeholder relationships and information constraints shape the nature of resource decision making.

REFERENCES

Anderson, J., Clement, J., and Crowder, L. V., 1999. Pluralism in sustainable forestry and rural development — an overview of concepts, approaches and future steps, in *Pluralism and Sustainable Forestry and Rural Development, Proceedings of an International Workshop,* 9–12 December, 1997, FAO, Rome, Italy.

Barnes, J. H., Jr., 1984. Cognitive biases and their impact on strategic planning, *Strategic Manage. J.,* 5(2), 129–137.

Bazerman, M. H. and Neale, M. A., 1992. *Negotiating Rationally,* Free Press, New York.

Becker, H. S., 1983. Scenarios: a tool of growing importance to policy analysts in government and industry, *Technol. Forecasting Soc. Change,* 23(2), 95–120.

Bhattacharyya, K. and Kumar, A., 1998. Beauty is in the process and not in the name: an alternative approach for participatory planning, *PLA Notes,* 31, 18–22, International Institute for Environment and Development, London.

Blythe, M. J. and Young, R., 1994. Scenario analysis: a tool for making better decisions for the future, *Eval. J. Australasia,* 6(1), 1–17.

Bocco, G. and Toledo, V. M., 1997. Integrating peasant knowledge and geographic information systems: A spatial approach to sustainable agriculture, *Indigenous Knowledge Dev. Monitor,* 5(2), 10–13

Borrini-Feyerabend, G., 1997. *Beyond Fences: Seeking Social Sustainability in Conservation,* Vol. 2, A Resource Book, IUCN, Gland, Switzerland.

Bossel, H., 1998. *Earth at a Crossroads, Paths to a Sustainable Future,* Cambridge University Press, Melbourne.

Brown, D., 1998. Participatory biodiversity conservation rethinking the strategy in the low tourist potential areas of tropical Africa, *Nat. Resourc. Perspect.,* 33(1–5), Overseas Development Institute, London.

Bunn, D. W. and Salo, A. A., 1993. Forecasting with scenarios, *Eur. J. Operational Res.,* 68(3), 291–303.

* The Swiss Development Cooperation is supporting CIFOR scientists and their partners in Madagascar, Indonesia, and Bolivia to test and evaluate the viability of future scenarios analysis in balancing stakeholder interests in biodiversity conservation settings.

Colfer, C. J. P., 1995. Who Counts Most in Sustainable Forest Management? CIFOR Working Paper 7, Bogor, Indonesia.

Datta, R., 1999. Seva Mandir: a learning organization, paper prepared at East-West Center–CIFOR Writing Workshop on Social Learning, East-West Center, Honolulu.

Decker, D. J., Krueger, C. C., Baer, R. A., Jr., Knuth, B. A., and Richmond, M. E., 1996. From clients to stakeholders: a philosophical shift for fish and wildlife management, *Hum. Dimens. Wildl.,* 1(1), 70–82.

Deshler, D., 1987. Techniques for generating futures perspectives, in *Continuing Education in the Year 2000. New Directions for Continuing Education*, Brockett, R.G., Ed., Jossey-Bass, San Francisco, 79–82.

Ducot, C. and Lubben, G. J., 1980. A typology for scenarios, *Futures*, 12(1), 51–57.

Duval, A., Fontela, E., and Gabus, A., 1975. Cross-impact analysis: a handbook on concepts and applications, in *Portraits of Complexity: Applications of Systems Methodologies to Societal Problems*, Baldwin, M. M., Ed., Battelle Memorial Institute, Columbus, OH, 202–222.

Edmunds, D. and Wollenberg, E., in press. A strategic approach to multistakeholder negotiations, *Development and Change.*

Fahey, L. and Randall, R. M., 1998. *Learning from the Future: Competitive Foresight Scenarios,* John Wiley & Sons, New York.

Farrington, J., 1996. Socioeconomic methods in natural resources research, *Nat. Resourc. Perspect.,* 9, Overseas Development Institute, London.

Fischhoff, B., 1988. Judgmental aspects of forecasting: needs and possible trends, *Int. J. Forecasting,* 7, 421–433.

Foran, B. and Wardel, K., 1995. Transitions in land use and the problems of planning: a case study from the mountainlands of New Zealand, *J. Environ. Manage.,* 43, 97–127.

Gigerenzer, G. and Todd, P. M., 1999. Fast and frugal heuristics: The adaptive toolbox, in *Simple Heuristics That Make Us Smart,* Gigerenzer, G., Todd, P. M., and the ABC Research Group, Oxford University Press, Oxford, 3–34.

Grimble, R. and Chan, M.-K., 1995. Stakeholder analysis for natural resource management in developing countries, *Nat. Resourc. Forum,* 19(2), 113–124.

Harrell, A. T., 1978. *New Methods in Social Science Research: Policy Sciences and Future Research*, Praeger, New York.

Holling, C. S., 1978. *Adaptive Environmental Assessment and Management,* Wiley International Series on Applied Systems Analysis, Vol. 3, Wiley, Chichester, U.K.

Huss, W. R. and Honton, E. J., 1987. Scenario planning: what style should you use? *Long Range Planning,* 20(4), 21–29.

International Institute for Applied Systems Analysis, 1979. Expect the Unexpected: An Adaptive Approach to Environmental Management, Executive Report 1, Laxenberg, Austria.

Johnson, B. L., 1999. Introduction to the special feature: adaptive management — scientifically sound, socially challenged? *Conserv. Ecol.,* 3(1), 10.

Jungermann, H. and Thüring, M., 1987. The use of mental models for generating scenarios, in *Judgmental Forecasting*, Wright, G. and Ayton, P., Eds., John Wiley & Sons, New York, 245–266.

Kahane, A., 1992. Scenarios for energy: sustainable world vs. global mercatilism, *Long Range Plann.,* 25(4), 38–46.

Kahn, H., 1965. *On Escalation: Metaphors and Scenarios,* Praeger, New York.

Lessard, G., 1998. An adaptive approach to planning and decision making, *Landscape Urban Plann.,* 40(1–3), 81–87.

Maarleveld, M. and Dangbégnon, C., 1999, Managing natural resources: a social learning perspective, *Agric. Hum. Values,* 16, 267–280.

Malafant, K. W. J. and Fordham, D. P., 1997. GIS, DSS and integrated scenario modelling frameworks for exploring alternative futures, in *Advance in Ecological Sciences,* Vol 1: *Ecosystems and Sustainable Development, Proceedings of a Conference,* Uso, J. L., Brebbia, C. A., and Power, H., Eds., Peniscola, Spain 14–16 October 1997, 669–678.

Margolis, R. and Salafsky, N., 1998. *Measures of Success*, Island Press, Washington, D.C.

McLain, R. J. and Lee, R. G., 1996. Adaptive management: promises and pitfalls, *Environ. Manage.,* 20(4), 437–448.

Millett, S. M., 1988. How scenarios trigger strategic thinking, *Long Range Plann.,* 21(5), 61–68.

Narayan, D. and Srinivasan, L., 1994. *Participatory Development Tool Kit*, World Bank, Washington, D.C.

Prabhu, B. R., Colfer, C. J. P., Venkateswarlu, P., Tan, L. C., Soekmadi, R., and Wollenberg, E., 1996. Testing Criteria and Indicators for the Sustainable Management of Forests: Final Report of Phase I, CIFOR, Bogor, Indonesia.

Robinson, J. B., 1992. Risks, predictions and other optical illusions: rethinking the use of science in social decision making, *Policy Sci.,* 25, 237–254.

Sapio, B., 1995. SEARCH (scenario evaluation and analysis through repeated cross impact handlling): a new method for scenario analysis with an application to the Videotel service in Italy, *Int. J. Forecasting,* 11(1), 113–131.

Schoemaker, P. J. H., 1991. When and how to use scenario planning: a heuristic approach with illustration, *J. Forecasting,* 10, 549–564.

Schoemaker, P. J. H., 1993. Multiple scenario development: its conceptual and behavioral foundation, *Strategic Manage. J.,* 14(3), 193–213.

Shindler, B., Steel, B., and List, P., 1996. Public judgments of adaptive management: a response from forest communities, *J. For.,* 94(6), 4–12.

Slocum, R. and Klaver, D., 1995. Time line variations, in *Power, Process and Participation — Tools for Change*, Slocum, R., Wichart, L., Rocheleau, D., and Thomas-Slayter, B., Eds., Intermediate Technology Publications, London, 194–197.

Steelman, T. A. and Ascher, W., 1997. Public involvement methods in natural resource policy making: advantages, disadvantages and trade-offs, *Policy Sci.,* 30, 71–90.

Stewart, J. T. and Scott, L., 1995. A scenario-based framework for multicriteria decision analysis in water resources planning, *Water Resourc. Res.,* 31(11), 2835–2843.

Tan-Kim-Yong, U., 1992. Participatory land-use planning for natural resource management in northern Thailand, Network paper 14b, Rural Development Forestry Network, Overseas Development Institute, London.

Taylor, B., Kremsater, L., and Ellis, R., 1997. Adaptive Management of Forests in British Columbia, Report, British Columbia Ministry of Forests, Vancouver, Canada.

van de Klundert, A. F., 1995. The future's future: inherent tensions between research, policy and the citizen in the use of future oriented studies, in *Proceedings of the Symposium, Scenario Studies for the Rural Environment,* 12–15 September 1994, Schoute, J. F. T., Finke, P. A., Veeneklaas, F. R., and Wolfert, H. P., Eds., Wageningen, The Netherlands, 25–32.

Van Huylenbroeck, G. and Coppens, A., 1995. Multicriteria analysis of the conflicts between rural development scenarios in the Gordon District, Scotland, *J. Environ. Plann. Manage.,* 38(3), 393–407.

Van Latesteijn, H. C. 1994. Scenarios for land use in Europe: agro-ecological options within socio-economic boundaries, in *Eco-regional Approaches for Sustainable Land Use and Food Production*, Bouma, J., Kuyvenhoven, A., Bouman, B. A. M., Luyten, J. C., and Zandstra, H. G., Eds., 12–16 December, ISNAR, The Hague, The Netherlands, 43–63.

Veldkamp, A. and Fresco, L. O., 1999. Exploring land use scenarios: an alternative approach based on actual land use, *Agric. Syst.,* 55(1), 1–17.

Wack, P., 1985a. Scenarios: uncharted waters ahead, *Harvard Bus. Rev.,* 63(5), 72–89.

Wack, P., 1985b. Scenarios: shooting the rapids, *Harvard Bus. Rev.,* 63(6), 139–150.

Walters, C., 1986. *Adaptive Management of Renewable Resources*, Macmillan, New York.

Wollenberg, E., Edmunds, D., and Buck, L., 2000. Using scenarios to make decisions about the future: Anticipatory learning for the adaptive co-management of community forests, *Landscape and Urban Planning,* 47, 65-77.

Wollenberg, L., with Edmunds, D. and Buck, L., 2000. Anticipating Change: Scenarios as a Tool for Adaptive Forest Management — A Guide for Practitioners, Working Paper, Center for International Forestry Research, Bogor, Indonesia.

Yin, Y., Pierce, J. T., and Love, E., 1995. Designing a multisector model for land conversion study, *J. Enviorn. Manage.,* 44, 249–266.

Section IV

Case Studies: Applications of Adaptive Collaborative Management Approaches

18 Community-Based Conservation Area Management in Papua New Guinea: Adapting to Changing Policy and Practice

Arlyne Johnson, Paul Igag, Robert Bino, and Paul Hukahu

CONTENTS

INTRODUCTION

The Crater Mountain Wildlife Management Area (WMA) in Papua New Guinea is an Integrated Conservation and Development Project (ICDP). It is currently testing the hypothesis that the establishment of viable business enterprises that *depend* on site biodiversity will lead to conservation of biodiversity, and sustainable use of the natural resources, by WMA landowners. Community-based projects that provide

0-8493-0020-7/01/$0.00+$1.50

socioeconomic incentives for local participation in conservation have been implemented as one method for establishing biodiversity conservation areas (Brandon and Wells, 1992; Western and Wright, 1994). Considerable debate at present exists about the effectiveness of ICDP methods to achieve conservation goals (Kramer et al., 1997). Since little systematic analysis of the methods and results of this approach have been conducted, we wanted to set up a monitoring system which would allow us to better evaluate whether socioeconomic development objectives being implemented in the Crater Mountain WMA may be resulting in increased biodiversity conservation action by natural resource owners. We will describe the methodology that we used to design and evaluate our project. We also discuss what project monitoring has revealed so far and give some examples of how this has changed our activities over time.

PROJECT SITE

The Crater Mountain WMA covers 2700 km^2 (Figure 18.1). In Papua New Guinea, 97% of the land is held in customary tenure, which gives landowners full ownership and control of all natural resources, with the exception of mineral reserves. WMAs are classified as multiple-use areas, and are declared by the national government upon request by the landowner. Owners submit a legal description of the boundaries, a list of the clan (family group) leaders who will sit on the local management committees, and determine the rules for natural resource use. Committees enforce rules, collect fines from violations, and can collect tax from enterprises operating in the WMA. In the Crater Mountain WMA, training and technical assistance for landowners in this endeavor is provided by the staff of a national nongovernmental

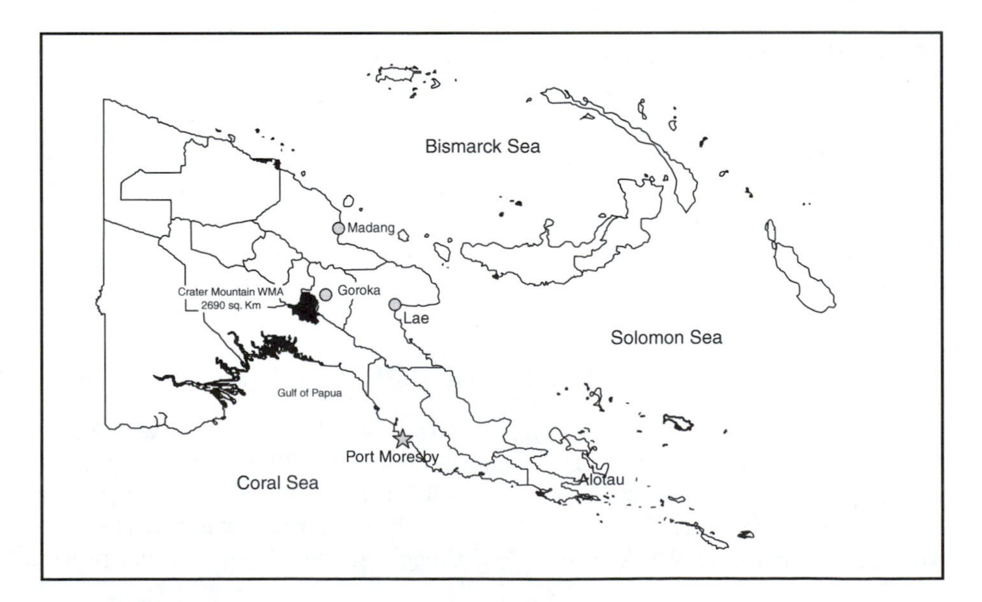

FIGURE 18.1 The Crater Mountain WMA in Papua New Guinea.

organization, the Research and Conservation Foundation (RCF), and the Wildlife Conservation Society (WCS). The project team includes biologists, small business developers, community development volunteers, scientists, and support staff.

The Crater Mountain WMA ranges from sea level to 3000 m in elevation. The altitudinal gradient includes lowland primary rain forest up to alpine scrub and grassland. Biodiversity is very rich. There are 221 species of birds and 84 species of mammals with numerous Papua New Guinea endemics (Beehler, 1993). Many plants, insects, and herptefauna are yet to be named.

The social landscape of the WMA is extremely complex. Land is owned by 22 clans of the Gimi and Pawaian language groups, who have settlements near four airstrips. There are no roads entering the WMA. Each clan manages its land independently of the others. Traditional rivalry and sorcery is ongoing between the clans. The population of the WMA is 3000 people, an average density of one individual per square kilometer. The landscape is mountainous with limited infrastructure for delivery of government services, health, and education to these remote areas. Average formal education of WMA residents is grade one.

Landowners are engaged in subsistence living through gardening, hunting, and gathering. Cash income has traditionally been earned through sale of coffee, and market sales of wildlife. Often the only other option seen by landowners for meeting growing cash and development needs is the selling of their raw minerals or forests to international interests.

In the late 1980s, to offset these growing cash needs, the Crater Mountain ICDP began assisting landowning clans to establish ecoenterprises. These are small business ventures that could potentially bring cash income without damaging the rich biodiversity of the area. There are now basic scientific field research and village guesthouse facilities, and village handicraft enterprises, based out of two airstrips (Figure 18.2). Traditional handicrafts are for sale in the villages or by mail order. Tourists are beginning to come to hike rugged WMA trails with local guides to see the unique flora and fauna. International and national scientists and students conduct research projects and training courses at the rustic field research facilities in the WMA.

METHODS

The expanse and complexity of socioeconomic and biological variables within the WMA are not atypical of many integrated conservation and development initiatives. It is often a challenging task for an ICDP team to select and prioritize the project resources on activities that have the highest probability of resulting in the desired goal, biodiversity conservation.

CONCEPTUAL MODEL

With the assistance of a methodology developed by the Biodiversity Conservation Network (Margolius and Salafsky, 1998), the authors' project team began by making a conceptual model (see Appendix A). The model serves as a pictorial representation

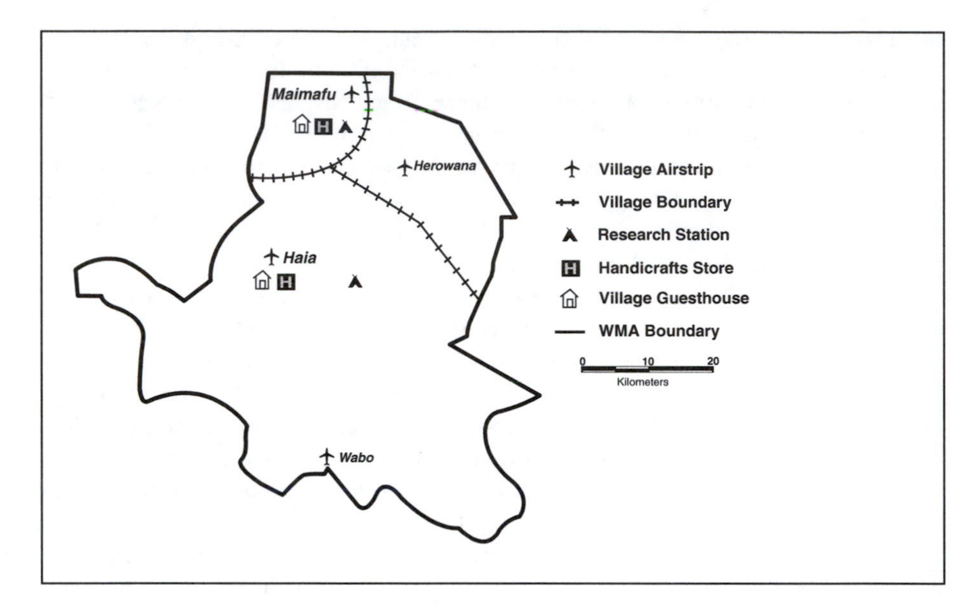

FIGURE 18.2 Ecoenterprises operating in the Crater Mountain WMA villages of Haia and Maimafu. © 1999 WWF.

of what the project team felt the site conditions in the WMA were (Ericho et al., 1999). It depicts the factors (in boxes) that the team thought were effecting the target condition of biodiversity (conservation). For example, an increase in the box marked, "human population," along with changes in the box marked, "cultural practices," were thought to be effecting a need for cash, which is indicated by the box marked, "cash needs." It was thought that this was, in turn, effecting the level of box marked, "market hunting," which was ultimately effecting the target condition of conservation of "biodiversity."

Upon completion of the model, the team identified key areas within the model where it thought its invention may have the highest probability of positively influencing factors that may, in turn, lead to biodiversity conservation. These "intervention areas" are indicated as circles on the model. The key assumptions that were made regarding the selection of the circled intervention sites were the following:

1. If cash income could be made from biodiversity remaining *in situ*, that it may serve to discourage other options by clans to "sell off" natural resources (e.g., nonsustainable wildlife market sales, large-scale logging, etc.).
2. Further education was needed if the ecoenterprises were to flourish because the rates of illiteracy among WMA residents were high. There was minimal experience by residents in the WMA with business ventures and the cash economy. There was a lack of awareness by landowners of the unique and potential monetary value of their natural resources if held *in situ*, in the WMA.

3. Local clan institutions required strengthening as human population and intrusion into the WMA for resource harvesting was increasing, while abundance of natural resources was reportedly decreasing. There had been a lack of support from either the local or national government to address these issues. Communities were expressing concern about how to collect information, make policy, manage, and enforce in this rapidly changing landscape (James, 1995).

PROJECT OBJECTIVES AND ACTIVITIES

The circled intervention factors were then formalized into four project objectives:

1. To increase the income of clans from the establishment of ecoenterprises;
2. To increase the capacity of WMA residents who work in the ecoenterprises;
3. To increase the number of decisions and actions that integrate monitoring results in the management plan;
4. To increase national involvement within the WMA as teachers and trainers, to build national capacity and to replicate the process, if successful.

Beginning in 1995, project activities were designed for a 3-year period that would move the WMA toward these objectives.

MONITORING PLAN

As we moved through the years with project implementation, how would we know if we were successful or not in achieving the objectives, and if these, in turn, had any effect on the goal? Was our hypothesis, and associated assumptions, correct? To monitor and evaluate our progress, we selected indicators that would work as "gauges" to measure the state and change in environmental and socioeconomic conditions at the site over time. For each indicator, we identified (1) the methods we would use to collect the data, (2) which project staff or community members would implement the method, and (3) when and where the monitoring would take place. This resulted in the Crater Mountain WMA Monitoring Plan (see Appendix B).

Project Goal: Conservation of biodiversity. To measure direct and indirect changes in the conservation of biodiversity, the project goal, we decided to monitor indicators of the direct use of plants and wildlife, as well as change in diversity and abundance of habitat and wildlife. For example, wildlife exported from the WMA is monitored by a village resident, the airstrip agent, who weighs passengers and cargo for each plane that exits. A data sheet is completed that identifies the wildlife species, the clan of the person who is sending the animal out, the method of collection, and the reason for export. In Papua New Guinea, domestic trade and sale of wildlife by national citizens is not prohibited.

Objective 1: Increase clan income from eco-enterprises that depend on site biodiversity. To measure changes in clan income, we monitor the extent and distribution of revenues to clans from ecoenterprises, and income from traditional cash crop and wildlife sales. We also monitor change in spending at the village trade store and market. One example of income monitoring is by village managers through receipt books for each enterprise, which record distribution of sales income to clan members. Another example is kilograms of cash crops monitored by the village airstrip agent as they are freighted out of the WMA by plane.

Objective Two: Increase clan capacity to participate in ecoenterprises that depend on site biodiversity. To measure changes in capacity, we monitor the number of participants and the skills taught at all village training events sponsored by the project. The application of skills can be measured through responses from visiting scientists and tourists on visitor questionnaires, and through the minutes of discussion and action taken at village meetings. Assessment of enterprise management is conducted annually by project staff and the village manager. Training events may include ongoing on-site training by resident staff, special workshops, and study tours to sites outside of the WMA.

Objective 3: Increase actions that use the monitoring program in management plan.

Objective 4: Increase national involvement as teachers and trainers. The last two objectives are also monitored through minutes from village meetings, a database of publication outputs, and records of training events led by community members, project staff and visiting instructors. Conservation action and project participation is tracked through the recording of minutes at village management committee meetings, written letters, and reports.

The project team began the design of the monitoring plan in 1995 and is just completing implementation now. What has been learned so far? Has this process been useful in guiding the project? Following are a few examples of preliminary monitoring results from two WMA villages. We will talk about how this has been used to modify and focus project activities.

RESULTS

The first objective of the project was to increase the income of the clans through establishment of ecoenterprise businesses. By 1997, the project had done that to various degrees. We found that in the lowland village of Haia, most income in 1997 was coming from the ecoenterprises, with a much smaller amount from sale of coffee and wildlife (Figure 18.3). In the mountainous village of Maimafu, we found that the average annual income was already much higher than in the lowlands, and that ecoincome is still only a minor part of the economy, as compared with sale of coffee (Figure 18.4).

If we look closer at the monitoring data from Haia village (Figure 18.5), we see that the biological field research station was the main source of cash income in 1997. We also learned something about clan involvement in the enterprise. The large triangle indicates that the research station is located on the land of clans 9 and 10. The small triangle shows that a tourism guesthouse and airstrip is located on the land of clan 7. Scientists enter Haia village by plane, overnight at the guesthouse, via the ground of clan 7. They then cross the lands of clans 6 and 8 on their 8-hour walk to the research station. Although the research business is owned by the entire

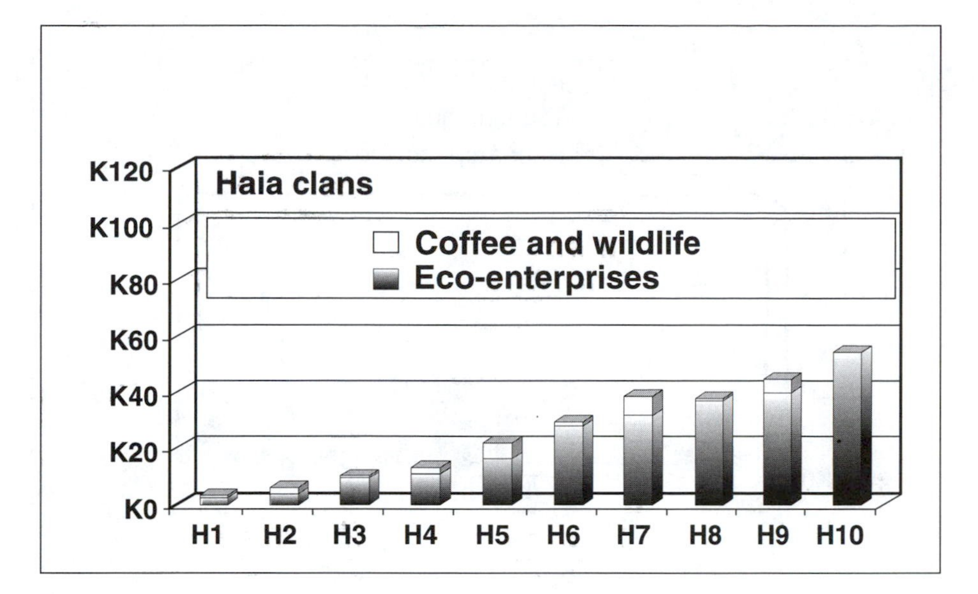

FIGURE 18.3 Annual cash income per capita per clan from ecoenterprises, wildlife, and coffee in Haia village in 1997. Individual clans are identified by the symbols H1 to H10. K is the symbol for the national currency, the kina. In 1997, one kina was equivalent to approximately 0.50 U.S. dollars.

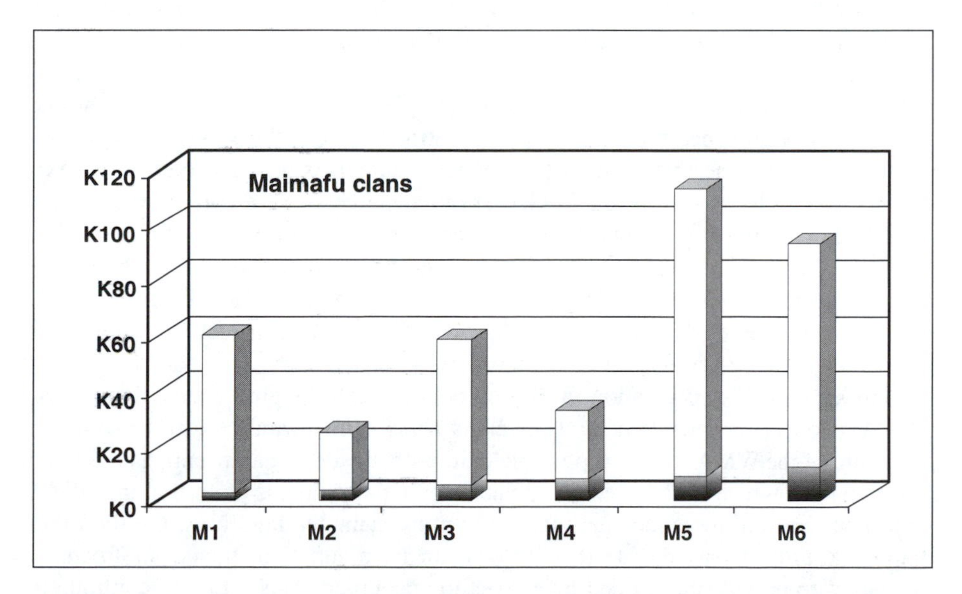

FIGURE 18.4 Annual cash income per capita per clan from ecoenterprises (gray) and coffee (white) in Maimafu village in 1997. Individual clans are identified by the symbols M1 to M6. K is the symbol for the national currency, the kina. In 1997, one Kina was equivalent to approximately 0.50 U.S. dollars.

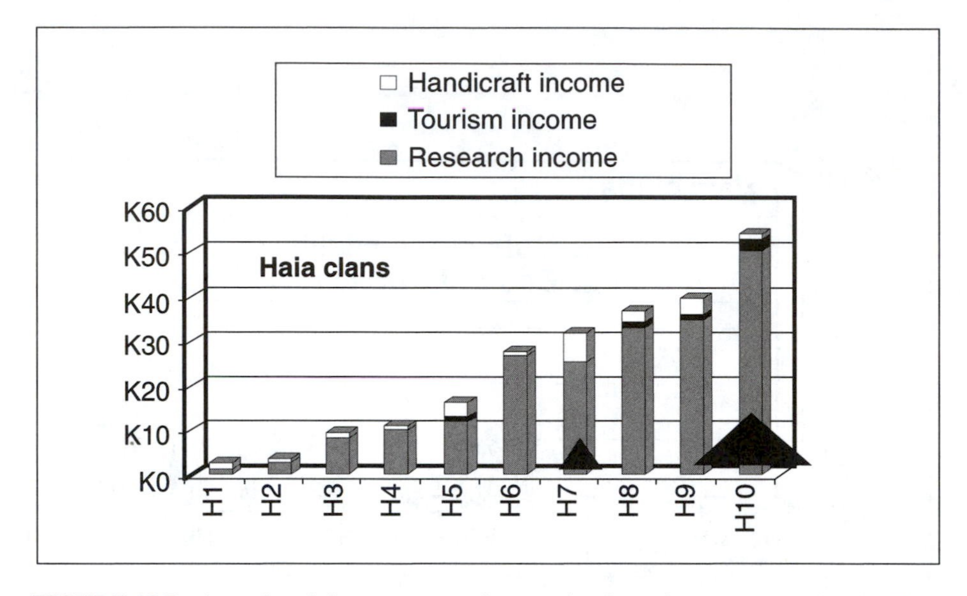

FIGURE 18.5 Annual cash income per capita per clan from three ecoenterprises in Haia village in 1997.

Management Committee of Haia, made up of all ten clans in the language group, economic monitoring has shown that it is primarily five clans (H6 to H10) that are currently benefiting the most. Because of traditional rivalry and sorcery, it appears that only the clans who are historical allies, or who have established marriage contact, have reasonably comfortable access to participation in employment at the research station.

What might these preliminary results tell about potential of these economic enterprises to influence biodiversity conservation? Figure 18.6 shows the approximate clan land ownership of the two villages. Haia clans 6 to 10 control approximately 90,000 ha of land in the WMA. If our hypothesis were correct, the project could be providing substantial economic incentives for conservation of one third of the WMA with the economic activity generated by the biological research station. If we want to provide similar incentives for other Haia landowners from clans 1 to 5, who own another third of the WMA, we may want to consider focusing other income-generating activities, like ecotourism, on their ground.

To support this speculation, a proposed large-scale logging concession (NFA, 1996) is now being considered by Haia clans 1 to 5, which would affect the southern one third of the WMA. It seems possible that, without additional incentives provided through increased alternative economic activity in the area, this portion of the WMA could be removed by landowners for other nonsustainable land uses. On the other hand, the project may decide that it does not have sufficient human or financial resources to provide the needed incentives to encourage clans 1 to 5 to continue to participate in the WMA. Perhaps the project should focus its limited resources on continued strengthening of the seemingly successful enterprise model that is operating on the lands of clans 6 to 10.

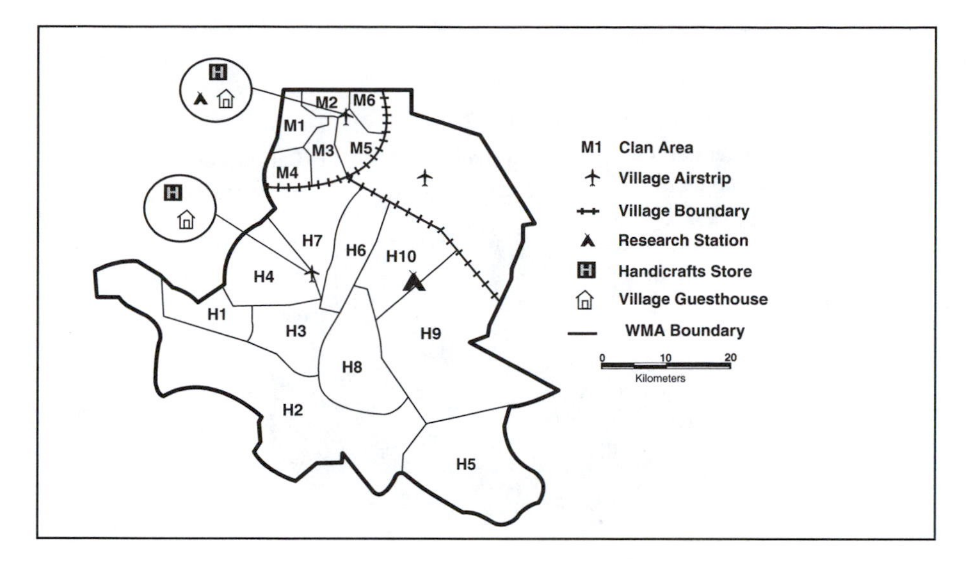

FIGURE 18.6 Map showing the present approximate land boundaries of clans M1 to M6, and H1 to H10, in the Crater Mountain WMA in relation to the location of ecoenterprises.

What will our monitoring eventually tell us about the initial hypothesis that a change in the use and abundance of natural resources may result as we observe a change in clan income from ecoenterprises? The results from the export and captive wildlife monitoring in the same two WMA villages, of Haia and Maimafu, over the last 9 months reveals that cassowaries (*Casuarius bennettii* and *C. casuarius*), large terrestrial birds, are the main species of wildlife that is being used (71%) (Figure 18.7). We found that almost 90% of these cassowaries captured were juveniles, which is not surprising since the adult birds are considered very dangerous for humans to handle. The other 29% of wildlife used during the time of this monitoring included an assortment of large mammals of the families Macropodidae (tree kangaroos) and Phalangeridae (cuscus), colorful parrots and cockatoos (family Psittacidae), and a hornbill (*Rhyticeros plicatus*). Landowners report that this current harvest of cassowary chicks is from clan-designated hunting areas within the lands they control in the WMA. Since little land mapping has been done, it is not clear if these designated nonhunting areas, which are set aside by each clan, are sufficient in size to serve as a viable population source to sustain this rate of harvest.

In response to the captive and export monitoring, we held a workshop with WMA management committees to conduct trend mapping (Ford et al., 1998) of the exploited species. In the exercise, residents described and charted how the availability of hunted wildlife had changed since the time of their grandparents. Workshop participants concluded that the hunting of cassowary, some tree kangaroos, and cuscus in most areas of the WMA is now nonsustainable. Residents thought that this was largely due to increasing human populations, a phenomenon that they said they had not had to manage for in previous generations. The most abundant remaining populations of the exploited wildlife were reported on the lands of the lowland Haia

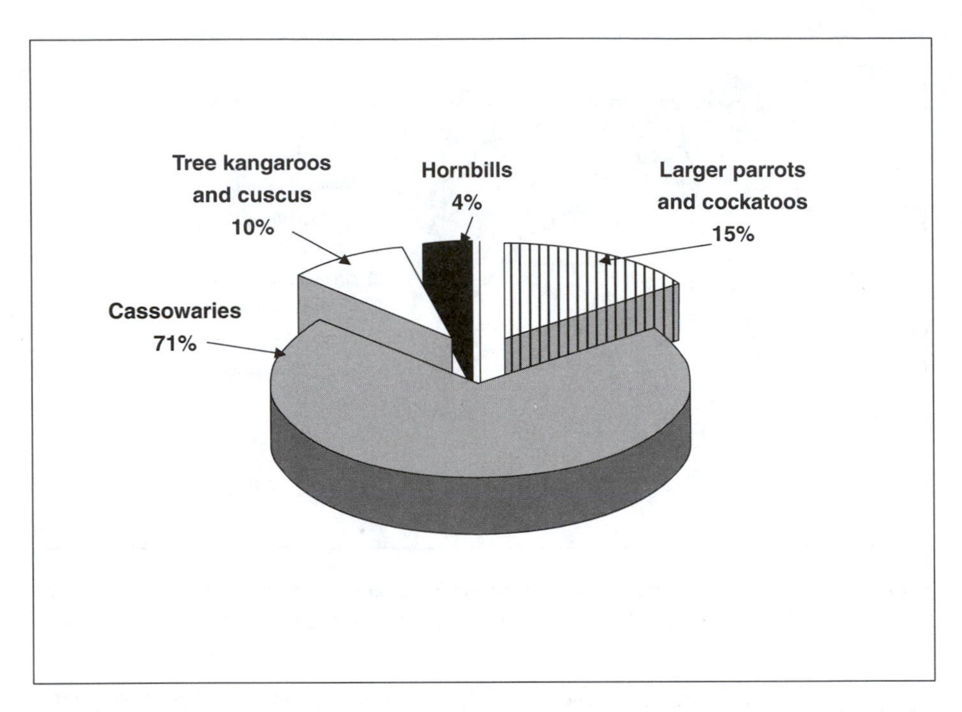

FIGURE 18.7 Exported and captive wildlife from the villages of Haia and Maimafu, in the Crater Mountain WMA, over a 9-month period from 1997 to 1998 (n = 104).

clans in the south half of the WMA where human population is still low in number, less than one individual per square kilometer (RCF, 1998).

Who is capturing and exporting this wildlife? Has the monitoring revealed any correlation between benefits from ecoenterprises and wildlife use? Figure 18.8 shows the percentage of the total Haia village wildlife used by each clan over 9 months as compared with their average ecoincome in 1997. This baseline shows a slight correlation between these two variables with the exception of clans 6 to 10. These two clans are benefiting from income received at the biological research station, but are using the same percentage of wildlife as other Haia clans, which are receiving much less.

There may be many reasons for this. One possibility may be the result of traditional practices. From the monitoring of village weddings and bride price paid, we know that clans 6 and 9 experienced a higher number of marriages than other clans in the last year. Exchanges not just of cash but also of wildlife, usually cassowaries, are very important in these ceremonies. This practice of wildlife use will likely not be changed by simply increasing cash income.

CONCLUSIONS

In response to preliminary monitoring results from 1997 and 1998, the project team may modify the focus of project activities and monitoring to include the following:

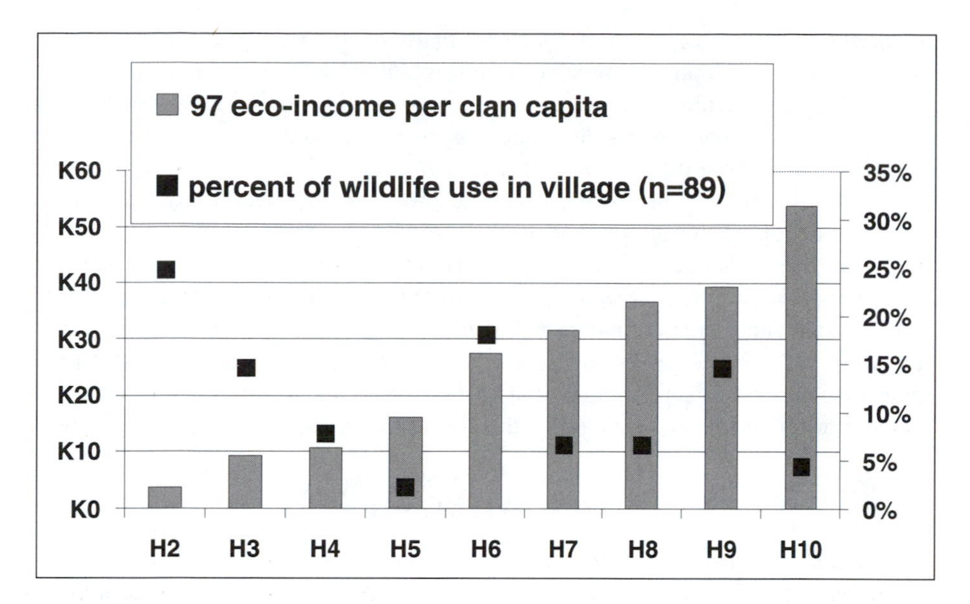

FIGURE 18.8 Average cash income per capita, per clan, in Haia village, as compared with the percentage of the total Haia village wildlife used by each clan over the last 9 months.

- To conduct further participatory planning with WMA communities to design and test traditional, and nontraditional methods, of regulating wildlife harvest that may be applied to the overexploited species identified by project monitoring in 1997;
- To conduct preliminary Global Positioning System (GPS) mapping of hunting and nonhunting zones on each clan's land to gauge the feasibility of present land use to provide for conservation of exploited species;
- To place an emphasis on monitoring the village consumption of wildlife for food, in addition to captive and export animals, to understand better the full extent of wildlife use in the WMA;
- To focus existing wildlife transects on the monitoring of heavily utilized wild cassowary populations;
- To focus community conservation education on discussions about the linkages between natural resource use and the viability of both WMA ecoenterprises and traditional subsistence livelihoods;
- To analyze available monitoring data collected on clan spending of cash income to understand change in WMA consumption as potentially related to change in ecoenterprise activity;
- Potentially to test training methods with interested families on the topic of household budgeting of cash income to meet cash needs.

This chapter has outlined the steps taken by an ICDP project team in developing an interdisciplinary monitoring program that is being used to test the hypothesis that implementation of selected socioeconomic development objectives will result in biodiversity conservation. The team found that, when working on a complex conservation and development initiative, a conceptual model was an essential tool for uniting an interdisciplinary, multinational project team in ongoing focused discussion and planning. Without the model, there was an increased potential to "get lost" or lose consensus about the steps to be taken next in the complex process. Although it is much too early for monitoring results to prove or disprove the hypothesis, the implementation of the monitoring program provided a framework from which to begin to assess change across a range of socioeconomic and biological variables that are operating in the protected area landscape. Baseline monitoring results have provided us with increased understanding of site conditions as well as the responses of WMA residents to project activities. The results have caused us to ask further questions, refine project activities, and focus monitoring methods, from within the context of the road map provided by the conceptual model.

ACKNOWLEDGMENTS

The authors acknowledge all their colleagues on the Crater Mountain Project management team. The Crater Mountain ICDP is a joint project between the Research and Conservation Foundation of Papua New Guinea and NYZS The Wildlife Conservation Society. The project is supported by the Biodiversity Support Program, a consortium of the World Wildlife Fund, the Nature Conservancy, and the World Resources Institute, with funding from the U.S. Agency for International Development.

REFERENCES

Beehler, B., Ed., 1993. A Biodiversity Analysis for Papua New Guinea, Papua New Guinea Conservation Needs Assessment, Volume 2, Biodiversity Support Program, Washington, D.C.

Brandon, K. E. and M. Wells, 1992. Planning for people and parks: designing dilemmas, *World Dev.,* 20(4), 557–570.

Department of Environment and Conservation, 1993. A biodiversity analysis for Papua New Guinea, in *Papua New Guinea Conservation Needs Assessment,* Vol. 2, Beehler, B., Ed., Biodiversity Support Program, Washington, D.C.

Ericho, J., Bino, R., and Johnson, A., 1999. Testing the effectiveness of using a conceptual model to design projects and monitoring plans for the Crater Mountain Wildlife Management Area, Papua New Guinea, in *Measuring Conservation Impact: An Interdisciplinary Approach to Project Monitoring and Evaluation,* Saterson, K., Margolius, R., and Salafsky, N., Eds., Biodiversity Support Program, Washington, D.C.

Ford, R., Lelo, F., and Rabarison, H., 1998. *Linking Governance and Effective Resource Management: A Guidebook for Community-Based Monitoring and Evaluation,* Center for Community-Based Development, Clark University.

James S., 1995. *The Crater Mountain Wildlife Management Area: Recommendations for Developing a Natural Resources Management Plan,* The Wildlife Conservation Society, New York.

Kramer, R., van Schaik, C., and Johnson, J., 1997. *Last Stand: Protected Areas and the Defense of Tropical Biodiversity,* Oxford University Press, New York.

Margoluis, R. and Salafsky, N., 1998. *Measures of Success: Designing, Managing and Monitoring Conservation and Development Projects,* Island Press, Washington, D.C.

NFA (National Forest Authority), 1996. *The National Forest Plan for Papua New Guinea,* May, Papua New Guinea Forest Authority, Port Moresby.

RCF (Research and Conservation Foundation) of Papua New Guinea, 1998. Minutes of 1998 Crater Mountain Wildlife Management Area Annual Landowners Meeting, July 22–24, Maimafu Village, Crater Mountain Wildlife Management Area, Papua New Guinea.

Western, D. and Wright, R. M., 1994. The background to community-based conservation, in *Natural Connections: Perspectives in Community-Based Conservation,* Western, D. and Wright, R. M., Eds., Island Press, Washington, D.C., 403–427.

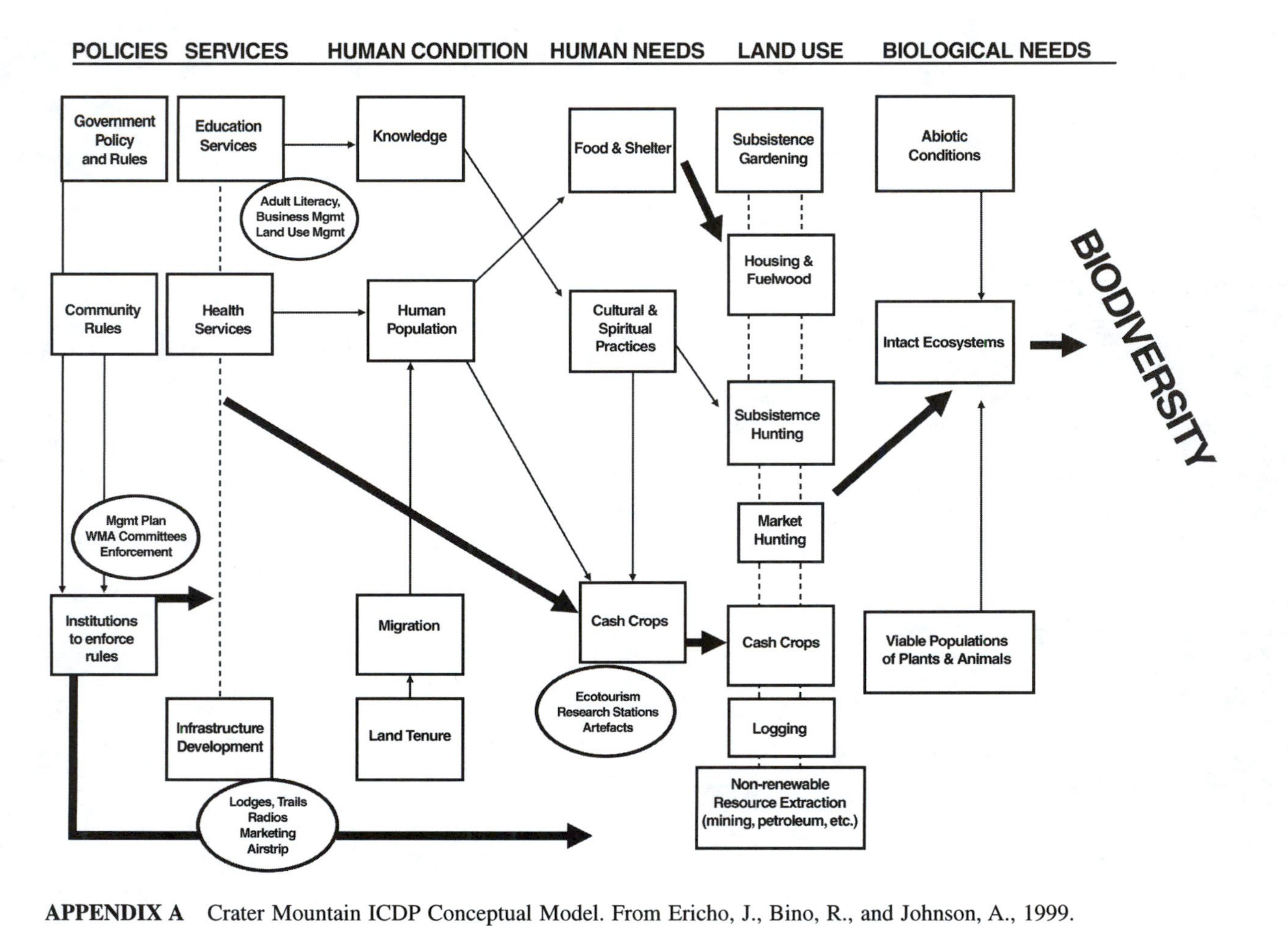

APPENDIX A Crater Mountain ICDP Conceptual Model. From Ericho, J., Bino, R., and Johnson, A., 1999.

APPENDIX B

Crater Mountain Wildlife Management Area Monitoring Plan

What is Monitored? (Indicator)	Why? (Indicator of…)	How? (Method)	When? (Frequency of data collection)	Where?	Who? (Staff serving as trainer of TLOs)	
					Staff	TLO (Trained Local Observer)
Bird species richness	Goal	Flyover counts	One morning and one afternoon per week	Airstrip	Biologist	Village resident
Bird species richness	Goal	Transect point counts	10 days in each of the months of February, June, and October	Three line transects near the village and three transects away from village	Biologist	Village resident
Consumption of wildlife	Goal	School diet survey	Last Friday of every month	Community school classrooms	Biologist	Community school teacher
Wildlife exported on planes	Goal	Plane survey	Each plane	Village airstrip	Biologist	Village airline agent
Captive wildlife in village	Goal	Survey checklist	Every 2 months	Village house lines	Biologist	Village resident
Wildlife signs	Goal	Line transects	10 days in each of the months of February, June, and October	Three line transects near the village and three transects away from village	Biologist	Village resident
Megafauna occurrences	Goal	Incidental observations	Ongoing	Field staff houses, research station, and guesthouses	All (residents, staff and WMA visitors)	
Plant use in handicraft production	Goal	Production records	Ongoing	Village handicrafts business	Business development officer	Village business manager
Area of cultivated land	Goal	Aerial photo flyover	Once every 3 years	Each village	Monitoring program coordinator	—
Size and location of clan conservation areas	Goal	GPS/description	Ongoing	Each clan	Biologist	Clan leader

APPENDIX B (continued)

Crater Mountain Wildlife Management Area Monitoring Plan

What is Monitored? (Indicator)	Why? (Indicator of…)	How? (Method)	When? (Frequency of data collection)	Where?	Who? (Staff serving as trainer of TLOs) Staff	TLO (Trained Local Observer)
Frog species richness	Goal	Visual encounter transects and leaf litter plots	One sampling every month	Wara Sera Research Station	Resident scientist	WMA resident(s)
No. of WMA businesses	Obj. 1	Business register	December of each year	Each village	Business development officer	Village business manager
Business income	Obj. 1	Income statement	End of each month	With each ecoenterprise committee (research, tourism, and handicrafts)	Business development officer	Village business manager
Business liabilities/assets	Obj. 1	Balance sheet	December of each year	With each ecoenterprise committee (research, tourism, and handicrafts)	Business development officer	Village business manager
Trade store sales by clan	Obj. 1	Sales records	Each day that trade store is open	Village trade stores	Business development officer	Trade store manager
Village market sales	Obj.1	Market survey	One market day each week	Village market	Business development officer	Village resident
Cash crop sales	Obj. 1	Freight records	Each plane	Village airstrip	Business development officer	Village business managers
Bride price	Obj. 1	Survey form	At each marriage	Village	Staff	Village resident
Handicraft sales	Obj. 1	Receipt books/activity records	Ongoing	Village handicrafts business	Business development officer	Handicrafts business managers
Research station sales	Obj. 1	Receipt books/activity records	Ongoing	Research stations	Business development officer	Research station managers

Tourism sales	Obj. 1	Receipt books/activity records	Ongoing	Village guesthouses	Business development officer	Guesthouse managers
Skills, education, and training	Obj. 2	Training records (event description, particapants, skills taught, levels achieved)	Each event	All training activities: natural resource management, monitoring, conservation education, business management, leadership, etc.)	All	—
Visitor feedback	Obj. 2	WMA visitor questionnaire	End of visit by tourist or researchers	Guesthouses and research stations	All	Business managers
Village committee capacity and action	Obj. 3	Minute keeping	Each meeting	Management and business committee meetings	All	Committee members
WMA publications	Obj. 3	Publications list	Ongoing	ICDP office	Monitoring program coordinator	—
Business practices standard	Obj. 3	Checklist	December of each year	With each ecoenterprise committee (research, tourism, and handicrafts)	Business development officer	Village business managers
Papua New Guinea teachers and trainers from within, and outside, the WMA, and the ICDP	Obj. 4	Training records	Each event	All training activities: natural resource management, monitoring, conservation education, business management, leadership, etc.	All	—

19 Integrating Biological Research and Land Use Practices in Monteverde, Costa Rica

Carlos F. Guindon, Celia A. Harvey, and Guillermo Vargas

CONTENTS

INTRODUCTION

Costa Rica has become known for its efforts to conserve its natural resources and has set aside 18% of its land area in national parks and reserves (Fogden and Fogden, 1997). A less-publicized, but increasingly important, aspect of conservation in Costa Rica is private reserves (Bien, 1997). The largest private reserve in the country (18,000 ha) is the Bosque Eterno de Los Niños (BEN), owned and managed by the Monteverde Conservation League (MCL), a local organization dedicated to the conservation of biodiversity within the Monteverde region. BEN forms part of a

0-8493-0020-7/01/$0.00+$1.50

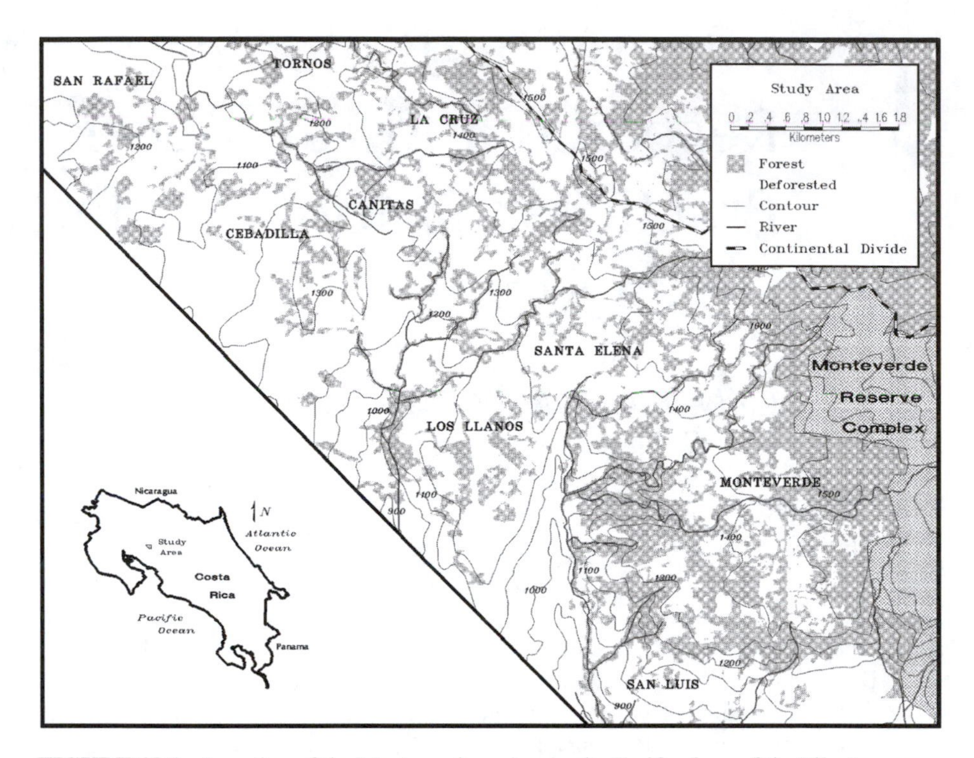

FIGURE 19.1 Location of the Monteverde region on the Pacific slope of the Tilarán mountain range in northwestern Costa Rica showing the communities adjacent to the reserve complex that participated in environmental education and reforestation activities. Forest cover is based on 1992 aerial photographs taken prior to the establishment of most of the planted windbreaks that now cover another 500 ha of the deforested area shown in this figure.

reserve complex that covers most of the upper slopes of the Tilarán mountain range and includes the private Monteverde Cloud Forest Biological Reserve (10,500 ha), administered by the Tropical Science Center; the Arenal National Park (12,123 ha); and a small reserve (310 ha) administered by a local public high school. This reserve complex provides a fascinating example of the challenges of integrating local, regional, national, and international visions of conserving and managing biological resources with many opportunities for applying adaptive collaborative strategies that attempt to fulfill conservation goals while addressing the needs of local communities.

Like most montane protected areas in Costa Rica, the Monteverde reserve complex primarily includes forest above 1500 m (Powell et al., 1995). Below this elevation lies an agricultural landscape with increasingly fragmented and reduced forest remnants within a matrix of pastures for dairy cows, and small plantations of coffee, bananas, and sugarcane (Figure 19.1). Primarily as a result of growth in ecotourism, generated by the presence of the reserve complex, many farms near the towns of Santa Elena and Monteverde are now being divided into lots for residential areas or tourist facilities. This change in land use tends to fragment and reduce forest habitats further with roads and power line rights of way.

Similar to other protected areas (Noss and Harris, 1986; Schonewald-Cox et al., 1992), as the Monteverde reserve complex becomes increasingly isolated due to the deforestation and fragmentation of the surrounding habitats, there is an ever greater need to manage the landscape to enhance the survival of regional flora and fauna (Powell and Bjork, 1994; Guindon, 1996). However, Monteverde differs from other protected areas in two significant ways: (1) most of the protected land exists within locally managed, private reserves; and (2) it has an astounding amount of biological and ecological information, generated by field research projects conducted within and around the reserve, that can be used to provide guidelines for conservation efforts (Nadkarni and Wheelwright, 2000). As a result of both local participation in land protection as well as a significant research base, the Monteverde region offers a unique opportunity to explore the challenges of integrating biological information into land-use practices. This chapter discusses briefly the importance of conserving forest fragments, remnant trees, and other natural habitats in the agricultural landscape for the conservation of regional biodiversity. Next, it highlights how biological information has been integrated into land-use practices through an adaptive and collaborative conservation strategy implemented by the MCL that includes environmental education, reforestation, and forest fragment protection. Finally, the chapter discusses the limitations and benefits of this approach to biodiversity conservation and highlights conclusions that have been drawn from the authors' experience in the Monteverde region.

IMPORTANCE OF MANAGING THE LANDSCAPE FOR THE CONSERVATION OF REGIONAL BIODIVERSITY

When the Monteverde region was settled during the 1940s and 1950s most of the forest was cleared to establish pastures for dairy farming (Burlingame, 2000). Smaller areas were set aside for coffee, bananas, and sugarcane. Since strong winds prevail for much of the year due to the trade winds passing over the continental divide (Clark et al., 2000), many areas of forest were left for wind protection. Other forest fragments were left for water protection, the provision of timber for construction, firewood, and fence posts, or on slopes too steep for farming (Guindon, 1988). Most farms are small (<50 ha) and have forest fragments that are less than 2 ha in size, although on some farms forest fragments reach 8 ha (Guindon, 1997). In addition to leaving forest fragments, most farmers also left remnant trees in their agricultural areas to provide shade for their cattle, timber, and other materials (Harvey and Haber, 1999). Today, most dairy pastures in the Monteverde region are dotted with a combination of relict trees from the original forest and pioneer trees that have recently colonized the pastures (Harvey and Haber, 1999).

Forest fragments, isolated pasture trees, and windbreaks are now recognized as important to maintaining the biological diversity of the region (Harvey et al., 2000). Independent research conducted since the 1970s has revealed that the Monteverde region hosts an incredible diversity of plant and animal life (Nadkarni and Wheelwright, 2000). Over 3000 species of vascular plants, including more than 700 species of trees (Haber, 2000); 658 species of butterflies (Stevenson and Haber, 2000); 161 species of amphibians and reptiles (Pounds, 2000); 425 species of birds (Young and McDonald, 2000); and 121 species of mammals (Timm and LaVal, 2000) have been recorded for the area. At least 10% of the plant species are endemic to the Tilarán mountain range (Haber, 2000) and 40% of the amphibians are endemic to the uplands of Costa Rica and northwestern Panama (Pounds, 2000).

A key component of the Monteverde landscape is its steep elevational gradient. This gradient results in a rapid change in species composition, particularly of plants (Haber et al., 1996; Guindon, 2000), and drives the seasonal movements of many vertebrate and invertebrate species that provide for the seed dispersal and pollination of many of the plant species. Of the 658 species of butterflies found at Monteverde, 360 are altitudinal migrants (Stevenson and Haber, 2000), as are 68 of the 425 bird species (Young and McDonald, 2000). Many of these migrating species depend on forest fragments, remnant trees, and other relict habitats in the agricultural landscape outside the reserve complex for resources, and the presence of these landscape features may partially mitigate the negative impact of fragmentation and habitat loss (Guindon, 1996; Stevenson and Haber, 2000; Harvey et al., 2000). If these forest habitats are reduced and fragmented further, plant and animal species may be lost and the dynamics of plant–animal interactions disrupted.

The Importance of Forest Fragments

Forest fragments in the Monteverde region have been found to contain a high diversity of trees: in a 10% sample of 30 forest fragments Guindon (1997) found 59 tree families, 130 genera, and 225 species, representing 31% of the tree diversity of the entire Tilarán mountain range above 700 m (Haber, 2000). Forest fragment size was the best predictor of tree species, genera, and family richness, with larger fragments containing greater diversity. After controlling for size, a significantly greater number of tree families, genera, and species were found within fragments at higher elevations. Although these data suggest that the protection of large forest fragments in the higher elevations should be a priority, forest fragments in lower elevations are also important because they contain a different set of species and provide resources, such as fruit and nectar, during different times of the year (Haber, 2000; Wheelwright, 2000). The conservation of trees and vertebrates may be mutually dependent on each other as nearly nine out of every ten of the 225 tree species found in the forest fragments produced vertebrate dispersed seeds with 66 percent primarily dispersed by birds and 21 percent by mammals (Guindon, 2000).

The forest fragments appear to play a critical role in the conservation of altitudinal migrant birds in the region, including some of the most spectacular species, such as the resplendent quetzal (*Pharomachrus mocinno*) and three-wattled Bellbird (*Procnias tricarunculata*), which migrate seasonally into the forest fragments to feed on fruit during times of the year when there is a scarcity of fruit within the reserve complex (Guindon, 1996; Powell et al., 2000). The use of forest fragments by these migrating species varies, depending in part on their ability or willingness to cross open space. For example, three-wattled bellbirds preferred forest fragments with high Lauraceae fruit availability and crossed open spaces to reach the forest fragments with the most Lauraceae fruit, even when these were small and isolated from the continuous forest of the reserve complex (Guindon, 1996). In contrast, resplendent quetzals and black guans (*Chamaepetes unicolor*), which also seasonally enter forest fragments to feed on Lauraceae fruit, were less willing to cross open space, favoring the larger forest fragments close to the continuous forest even when more fruit was available in smaller, more-isolated fragments. Further loss, degradation, and isolation of forest fragments could not only decrease the diversity and abundance of altitudinal migrants, but decrease the seed dispersal and regeneration of many plant species. It could also negatively impact the local tourist economy, which depends to a large extent on people coming to observe spectacular birds like the quetzal and bellbird (Powell and Bjork, 1994).

The Importance of Remnant Trees

Another landscape feature that plays an important role in the conservation of biodiversity is the presence of scattered trees in pastures. The diversity and densities of remnant trees within pastures surrounding the Monteverde reserve complex is surprisingly high. In a survey of 237 ha of pasture, a total of 5583 trees of 190 tree species (mean density of 25 trees/ha) were found (Harvey and Haber, 1999). The 190 tree species represented approximately 60% of the tree species occurring in the study area. Many of these trees were primary forest species, accounting for 33% of the total stems and 58% of the species.

In addition to representing an important component of the biodiversity of the region, remnant trees greatly enhance the vegetational and structural complexity of the farms and provide habitats and resources for other species. Over 94% of the trees found in pastures at Monteverde provide fruits for birds, bats, or other animals and many of the most common remnant trees (e.g., *Acnistus arborescens*, *Sapium glandulosum*) provide fruits for a large number of bird species (Harvey and Haber, 1999). Pasture trees may be particularly important food sources for animals because isolated trees often produce larger or more frequent fruit crops than their forest conspecifics. In addition, these trees may serve as "stepping-stones" that facilitate the movement of birds within the agricultural landscape, thereby partially alleviating the impacts of habitat fragmentation and deforestation (Date et al., 1991; Guevara et al., 1998). In addition, the remnant trees attract frugivorous birds, bats, and other animals into pastures, thereby increasing the dispersal of tree seeds into the pastures and facilitating forest regeneration if pastures are abandoned (Harvey, 1999).

Although the density and species richness of trees in pastures in Monteverde is fairly high, there is evidence that this diversity will decrease in future years unless concerted efforts are made (Harvey and Haber, 1999). Many of the primary forest trees show no signs of regeneration within the pastures. This lack of primary forest saplings does not appear to be due to the inability of trees to produce seeds or to the failure of seeds to germinate; instead, it appears to be due to continual grazing by cows and the weeding of pastures by farmers. In addition, many of the primary forest trees occur in low densities, with 62% of the trees being represented by ten individuals or fewer. These low densities make it easy to eliminate entire species from the pastures by simply cutting a few trees. As a consequence, efforts to conserve biodiversity within the region should include the maintenance of trees within pastures, as well as the protection of forest fragments.

A STRATEGY TO INTEGRATE BIOLOGICAL INFORMATION INTO LAND-USE PRACTICES

HISTORICAL BACKGROUND

The Monteverde region has undergone a radical transformation during the last 30 years from an agricultural frontier to a world-renowned conservation area and ecotourism site. It has also witnessed the evolution of top-down, externally driven conservation strategies to strategies that are more collaborative and responsive to local landowner needs. In the late 1960s, biologists George and Harriet Powell were the first to see the danger of the region losing its wealth of biological resources as the agricultural frontier advanced up both the Pacific and Atlantic slopes (Burlingame, 2000). In 1972, with the support of only a few local residents and the Tropical Science Center located in the country's capital, they created the Monteverde Cloud Forest Reserve. At this time, most local landowners were intent on developing their farms and advancing the agricultural frontier, leaving forest only on steep slopes, around springs, or for wind protection. Although the community of Monteverde was an exception, setting aside 554 ha of forest on the top of the continental divide to protect the headwaters of its principal water source, they were not concerned about protecting the biological diversity of the region.

Until the late 1980s the majority of local farmers did not understand or value the creation of the Monteverde Cloud Forest Biological Reserve. However, as people observed how researchers and tourists came from all over the world to admire the beauty and wonders of the cloud forest, many became more interested in learning about these resources. Tourism began generating new jobs and activities for local people and many of these activities, such as guiding and the production of arts and crafts related to nature, required learning about biodiversity and its conservation.

Growing local interest in the conservation of the biological resources of the region, combined with concern about the continued degradation of Pacific slope habitat outside the Monteverde Cloud Forest Reserve, resulted in the founding of the MCL in 1986 by a group of resident biologists and local farmers (Burlingame, 2000). The goals of the organization included the conservation, protection, and recovery of natural resources; the improvement and protection of the physical, biotic, and cultural environment; and the search for a healthy balance between humans and nature (Burlingame, 2000). To achieve these goals, the MCL devoted resources not only to the purchase of forested land (creating BEN), but also to environmental education, reforestation, and the protection of forest fragments in communities adjacent to the reserve complex.

ENVIRONMENTAL EDUCATION

Backed by the directives of its statutes, the MCL environmental education program took a broad approach, including cultural values and social issues, and targeted a diverse audience of school children, teachers, and landowners (Burlingame, 2000). One key component of its education approach was to assist local communities identify their principal environmental concerns, rather than impose the concerns of the MCL. As community concerns were addressed, such as waste management and agricultural chemicals, a dialogue was opened and other issues could be discussed such as the importance of native habitats for the maintenance of the biological diversity of the region.

A second key component of the education program was its emphasis on collaboration with schools, cooperatives, milk producers, and other organizations, creating opportunities to share concerns about environmental issues. This approach resulted in shared activities, such as the school environmental award competition, which evolved into the annual celebration of the "ecological day," the production of organic fertilizer from coffee hulls, the establishment of windbreaks, and a recycling program. The collaborative approach and efforts of the environmental education program increased community awareness about conservation issues and provided communities with the support necessary for the successful establishment, expansion, and evolution of activities such as reforestation.

Reforestation with Windbreaks

The MCL reforestation program started small with the help of volunteers who experimented with planting a few native tree species in windbreaks and degraded pastures. It soon expanded into a large-scale project as funding became available in 1988 (Burlingame, 2000). Although the project proposal incorporated existing biological information on forest fragments and altitudinal migrants, proposing the establishment of corridors by reforesting degraded Pacific slope habitat, it did not contain a clear strategy on how to gain landowner support for implementing these land-use changes. Using the support and collaborative approach of the environmental education program, farmers were consulted about their reforestation needs and presented with the options for receiving financial assistance to subsidize the cost of planting trees. During this process, it became evident that farmers were not interested in reforesting large areas with plantations to establish corridors. Instead, their first priority was to establish windbreaks to protect their cattle from the strong trade winds. Although windbreaks could potentially provide some habitat for wildlife, or serve as corridors, they were clearly of less conservation value than larger-scale reforestation plots because of their small size and narrow width. However, the MCL program agreed to adapt their reforestation program to incorporate the establishment of windbreaks as a first step to involving farmers in reforestation.

The next challenge was to convince government officials to allow the use of government subsidies for planting narrow strips of trees in windbreaks using native species rather than in one-hectare blocks using the exotic species required by the government incentives program. After discussions with the MCL, government officials agreed to the planting of trees in narrow strips to form windbreaks but did not want to risk planting untested native species. Landowners were also hesitant to incorporate native species as exotics such as eucalyptus (*Eucalyptus diglupta*), pine (*Pinus caribaea*), and cypress (*Cupressus lusitanica*) had already been planted locally and accepted as species to plant in windbreaks or around buildings. Many members of the MCL were very skeptical about the value of planting windbreaks with exotic tree species. However, the directors of the environmental education program and the reforestation program were convinced of the importance of taking into account the interests of the landowners even if it meant potentially sacrificing the biological value of the project. As a result, the reforestation program established windbreaks consisting primarily of exotic species such as cypress, whistling pine (*Casuarina equisetifolia*), rose apple (*Syzygium jambos*), and loquat (*Eriobotrya japonica*).

The reforestation program was very effective in obtaining landowner participation as it addressed a landowner need resulting in the planting of over 1000 windbreaks covering a total of 500 ha. These windbreaks generally consisted of two to four rows of trees, and averaged 6 m in width and 60 m in length (Harvey, 1999). Areas around springs were also fenced and reforested in three communities concerned about the protection of their water sources.

Landowners were given direct input into the development and execution of the reforestation program through representation on the reforestation committee. The establishment of an annual reforestation day was also very effective in stimulating and maintaining landowner interest, as it provided recognition to landowners with the best projects. The annual event took place in different communities each year and was conducted in a festive atmosphere that respected and promoted local cultural practices. Landowner knowledge and experience were given value by consulting with them on how and when to plant trees, where to place windbreaks, how to establish living fence posts, and what native tree species to use for seed sources. Field days allowed farmers to learn about projects on other farms and exchange information.

An unexpected obstacle that the MCL encountered in the reforestation program was that the large demand for fence posts to fence the windbreaks against cattle entry resulted in farmers cutting native tree species from their forest fragments. To reduce this impact, various alternatives to the use of split wooden posts were discussed and incorporated into some of the later windbreaks. These included the use of electric fences with metal stakes with insulators, living fence posts, and concrete posts.

As the reforestation program evolved and expanded (1989 to 1995), seeds from more and more native species were collected and tested in the nursery for their germination and growth. One native species with a very localized distribution, "tubú" (*Montanoa guatemalensis*), showed great promise and was soon incorporated into most of the windbreaks in the area. It grew quickly even on fairly poor soils providing a windbreak for other, slower-growing species. Because local landowners were already familiar with it as a species that could provide good long-lasting fence posts, they readily adopted it as a windbreak species. Over 40 other native species were tested, but were incorporated into the windbreaks on a smaller scale, as they did not show as much promise for the quick establishment of wind cover or timber production.

Protection of Key Forest Fragments and Biological Corridors

When funding for the reforestation project ended in 1995, additional funding was obtained to implement the protection of forests on farms and the establishment of biological corridors, a project known as "bosques en fincas." This project broadened the geographical scope of the previous reforestation program to include farms and communities bordering the reserve complex on the Atlantic slope, as well as farms on the Pacific slope that had previously been involved in the reforestation program.

The experience developed over 5 years of establishing windbreaks under the reforestation program and participating in environmental education program activities was critical in preparing landowners to take this next step of protecting their forest fragments and reforesting and fencing gaps in proposed biological corridors (C. Guindon, personal observation). Research had also provided further evidence of the importance of forest fragments and windbreak corridors to the maintenance of biological diversity on the Pacific slope (see discussion of the research of Guindon; Nielsen and DeRosier; Harvey; and Harvey et al.). To achieve its goals, the bosques en fincas project reforested with native species and fenced forest fragments (using concrete and living fence posts) to prevent their degradation by cattle.

The bosques en fincas project was unique in that it integrated the collection of biological data with education and extension by using a team of biologists, extensionists, and environmental educators. This team approach proved to be very effective in creating dialogue and collaboration both within the MCL and outside the organization (C. Guindon, personal observation). Including biologists on the team assured the quick transfer of the information generated on birds, butterflies, and trees within forest fragments to educators and extensionists who could then transfer it to landowners, schoolchildren, and community organizations. Further collaboration was created by having researchers participate in some of the education and extension activities, such as visiting schools and community groups or working in the tree nurseries, while educators and extensionists, as well as some farmers, accompanied researchers in the field.

Another important adaptation of this project was that instead of the MCL owning and operating the tree nurseries, two local women's groups were trained in the establishment and operation of tree nurseries. These women and their families increased their incomes while gaining administrative skills that allowed them to maintain the two tree nurseries even after the bosques en fincas project ended. Additionally, they learned about the fruiting and flowering of native trees as well the collection, germination, and planting of the seeds. They also learned about tree distributions and abundance and their natural pollinators, seed dispersers, and habitat requirements.

BENEFITS OF THE STRATEGY

The MCL approach of integrating environmental education, forestry extension, and field biology, with its responsiveness to landowner needs and its collaboration with other local organizations and institutions, has had a positive impact on the conservation of biodiversity within the region.

Environmental education and extension activities were effective in facilitating discussions among landowners, biologists, and extension agents on the regional importance of windbreaks, forest fragments, and forest corridors. It also allowed local families the opportunity to acquire a better understanding of the interactions between protected areas and their forest fragments, tree plantations, and agricultural land

(G. Vargas, personal observation). Perhaps most importantly, these discussions and activities provided farmers and their families a chance to participate in the decision-making process related to the conservation and management of regional biodiversity.

The reforestation program greatly expanded landowner interest and knowledge in tree planting. Many landowners became more interested in planting native trees, as they were able to directly compare the growth of native species with the growth of exotics in the plantations. Landowners also became more interested in native species as a result of learning about their function of providing fruit for wildlife species such as the resplendent quetzal. Additionally, the reforestation program generated a vast amount of knowledge about the collection, germination, and establishment of native trees, which continues to be used locally as well as at other sites within Costa Rica (O. Varela, personal communication).

Landowners have benefited directly from windbreaks through increased milk production by reducing the physical stress on cattle from wind and by reducing pasture desiccation during the dry season. The windbreaks have also started to provide posts and poles for farm use, thereby reducing the extraction of these materials from forest fragments or remnant trees. Some farmers have also used the protection of the windbreaks to diversify their farm output by planting vegetable gardens and crops in the areas adjacent to the windbreaks.

In addition to their agronomic value, the windbreaks appear to be playing important roles as habitats and corridors for plant and animal species, despite consisting primarily of exotic species (Harvey, 1999; 2000). Research has shown that the windbreaks provide habitat for the regeneration of over 90 native tree species, including many primary forest species, thereby increasing the diversity of trees within the pasture landscape (Harvey, 1999). Another study by Nielsen and DeRosier (2000) found 68 species of birds using natural windbreaks and 52 species using planted windbreaks. Windbreaks that were directly connected to forest patches had a greater number of birds using them than windbreaks that were separated from forest fragments by at least 20 m. This is evidence that the windbreaks may serve as corridors that allow forest birds to move across the agricultural landscape. The preference of birds for windbreaks that are connected to forests was reflected in the fact that connected windbreaks had significantly higher densities and species richness of bird-dispersed tree seedlings than windbreaks that were isolated from forests (Harvey, 2000).

The establishment of protected forest fragments and biological corridors was able to achieve simultaneously goals of conservation and the development of sustainable practices of production on the farms with agroecotourism, windbreaks, water protection, fence posts, firewood, and lumber. At the same time, farmers were able to receive economic benefits to compensate them for the short-term reduction in income from land designated for reforestation or protection.

Landowners were made aware of the importance of forest protection and the establishment of biological corridors for the conservation of regional biodiversity, as well as for soil/water conservation. Knowledge of the cost and labor of planting trees increased farmers' appreciation for the services provided by seed dispersers such as quetzals, toucans, and bellbirds, and the negative impacts of cattle on the natural regeneration of desired timber species within their forest fragments.

LIMITATIONS OF THE STRATEGY

Although the strategy of providing environmental education to local people, promoting reforestation, and protecting forest fragments has had a positive impact on the conservation of biodiversity within the region, its potential impact has been limited by a lack of continued funding for conservation activities and the failure of projects to provide follow-up to conservation initiatives. Most of the MCL projects have been funded for only 2 to 4 years with limited options for continuing activities and achieving long-term goals. In many cases, the public and private institutions collaborating in the projects have not been oriented toward strengthening the decision-making capacity and self-initiative of communities. Consequently, when projects end, communities are often incapable of continuing or replicating the activities initiated by the projects. Finally, the projects, even when receiving government funding, have had a limited ability to establish legally binding commitments for conservation on privately owned farms. The contracts that have been established for windbreak and forest protection incentives have been for a relatively short-term period of 3 to 5 years and do not assure long-term protection of the biodiversity of the region.

The lack of continuation means that despite the keen interest in reforesting additional areas with windbreaks and in protecting additional forests from cattle, there is currently little funding to support these activities. In addition, the tree nurseries owned by the MCL that produced plants for reforestation efforts on the Pacific slope have been partially or completely closed, making it difficult for some farmers to obtain planting materials. As a result, there are few trees available for farmers interested in planting additional windbreaks or widening existing ones.

Another limitation of the conservation strategy has been the failure to provide ongoing technical assistance on the management of established windbreaks. Many of the windbreaks are now more than 10 years old, with trees that could be harvested for timber or fence posts. There is no clear idea, however, on how best to harvest trees without destroying the capacity of the windbreaks to provide wind protection. The lack of technical support has also led some farmers to neglect their windbreaks and let the fences fall into disrepair, resulting in these windbreaks becoming severely degraded by cattle and no longer fulfilling either conservation or farm productivity goals.

A final negative result of the way the current strategy has been implemented is that some farmers who participated in short-term projects feel "betrayed" by the lack of project continuation and become less supportive of the organization. These sentiments may arise in part from a lack of clearly communicating the time period of project funding and the ability of the organization(s) to provide continuity after the original funds are expended.

CONCLUSIONS

The author's research and that of others has indicated that habitat remnants on farms surrounding the Monteverde reserve complex are important to maintaining the biological diversity of the region (Harvey et al., 2000). The experience of the MCL and others with helping landowners and local communities incorporate this information into their land-use practices has been mixed. One successful aspect of this approach has been identifying the needs of farmers for windbreaks, fencing, and watershed protection and adapting short-term conservation objectives to channel resources to address these needs. Addressing farmer needs in ways that value and respect their knowledge and abilities led to the participation of most farmers in the reforestation program. Many of them subsequently supported regional conservation initiatives to create biological corridors and protect forest fragments.

However, accomplishments can be short term as landowners may not always value the biological resources on their property enough to maintain them without receiving incentives, properties may be subdivided and change hands, and organizations change programs and priorities. Tourism provides an example of how the solution to one problem can lead to the creation of others, which in turn require new solutions. Tourism has become a key factor in land-use changes around the Monteverde reserve complex. Many of the farmers with land close to the protected area who planted windbreaks, protected their forest, and established corridors are now subdividing their property and selling off house lots. These changes involve new stakeholders, landowners who are not farmers and have small house lots. Thus, new strategies and tools must be developed in collaboration with these landowners to minimize their impact on the regions biological resources. One new tool being applied in the Monteverde region is conservation easements. The easements are being used to assure the long-term maintenance of biological habitat, including biological corridors, while assisting landowners with neighborhood planning. They can be very slow and tedious to establish, but the process provides a forum for developing a regional perspective on how to integrate biological and social issues, and offer the chance to protect biodiversity over the long term.

Funding for the long-term restoration and protection of habitat outside of the reserve complex is important as short-term projects lack continuity. Assuring long-term support will require collaborating with new stakeholders and developing new strategies. The new stakeholders in tourism are beginning to support some conservation efforts, such as annual bird counts, a bellbird conservation project, and the annual ecological day activities. Local natural history guides are donating their time to produce, distribute, and plant seedlings of local tree species. The MCL has successfully negotiated an environmental service fee with a hydroelectric company in return for protecting the watershed that supplies the water for its generating plant. An increasing number of farmers are receiving environmental service fees from the government for protecting forests on their farms. Another local organization, the Monteverde Institute, with income generated through courses given in tropical biology, is supporting the easement program and facilitating the channeling of funds for a land trust.

The experience in Monteverde suggests that conservationists must be patient and willing to set aside short-term goals to take into account the needs of landowners. It also requires that biologists and landowners discuss their goals and concerns and seek solutions that simultaneously, to the best degree possible, address conservation and productivity objectives. Communication and information exchange is critical to this process. Once landowners know that they are being listened to, they are more willing to listen to alternatives. Organizations must be willing to modify their programs and projects to adapt to the ability and willingness of landowners to incorporate changes in their land-use practices and to take advantage of unexpected opportunities for conservation, if they are to be successful in conserving biodiversity within private farmlands.

Even the best-designed reserves will not incorporate the habitats and resource needs of all of the biological diversity of a region. As a result, it is imperative that people learn to manage resources in an adaptive and collaborative manner. A major step taken by the MCL in its reforestation, environmental education, and research programs has been their evolution into more integrated and collaborative experiments in the management of natural resources.

ACKNOWLEDGMENTS

The authors thank the many farmers who have participated in the projects and allowed for the research and extension opportunities described. Also thanked is the Monteverde Conservation League for its support of the projects, as well as the many agencies that funded these projects: The Canadian International Development Agency, World Wildlife–Canada, The National Fish and Wildlife Foundation, and The RARE Center for Tropical Research. Finally, the authors thank the Monteverde Institute and the Estación Biológica for providing logistical support and facilities for biological research in the area.

REFERENCES

Bien, A., 1997. *Boletín de la Red de Reservas Naturales*, Número 1, Febrero.

Burlingame, L. J., 2000. Conservation in the Monteverde zone, in *Monteverde: Ecology and Conservation of a Tropical Cloud Forest*, Nadkarni, N. M. and Wheelwright, N. T., Eds., Oxford University Press, New York, 351–375.

Clark, K. L., Lawton, R. O., and Butler, P. R., 2000. The physical environment, in *Monteverde: Ecology and Conservation of a Tropical Cloud Forest*, Nadkarni, N. M. and Wheelwright, N. T., Eds., Oxford University Press, New York, 15–33.

Date, E. M., Ford, H. A., and Recher, H. F., 1991. Frugivorous pigeons, stepping stones, and weeds in northern New South Wales, in *Nature Conservation 2: The Role of Corridors*, Saunders, D. A. and Hobbs, R. J., Eds., Surry Beatty & Sons, Chipping North, Australia, 241–245.

Fogden, M. and Fogden, P., 1997, *Wildlife of the National Parks and Reserves of Costa Rica,* Editorial Heliconia, Fundación Neotrópica, Costa Rica.

Guevara, S., Laborde, J., and Sanchez, G., 1998. Are isolated remnant trees in pastures a fragmented canopy? *Selbyana*, 19(1), 34–43.

Guindon, C. F., 1988. Protection of Habitat Critical to the Resplendent Quetzal (*Pharomachrus mocinno*) on Private Land Bordering the Monteverde Cloud Forest Reserve, M.S. thesis, Ball State University, Muncie, IN.

Guindon, C. F., 1996. The Importance of forest fragments to the maintenance of regional biodiversity in Costa Rica, in *Forest Patches in Tropical Landscapes*, Schelhas, J. and Greenberg, R., Eds., Island Press, Washington, D.C., 168–186.

Guindon, C. F., 1997, The Importance of Forest Fragments to the Maintenance of Regional Biodiversity Surrounding a Tropical Montane Reserve, Costa Rica, Doctoral dissertation, School of Forestry and Environmental Studies, Yale University, New Haven, CT.

Guindon, C. F., 2000. The importance of Pacific slope forest for maintaining regional biodiversity, in *Monteverde: Ecology and Conservation of a Tropical Cloud Forest*, Nadkarni, N. M. and Wheelwright, N. T., Eds., Oxford University Press, New York, 435–437.

Haber, W. A., 2000. Plants and vegetation, in *Monteverde: Ecology and Conservation of a Tropical Cloud Forest*, Nadkarni, N. M. and Wheelwright, N. T., Eds., Oxford University Press, New York, 39–70.

Haber, W. A., Zuchowske, W., and Bello, E., 1996. *An Introduction to Cloud Forest Trees: Monteverde, Costa Rica,* Impresión Comercial, La Nación S.A.

Harvey, C. A., 1999. The Colonization of Agricultural Windbreaks by Forest Trees in Costa Rica: Implications for Forest Regeneration, Ph.D. dissertation, Department of Ecology and Systematics, Cornell University, Ithaca, NY.

Harvey, C. A., 2000. Windbreaks enhance seed dispersal into agricultural landscapes in Monteverde, Costa Rica, *Ecol. Appl.,* 10, 155–173.

Harvey, C. A. and Haber, W. A., 1999. Remnant trees and the conservation of biodiversity in Costa Rican pastures, *Agrofor. Syst.,* 44, 37–68.

Harvey, C. A., Guindon, C. F., Haber, W. A., DeRosier, D. H., and Murray, K. G., 2000. The importance of forest patches, isolated trees and agricultural windbreaks for local and regional biodiversity: the case of Monteverde, Costa Rica, in *XXI IUFRO World Congress,* 7–12 August 2000, Kuala Lumpus, Malaysia, International Union of Forestry Research Organizations, Subplenary sessions, 1, 787–798.

Nadkarni, N. M. and Wheelwright, N. T., 2000. Eds., *Monteverde: Ecology and Conservation of a Tropical Cloud Forest*, Oxford University Press, New York.

Nielsen, K. and DeRosier D., 2000. Windbreaks as corridors for birds, in *Monteverde: Ecology and Conservation of a Tropical Cloud Forest*, Nadkarni, N. M. and Wheelwright, N. T., Eds., Oxford University Press, New York, 448–450.

Noss, R. R. and Harris, L. D., 1986. Nodes, networks, and MUMs: preserving diversity at all scales, *Environ. Manage.,* 10, 299–309.

Pounds, J. A., 2000. Amphibians and reptiles, in *Monteverde: Ecology and Conservation of a Tropical Cloud Forest*, Nadkarni, N. M. and Wheelwright, N. T., Eds., Oxford University Press, New York, 148–171.

Powell, G. V. N. and Bjork, R. D., 1994. Implications of altitudinal migration for conservation strategies to protect tropical biodiversity: a case study of the Resplendent Quetzal *Pharomachrus mocinno* at Monteverde, Costa Rica, *Bird Conserv. Int.,* 4, 161–174.

Powell, G. V. N., Bjork, R. D., Rodriguez, S. M., and Barborak, J., 1995. Life zones at risk: gap analysis in Costa Rica, *Wild Earth*, 5, 46–51.

Powell, G. V. N., Bjork, R. D., Barrios, S., and Expinoza, V., 2000. Elevational migrations and habitat linkages: using the Resplendent Quetzal as an indicator for evaluating the design of the Monteverde reserve complex, in *Monteverde: Ecology and Conservation of a Tropical Cloud Forest*, Nadkarni, N. M. and Wheelwright, N. T., Eds., Oxford University Press, New York, 439–442.

Schonewald-Cox, C., Buechner, M., Sauvajot, R., and Wilcox, B. A., 1992. Environmental auditing: cross-boundary management between national parks and surrounding lands: a review and discussion, *Environ. Manage.,* 16, 273–282.

Stevenson, R. and Haber, W. A., 2000. Migration of butterflies through Monteverde, in *Monteverde: Ecology and Conservation of a Tropical Cloud Forest*, Nadkarni, N. M. and Wheelwright, N. T., Eds., Oxford University Press, New York, 118–119.

Timm, R. M. and LaVal, R. K., 2000. Mammals, in *Monteverde: Ecology and Conservation of a Tropical Cloud Forest*, Nadkarni, N. M. and Wheelwright, N. T., Eds., Oxford University Press, New York, 223–235.

Wheelwright, N. T., 2000. Conservation biology, in *Monteverde: Ecology and Conservation of a Tropical Cloud Forest*, Nadkarni, N. M. and Wheelwright, N. T., Eds., Oxford University Press, New York, 419–432.

Young, B. E. and McDonald, D. B., 2000. Birds, in *Monteverde: Ecology and Conservation of a Tropical Cloud Forest*, Nadkarni, N. M. and Wheelwright, N. T., Eds., Oxford University Press, New York, 179–204.

20 Linking Geomatics and Participation to Manage Natural Resources in Madagascar

Richard Ford and William J. McConnell

CONTENTS

INTRODUCTION

Environmental planners now widely acknowledge that community-based natural resource conservation in ecologically fragile zones is unlikely to occur unless rural farmers have reliable economic alternatives to traditional, extensive agricultural practices (Margoluis and Salafsky, 1998). In Madagascar, the creation of protected areas, particularly tropical forests, has limited the livelihood options of the rural communities adjacent to these forests, whose access to potential farmland and forest resources is constrained. Meanwhile, farmers continue to suffer the effects of natural resource degradation resulting from logging, mining, and decreasing land availability (Keck et al., 1994).

0-8493-0020-7/01/$0.00+$1.50

To improve the well-being of farmers and to establish a basis for long-term forest conservation, planners in Madagascar adopted an approach in the early 1990s that combined rural development activities with natural resource conservation techniques. As part of the first phase of the country's National Environmental Plan (PE1), USAID funded Integrated Conservation and Development Projects (ICDPs) in six of Madagascar's protected areas under an umbrella program entitled Sustainable Approaches to Viable Environmental Management (SAVEM). The ICDPs were designed to reduce deforestation by providing solutions to the economic marginalization of communities adjacent to the rain forest enclosures.

This chapter describes the evolution of one of the initiatives in conservation and development that was carried out in Madagascar in the mid-1990s. It focuses on an ICDP then operating in the Andasibe/Mantadia Protected Area Complex, located in the hilly east-central region of the country. Known as APAM (Aires Protegées d'Andasibe/Mantadia, or the Andasibe/Mantadia Protected Areas), the project included several institutional partners.* The map (Figure 20.1) provides details. Two of the partners, the Clark University Program for International Development and Sampan' As Mamba Ny Fampandrosoana (SAF/FJKM), a church-based Malagasy nongovernmental organization (NGO), established a foundation for long-term natural resource conservation through a process of community-based, rural development in villages surrounding the APAM. Development practitioners from the APAM project facilitated early stages of an economic transition from traditional livelihood strategies based upon extensive, shifting cultivation to practices that are more viable, given a permanent constraint on access to new lands. It was presumed that land-use intensification would lead both to improvements in the quality of farmers' lives and physical environments and to the likelihood of long-term conservation of the local forest.

This study demonstrates these transitional stages with examples from one participating village, Vohibazaha. It highlights how an adaptive, social learning approach helped to create the conditions for the intensification process to gain momentum. The case underscores the effectiveness of applying geomatics** within a participatory planning framework for catalyzing joint action toward local land-use intensification. This social policy experiment in participatory planning methods was driven by the hypothesis that a U.S. university, a local NGO, and villagers in Vohibazaha, within a very limited budget, could accomplish the following:

1. Develop information from locally derived data and convert these into community-based action plans that the village would support and implement;
2. Expand the trust gained from the local action into larger-scale development projects;

* The full list of USAID-supported participating institutions includes the primary contractor, VITA (Volunteers in Technical Assistance); three sub-contractors, SAF/FJKM, Clark University, and Tropical Forestry Management Trust (TFMT); and four sponsoring agencies, including: (1) the Département d'Eaux et Forêts (DEF), (2) L'Association Nationale pour la Gestion des Aires Protéges (ANGAP), (3) the Grant Management Unit of USAID, and (4) Partners Acting and Cooperating Together (PACT).
** Geomatics refers to a suite of related technologies including global positioning (GPS) receivers, the interpretation of airborne and satellite remote sensing imagery and geographic information systems (GIS).

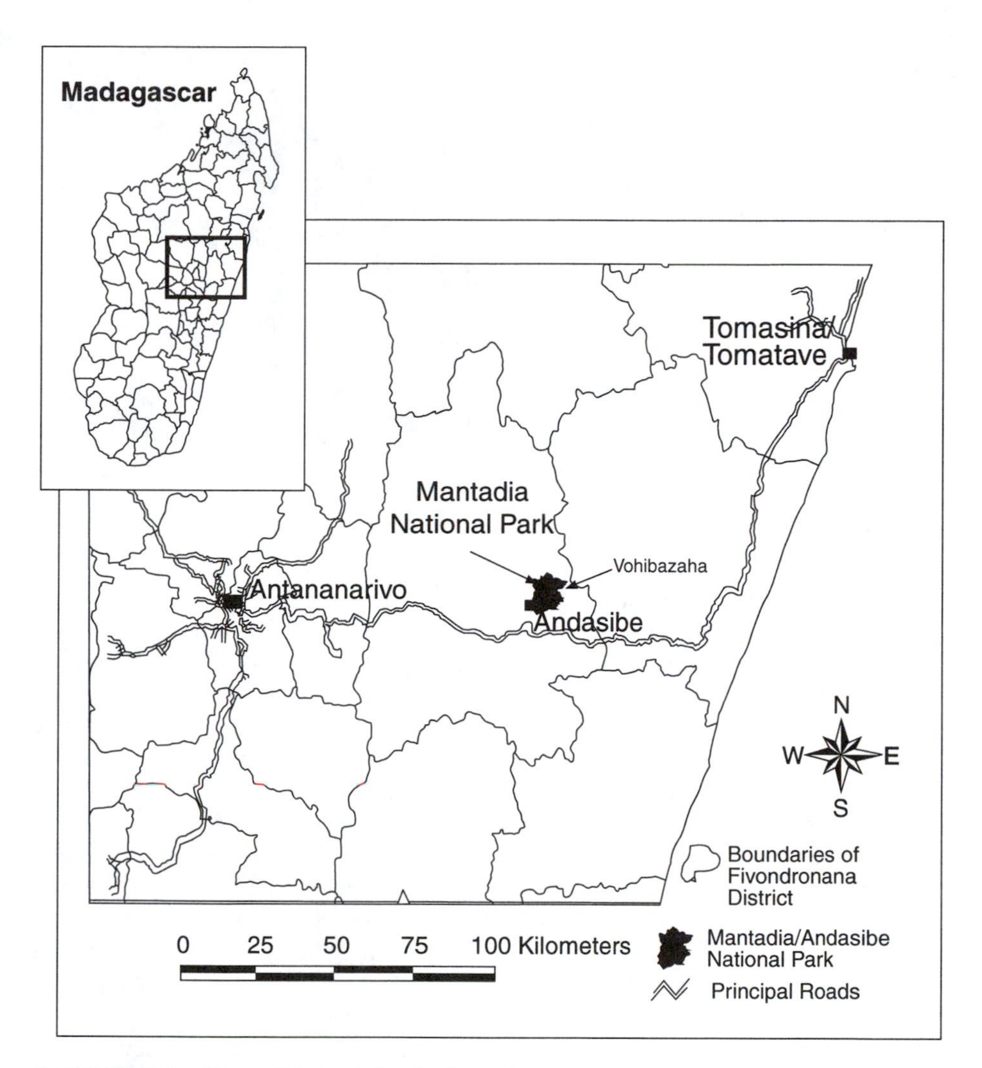

FIGURE 20.1 Map of Mantadia/Andasibe and environs.

3. Use the trust and database to undertake community-based conservation planning;
4. Scale up the database for community-based local and Geographic Information System (GIS)-based regional monitoring of land cover change;
5. Maintain the role of farmers and local resource users as central to the planning and implementation process in both conservation and development.

REGIONAL SITUATION

The AMPA complex comprises two neighboring forests. The first, the Analamazaotra Special Reserve, includes 810 ha of secondary forest, regenerated since the early 1900s when the forest was cleared by the colonial state for railroad construction,

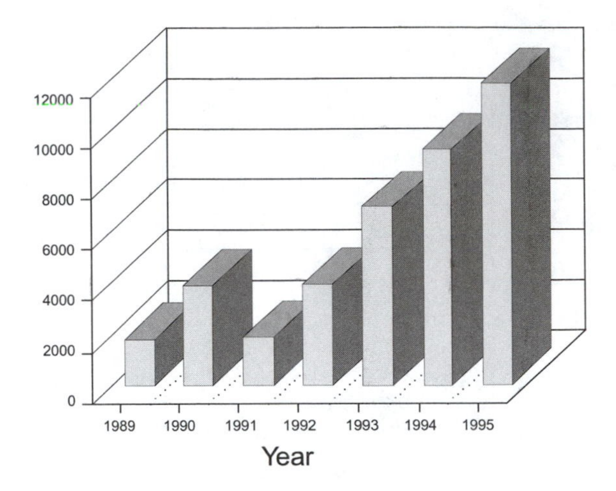

FIGURE 20.2 Visitors to the Andasibe Special Reserve.

locomotive fuel, precious hardwood extraction, mining, and by Malagasy farmers for rice cultivation. The special reserve contains a variety of rare wildlife species, most notably the *Indri indri* lemur, and draws a high percentage of foreign tourists because of its accessibility by rail or highway. The Mantadia National Park, established in 1991, opened to tourism in 1996. It consists of 10,000 ha of endemic primary rain forest in which campsites and trails of different grades of difficulty for tourists have been created by the APAM conservation team.*

These protected areas are among Madagascar's most popular for foreign travelers. Situated 2½ hours east of Antananarivo, the capital, off a well-paved highway, the special reserve and national park are easily accessible and a natural stop for tourists heading for the coastal city of Tamatave. The special reserve has enjoyed a steady annual flow of tourism (Figure 20.2), which is likely to increase with the opening of the Mantadia National Park. As of 1995, the reserve had the largest number of visitors of any protected area in Madagascar.

Until recently, the entrance to the Mantadia forest was virtually inaccessible to project staff and outside researchers, as the only road leading to the park was built by one of two graphite mining companies in the region. APAM and the company owners have since agreed upon mutual use of the road to allow for tourism in the park. These graphite companies have been operated by two French families since the 1920s and employ approximately 300 to 400 Malagasy laborers. With the formation of the national park, the mine companies can no longer explore for new sources of graphite. They are therefore investing in the high-volume tourist industry that the national park is expected to attract.

Another forest-based enterprise in Andasibe is the Complexe Industriel du Bois d'Andasibe (CIBA), which is a state-subsidized logging firm hired to cut railroad ties for regular railway maintenance. Researchers estimate that CIBA furnishes

* Much of the material on the region and project setting is based on: Ndranto Razakamarina et al., 1996.

between 60,000 to 120,000 railroad sleepers per year.* In addition, CIBA manufactures furniture and planks. Wood comes from concessions to the north and west of the Mantadia National Park. In the 1980s, the timber mill employed as many as 400 people, but by the mid-1990s, the workforce had decreased to approximately 100. Andasibe is home to a total population of 12,000 people, the majority of whom are migrants from other regions of the island. This canton and the surrounding cantons of Ambatovola, Beforona, and Moramanga have a population density ranging from 10 to 20 inhabitants per square kilometer (Keck et al., 1994)

This chapter examines the suite of development activities that have occurred in 1 of the 13 villages adjacent to the protected areas and participating in the APAM ICDP. It highlights the opportunities afforded through village-based, participatory development to reach the goal of natural resource conservation in the region.

LOCAL SITUATION

Vohibazaha (pronounced *Vo-ee-bá-za*) is a village of particular interest because it is located immediately adjacent to the Mantadia National Park. Residents have "illicitly" practiced *tavy*** in the national park. They are aware that farming in the park is prohibited, but they claim that their ancestral territory was unfairly expropriated by the state during the delimitation of the park's boundaries. L'Association Nationale pour la Gestion des Aires Protéges (ANGAP), the national protected area planning and management agency, has attempted to forge a compromise with villages surrounding the park by conceding a buffer zone, wherein farmers who have already begun to cultivate may remain on their land if they practice intensive farming techniques, such as agroforestry. Because of the contested land issue and the pervasive practice of *tavy*, the area became a high-priority zone for the APAM project.

From Andasibe, the journey to the remote villages is time-consuming and, for outsiders, adventuresome. The nearest train station is at Ambatovola. From there one must walk along the railway track for another hour, traverse the Sahatandra River on a bamboo raft — until APAM helped villagers to erect a footbridge — and hike another 3 km on often slippery footpaths to the permanent residential core of Vohibazaha.

As of the mid-1990s approximately 640 people lived in Vohibazaha. Village populations in this region consist of several extended families, each headed by one or more clan patriarchs called a *tangalamena*. Collectively, the *tangalamenas* administer village affairs: they mediate conflicts, preside over rites and celebrations, allocate land to extended family households, disseminate important news, and officially welcome visitors. Several of the *tangalamenas* have become actively involved with development planning in collaboration with APAM personnel. In some villages, a *tangalamena* has taken on the role of president in the local village development committee, formed after Clark University and SAF/FJKM representatives conducted their initial participatory rural appraisals.

* See Priya Shyamsundar, 1993, and Pierre Berner, 1993. Shyamsundar estimates 60,000 sleepers per year are manufactured by CIBA, whereas Berner estimates 120,000 per year.

** *Tavy* is the traditional slash and burn system for growing hill rice in Madagascar.

Many residents of Vohibazaha trace their origins to a more distant, now vanished village that has gained a mythological status. In the late 19th century, people moved from their ancestral village, pushed out by a bellicose king, to inhabit this area near the Volove and Sahasarotra Rivers. This initial group was later joined by several subsequent groups of immigrants, explaining the village's name, Vohibazaha, from the Malagasy phrase the "place of the foreigner." Vohibazaha has a semifunctional primary school and a fairly well attended Anglican Church, an institution that brings occasional development resources from the outside. A small shop in the village intermittently stocks cooking oil, salt, matches, rum, and "food to go with rice" (*laoka*), such as dried fish, beans, or eggs. However, many of these supplies are meager, expensive, and frequently unavailable. During the rice-growing season, from late October to June, the wooden houses of the villages are mostly deserted, and residents live on their rice plots sowing, weeding, chasing away birds, tending to interplanted crops, and eventually harvesting the reddish, flavorful hill rice.

Residents of the area rely on rice as their subsistence and primary cash crop. Throughout the island wide, rice for the Malagasy people is a mainstay of dietary custom and cultural tradition. The Betsimisaraka of this region assess a household's quality of life by the amount of rice it harvests. It is cause for utmost anxiety and a sign of poverty if a family cannot eat rice three times a day. Unfortunately, a majority of farmers do not harvest enough rice to meet subsistence needs year round. Each season, men sell a portion of their household's harvest in the markets of Andasibe and Moramanga to acquire cash for clothes and other necessary items. Yet in the months before harvest, the lean months, a large proportion of households are forced to buy back rice at a much higher price. Wealthier households tend to invest in zebu cattle and in expansion of their rice fields, since they have income to hire friends and other family members to labor on the fields when necessary.

By the time the APAM project began in Vohibazaha, considerable agricultural encroachment into the Mantadia National Park had already occurred. Although figures are difficult to confirm, village elders of Vohibazaha estimated that as many as half of the village households were practicing *tavy* inside of the Mantadia National Park before the boundary was demarcated. SAF/FJKM proposed a plan under which farmers would be allowed to continue cultivating their fields in a "buffer zone" within the park, on the condition that they apply a cluster of soil and water conservation techniques, such as planting rows of vetiver grass and leguminous trees along ground contours, to improve the quality sufficiently to enable permanent cultivation, thereby negating the need for clearing new fields in the park. Farmers are not necessarily willing to invest their time and energy into locally untested intensive agricultural techniques, particularly if it is intended to save the forest for a lucrative state-run tourist industry.

The task of APAM, then, was to assure that farmers saw their investment in the conservation effort bringing material gain to their families and their community. Since the benefits of agroforestry take several years to manifest, rural development activities in villages are an expeditious means to improve livelihoods, transfer technology and skill, strengthen the leadership capacity of community members, and establish good relations between conservationists and farmers.

The APAM project began with an appreciation for four pertinent conditions:

1. The practice of *tavy* was widespread and had been a source of livelihood since time immemorial. Although there was some evidence that paddy rice had been cultivated in the area, it was clear that *tavy* was the dominant and preferred food crop.
2. Hostility toward the park was considerable. Park planners had arrived in the area in the mid-1980s and demarcated the boundaries of the park without consulting with the local residents. The borders were selected with reference to the location of the different forest species, not on the basis of the villagers' needs or priorities.
3. Because villagers had not been consulted in creating the park, there was a perception on the part of the community that the park would be of little if any benefit to the farmers.
4. Poor planning and weak communication had created an extremely tense situation in which the villagers saw the park as a force threatening their livelihood systems and park officers saw the farmers as irresponsible and ecologically destructive.

The problem confronting the park and the people was whether a shared land-use plan could be devised that would enable both parties to meet their objectives — could there be a win–win solution? A special feature of this project was experimentation with geomatics as planning and monitoring tools to strengthen initiatives in conservation and development.

CRITERIA FOR USING GEOMATICS IN THE APAM PROJECT

The application of geomatic technology in the APAM project had two related components. First was the extension of sophisticated remote sensing and GIS technologies in use at the national scale to the local park office. The second involved the use of simple, low-cost air photograph interpretation and computer-mapping technology to bring the knowledge and priorities of the farming communities into the planning process on an equal footing with the information and goals of the National Park Service.

In contemplating use of geomatics, the project developed several criteria for selecting when and how to use it. These considerations included:

Are participation and transparency improved? Geomatics should be used to expand and enhance the participation of the people and to increase the transparency of project implementation and management.

Is it cost effective? Geomatics should be used only if it is cost-effective and provides benefits to the people, proportional to the cost of setting up and operating the system.

Will its use increase opportunities for stakeholders? Geomatics planners should place priority on helping villagers and park managers to generate new options for improving development and conservation.

Can it be sustained? Can printed maps or other products from the geomatics exercises be left with park and village officers to assist in monitoring changes in land use cover?

Is it stakeholder driven? Do all stakeholders see reason to use the geomatics?

Can the information gathered be used in coercive way? While geomatics and databases are neutral, their applications may not be. Can the information on land use be used in ways that would adversely impact either park or people and, if so, can the geomatics staff safeguard against such uses?

Is the scale appropriate? Can the information be assembled in ways that will be useful to land users at local levels as well as policy officials at regional and national levels?

FROM DEVELOPMENT TO CONSERVATION: FIVE PHASES OF INTEGRATION

Several formal steps form the core of the rationale for developing plans for integrated conservation and development. The process for communities to incorporate new and more-conservation-minded livelihood strategies requires the provision of economic alternatives to slash-and-burn rice cultivation as well as the formation or strengthening of community institutions involved in natural resources management. SAF/FJKM and Clark University, the partners responsible for the community development component of the APAM project, facilitated five phases of integration between the park staff and villagers in the process of shifting farmer livelihood practices toward more environmentally and economically sustainable forms.*

1. ***Engagement*:** Using community-based methods such as Participatory Rural Appraisal (PRA) in the initial phase of planning for conservation and development, community members become immediately and actively engaged in the integrated planning.
2. ***Development Implementation*:** Based on communities' action plans, derived from PRA activities, microprojects are launched in villages as soon as possible. Funding, materials, and labor are provided jointly by community members and participating institutions.
3. ***Expanded Projects*:** Project staff introduce possibilities for new projects in villages that were not previously proposed in the community action plans. This is a means to stretch the vision of community residents; it allows people to experiment with new livelihood pursuits without great personal risk, since the project provides technical and partial financial assistance.

* The authors are indebted to David Richards, who led the project design team, for conceptualizing these transitional stages.

4. ***Agricultural Intensification and Conservation Plans*:** Project staff members introduce alternatives to existing agricultural practice (*tavy*), such as agroforestry, that benefit the community and the resource base. It is important to act on the momentum and enthusiasm generated from the microproject activity. In the case of APAM, project staff worked with community members to create formal plans for resource conservation and agricultural intensification.
5. ***Communities as Partners in Protected Area Management*:** The final stage of APAM and similar projects stabilizes the resource management practices in the protected forest and surrounding territory. Phase five calls for the project staff and participating village communities to facilitate durable partnerships between and among stakeholders. In the case of APAM, this final phase worked while the project was in operation and continues to work on site.

Having agreed on the nature of the problem and a set of steps for solving it, the team determined that the first major planning step was to find out what the people knew and what they felt were their highest priorities. The project staff did not begin by constructing a large database of economic, ecological, and social data of the area. Some baseline data were already available from the original park planning exercises in the late 1980s and early 1990s, including a set of air photographs taken in 1991. There were also topographic maps prepared in the 1960s from air photographs acquired in the waning days of French colonial rule. Instead, the team began the process of engagement with the communities.

Phase 1: Engagement

Sustainable conservation and development require local institutions to support new approaches. These institutions would become the driving force of the APAM local planning and action — the core elements in the process of initial engagement. To work effectively, these local institutions require systematic and structured methods to gather data from community and village residents. It is also important to shape these data into plans and courses of action that the majority of the village will support. With these needs in mind, the team joined with villagers and used PRA to gather and analyze microinformation about the communities adjacent to the park.

PRA provides tools to manage participatory approaches to rural development.* The method helps rural communities to support activities that they design and implement. It strengthens local leadership and institutions, and helps integrate sectors at the community level related to natural resource management. PRA helps to build collaboration among different agencies external to the community as well as within the community. In the case of the Mantadia villages, PRA has opened doors to

---

* For additional information about use of PRA in Madagascar see Introduction to PRA (Program for Internal Development, 1989) or *Analyse Participative en vue de la Réduction de Pression sur une Aire Protégé* (Reveley et al., 1993). Both of these booklets are available from SAF in Madagascar or Clark University in the U.S.

community leaders and institutions. It has helped to build trust between park and people. It has been an important tool for creating partnerships between outside agencies such as SAF and village institutions.

Three assumptions form the basis of PRA:

1. ***Local Knowledge*:** Farmers have knowledge and information, but it can be more effectively communicated to outside institutions if organized differently.
2. ***Community Institutions*:** Villagers have resources that can be more effectively mobilized.
3. ***Attracting Outside Help*:** Outside resources are available, but are helpful only if applied in the context of village-identified priorities.

PRA uses a variety of innovative data-gathering techniques such as sketch maps, transects, seasonal calendars, trend lines, time lines, institutional diagrams, livelihood maps, resource access ranking, and options assessment ranking. Further, PRA calls on the Rapid Rural Appraisal criteria of productivity, sustainability, equitability, and stability to find solutions to problems the community identifies as its most severe. The solutions are then organized into a Community Action Plan (CAP) in which specific community groups make commitments to carry out particular tasks.

Gathering and analyzing data for PRA rely on visual data collection tools — charts, tables, and graphs, reaching out to many social, ethnic, gender, class, and age groups within a community. Most data are collected, analyzed, and ranked in large group meetings. Data are left with the community for further analysis, ranking, action planning, and monitoring. In Mantadia, this approach has yielded a rich and introspective view of community needs and resources. Community residents and project staff have used the data to construct action plans, integrating technical, social, ecological, and managerial considerations. More recently, communities and project staff have joined together to develop indicators and community logbooks to monitor the progress of the development activities.

In Vohibazaha, the village and team collaborated to gather data and rank problems. It is instructive to note the problems that the village identified, and to understand how present-centered events influenced the ranking. During the weeklong data collection and analysis in Vohibazaha, a 5-year-old child became ill. The child's father began to worry as the illness deepened. He decided to seek medical help and began the 4-hour walk to the train station and then the sometimes 36-hour wait for a train. The child died in his father's arms about 2 hours from the train station. The PRA exercises were suspended for 3 days while the entire village mourned the loss of one of its young. Vohibazaha's ranking of its most serious problems, ordered from most to least severe, was:

1. Health problems of the people
2. Conflicts and land rights in the park
3. Commercial linkages with the outside world
4. Need for increased agricultural production and food production
5. Livestock management
6. Social problems, especially among the village's young people

TABLE 20.1
Vohibazaha Community Action Plan

Action	Materials/Follow-Up	Responsibility
Build health clinic	• Obtain labor, wood, nails, roofing	• The community
Provide medicine	• Keep supply in village	• External organization
	• Explore uses of products from the forest	• The community
Identify medical staff	• Arrange periodic visits by external health practitioner	• External organization
		• The community, after person is trained
	• Train local village health worker	
Offer local classes in health education	• Use new clinic and trained village health worker	• The community
		• External organization
Provide technical health services	• Recruit outside staff	
Negotiate park boundaries	• Organize meeting with park director	• The community
	• Write letters	• The community
	• Meet with park commission	• The community
	• Create village committee	• The community
Create commercial linkages with outside businesses	• Think about improving the trail and river crossing	• The community
	• Continue discussion at next PRA meeting	

Park, project, and village leaders used this ranking to consider what could be done to address local problems. After extended discussions, the community adopted a community action plan (Table 20.1).

With an action plan in place and several community groups committed to its implementation, Vohibazaha was ready for Phase 2, implementing small-scale projects. At this point in the process, it was clear that the villagers were assuming ownership and responsibility for the small-project activity envisioned in the engagement process. It was equally clear that without outside help in medical assistance, technical training in health care, and overall management that little would happen to implement the plan.

Phase 2: Implementing Initial Development Activities

Several concepts are important for implementation. There is need to manage the local contributions of time and material. There is equal need to seek out the external support from park gate receipts, NGOs, government agencies, and donor organizations. To keep the momentum of the five-stage integration moving, it was also necessary to look for ways that some portions of the CAP could be accomplished in a relatively short period of time.

Health

To address the community's first priority, improved health services, the village identified two options. The first was to organize a village pharmacy to sell basic medicines for a small price. The second was to seek ways that a doctor or health practitioner might visit the village on a periodic basis. The village pharmacy was planned and implemented within a few months. One resident was familiar with basic health principles and agreed to manage a shop for medicines; SAF found funds to subsidize the cost. Basic pills such as aspirin, malaria control, and simple tablets for coughs, fever, and colds became available at the village pharmacy. The village health group also worked with SAF to arrange for monthly visits by a doctor, based in a nearby town. SAF helped with fund-raising to pay a portion of the costs. Villagers contributed a small amount to pay for the medicines as well as the costs of the doctor's time.

Forest Access

Despite a clear legal prohibition on cultivation inside the park, local park officials sought a means to allow farmers to continue using fields that had been in production prior to the beginning of the APAM project. Continued use of the fields inside the park depended on the farmers not expanding their fields farther into the park. SAF/FJKM drafted a plan stipulating conditions under which cultivation could continue. To avoid expulsion from the park, groups of farmers agreed to implement several dozen soil and water conservation measures. The project supervised and paid farmers for their labor during initial training on demonstration plots, with a plan to compare yields with those obtained under traditional management. It was expected that the improved yields would convince the farmers to adopt the "improved practices." The park director agreed to submit the contract to the National Park Service for ratification.

Food Security

Declining soil fertility, the rising cost of purchasing rice, and constrained access to land caused by creation of the park have combined to create agricultural problems for most of the villages in the region. Vohibazaha is no exception. One of the major goals of APAM has been to help villagers ensure food security without expanding their agricultural lands. Longer-term discussions began, based on the priority of the park for intensified agricultural production and of the community for greater food security. As a result of several meetings and negotiations, villagers agreed that one way to help them with health and commercial links was to focus on building one or more village grain storage facilities, and to establish a fund for the purchase of rice.

Most farmers would normally carry a portion of their recently harvested rice 4 to 5 hours along the difficult trail to the nearest merchant, in Ambatovola. The trek includes crossing the Sahatandra River on a precarious bamboo raft. There they would sell their rice at a low price — because many farmers in the region were bringing their rice to the same merchant during harvest season — and use the cash to buy needed clothing, tools, and other durable goods. After 6 months, when the rice they kept at home was finished, they would walk back to the merchant and buy rice at a dramatically increased price. The winner in the transaction was the merchant; the losers were the farmers and their families.

SAF field staff negotiated an agreement in which the project would establish a "revolving fund" that the village development committee would use to form a rice storage cooperative. Villagers built the granary themselves, at no cost to the project. The "revolving fund" enabled them to buy rice at harvest time that was then resold at a modest profit in the months leading up to the next harvest. Unfortunately, following the sale of the initial stock of rice, the rice storage cooperative was disbanded when the "revolving fund" was reportedly shared out among the cooperative leadership.

They also began conversations about small-scale poultry management as well as fishing in the rivers in the region.

Phase 3: Stretched Vision

Whereas the initial conversations and actions in Phases 1 and 2 had taken about 18 months, the process moved more quickly in Phase 3. Informal discussions had been building in the village. One conversation is of particular importance.

The villagers' original vision was that improving the trail could increase external access. That would make it easier to carry rice to the merchant in Ambatovola. Simply facilitating transport in and out of the village, however, would make no major change in the role of the external merchant nor did it free the farmers from their economic dependence on a system in which they would lose money every time. In spite of this dilemma, it was clear that improved access could help in marketing products resulting from agricultural intensification, such as citrus, coffee, beans, and avocado.

SAF assisted farmers on trail improvement by providing limited funds for hand tools. At the same time, they began discussions about other ways to strengthen links with the outside. Farmers suggested many things, some quite expensive and impractical. However, many talked about replacing a bridge across the Sahatandra that had been washed away in a severe flood some years ago. It was a village asset that had previously worked, but that the community could not maintain by itself. The government had lost interest in helping the isolated communities in the rain forest so had never replaced the bridge. The bridge conversation became the first step in formal consideration of an activity fully beyond the scope of the original action plan and one that would require a significant effort by the farmers, and by park and project staff. It would fulfill a major goal that the community had identified, but would do it in a way the villagers had not considered. It was a conversation that was dependent on the trust, accomplishment, ownership, and self-confidence that had emerged from Phage 2 of the APAM five-level strategy for integrating conservation and development.

Farmers agreed that the bridge would be an important boost to their planning, so long as they could get some outside help. Again, SAF became involved. They organized meetings with park staff, the APAM project group, and eventually an engineering firm in a nearby town. APAM reviewed the proposal and in particular asked what the community was prepared to contribute to the construction of the bridge.

All the stakeholders agreed on the bridge construction plan. It called for contributions from the community as well as from the park gate receipts fund. When the work was completed and all the figures tallied, the community had provided 1917 person-hours to gather 3.6 tons of sand and 15 m^3 of stone and gravel. They also cut enough planks, 5-cm thick, to form the bed of a footbridge 1 m wide and 64 m long. The breakdown of these figures included 333 villager workdays to clear paths, 61 workdays to break rock, 242 workdays to mix cement, and 355 workdays to assist in assembling the metal work for the bridge. The work the community provided would have cost U.S. $3000 had it been necessary to hire or buy the materials.

The park fund contributed $10,000 that paid for the cement, metalwork design and cutting, and the transportation of these materials. The funds also paid for engineering feasibility studies and siting as well as the overall design of the bridge. In the last few weeks, the engineer was so impressed with the energy and competence of the villagers' work that he turned back $1000 of his fee to reduce the overall cost of the bridge.

Records that the president of the village development committee kept of hours worked and materials produced documented that every family in the village contributed to the process. It was a time of hard work but of much community enthusiasm. The spirit and energy that the bridge construction created extended into discussions about a conservation strategy.

Phase 4: Intensification and Conservation Planning

At this stage, geomatics became important in village land-use planning. Stretching visions required a means for farmers to see beyond their present situation. Geomatic techniques offered the means to help.

Throughout the discussions, the SAF/Clark team had relied largely on topographical maps and farmer testimony to plan projects — granaries, the village pharmacy, agroforestry, and crop intensification. At this point, a geomatics specialist from Clark University used low-altitude 35-mm aerial photographs of the region commissioned by ANGAP in 1996 to prepare a high-resolution, true-color photographic mosaic of the village lands and adjacent portions of the park. Project personnel took the photographic mosaic, along with inexpensive mylar overlays and grease pens, to the village. With no instruction or guidance, a group of farmers interpreted the photographic mosaic using the mylar overlays (Figure 20.3). Men and women joined together to plot familiar landmarks, including watercourses and trails, and identified each of the recently harvested *tavy* fields that had just been burned prior to the photographs being taken. The villagers easily identified specific sites in the photographs. Quite unexpectedly, the farmers also identified several gardens (*potro*), about which the project staff previously knew little, since they tend to be nestled in small valley bottoms, often far from the main path leading from the village center to the park boundary.

In subsequent discussions, it became apparent to the project staff that *tavy* cultivation in the community was organized around a complex set of rules of access. In the end, three types of land tenure were described. The first concerns traditional clan holdings in the area immediately surrounding three tombs located on hilltops

FIGURE 20.3 Villagers create land use map from photographic mosaic of their community.

surrounding the residential core of the village. Access to these *sembontrano* lands is obtained through a request to the appropriate *tangalamena*, requiring a ritual cattle sacrifice and the approval of the clan leadership each year. Outside of the *sembontrano*, access to fallow land, called *jingeranto*, is controlled by the person who originally cleared the land, although access is rarely denied to close relatives. Finally, access to fallow lands *inside the park* is completely at the discretion of the person who originally cleared it. Although this is the most fertile land in the village, it was cleared illicitly, and farmers are reluctant to implicate themselves in this situation.

In this process, the project staff learned a great deal about land-use practices in a community in which they had been working for 3 years. There are two possible explanations. One was the photographic mosaic. Farmers had no trouble finding the rivers, streams, valleys, and hill slopes that were their homes and fields. They could easily map, using their own language and symbols, where they had practiced *tavy* and also where they had grown commercial crops including coffee, beans, avocado, and citrus. Second, the farmers were forthright and candid in talking about these practices in ways they had not done in the initial meetings and planning sessions. In part, this was because they were faced with the fact that the project held quite clear evidence of their cultivation practices. In addition, however, the community recognized that SAF had delivered what they said they would on the small projects, thereby winning the trust and confidence of the villagers.

Project staff learned several important things as a result of the mapping exercise. First, the complex nature of the system of land tenure in the community had previously been unclear to project staff, most of whom were from other parts of the country and had spent limited time in the village. As a result, rather than building upon these preexisting institutions, four new "user groups" had been formed to sign the draft contract for the use of the buffer zone. The new groups cut across a set of local institutions developed over many generations. Second, the project staff learned that gardens were much more extensive and diverse than had previously been realized, and that significant intensive, permanent agriculture was already being practiced. Rather than building upon the existing successes in gardening, the project had introduced a panoply of soil and water conservation measures from an agroforestry manual developed in other countries.

The photographic mosaic analysis opened several new points of discussion, points essential to expanding the vision of the project and the community. This new information became a fundamental step toward initiating discussions about conservation planning including considerations of moving more quickly to more sedentary and commercial agricultural practices.

Phase 5: Comanagement Plan

The local park office had simultaneously been developing its own geomatic database of the evolution of land cover in the periphery of the park, using Global-Positioning System (GPS)-assisted surveys, as well as aerial photography from 1957 and 1991, in conjunction with satellite imagery from 1993 (Figure 20.4). The result was a land cover history, showing major deforestation in the first period (1957 to 1991), and relatively minor deforestation thereafter, particularly along Vohibazaha's boundary with the park. For this reason, Vohibazaha was seen by the project as a good partner, and it was treated as a model for collaboration. There are several possible explanations for the slowing of deforestation by Vohibazaha's farmers since 1991.

First, in the 1950s there was a great deal of commercial logging underway in the area. It is possible that the major changes in forest cover took place in the late 1950s, after the air photographs but long before the proclamation of the park boundaries. A second possibility is that gradual expansion of *tavy* was progressing persistently through to the park demarcation but stopped as a result of intensive surveillance by the park's "Guards Forestiers" who deployed GPS receivers to accomplish detailed mapping of illicit *tavy* within the park. A third, and one would like to think valid, hypothesis is the belief that the negotiations among park officials, the APAM project, and the community — discussions begun in 1992–93 — that opened the conversations to negotiate a conservation plan had an early impact on farmer behavior. During the 4 years of development/conservation discussions, park officials allowed farmers to continue using fields inside the park that had previously been cleared. However any new clearing was strictly forbidden. Given the confidence built up through the small development projects and the photographic and map data that the farmers could see, it was not difficult to draw up a conservation plan.

In their draft management plan, park officials had proposed classifying the park into three levels of protection. One zone would be off limits to all people and be protected as an exclusion reserve; the second would be available for research but prohibit tourists; the third park zone would be open to tourists who would be entering

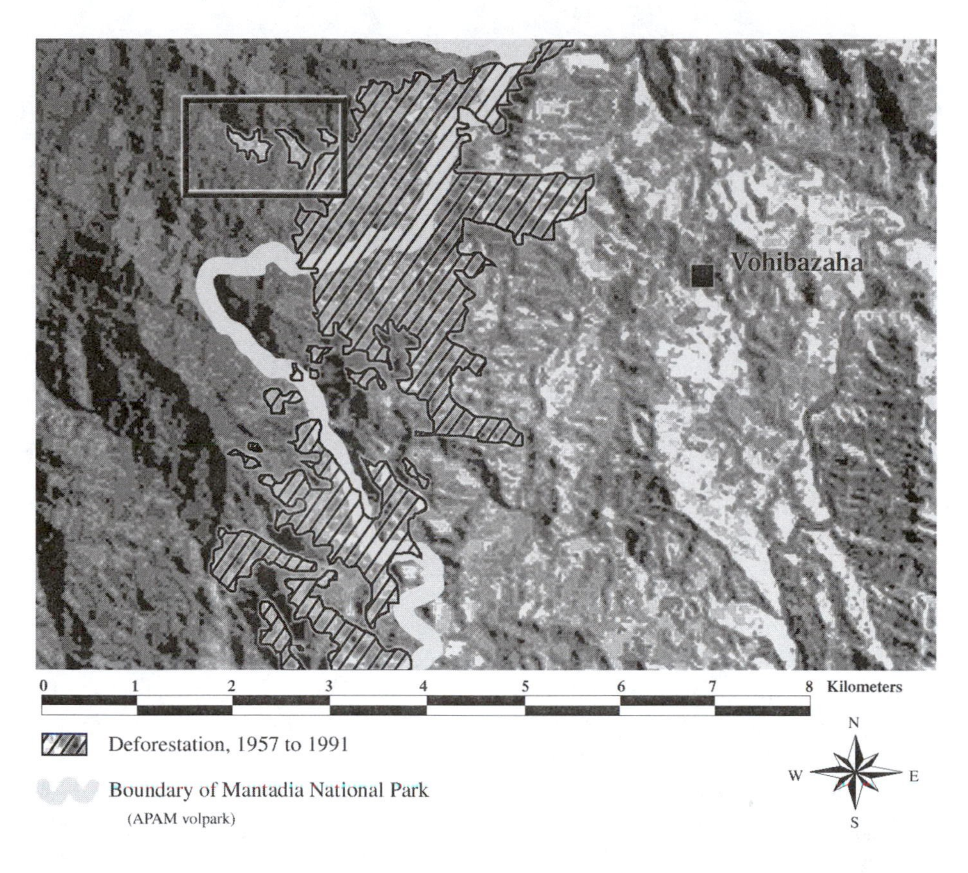

Figure 20.4 Forest cover change in the vicinity of Vohibazaha, 1957 to 1996. The residential core of Vohibazaha is shown amidst the community's farmland, limited to the east by the Sahatandra River and to the west by the Mantadia National Park. The hashed region indicates forest that was cleared for farming over about four decades. Some of this clearing constituted encroachment into the park. The change in forest cover was assessed through manual stereoscopic interpretation of panchromatic aerial photographs flown in 1957 (mission R47-S47, frames 057 and 058) and 1991 (mission 108/400, frames 056 and 057.) The backround image, from a Landsat Thematic Mapper satellite image (WRS2 158/073) acquired in September 1993, shows that forest cover remained stable between 1991 and 1993 (see box.) This stability was confirmed by analysis of low-altitude true color stereoscopic aerial photographs commissioned by the National Park Service (ANGAP) in 1996. Four of these photographs were scanned at a nominal ground resolution of approximately 1.2 meters, and processed to produce a 1:8,000 scale georeferenced mosaic that was used by community members to map land use, as shown in Figure 20.3. The satellite image and 1996 aerial photographs were provided courtesy of USAID/SAVEM.

from the main gate, located on the west side of the park, opposite Vohibazaha's park perimeter. The people's portion of the plan was an agreement that they would not expand any farther into the park as long as the park continued to invest 50% of park entrance revenues in joint development projects. The granaries, health, agroforestry, and access efforts had created a positive attitude sufficient to persuade the farmers that alternative livelihoods were possible.

The draft management plan was then submitted through ANGAP headquarters in Antananarivo (Tana), to the Ministry of Waters and Forests, which responded that it did not have the legal authority to allow cultivation inside a national park. Authorizing such cultivation would require redesignating that portion of the park, which would only be acceptable if the park could be expanded by equal amounts in other directions. The villagers, project staff, and park director were disappointed with this news.

Given the failure to gain ratification of the draft contract, local park officials were forced to look closely at the results of the SAF/FJKM initiative to promote soil and water conservation in the proposed buffer zone. By the end of the agricultural season, it was clear that the demonstrations had not been carried through to fruition, with farmers unable or unwilling to commit the required efforts. Park officials felt they had no choice but to disallow further use of the lands inside the park, amounting to an expulsion of the farmers from the village's most fertile lands. Farmers appealed to the *tangalamena* responsible for the village's oldest and largest *sembontrano*, who agreed to relax the traditional requirements of cattle sacrifice. Unfortunately, these were some of the least fertile lands in the village, and produced poorly.

All of the work of the previous 5 years seemed lost. Yet the director was not daunted, and the community did not give up hope. With the approbation of the village leadership, the park director used the community's maps as evidence of the farmer's good faith in negotiating a jointly acceptable future in order to press his case with the Ministry of Waters and Forests. Although there has yet to be a final decision on the redesignation question, the last news was hopeful, and the maps appear to have been effective in persuading a reconsideration of the national legislation governing the issue.

It has been 3 years since the joint conservation plan was implemented. The bridge is still intact with some maintenance given to the planking, although more would be helpful. Park officials state that the park continues to provide shared funds to the community for its continuing work in implementing small projects. Two new hotels constructed to accommodate the expected increase in park visitors are still open and seem to be at least holding their own.

IMPLICATIONS FOR ADAPTIVE COLLABORATIVE MANAGEMENT

Findings that emerged from the experiences of Mantadia and Vohibazaha are pertinent at several levels. There are four overall observations. The first considers the nature of collaboration. Villagers and farmers know a great deal and will share this information if they perceive how it will lead to more productive or sustainable land use. That many villagers throughout Africa have little formal education does not diminish contributions they can make toward responsible social, economic, and natural resource management. Rural residents will thrive and contribute to partnerships between local and external agencies if they see that such cooperation improves their livelihoods and their ecosystems. To form authentic partnerships requires meeting villagers on their own terms through processes of engagement such as those outlined in this chapter. Residents of Vohibazaha distrusted the park officers as well as the NGO and APAM team staff at the beginning. They assumed that the purpose of the initial visits was to impose even greater restrictions on their land-use practices. If there is to be adaptive and collaborative planning and management of natural resources, one needs first to establish the rapport within which such collaboration can take place. Use of participatory planning tools worked effectively in Vohibazaha to build trust.

A second finding considers the relationship of outsiders to the process. Even though the participatory tools worked well to systematize and organize what farmers already knew, it became clear that the farmers did not know everything they needed to about their environment or production methods. Nor did they necessarily have the tools and skills to analyze or compare trends in their own ecological or livelihood situations. External help such as developing the photographic mosaic, selecting and using indicators, or introducing visual planning tools, described briefly in this chapter, proved to be invaluable in setting goals, monitoring changes, and mobilizing resources. The Vohibazaha experience clearly documents that each stakeholder had significant and lasting things to contribute to conservation and development. It is important to develop and maintain linkages between insiders and outsiders.

A third lesson relates to integrating resource conservation and development. Although conservation is important to all villagers, it is unlikely to be their highest priority. Strategies of engagement that link conservation and development and clearly establish interactive relationships between the two are a necessary ingredient if interventions to protect long-term biodiversity are to be sustained. The concept runs parallel to Norman Uphoff's phrase, "assisted self-reliance," in which insiders and outsiders play important roles. The prospect of inside–outside linkages evolving into long-term collaboration, such as the ANGAP agreements between villagers and the park, suggests two lessons: that collaborative and integrated models can be sustained if they respond to locally identified needs, and that continual fine-tuning and amending of action plans and agreements are needed to sustain the agreements

Finally, comanagement can work at both national and local scales as well as making connections between the two. The apparent "success" in limiting deforestation on the borders of the Mantadia National Park suggests that the villagers were prepared to abide responsibly by their agreement. When national ministry officials reversed the agreement, local park officers and the people used their local data — some assembled through use of geomatics — to renew a dialogue with the national ministry. In part, this dialogue could take place because the local data were well organized and systematically presented, using geomatics as well as some of the data collected earlier through community participation. A climate in which local–national interaction can take place suggests that adaptive collaborative management can be a valuable tool for planning and decision making at several levels.

Findings more specific to the methods employed in Mantadia/Vohibazaha are also important. Some relate directly to combining geomatics and participation. Even though the role of geomatics was limited and the total cost of the time and materials well below $5000, the images provided critical assistance in several ways. First, farmers worked well with the air photographic mosaic. The pictures proved to be invaluable to open conversations for Phases 4 and 5 on the complex issues of planning for conservation and development. It is not clear whether the farmers would have been as forthcoming had the team used the photographs at the very beginning of the process. The point is that using them midway through the planning phases worked effectively. Second, use of the air photographs helped villagers to acquire ownership of collaborative planning. Because they shared ownership of the planning, probabilities increased that multiple ownership of the plan would follow. Third, use of geomatic tools enabled all parties to scale up from one local region to a national information base. In the case of Mantadia, geomatics-based maps were of direct use to influence national policy. Finally, geomatics helped the community move to new levels of detail not available through the conventional engagement tools used in the initial phases.

There are additional findings. Comanagement planning was an important element in limiting forest degradation between 1991 and 1996. Land tenure and land use are complex in the Mantadia area. Knowledge about the role of ancestral lands became an important factor in planning and limiting the expansion of *tavy*. Although this information was not forthcoming in the initial stages of planning, it emerged later because of confidence established in the earlier stages and the effectiveness of the visual planning tools supported through geomatics. For another point, agricultural intensification was already present in the community, involving the production of commercial crops such as coffee, beans, avocado, and citrus. Community-based planning helped to make this process more visible. A third lesson notes that illicit deforestation had been occurring outside of the jurisdiction of traditional local land tenure institutions. It is essential for project staff to know and understand details of such local institutions, and to build upon them in devising new land management strategies. Traditional land tenure institutions proved flexible in times of crisis, for example, in allowing emergency access to *sembontrano* lands following the expulsion of the user groups from the proposed buffer zone in 1997. Their involvement in many other situations could be equally important.

CONCLUSIONS

In conclusion, it is helpful to return to the chapter's initial premise. Combining participatory and geomatic tools enabled the APAM project to engage village groups to gather data, determine the community's highest-priority needs, design development action plans, select indicators, and implement the plans. The process further enabled the multiple stakeholders to move beyond development to conservation planning and management. This five-phase process of planning appears to be an effective way to introduce adaptive collaborative concepts that can function at local and national levels. It will be many years before all of these findings can be verified. During the interim, additional experiments that combine participatory strategies of engagement and negotiation with geomatics are worthy of consideration.

REFERENCES

Berner, Pierre, 1993. Conservation-Based Forest Management Development Strategies to Reduce Pressure on the Analamazaotra-Mantadia Complex, Madagascar. Unpublished document, prepared for APAM and Tropical Forestry Management Trust (TFMT), Gainesville, FL.

Keck, A., Sharma, N. P., and Feder, G., 1994. Population Growth, Shifting Cultivation, and Unsustainable Agricultural Development: A Case Study in Madagascar. The World Bank, Washington D.C.

Margoluis, R. and Salafsky, N., 1998. *Measures Of Success: Designing, Managing, And Monitoring Conservation and Development Project,* Island Press, Washington, D.C.

Program for International Development, Clark University and National Environment Secretariat, Ministry of Environment and Natural Resources, Kenya, 1989. *Introduction to PRA.* Clark University, Worcester, MA. (Available in 12 languages including Malagasy, French, and Kiswahili.)

Razakamarina, N., Ford, R., Sodikoff, G. M., Wood, S., Toto, E., and Laris, P., 1996. *Negotiating Conservation: Reflections on Linking Conservation and Development in Madagascar: Interim Thoughts on Collaborative Approaches to Sustainable Livelihoods and Resource Conservation Around the Andasibe/Mantadia Protected Areas, Madagascar,* unpublished document, Clark University and APAM, Worcester, MA.

Reveley, P.S., Rajaona, J., Rasamison, F., Ramambasor, H., Rabarison, H., Razakamarina, N., and Razafindrakotohasina, N., 1993. *Analyse Participative en vue de la Réduction de Pression sur une Aire Protégé*, SAF, Andasibe, Madagascar and Program for International Development, Clark University, Worcester, MA.

Shyamsundar, Priya, 1993. Socio-buffering Around the Mantadia National Park: Problems and Possibilities, unpublished graduate thesis, Duke University, Durham, NC.

21 Facilitation, Participation, and Learning in an Ecoregion-Based Planning Process: The Case of AGERAS in Toliara, Madagascar

Paul D. Cowles, Soava Rakotoarisoa, Haingolalao Rasolonirinamanana, and Vololona Rasoaromanana

CONTENTS

0-8493-0020-7/01/$0.00+$1.50

INTRODUCTION

The purpose of this chapter is to characterize and assess key aspects of the pilot implementation of a collaborative ecoregion-based planning process called AGERAS, in Toliara, Madagascar. This unique process has arisen in response to the need to address widespread and severe biodiversity loss in a coordinated fashion at a broader ecoregional or landscape level. While protected area management is critical to biodiversity conservation, the Malagasy example highlights the importance of having other approaches to dealing with threats to biological resources outside protected areas. The long-term conservation of these resources calls for adaptive, collaborative approaches to dealing with threats to biodiversity at broader geographic scales. The AGERAS process uses participatory planning committees made up of representatives from local, communal, and regional levels to analyze environmental problems and identify and implement strategies and actions capable of responding to those problems. The AGERAS pilot implementation in Toliara, Madagascar provided examples of how to facilitate broad stakeholder involvement while ensuring that the process leads to learning and adaptation. This chapter highlights some of the key lessons that have been learned concerning the facilitation of the process, the use of planning committees to implement the process, and the role of action-research as a tool for learning.

The case study is developed from the perspective of key actors who worked together to implement the AGERAS process in Madagascar. The group includes individuals who were involved directly in the Toliara pilot implementation, a lead investigator of the overall AGERAS pilot evaluation, the technical assistant to the AGERAS process, and the director of the National AGERAS Unit. The authors hope the lessons that have been distilled from a blend of experience and perspective may be of some value to the process as it is currently being implemented in four other regions of Madagascar, and to initiatives elsewhere that seek to balance biodiversity conservation and local livelihood goals through adaptive collaborative management (ACM).

MADAGASCAR, NEAP, AND AGERAS

The people of Madagascar are searching for new ways to interact with their environment. The fourth largest island in the world, Madagascar was once contiguous with the African continent. Around 100 million years ago, when it separated from Africa, its flora and fauna began evolving along different lines than those on the continent. Today, thanks to this isolation Madagascar possesses one of the most unique assemblages of floral and faunal biodiversity in the world (Table 21.1). At the same time Madagascar is among the poorest of countries (Table 21.2). There is a desperate need for economic development and stability, including, for many of the

TABLE 21.1
Madagascar's Biodiversity

- High level of endemism:
  - 80% of plant species
  - 95% of reptiles
  - 99% of amphibians
  - Nearly 100% of primates (Lemurs)
  - Higher order endemism (genera and families) is common
- Forest cover of Island through time:
  - Nearly total in prehistoric times
  - 25% of total surface in 1950
  - 20% in 1972
  - Less than 15% today

Source: World Bank (1996) and Langrand (1995)

TABLE 21.2
Madagascar's Economy

Per capita income down 40% over last 20 years to $230
75% of the population lives in poverty, most (90%) in rural areas
Natural resource based:
  Agriculture, livestock, fisheries, and forests: 31% GDP, 60% of exports, and 70% of work force
  Natural resource–based tourism is growing sector
Economic cost of soil loss and siltation equals 5 to 15% of GDP
Advances are being made
  Lower import tariffs
  No commodity subsidies
  Improved rice harvest 1994/95
  Rational pricing of energy

Source: World Bank (1996).

island's rural poor, basic food security. The interaction of these disparate realities, along with other political and cultural root causes, has created a complex and vicious cycle resulting in the destruction of Malagasy biodiversity and continued socioeconomic decline.

Madagascar's ecosystems range from humid and arid tropical forests to ecologically critical mangroves and coral reefs, all with a high level of endemism (Nicoll and Langrand, 1989). The use of these ecosystems is exploding in a relatively unchecked fashion with vast interconnected consequences for nature and society. For example, forests are disappearing from Madagascar at an alarming rate, mainly due to *tavy* or *tetikala* (slash-and-burn agriculture) (Langrand, 1995), but also due to fire and illicit or uncontrolled commercial and private timber harvest and mining. If forests continue to disappear at historic rates it is likely that in the next 25 to

30 years natural forests will exist only on very inaccessible steep slopes (Green and Sussman, 1990). The consequences of this deforestation and extreme fragmentation of ecosystems would likely result in massive extinctions of plants and animals that exist nowhere else in the world. These are biological resources that could be of economic importance to the country, through potential pharmaceuticals and other biodiversity-based products. The loss of the ecological functions these forests perform would also worsen an already severe agricultural crisis through soil loss and impoverishment, further complicating Madagascar's quest for economic improvement. It is clear that the country is fast approaching what could be the last chance to find solutions that can both conserve biodiversity *and* ensure the sustainable development of the Malagasy people.

To save its unique fauna and floral diversity in the face of these challenges, the government of Madagascar was the first in the African region to establish a National Environmental Action Plan (NEAP) in 1989. The NEAP was designed to encourage the reconciliation of humans and their environment (Government of Madagascar, 1998). This ambitious plan was to be implemented over a period of 15 years covering three phases. The first phase (EP1) took a project approach to resolve a range of critical environmental issues, and focused on the creation of a protected area management system and the institutions needed to address environmental issues. The second, and current phase (EP2), focuses on the "regionalization" of participatory environmental management through the use of a programmatic approach that shares responsibilities between multiple actors and builds on the accomplishments of the first phase. The third phase, slated to begin in 2002, will focus on ensuring that environmental concerns become an integral part of the macroeconomic development of the country.

The main implementation mechanisms during the first phase of the NEAP were the Integrated Conservation and Development Projects (ICDPs) and the "mini-projects," which involved small, community-based natural resource management actions such as irrigation dams, reforestation, etc. Although these two mechanisms did have limited successes, they were not able to generate the impact needed to stem the level of environmental degradation taking place in Madagascar. Multiple evaluations of the EP1 activities were carried out and some key lessons began to emerge. One of the more critical lessons was that the ICDPs and miniprojects had too narrow a focus in trying to address extremely complex environmental problems such as deforestation, erosion, etc. This narrow focus pertained to geography as well as to the extent to which stakeholders at local, regional, and national levels were involved in the projects (World Bank, 1996; McCoy and Razafindranibe, 1997). These and other lessons (Table 21.3) led to the creation of three environmental components specifically for the EP2.

1. AGERAS (Appui à la Gestion Environnementale Régionalisée et à l'Approche Spatiale) was created as an iterative planning process to broaden participation, foster synergy between a range of different stakeholders and sectors, and improve management of the environment, especially biological diversity. Key to the process is the development of an analysis of environmental problems that takes into account root causes at multiple levels: local, regional, national, and international.

TABLE 21.3
Selected ICDP Lessons Learned

Conservation-linked development efforts need to give more consideration to the underlying causes of biodiversity threats and need to take a broader geographic view of them. Threats should not be assessed only within the peripheral zone of a protected area since they can sometimes originate from much farther away.

Collected a wide range of data for analysis. Gather information from communities, other agencies, and organizations in the region.

Seek "Quality" participation from stakeholders. Don't rush through community involvement. Establish relationships with traditional and official community authorities.

Engage participants in activities in such a manner that they take responsibility for the activity to increase community ownership and improve chances for the long term sustainability of an activity.

Increase collaboration with local authorities, government services, and local NGOs when undetaking field activities.

Build lasting relationships between stakeholders and help them to build a common vision.

Source: McCoy and Razafindranibe (1997).

2. The GELOSE (Gestion Locale Securisée) was created and adopted in Malagasy law as a way of increasing the rights and the responsibilities of local communities to manage natural resources at a local level.
3. FORAGE (Fonds Regional d'Appui à la Gestion de l'Environnement) was created as a fast-reaction grant fund to aid in implementing actions that could have important environmental impacts.

These three elements were envisioned as independent mechanisms that would nonetheless work together to enhance natural resource management in Madagascar. For example, it was expected that FORAGE would provide funds for some of the actions identified by the AGERAS process. GELOSE was a potential tool that AGERAS participants could use to improve management of natural resources at a local level.

The implementation of AGERAS as a component of the EP2 came in two phases. The first was a 1-year pilot phase (1996 to 1997), funded by the United Nations Development Program (UNDP), to develop and adapt the process to the cultural, social, and institutional realities of working in the diverse regions of Madagascar. This pilot phase was also an opportunity to test the FORAGE funding mechanism. After the pilot phase, the AGERAS component went into full swing with the development of a National AGERAS team based in the National Office of the Environment and establishment of AGERAS facilitation offices in the three regions, where the pilot phase was implemented: Toliara, Fianarantsoa, and Lake Alaotra.

These regional AGERAS offices were tasked with following up on pilot activities, establishing or reinforcing multi-institutional planning committees, and facilitating the committees through the AGERAS process (Figure 21.1). This chapter focuses on the particular case of Toliara during the pilot phase because it provides the broadest range of lessons relevant to the ACM of threatened biodiversity.

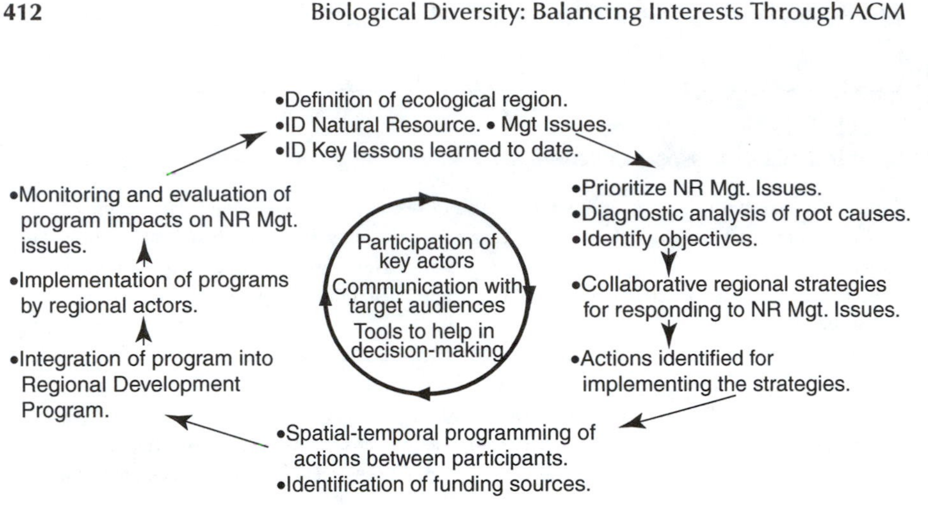

FIGURE 21.1 Elements of the AGERAS process.

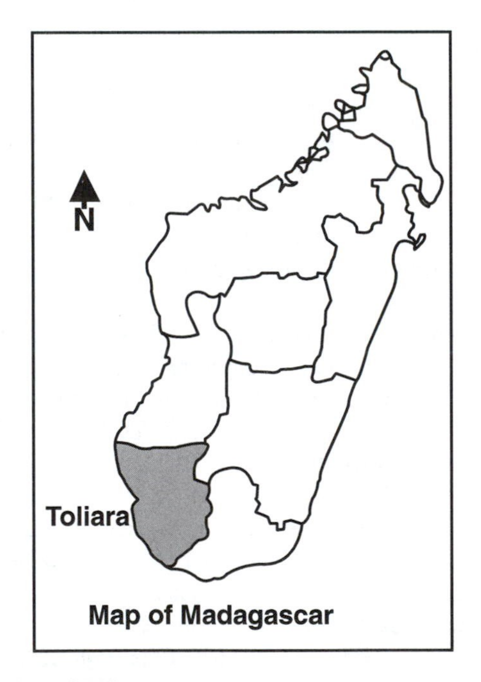

FIGURE 21.2 Location of Toliara.

AGERAS PILOT PHASE IN TOLIARA

The AGERAS pilot phase in Toliara region (Figure 21.2) began in November 1996. A Malagasy Development nongovernmental organization (NGO) called TAMBIRO was hired by the National Office of the Environment to facilitate the process in Toliara. The TAMBIRO team was experienced in participation and consultation processes as a result of its earlier involvement in biodiversity conservation-oriented environmental planning activities undertaken in Phase I of the NEAP.

The role of TAMBIRO was to act as a neutral facilitator of the AGERAS process. This involved helping interested stakeholders at local, communal, and regional levels to organize themselves, and providing participatory processes for stakeholders to analyze environmental problems in the region and come up with shared solutions. The facilitator works as a catalyst, providing tools and information and acting as a communication hub to help stakeholders work through the process. It is important to note that the facilitators must remain neutral; their interest is centered on implementing the process so that it serves the needs of the stakeholders.

The Toliara Region

Toliara is an arid to semiarid region characterized by dry, dense tropical forests bordering on the spiny forests of the south. Mangroves and coral reefs are also important to the ecology of the region. The area has a high level of endemism especially among reptiles and amphibians. There are three predominant cultural groups in the region: Mahafaly in the south, Bara in the west, and the mix of Masikoro, Vezo, and Mikea in the north part of the region. The Mahafaly and Bara tribes depend on grazing and subsistence agriculture, generally slash and burn. The Masikoro tribe also depends on livestock and small-scale slash-and-burn agriculture, whereas the Vezo rely mainly on traditional fishing and the Mikea depend on hunting and gathering of forest products. It is interesting to note that the Mikea are the only forest-dwelling, hunter gatherer-based culture in Madagascar. There is intense migration into the region, both seasonally and permanent, from the south. The presence of a large port in the city of Toliara has a significant impact on the economy and the use of natural resources in the region.

The remaining indigenous forest blocks in this region are found surrounding the city of Toliara and in the northern part of the region, the Mikea forest. The Toliara forest is threatened by both charcoal and slash-and-burn exploitation, while the Mikea forest is threatened mainly by slash-and-burn agriculture focused on corn production for export.

The main crops in the region are corn, cotton, cassava, and beans, but cotton is the only crop produced on a large commercial basis. The others are produced by small farmers in the north and west part of the region. The southern part of the region is not fertile because of its high lime content. After cotton, corn is at present the most profitable crop because it is exported to Reunion and Mauritius Islands for use as swine feed.

Methodology Used by TAMBIRO

Faced with the complexity of the environmental problems in this region, three steps were taken by the TAMBIRO team to facilitate implementation of AGERAS.

The first step was the creation of a regional-level task force charged with implementing the process. This task force was open to anyone interested in participating and was made up of over 60 people from government (elected and appointed),

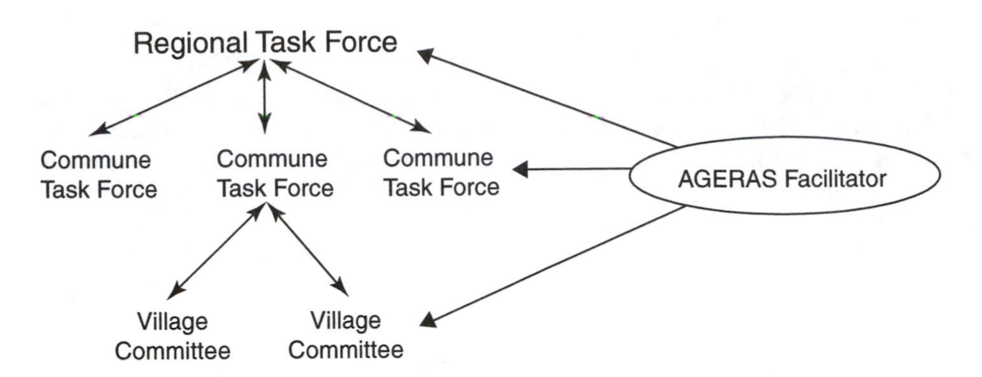

FIGURE 21.3 Organizational chart of regional-level task force.

private sector, voluntary associations and NGOs, development projects, and government technical services (Water and Forests, Agriculture, etc.), as well as researchers. Although the official group was large, the regularly participating membership of the task force varied from about 30 to 35 people.

The second step, and a key innovation in the approach was the development of an extensive network of consultative groups at the village and commune levels. These groups fed information into the regional-level task force, as illustrated in Figure 21.3. These groups were used to aid in the analysis of the problem of deforestation in Toliara. Information about the issue flowed from the village level toward the regional level where it was consolidated and redistributed back to the lower-level structures. This structure gave the AGERAS facilitator the ability to encourage a two-way exchange of information instead of a strictly top-down or bottom-up approach.

The third step was the actual implementation of the AGERAS process involving the development of strategies and actions based on an in-depth analysis of the causes of deforestation. However, two factors led to a modification of this part of the process. The first was that, while the facilitators wanted to have an impact on deforestation, the fact remained that this was a pilot project in place for only 1 year and therefore unable to make long-term commitments to strategies and actions. A second factor was that the TAMBIRO team felt that the information they had gleaned through the analysis was incomplete. Because of these factors, it was decided to use an action-research type approach to implementing a few discrete actions. Although it was hoped that these actions would have an impact on deforestation, their implementation was designed so that they would provide further information on the causes of deforestation and the best way to implement responses. It was hoped that this approach would both provide a better understanding of the deforestation problem and provide examples of activities that can have an impact on the root causes of deforestation.

Deforestation Problem Defined

One of the key steps in the Toliara AGERAS process was helping the task force network to develop a common understanding of the underlying causes of deforestation in Toliara through an in-depth analysis. This required a significant amount of time, because of the need to exchange information among groups in the network, but it gave the network a common focus for the planning process. The following is a summary of the findings of the analysis.

The environmental problems in Toliara region were categorized according to two questions. (1) Is forest degradation due to slash-and-burn activities for corn production? This problem affects both the forest surrounding the City of Toliara and the Mikea forest. (2) Is forest degradation due to charcoal production that affects the forest surrounding Toliara City in particular?

Concerning charcoal production, the main impact is the disappearance of precious woods such as *Geospyrus* sp. and *Dalbergia* sp. These dense, hardwood species are available only in the remaining primary forest surrounding Toliara. They are highly valued for charcoal production because of their long, slow burn rates. Since the species have been exploited for many years, charcoal producers are now forced to travel farther and farther from Toliara to find suitable stands. This, in turn, forces the producers and their workers to establish temporary settlements that need to produce food from slash-and-burn agriculture. This accelerates forest destruction and has led to the loss of approximately two thirds of the original forest cover around Toliara. The main markets for charcoal are household and restaurant use, as well as use for the production of kiln-fired bricks and lime.

Corn production contributes enormously to the degradation of the Mikea forest as well as to the loss of forest surrounding Toliara. Corn, while being an important food source in the region, is particularly valuable as an export to Reunion and Mauritius where it is used as a livestock feed. In general, it is felt that corn from newly deforested land is of a higher quality, both in terms of taste and nutrition, than corn grown on existing agricultural land. This is largely due to the lack of knowledge and materials needed for the appropriate management of existing farmland. The high price paid for export-quality corn and the indeterminate nature of land tenure in forested areas encourage migration into the area by the Antandroy, Mahafaly, and Bara tribes. This in-migration causes residents in the area to feel the need to "make their claim" to forested areas by clearing them and planting corn. These factors along with a complete lack of appropriate regional development policy have worked together over the last 10 to 15 years to accelerate forest destruction. Today, only half of the Mikea forest's original 40,000 ha remains.

Strategies and Actions Developed

Based on this analysis, the task force network set out to develop potential strategies and actions that could reduce pressures on the forests from charcoal and corn production. Four strategies were developed, which are presented below with example actions.

1. To develop new sources of charcoal and to provide alternative financial opportunities for charcoal producers.
   - Action: Promote reforestation by charcoal producers in collaboration with the regional office of forest and water department.
   - Action: Promote improved charcoal production kilns that produce longer burning charcoal.

2. To encourage use of improved technologies and new energy sources to reduce demand for charcoal.
   - Action: Diffusion of a new type of cookstove to households and restaurants in collaboration with a regional energy project of the World Wildlife Fund (WWF).
   - Action: Promote alternative sources of combustion for brick and lime production (e.g., plantation woods, such as eucalyptus, or peat from local bogs).
3. To provide development activities in the immigrant "departure zones" to encourage sedentarization among potential immigrants.
   - Action: Create wells for irrigation, potable drinking water, and water for cattle in the deep south of the region.
   - Action: Create ecotourism zones for income generation activities linked to biodiversity conservation.
   - Action: Repair communal roads to improve access and the trade of goods in the region.
4. To encourage improved agricultural techniques in already deforested zones to promote permanent settlement and increase income from this land.
   - Action: Create/rehabilitate irrigation channels in the deforested zones.

Action–Research Approach Adopted

At this point in the normal course of the AGERAS process the facilitators would work with the task force network and various donors or NGOs to develop proposals for the implementation of these strategies and actions. As stated above, however, the pilot phase was not able make this long term commitment to the process. Although it was expected that a full-time AGERAS team would be deployed in Toliara in the near future, it was not a given. Additionally, there was a certain level of uncertainty among the facilitators and the stakeholders regarding the accuracy of the analysis and therefore the effectiveness of the actions in slowing deforestation. Based on these issues, it was decided that a few of the above actions would be implemented in discrete areas in an action-research approach. The actions selected included irrigation canal rehabilitation, well digging, and community forest management.

The actions were to be funded through the pilot implementation of the above-mentioned FORAGE mechanism, which was controlled from the National Office of the Environment, based in the capital. The facilitators selected contractors to implement the actions through a competitive bidding process. In each community where actions were to be implemented, the facilitators also hired observers to collect qualitative information related to the following types of questions:

- What is the organizational capacity of the community involved?
- Who in the community is likely to facilitate or block implementation?
- Who is going to participate or not participate and why?
- Does the action have an impact on the biodiversity threat (i.e., migration, charcoal or corn production)?
- What is the role of women in the system?
- How are conflicts resolved?

The observers were most often students. Multiple observers were engaged in each community to help validate results.

Although the implementation of the actions had certain problems, as identified below, the approach did have value as a learning tool for the facilitators and the stakeholders. The main lessons learned from the actions were related to the complexity and challenging nature of implementing even small projects in communities. Some of the questions, such as impacts on biodiversity threats, were difficult or impossible to assess given the brief time period for the pilot phase. However, the facilitators and members of the task force network emerged from the experience with a much better understanding of community dynamics and capacities and the potential for conflict where these types of actions are being implemented. This prepared them for the challenges that would come with the full implementation of the process once the full-time AGERAS team was in place.

CHALLENGES DURING THE PILOT PHASE IN TOLIARA

The pilot implementation of AGERAS in Toliara presented many challenges. The following is a discussion of some of the key issues that emerged that required considerable adaptation and learning by the facilitators and participants in the process.

Implementation of Actions

A key issue facing the pilot phase was that the funding mechanism to be used for implementing actions (FORAGE) was completely untested. The administrative procedures were extremely cumbersome and the bureaucracy involved in approving and releasing funds was very slow. These administrative problems kept some of the planned actions from being implemented. This had the effect of discouraging both the facilitators and the stakeholders who had built up high expectations based on the work that they had accomplished together.

Conflict Management

Managing conflicts among stakeholders proved to be one of the more time-consuming aspects of the implementation of the pilot phase (TAMBIRO, 1996; 1997). Most conflicts surrounded the identification of strategies and the prioritization of actions to be implemented. The facilitators had to mediate conflicts and facilitate negotiations around two main types of issues: conflicts between landlords and laborers over land use and perceived conflicts between private sector and NGO interests (TAMBIRO, 1997).

Once the process began to focus on the need to improve agriculture techniques used by small farmers, the large landowners felt that they might lose access to the cheap labor they had been using to work their own lands and that they might have to contend with more competition from small landholders. The facilitators had to work hard to mediate between the stakeholders to find an acceptable approach to this activity. Eventually, it became clear to the large landowners that it was not in their best interests to block assistance to small farmers and that the likely impacts on the labor and agricultural markets would not be as severe as expected.

Private sector interests proved difficult to involve in the AGERAS process. This was due mainly to their perception that this type of process could not benefit them. The facilitators devoted a lot of time to convincing these stakeholders of their long-term interest in participating in the process. Once a few members of the private sector became involved, they found themselves viewed as the enemy by many of the NGO stakeholders. There was a general perception by the NGOs that the private sector, because of its links to commerce in charcoal and corn, was the root cause of the problem. The private sector for its part felt that the NGOs were attempting to stifle their efforts to make a profit. Eventually, the facilitators were able to mediate an uneasy truce between these groups, which allowed the NGOs to focus on improving the livelihoods of communities and the private sector to focus on the long-term benefits they could gain from more sustainable land management in the area.

Education and Communication

A major task for the facilitators was to develop reliable data, and to share that information with the task force network. The challenge essentially was to tailor the form and content of relevant information to the respective stakeholders. There were large disparities in education and cultural perceptions among the stakeholders. At the same time, bringing everyone to a more or less equal footing in terms of an understanding of the analysis was a necessity. Needless to say, this also slowed implementation of the process and required extensive education and communication skills on the part of the facilitators.

Gender Issues

The Toliara region is well known for its highly patriarchal society, which excludes woman from any form of decision making (Rakotoarisoa, 1997). Involving women in the process was an important point of contention with which the facilitators had to deal. Little headway was made, however, partly because of the low education level of many women in the communities. The gender participation issue in this region is still highly controversial. Educated women at the regional level may overcome it to an extent through active participation, where acceptance of their ideas is more common.

The Perception of Helping Only People Who Break the Law

One of the striking conclusions reached by the facilitators was an appreciation that the process, like many conservation-linked development projects, was in effect serving the interests of those engaged in nonsustainable resource exploitation. At the same time, it was viewed as contributing comparatively little to strengthen the position of groups who were trying to exploit natural resources sustainably (Rakotoarisoa, 1997). A similar conclusion was reached in evaluating the Madagascar ICDPs (McCoy and Razafindranibe, 1997). In effect, because of the focus on biodiversity threats, the projects tended to work mainly with people who were exploiting forests illegally while ignoring their law-abiding neighbors. To some extent, the facilitators were able to address this issue by using broad stakeholder involvement. But it is a distortion of the conservation program approach that must be dealt with if more effective and long-term biodiversity conservation is to be achieved.

INSIGHTS AND FUTURE DIRECTIONS

The following insights generated from the pilot experience may be applied to aid in the improvement of the AGERAS process as it is being implemented currently. These may also be of use to other conservation initiatives using broad-based stakeholder involvement.

USE OF FACILITATORS

For the AGERAS process the presence of an on-site, full-time facilitator enhanced implementation of the process. Oftentimes in Madagascar, committees are formed, especially at the regional level, without providing for a facilitator. These groups tend to operate without focus and often fail within a short period of time. By paying attention to the experience with the use of neutral facilitators in Toliara, there is hope to improve the continuity and effectiveness of group processes.

First, outside facilitators who understand the context and culture and are able to bring new practical vision and tools essential to the process. The facilitators need to have the skills to be effective in communicating or translating between different groups and interests that may have differing education levels and cultural practices.

Conflict management skills that are sensitive to traditional cultural approaches to conflict are essential to the facilitation role. The facilitators often found that the only way of progressing from one step of the process to the next was by sorting out and dealing with the conflicts that were blocking progress. To be credible in this role and effective generally, the facilitator must be neutral. A facilitator's power lies in people's trust that he or she is not interested in how a given issue is resolved, only that it is resolved in a way that is fair to all stakeholders.

Facilitators should focus first on building relationships and getting people to work together within the task force network, then help the group to improve technical analysis. The most difficult part of any participatory process is getting people from different and sometimes conflicting sectors to work together. At the same time, this is extremely important to the long-term success of the process. Too often, processes can focus on trying to get the best data rather than developing the relationships that will be necessary to make hard decisions about natural resource uses. It is better to work with the best available knowledge, while building these working relationships, even though the scientific quality of the information may be suspect. Later, better science can be brought into the process.

USE OF TASK FORCE NETWORKS

The three-tiered consultative task force network used in Toliara proved to be effective in ensuring information flow and encouraging understanding between different groups within the region. The following insights were generated from the experience with these groups.

Helping task forces to self-evaluate their function and effectiveness leads to more effective participation and ownership in the process. After 1 year, the facilitators initiated a self-evaluation of the task forces covering all aspects of the operation. This allowed the groups to understand better their own dynamics and to improve their decision making. Based on these types of self-evaluation, task forces can be targeted for capacity building in areas such as management of information, analysis, programming and planning, and monitoring and evaluation. These groups can be ready-made classrooms and are often hungry for new skills and techniques. Also, the multidisciplinary nature of the task forces ensures that new skills are spread through many different sectors within the region.

It is important for the process and the networks to begin small and build toward broader ecoregional scales. The AGERAS process in Toliara began by focusing on specific areas threatened by overexploitation within a regional context. This was much easier for the committees to grasp than starting with a focus on the entire region. Since completion of the test phase, the regional task force has recognized that it does not have a sufficiently broad base of knowledge about the region and has instituted a larger-scale analysis of the environment at the regional level.

Use of Action-Research Approach

The action-research activities did succeed in providing valuable information to the process on the human dynamics of the deforestation issue in Toliara, as well as the complexity of implementing projects at a community level. The AGERAS team has gleaned a number of insights from the use of action-research in Toliara.

First, clear linkages must be explicitly described between the actions and the root causes of the environmental issue. This allows for the development of clear assumptions and objectives that can then be monitored and adjusted if necessary. Without these linkages, it is difficult to demonstrate clearly why one development activity might be more desirable than another.

The use of an independent village-based observer to follow implementation qualitatively and quantitatively was valuable in understanding the dynamics of the issue at the local level. The observer paid attention to how people did or did not organize themselves, which groups were marginalized by the action or had an interest in blocking it, and what other actions would be needed to ensure sustainability of positive outcomes.

Indicators of behavioral change that are capable of being followed easily and providing usable information quickly are more useful in the short term than ecological impact indicators. In other words, it is easier and more indicative of conservation progress to follow how many hectares of forest are lost to slash-and-burn agriculture rather than the distribution or density of certain animal or plant species.

Finally, rather than performing just a few action-research type activities to improve the analysis, *all* the actions emerging from the AGERAS process should be "learning opportunities" tied to our assumptions of the root causes of the environmental issue. This more "adaptive management" type approach to actions not only would improve the information base, but should also provide AGERAS a way of better ensuring an effective, positive impact on the environmental issue.

These insights and ideas are currently being incorporated into the AGERAS process in the four other regions where it is being implemented, as well as in Toliara. The Toliara AGERAS process is the farthest advanced of these five ecoregions. Now in its second year of implementation, the process has expanded its view of environmental problems to include a much larger landscape, including marine and terrestrial ecosystems. Additionally, the AGERAS unit in Toliara has recently teamed up with the new WWF/U.S. Spiny Forest Ecoregion-Based Conservation Project to ensure a better mix of science and participatory planning as responses to critical environmental problems are developed.

CONCLUSIONS

This chapter has looked at the pilot implementation of the AGERAS process in Toliara, Madagascar. This collaborative landscape-level planning process came about as a response to threats to biodiversity that were not being dealt with through standard protected area management approaches, such as that of ICDP. The process requires broad-based stakeholder involvement and an adaptive approach that allows for learning and change. The Toliara pilot project provides an example of how this type of process can be implemented, using a neutral facilitator, an extensive network of public task forces, and an action-research approach to learning while doing.

In looking at the overall success of the pilot project, it was really the use and interaction of all three of these aspects that allowed the process to deal with the numerous obstacles that it faced.

- The presence of a facilitation team ensured that the process had a shared vision for where it was going and a neutral third party to help manage conflicts as they arose.
- The three-tiered task force network gave the process legitimacy among stakeholders and a structure to allow back-and-forth communication across the full range of types of stakeholders involved in the issue of deforestation.
- Action-research provided a feedback mechanism by which participants could learn more about the dynamics of deforestation and the implementation of actions to change people's behaviors.

At the same time the pilot implementation underlined many of the perennial problems encountered in conservation and development initiatives. Bureaucratic bottlenecks blocked implementation of actions. Severe conflicts over natural resource use between different stakeholders slowed the process and frustrated participants. Local women had very little voice in how the process progressed, and law-abiding local people often felt ignored while the process focused on trying to help those who were illegally clearing forest.

Conservation and development initiatives such as AGERAS will have to begin to deal with these types of issues more successfully if significant progress is to be made. Facilitators will have to be more creative and skillful in dealing with conflicts involving disadvantaged groups effectively. Information flow between the task forces must be improved and the range of stakeholders involved should be constantly reassessed and increased if necessary. Finally, action-research-type feedback loops must begin to provide good information on the effectiveness of actions to curb biodiversity loss.

The AGERAS process represents an alternative to more traditional conservation strategies used in Madagascar. If the process can continue to involve a broad spectrum of stakeholders and continue to help participants learn and adapt based on lessons learned, it might well contribute to a future for Madagascar that includes a healthy, diverse ecology *and* economy.

ACKNOWLEDGMENTS

This case study was funded by the U.S. Agency for International Development through the MIRAY Program, implemented by Pact, Inc. The ideas and opinions expressed herein are solely those of the authros and do not necessarily reflect those of USAID, nor of Pact, Inc.

REFERENCES

Conservation International, 1995. Rapport de l'atelier Scientifique sur la Priorisation des Sites pour la Biodiversité de Madagascar, PRIM/GEF.

Green, G. M. and Sussman, R. W., 1990. Deforestation history of the eastern rain forests of Madagascar from satellite images, *Science,* 248, 212–215.

Government of Madagascar, 1998. Charte de l'Environnement.

Langrand, O., 1995. The effects of forest fragmentation on bird species in Madagascar: a case study from Ambohitantely Forest Reserve on the central high plateau, Master's thesis, University of Natal, South Africa.

McCoy, K. L. and Razafindranibe, H., 1997. Madagascar's Integrated Conservation and Development Projects: Lessons Learned by Participants. Project Employees, Related Authorities, and Community Beneficiaries, USAID, Madagascar.

Nicoll, M. E. and Langrand, O., 1989. *Madagascar: Revue de la Conservation et des Aires Protégées,* World Wildlife Fund, Gland, Switzerland.

Rakotoarisoa, S. V., 1997. Rapport sur l'évaluation par les bénéficiaires des trois régions de la Phase Test d'AGERAS: Alaotra-Fianarantsoa-Toliara, PNUD/ONE AGERAS.

TAMBIRO, 1996. Rapport d'avancement du Processus Ageras, Analyse diagnostic, Tambiro.

TAMBIRO, 1997. Rapport d'avancement du Processus Ageras, Strategies et actions, Tambiro.

World Bank, 1996. Madagascar: Second Environmental Program, Project Document.

22 Reclaiming Ancestral Domains in Pala'wan, Philippines: Community-Based Strategies and Perspectives on Adaptive Collaborative Management

Maria Cristina S. Guerrero and Eufemia Felisa Pinto

CONTENTS

0-8493-0020-7/01/$0.00+$1.50

INTRODUCTION

The island of Pala'wan, which is located west of the main chain of the Philippines, is one of the country's largest national protected areas; its biodiversity is also one of the most vulnerable. Pala'wan is also recognized internationally by the United Nations and the World Conservation Union (IUCN) as a conservation hotspot. Through national legislation in 1992, Pala'wan is governed within a comprehensive environmental protection and sustainable development framework. The Strategic Environmental Plan for Pala'wan (hereafter referred as the Pala'wan Plan) is a novel framework that attracted significant donor support and buy-in from the local government and general public despite the challenging implications of its implementation.*

The responsive political climate in Pala'wan is an extension of the national wave of democratization and devolution after the 1986 People's Power Revolution that deposed then President Ferdinand Marcos. Strong public demand for government accountability, local participation, and the rights of indigenous people with respect to conserving Pala'wan's biodiversity were converging. These were complemented by an environmental movement in the province backing legislation to address threats to environmental degradation.

The challenges before the Pala'wan Plan are prodigious because of conflicts among multiple actors over the use and management of resources found within environmentally critical forest, mines, and agricultural areas, small islands and mangrove areas, and tribal ancestral domains that traditionally spanned terrestrial and/or coastal territories. Besides conservation interests, various economic interests are at play in Pala'wan within a context of unequal power. Addressing these interests is even more complicated by the question of decision making and implementing authority and by the extent to which marginalized indigenous communities can influence the demarcation and management of Pala'wan's environmentally critical areas. This chapter illustrates how this latter concern might be addressed through adaptive collaborative management (ACM).

Two indigenous communities in the Municipality of Rizal in Pala'wan, Philippines are featured here. Both adopted provincial and/or national policy frameworks to meet their livelihood needs. In the course of following the guidelines set forth by these frameworks for ancestral domain delineation and management, the Pala'wan** indigenous communities in the municipality of Rizal (Figure 22.1) apply strategies and tools that employ iterative planning and management processes and a level of collaboration with external groups. These adaptive and collaborative strategies provide the opportunities to bring further legitimacy and effectiveness to the communities' management control over their ancestral domains; cultivate fluid community-based processes and more informed community decision making; build comprehen-

* The plan, which was legislated by special law, localized the responsibility for planning, management, regulation, and utilization of natural resources for the province, the first such devolution of natural resource governance in the country. This would have otherwise been the mandate of the powerful Department of Environment and Natural Resources.

** The term *Pala'wan* refers to the indigenous group as differentiated from Pala'wan, the name of the province.

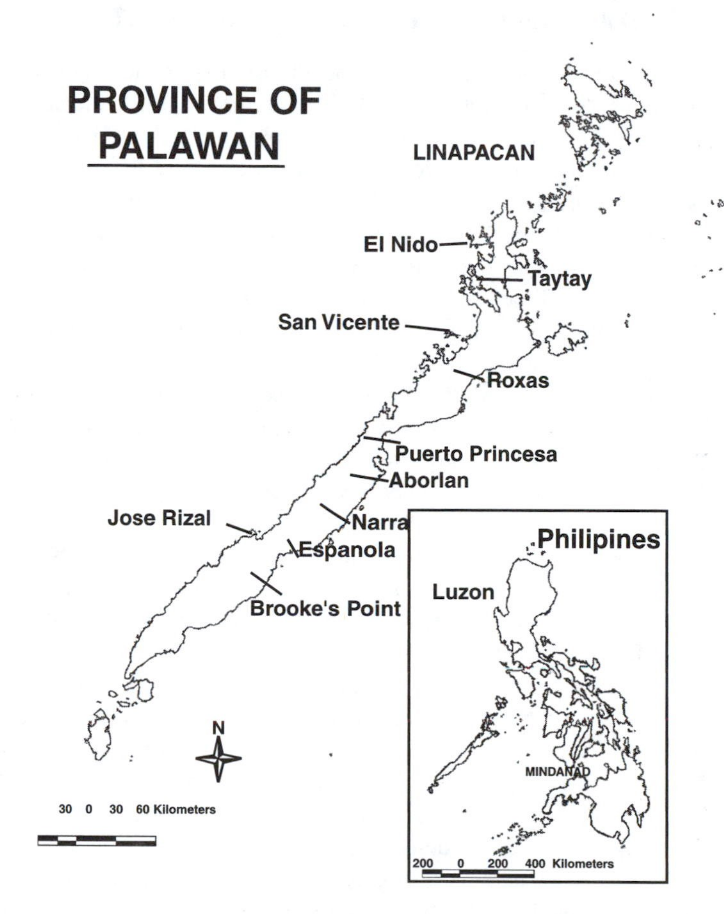

FIGURE 22.1 Map of Province of Pala'wan, the Philippines.

sive institutional support and capacity-building mechanisms for indigenous communities; strengthen the economic viability of community-centered livelihoods; and mitigate bureaucratic implementing rules and regulations for ancestral domain and/or environmentally critical areas delineation and management.

The authors draw the conclusion that institutional and policy support for community-based initiatives are critical for indigenous communities to engage effectively in the process of delineating and managing ancestral domains across the island province. Institutional and policy support includes technical, funding, and legal assistance, leadership and capacity-building, and community-organizing and strengthening; all these influence the parameters for ACM within ancestral domains. Furthermore, equitable power-sharing arrangements and flexible intracommunity collaboration mechanisms must be built into the strategies and tools for ancestral domain management to create and sustain positive impacts on indigenous community development, as well as on the management of Pala'wan's environmentally critical areas.

ANCESTRAL DOMAIN POLICY FRAMEWORKS AND PALA'WAN COMMUNITY ENGAGEMENT

Pala'wan is considered the last frontier of the Philippines, mainly because of its geographic isolation. Even though it is relatively removed from the mainstream political and economic development of the Philippines, Pala'wan itself has been developing rapidly in the last two decades. The integrity of Pala'wan's natural resources has been facing serious threats from commercial logging operations (up to 1992); settlement by impoverished lowlanders and migrants from other island-provinces in the Philippines; small-scale farming and subsistence agriculture; mining activities; and destructive commercial fishing practices. The fires that ravaged Pala'wan forests in 1998 due to the El Niño phenomenon significantly reduced its forest cover by at least 15,000 ha.*

The allure of Pala'wan to environmentalists and policy makers is reflected in the numerous measures to preserve its frontier character through a series of policy actions, nationally and internationally. In 1967, Pala'wan was declared a National Game Refuge and Bird Sanctuary.** Thereafter, a presidential decree also declared the province a Mangrove Reserve.*** By 1992, 763,000 ha of the 985,000 ha classified by the government as National Protected Areas were located in the Pala'wan province, including the St. Paul Subterranean River National Park, El Nido Marine Reserve, and the Tubbataha Reefs National Park.**** Internationally, Pala'wan was designated as a Man and the Biosphere reserve in 1990.

In 1979, the Pala'wan provincial government developed the Pala'wan Integrated Area Development Project***** that set sustainable development and environment protection goals for the province. This project later spawned the formulation and legislation of the Strategic Environmental Plan for Pala'wan under Republic Act 7611 in 1992. The Pala'wan Plan primarily provides the formation of an Environmentally Critical Areas Network, where terrestrial, coastal/marine, and tribal ancestral lands are designated as distinct and special zones for protection and regulated use. The Pala'wan Council for Sustainable Development, which was created by virtue of the Pala'wan Plan as a distinct policy-making body, governs the province and its natural resources.

---

* In the 1990s, the forest cover of Pala'wan was the highest in the entire country. Statistics varied, however, from 40 to 54% of the total land area of the province.

** Pala'wan is home to 232 wildlife species, of which 11 nonflying mammal species and 14 bird species are endemic to the province. These species comprise 23% of the entire fauna in the country (Eder and Fernandez, 1996).

*** Pala'wan's mangrove forest is the second largest in the country.

**** This was through the 1992 National Integrated Protected Area Systems Act (NIPAS), which mandates the "classification and administration of all designated national protected areas to maintain essential ecological processes and life support systems, to preserve genetic diversity, to ensure the sustainable use of resources in these areas, and to maintain their natural conditions to the greatest extent possible."

***** The Pala'wan Integrated Area Development Project provided the institutional support for a range of activities: upland agricultural extension work, fisheries support, land classification and titling, environmental awareness and education, infrastructure development such as building irrigation, waterworks, and road systems, access to credit, health services, and other related components. The project was funded by the European Economic Community (later the European Union) and the Asian Development Bank.

The council does not govern alone, however. It operates within a context of other local and national institutions that also manage Pala'wan's environment and natural resources. The council must work within the controversial "power" dynamics with the Department of Environment and Natural Resources (DENR), the primary government body responsible for the conservation, protection, management, and development of the country's environment and natural resources. The council also must address the clamor of the indigenous communities in Pala'wan, who, through the DENR, are acquiring legal rights to occupy, manage, develop, and use their ancestral domains in the province as a consequence of a nationwide shift in land and forest policy* that became more inclusive and cognizant of indigenous peoples' rights. This situation implies that there could be conflicts in terms of who wields the higher authority and how these institutions could work together or engage with each other, if at all.

From the Boundary Delineation to Securing the CADCs in Rizal, Pala'wan

A certificate of Ancestral Domain Claim (CADC)** was awarded by the DENR to two Pala'wan*** indigenous communities in the Municipality of Rizal in southern Pala'wan in 1997. These were among the few that were granted in the province. This CADC has gone through several permutations for delineation, actual survey, and recognition of the ancestral domain claims and for the formulation and implementation of the community's management plans. The two cases clearly illustrate the key drivers of ACM and provide lucid insights and recommendations for promoting community-based natural resource management and ancestral domain delineation and management.

Community organizing support for the identification of the ancestral domain of the Pala'wan indigenous communities**** in the villages of Campung-Ulay and

* This pivotal shift started in 1982 when the DENR mounted a flagship Integrated Social Forestry Program to address the problem of massive deforestation in the country and to imbue more people-oriented strategies in forest management. Since then, forest policies in the Philippines progressively included provisions that recognize indigenous communities and their rights to social justice and to their ancestral domains. In addition, the 1987 Philippine Constitution bolstered the recognition of ancestral domains and the reinstatement of earlier laws that have been overlooked, such as the 1909 Carino decision recognizing native titles; the 1952 Public Land Law whereby all lands occupied by Filipino citizens before July 1946 are subject to ownership claims; and the Ancestral Lands Decree in 1974, whereby agricultural lands are declared alienable and disposable after they have been occupied and cultivated by indigenous peoples for at least 10 years before the decree took effect. These laws were used as legal bases for indigenous rights advocacy and the corresponding ancestral domain policies and programs.

** CADCs are legal instruments provided by the DENR to indigenous communities that certify their prior claims of possession, occupation, and use since time immemorial, of the land and natural resources, including all adjacent areas, that traditionally belonged to them and are necessary to ensure their economic, social, and cultural welfare.

*** The Pala'wan are the second largest indigenous group in Pala'wan. They are a traditionally shifting cultivation people inhabiting the hinterlands of the municipalities of Quezon, Rizal, Brooke's Point, Espanola, and Bataraza in southern and southwestern Pala'wan.

**** The Pala'wan ancestral domain covers the mountain range of Mt. Mantalingahan, Pala'wan's highest peak at 2086 m. This major mountain range is one of the identified priority areas for protection according to the Pala'wan Plan.

Punta-Baja in Rizal was first mobilized in 1989 by an indigenous peoples' federation called NATRIPAL (United Tribes of Pala'wan). NATRIPAL and their assisting organizations recommended that the DENR designate the municipality of Rizal as one of the three national pilot areas for the implementation of the CADC program of the department. In 1993, the Provincial Special Task Force on Ancestral Domains, which was the provincial implementing body of the DENR administrative order for CADCs, was formed. The creation of the task force signaled the launching of the CADC program in the province. It also provided a platform to launch the identification and delineation of tribal ancestral lands as a distinct component of the Pala'wan Plan Environmentally Critical Areas Network. With a joint mandate from the DENR and the council, the task force supervised the information campaign and the succeeding steps in the delineation of identified ancestral domain claims in the province.

In tracing the steps leading up to the awarding of the CADC to the Pala'wan of Campung-Ulay and Punta-Baja by the joint DENR-Council CADC program, three related developments are noted. First, information dissemination, legal orientation, awareness building, and other documented activities spurred community organizing, feedback, and dialogue among the Pala'wan and other parties with additional claims. Second, the commitment and cooperation of the task force were contingent upon the availability of financial, logistical, and technical support for the delineation of the Pala'wan ancestral domain claim. Political will in the task force collaboration was critical in order to produce tangible outcomes, such as the actual survey of the ancestral claims. Third, a series of field activities and adaptation with the use of mapping and facilitation techniques provided the tools for the Pala'wan to participate and be better informed in making decisions and negotiating the boundaries and extent of their ancestral domain claims.

Information, Documentation, and Feedback

As a first step, the task force coordinated the CADC information campaign. Primers about the DENR CADC program and its guidelines were produced in Tagalog and were distributed at community orientation sessions. An information session was conducted in each of the 11 villages in Rizal including Campung-Ulay and Punta-Baja. These sessions were important in clarifying what the program was about and served also as vehicles for organizing the community to gather and document the proofs of their ancestral claims. After the information session, they conscientiously produced the evidence that could support their ancestral domain claims, including community sketch maps, community census, testimonials from elders, photographs of sacred grounds, planted fruit trees, burial grounds, and historical accounts, which they later presented to the task force.

The information sessions were attended by a majority of the Pala'wan in the villages, some of the lowland migrants living in the villages, and some elected local government officials. These sessions provoked questions and concerns from lowland migrants, concession holders, and local government officials. The lowland migrants were concerned that they would be evicted. Local government officials were concerned that the CADC program either exempted the indigenous Pala'wan from acceding to local government rules or it allowed them to refuse assent. From the outset, the forest product and mining concession and pasture license holders were strongly opposed to the CADC program because they anticipated that the Pala'wan would also claim their concession areas as ancestral domain.

To address the concerns, the task force played its role as facilitator as effectively and objectively as possible.* The task Force responded to letters of inquiry, written complaints, and counterclaims through official correspondence that explained, for example, that lowland migrants would not be evicted unless there is sufficient evidence that they acquired the land that they occupy through stealth or by force. The task force also convened follow-up meetings and public discussions to clarify the CADC program guidelines and to present the available documentation or proofs of claim of the Pala'wan that showed that their ancestral claims had merit. Eventually, the abundant evidence presented and the decision to respect and follow the CADC guidelines broke the impasse that challenged their ancestral claims.

Support for Collaboration

The widely dispersed nature of the Pala'wan communities in the 11 villages, the sizable number of migrants in the villages, and the incidence of overlapping claims across the villages led to a contentious and drawn-out process of boundary identification and validation of the Pala'wan claim. Originally, the Pala'wan ancestral domain was a contiguous claim that spanned the 11 villages (*barangays*). The size of the Pala'wan claim was first estimated at 70,000 ha, 64% the size of the entire municipality. The immensity of the claim became a focal point of conflict and had transaction costs in terms of broad community support.

After the first information session, the gathering of proofs for the Pala'wan claim in the 11 villages was supported by PANLIPI, a legal assistance organization working closely with NATRIPAL. The PANLIPI resources were very limited despite the large area to cover. In accordance with the guidelines of the CADC program, the Pala'wan claim had to be first field-validated by the task force before the actual survey could be started. As the steps proceeded, logistical and financial requirements to support the task force increased. Similarly, boundary conflicts became more serious, which led the municipal government of Rizal to refuse to endorse the Pala'wan claim. After 2 years the task force reached a deadlock. Their political will was weak. In the end, for multiple reasons, the DENR could not sustain the funding for the Pala'wan ancestral domain delineation as a national CADC pilot site. Consequently, the lack of funding support also hampered the flow and timing of the task force's field activities.

Boundary Identification and Negotiation

Given the standoff, the task force recommended a change in delineation strategy. Instead of applying for an 11-*barangay* contiguous claim, the task force advised the

* The process was stalled because of strong objections of the local government about the delineation of the Pala'wan claims. For more than a year, the Pala'wan claims were pending because the task force failed to take command of addressing the counterclaims and other issues and concerns of other actors.

Pala'wan to redefine their strategy into separate village claims. Logistically and politically, the task force found it easier to process separate claims. The Pala'wan community organizations in Campung-Ulay and Punta-Baja, namely, CAMPAL (Association of Trustworthy Pala'wan) and PINPAL (Unified Pala'wan), were the first two to resubmit their CADC application to the task force, as they were recipients of NATRIPAL funding for CADC delineation, resource protection, and community livelihoods.* The task force was then able to proceed with the field validation and survey for Campung-Ulay and Punta-Baja with these counterpart funds.

Another year of community and task force meetings, fieldwork, negotiations, and public dialogue around boundary issues elapsed. These activities became a part of a NATRIPAL-managed program, which had an overall goal of promoting the capacity building of the federation and selected indigenous community organization members, such as CAMPAL and PINPAL. The program had the mandate to facilitate the development of community enterprises out of nontimber forest products, particularly rattan, Manila copal, and honey in Campung-Ulay and Punta-Baja. It also supported the two communities in related sociopolitical and cultural issues including the security of legal access to use and community management of nontimber forest products and other natural resources.

The program was central to relationship building between the task force and the communities. The program staff also championed in the task force the timely follow-up and resolution of conflicts and the consistent application of community-inclusive activities and field-validated boundary delineation. The fieldwork entailed the site inspection and mapping of "conflict areas," investigation of field activities, and finally the actual survey using global positioning system (GPS) units by a team consisting of mostly Pala'wan. At least two Council and DENR staff provided technical support in the survey and field validation. The first acted as team leader in the survey and the latter acted as the team leader for the field validation. A NATRIPAL community organizer provided documentation support. The maps were very useful in the actual negotiation over what to do with the "conflict areas"** or the areas with overlapping claims.

CAMPAL and PINPAL leaders organized community meetings whereby the validation and survey team were given opportunities to walk through their findings,

* NATRIPAL and its partners, WWF-Philippines Program, PANLIPI and the Tribal Filipino Apostolate, received $0.6 million funding over 3 years (1995 to 1998) to implement the program entitled "Community-based Conservation and Enterprise Program for Indigenous Communities in Pala'wan, Philippines." Funding for this program was from Washington, D.C.–based Biodiversity Conservation Network (BCN). BCN is a USAID-recipient consortium program of the World Resources Institute, World Wildlife Fund, and the Nature Conservancy. The program was hinged on a donor-driven hypothesis that if local communities receive sufficient benefits from a bidoviersity-linked enterprise, then they will act to conserve the resources it depends on. In the Pala'wan context, and elsewhere in the Philippines, secure resource tenure and access is a prerequisite to setting up a biodiversity-linked enterprise. The program used the CADC model to test this hypothesis.

** These "conflict areas" included a 600-ha World Bank–funded coconut seed garden project that was fully supported by the local government in Campung-Ulay and the Rizal municipal government, the village improvement and expansion project of the local government in Campung-Ulay, and the nontimber forest product concessions in both villages. With the exception of the nontimber forest product concessions, the Pala'wan decided to excise the first two areas from their ancestral domain claim.

present their survey map, and discuss conflict scenarios with the other community members. The task force also organized two public meetings to share opinion, clarify the guidelines, and interpret legal terms. Over a period of 3 years, CAMPAL and PINPAL, with some external support, amassed the critical information needed to decide on the size and extent of their claim. In addition, the combination of field validation, survey, community meetings, dialogues, and community discussions spawned an iterative and dynamic process that was highly favorable to a negotiated solution to boundary conflicts and overlapping claims.

SUSTAINING MEANINGFUL COMMUNITY PARTICIPATION

When the task force decided to process the Pala'wan claim separately, NATRIPAL and PANLIPI perceived this to be a divisive and compromising strategy. The 11 communities were unhappy about the situation. However, with the exception of Campung-Ulay and Punta-Baja, the other communities realized that they had limited options and resources to suggest alternatives.*

Because securing a CADC was to provide them with a legal basis to establish forest-based livelihoods, CAMPAL and PINPAL leaders mobilized house-to-house visits and community meetings to proceed with the reapplication. At 3 years into the delineation process, community organizing was critical to resolve the Pala'wan's loss of confidence in the task force. From 1995 to 1997, NATRIPAL funding provided the means for CAMPAL and PINPAL leaders, with community organizing support of NATRIPAL (e.g., facilitation of community meetings, scenario-building, and discussion), to recreate a momentum of trust among the Pala'wan in the CADC process. They rejuvenated the CADC information campaign in Campung-Ulay and Punta-Baja and retraced the delineation steps to where there was contention among task force members and other actors with counterclaims. More Pala'wan participated in the field revalidation and survey, with the provision of a minimal stipend to compensate for lost income on the days they worked with the validation and survey teams. Once engaged in the fieldwork, the Pala'wan were able to offer their opinion, ask critical questions, and state their position, particularly about the "conflict areas" within their ancestral domain claim.**

* There was no strong dissent about this decision. A factor was that all communities were encouraged to resubmit their CADC applications. Those who resubmitted are included in the priority list for future ancestral domain delineation, once funding is available. In addition, a few of the remaining communities have active leaders (and assisting organizations) in place who continue to mobilize support for their community for CADC delineation, as well as livelihood and other basic services while the CADC process is stalled.

** For example, after mapping and inspecting a forest product concession held by a non-Pala'wan, CAMPAL firmly decided to keep its claim to the area. It chose not to excise this area from its claim, not only on the basis of ancestral domain rights, but also on the basis of an important finding from field validation. CAMPAL discovered that it had not been doing rattan reforestation in their rattan concession; by mistake it was doing it within the boundaries of the adjacent non-Pala'wan-managed concession, which, in the process, was benefiting from the fulfillment of the concession holder's responsibilities to reforest the areas harvested.

The DENR finally awarded the CADC to the Pala'wan of Campung-Ulay and Punta-Baja in 1997. At this time, the task force had already championed two* other CADCs in the province and there were at least eight more CADC applications with available funding support that were at different stages in the delineation process.** The two Rizal experiences strengthened the task force and helped to clarify the roles within the multisectoral group. The two cases also helped to define where technical assistance (e.g., GPS delineation of boundaries and survey) or community organizing support (e.g., information campaign, conflict and boundary negotiations) are most needed in the delineation process.*** After this, the Pala'wan communities in Campung-Ulay and Punta-Baja braced themselves for an equally, if not more, challenging process of preparing their ancestral domain management plans and actually implementing and monitoring them. The next section describes how the Pala'wan took more responsibility for ancestral domain management. It also suggests an expanded view about community participation and the implications for sustaining it.

STRATEGIES AND TOOLS FOR ANCESTRAL DOMAIN MANAGEMENT

A main responsibility of CADC holders is to submit an ancestral domain management plan that defines in concrete terms how holders will develop, utilize, and manage their ancestral domains and the natural resources found within them. This section illustrates the adaptive management strategies and tools used by the Pala'wan communities in their ancestral domains and the various outcomes of these strategies.

The Pala'wan of Campung-Ulay and Punta-Baja had already completed their management plan by the time they had the ceremonial awarding of their CADC in 1998. A collection of tools, namely, participatory appraisal methods, resource inventories, and geographic information technology, was used in the preparation of the plan. The mangement plans were adapted in the past 2 years in an amendment process. Amendments proved necessary for various reasons including natural and human-enhanced disasters (forest fire and a need for forest regeneration), the need for additional forest protection measures, and economic concerns.

Adaptation of the plan is built into the planning process, particularly since DENR requires annual work plans. However, these communities have modified their plans beyond the basic yearly requirements by periodically incorporating new principles

---

* Two more have been approved by the DENR in Manila but have not been awarded to the communities to date because of the lack of approval from the local government.

** Where possible, NATRIPAL and PANLIPI and other support groups still continue to lobby for the resourcing of the nine other community CADC applications in Rizal. It continues to be a struggle, but it is important to note that the cause has not been abandoned.

*** There were significant differences in the community organizing and project involvement history of the two sites that had an impact on the extent to which resources had to be invested for community organizing and CADC facilitation in the two communities. However, it is beyond the scope of this chapter to delve into the contrasting institutional dynamics of the communities and the challenges of community organizing and facilitation within the context of project partnerships. From the point of view of the application of adaptive and collaborative strategies within available policy frameworks, as illustrated in this chapter, the differences in outputs and outcomes are not sufficiently significant to merit a comparative discussion of the two experiences.

and ideas into their plans. Admittedly, the Pala'wan expedited the preparation of their management plans to secure extraction rights to gather and transport nontimber forest products within the domain. They did so, however, with the knowledge that their management plans could also be amended later with the sanction of DENR.

The Pala'wan were able to reclaim their ancestral domains on the basis of their social and cultural rights; their CADCs were granted both on principles of social justice and as a mechanism toward wealth distribution. The Pala'wan emphasized sustainable income-generating and capacity-building activities as crucial strategies to protect further the critical areas in their homelands.

Ancestral Domain Management Planning Process

In 1998, one certificate was granted over an area covering two *barangays*, in Rizal, Pala'wan. Since it was recognized that two distinct groups managed each village resource base, each group developed its respective management plan. The groups coordinated, however, in clarifying the physical boundaries between their geographic areas and in setting the policies and procedures to transport products outside the ancestral domain.

Although the Pala'wan have taken the lead in preparing the plan, the assisting organizations facilitated the process and they included collaborating agencies as observers. This strategy provided legitimacy and added technical assistance to the process.* Plans were validated in a series of community meetings and discussions. Some 70% of Pala'wan from Punta-Baja and Campung-Ulay prepared the ancestral domain management plans over a 1-month period in 1997 and finally affirmed by the DENR in March 1998. The formulation of the management plan was expeditious for a number of reasons. First, there was little interest in both communities to prolong the CADC process because it had already dragged on for over 3 years. At this stage, the main interest of the community was to receive concrete benefits from the CADC. Second, community organizers and facilitators who assisted the community leaders in overseeing the planning process worked and live in the communities for an extended period. Third, the use of Participatory Resource Appraisal (PRA) techniques facilitated the documentation and processing of community ideas, strategies, and priority information that eventually went into the management plan (Yupracio, 2000; Limsa, 2000).

The management plan includes a Pala'wan ethnography illustrating their cultural history and customary law. Seasonal calendars are interpreted in the plan to document their primary activities as well as the length and timing of these activities. Traditional gathering practices of nontimber forest products (NTFPs) form a major part of the management plan. NTFPs as well as other natural resources that had subsistence and/or commercial use are plotted on the resource map. The traditional political structure of the Pala'wan are also described and it conveys the hierarchy of authority within the CADC. The plan also includes a stakeholder diagram that shows relationships among the CADC holders, other agencies, and sectors.

* Since the plan does not need to be approved by any government agency and only affirmed/accepted, then added fear and threat of manipulation or denial of plans are minimized.

Adaptive Planning

The management plans were completed in a relatively short time. The community members and DENR officials recognized that there was a need to add to the original plan after the destruction caused by the forest fires in April 1998. Subsequently, a second round of amendments was executed in October 1999. The irregularity of the amendments reflects how adaptations are made based on experiences and major events and not set within an annual or regular time frame.

The community chose to use the amendment process to update its management plan, following what it thought was the most expeditious and legal path. Legalizing the changes and providing them to the public would be a process that other stakeholders and government agencies could not challenge because of its voluntary and transparent nature.

The amendments included changes in penalties for violations of the stated laws of the management plan since those of the original plan appeared too strict and unrealistic. The amendments included very relevant and timely policies on bioprospecting, the responsibility of the indigenous population in monitoring and penalizing illegal logging and harvesting activities, and a moratorium on the catch of endangered bird species. The amendment also included sections on other products that the community could benefit from, such as dead standing trees and *kulba* (a material used for planting orchids).

These amendments (Table 22.1) were submitted to the municipal office of the DENR and to community members and assisting organizations. They were not questioned. The amendments have been accepted and are enforced like other provisions of the original plan. Interviewed CADC holders said they would amend their CADC again when the need arises. They are also prepared to amend the plan if existing policies need revision or if new policies need to be developed based on changing circumstances.

Planning for Sustainable Livelihoods

The amendment process lends well to the communities' transition from mere CADC recipients or beneficiaries to responsible and effective ancestral domain managers. As the communities develop more experience in implementing concrete livelihood strategies within their ancestral domain, the management plan is amended to expand the institutional and economic dimensions of the plan. In particular, the plan is amended to reflect livelihood opportunities and the community's corresponding strategies, expanding leadership and membership roles, and community monitoring issues.

Rattan, honey, and almaciga resin* are examples of NTFPs that are the primary source of livelihood for the Pala'wan of Rizal. For generations, official utilization and trade of these concessions were in the hands of migrant concessionaires. Compensation for harvest has been low and debts accumulated (Warner, 1979). The CADC gave the indigenous communities the long-awaited right to preferential use

* Almaciga resin is a natural resin from the prime hardwood species *Agathis philippinensis* used for boat sealants, adhesives, paints, etc. This product is exported mainly to Europe.

TABLE 22.1
Ancestral Domain Management Plan Amendments, October 7–8, 1999

I. Process for making amendments to the ADMP
II. Reviewing and revising penalties/punishments to harvesting almaciga resin
III. Reviewing and revising penalties/punishments on rattan harvesting
IV. Policies on the use of salvaged logs and dead standing trees
V. Policy/process to prevent the entrance of migrants into the ancestral domain
VI. Policy/process to expel migrants in the ancestral domain that have entered against the rules of law
VII. Responsibility of the community members if they observe illegal activities within the ancestral domain (whether the illegal activity is performed by a migrant or indigenous community member)
VIII. Penalty/punishment to a leader that has not complied with the policies of the ADMP
IX. Plans and policies concerning quarrying and small-scale mining
X. Process of accepting new projects in the ancestral domain
XI. One year plan of activity of the community
XII. Recognized games within the ancestral domain
XIII. Process in obtaining decisions from the council of elders
XIV. Process for those that temporarily want to stay in the ancestral domain
XV. Zoning of natural resources for long-term benefit
XVI. Moratorium on catching birds
XVII. Policies on other indigenous community members of Pala'wan, Tagbanua, and Batak who would like to stay in the ancestral domain and/or avail of the benefits of the natural resources that can be found within the ancestral domain
XVIII. Indigenous peoples from outside the ancestral domain that would like to benefit within ancestral domain
XIX. Policies on lowland (irrigated) rice farming within the ancestral domain
XX. Policies on bioprospecting

Note: Translated from the Tagalog original by the authors.

Source: CAMPAL, 1997.

of these resources.* In addition, the radical opinion drafted by the DENR legal affairs unit, stating that the management plan could serve as a permit for all minor forest products,** gave an incentive for the Pala'wan to prepare sustainable harvesting plans for different products found in their ancestral domains. This meshed well with the multiproduct approach of livelihood of the indigenous communities. In addition, this could take pressure off already heavily extracted products.***

To build both economically viable and environmentally friendly enterprises, a biological inventory of the top seven NTFPs was conducted for both sites. This gave recognition and scientific support to the proper tapping techniques utilized by the

* Previously, Manila copal or almaciga concessions, for example, were put to bid in a process requiring cash bonds, rigid business plans, and other strict requirements, which indigenous groups found overwhelming and impossible to comply with.

** A legal opinion released by the DENR in July 1997 clarifies that a management plan qualifies as a legal resource extraction instrument.

*** Indigenous fruit especially seem promising, but these are still being developed.

Pala'wan of Manila copal or almaciga (*agathis philippinensis*) resin. The results of the inventory showed no scarcity of resin in the ancestral domain (NATRIPAL, 1998), which reflected that the Pala'wan were using sustainable methods of collecting resin. On the other hand, with the decrease in rattan production, zones were established for rattan production and regeneration. Local data were used to prepare a computer-generated image to aid in planning and management efforts. The biological inventory became the basis of computing a realistic annual allowable cut for rattan that replaced the "table" survey and inventory conducted by the DENR. The biological inventory also identified two major NTFPs, *anibong* and *pandan*, for future product development.

Community-Based Management Units

To operationalize the plans for the ancestral domain, CAMPAL and PINPAL, with the assistance of NATRIPAL, developed and strengthened community structures. Particularly important are the enterprise development unit, paralegal team, and *manungukom* or the Council of Elders, and officers.

Given the turnover of natural resources from government to communal and community management, it was seen as important to set up an economic structure to handle the transactions that were once coordinated by an elite group of traders. Through the NATRIPAL project, an area servicing unit was established to handle functions such as credit and savings, merchandising through a consumer store, and sales of NTFPs. With the phaseout of the project and the realization of large overhead expenses, the operations in Campung Ulay still consist of a small store with a business manager for NTFP sales. Punta Baja runs operations through a treasurer and business manager.

Based on feedback from evaluations, one of the most effective mechanisms of protection of the ancestral domain was through the paralegal team. This team was handpicked from the community to serve as the guards of the CADC and enforcers of community policies. These teams were provided paralegal training, including training on seizure and arrest, boundary protection and monitoring, and formulating sworn statements, and conducting citizen's arrests on illegal activities.

Finally, operations within the ancestral domain are still monitored and must pass through the decision of the *mangungukom* or the traditional Council of Elders. Even though traditional leaders were commonly called upon for issues pertaining to customary laws, for example, on marriage rites, their roles were expanded to address conflict over economic and environmental concerns as well. The thinking behind this was that critical issues should still be handled by a handful of wise leaders. Leadership training has been conducted to clarify roles with elders and operational staff. Those entering the CADC or wishing to start a project in the CADC must first seek approval from the *mangungukom* (see Figure 22.2).

Results of Community-based Management

Elders, officers, and members interviewed in July 2000 unanimously saw many positive results from the delineation and transfer of management of ancestral domain

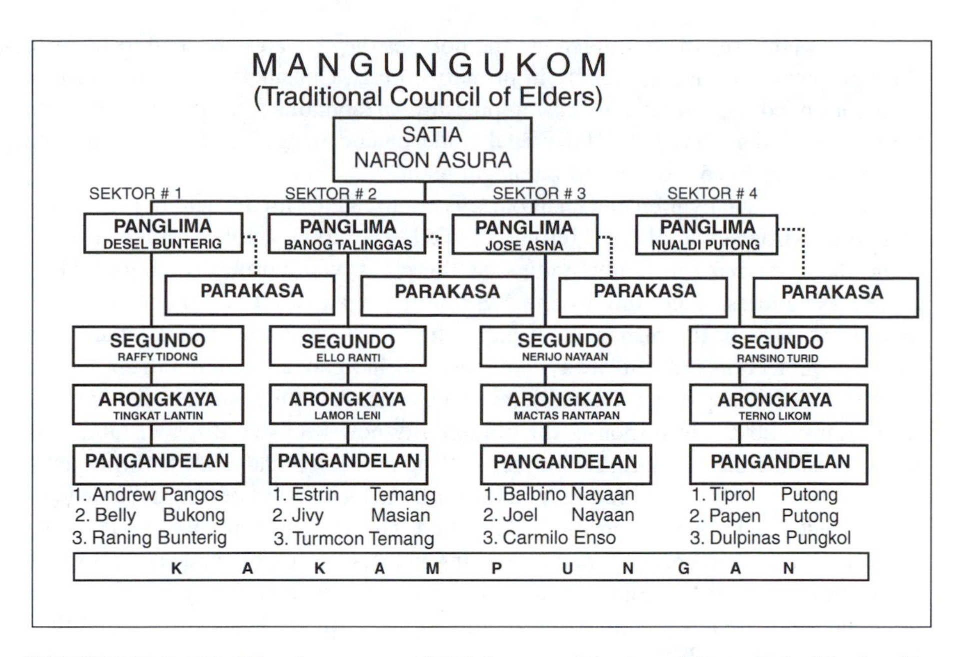

FIGURE 22.2 Traditional structure of Pala'wan community of Punta-Baja, Rizal, with special attention to the council of elders and hierarchy in various sectors of the ancestral domain. (From PINPAL, 1997.)

to the Pala'wan. With the CADC, these Pala'wan feel they are more recognized than in the past when officials only visited them during election time. In earlier years, the Philippine Coconut Authority (a quasi-government body) was pushing projects in the area without any consultations, resulting in many conflicts in the area.* After the awarding of the CADC, the community was first contacted before the implementation of a government project on tree farming (Nasura, 2000).

Also, Pala'wan CADC holders remember the recent time when migrant traders and NTFP concession holders dictated price and volumes gathered from the forests. At that point they had difficulty controlling the overextraction that was taking place and were mere laborers on their own lands. Now, they have the right to transact their business with their chosen business partners and receive better prices for their products than the previous concessionaires provided. The Pala'wan now have a voice in natural resource protection and they are also able to control the influx of migrants and CADC entrants (Taom, 2000).**

* Once the conflict almost resulted in a skirmish with the Pala'wan community threatening to fight using its traditional *supok* or poisoned blow darts.

** With their new rights to protect the domain, they have been able to apprehend illegal loggers and illegal almaciga resin harvesters. Early in 1998, Roger Majid, a 21-year-old Pala'wan stopped a truck carrying illegal lumber within the ancestral domain. As a result, the mayoral candidate that owned the truck had to give up his candidacy for fear of a scandal (Dalabajan, 1999). Just months earlier, Pala'wan of the neighboring village of Campung-Ulay confiscated 900 board feet of illegally cut *kamagong*, premium, endemic Philippine hardwood (Dalabajan, 1999).

They were also able to implement the policies they have established to regulate slash-and-burn farming, especially in primary forests. Considering the large effects of uncontrolled fires and the increasing pressure on land, this was also an adaptation to previous indigenous law. The illegal trade of endangered birds has also been diminished thanks to a dedicated paralegal team.

Although illegally prepared charcoal was confiscated with the help of the municipal and environmental officer (Terong, 2000; Lania, 2000) implementing management plans was risky for communities, as the plans were not always respected by other stakeholders. According to one community organizer, it is difficult to implement the CADC or the management plan when local government units charged to assist in protection did not always support the indigenous communities' seizure activities. For example, confiscated lumber was released at the whim of the mayor's staff (Limsa, 2000). In response, the community now keeps confiscated lumber in its possession until it is sure the lumber will be processed and dealt with properly.

Intracommunity collaboration remains problematic as roles of traditional leaders are expanded and new relationships established. The management plan is still something new to other indigenous members living in the ancestral domain and is difficult to disseminate to many people spread out across a sizable area. The members of the *mangungukom* are still familiarizing themselves with their expanded role and their new responsibilities. Even though community management is the ideal, some community members have long-term relationships with and debts to outside traders. The failure to pay these debts makes it difficult for community enterprises to take off.

Management of enterprises is also a sensitive and tricky issue since this involves technical skill, recording, and business sense. The structure and composition of enterprise teams also seem to be changing as members test their roles and some give way to others. The influx of new cash into the community is also a cause of intrigue and, at times, doubt. With the phaseout of community-based NATRIPAL staff, and focus on less-developed areas, the associations of PINPAL and CAMPAL continue to need institutional and technical support in operations within the CADC.

It is also in the management of community-based enterprises that the ability of the Pala'wan to shift unequal power arrangements in their favor is tested. There is great risk that the turnover of forest product concessions to the communities will result in animosity of previous concession holders toward the indigenous community. It is suggested that assisting organizations provide a venue or operational framework where previous concession holders can still have a role (although less prominent) in the trade chain. For example, to reduce conflict, a benefit-sharing scheme in the use and management of forest product concessions can be developed to facilitate continued yet balanced relationships between the Pala'wan and the previous concession holder.

Exploring benefit-sharing schemes may be necessary considering the recognized weakness of internal community structures. It has been recognized that even if community institutions are active, they are still organizationally fragile and easily challenged. This is especially true for enterprise functions. The success of such a program, however, hinges on the potency of these community structures. In the transition period to community management, "friendly economic transaction" agreements may be necessary so that economic functions are not affected as the community builds capacity. How such relations of trust will be built considering past decades of unequal relations is still a difficult question to answer, but assisting organizations can act as mediators in forging such agreements.

Finally, the size of the CADC and lack of clarity of its borders makes it difficult to process borderline cases of illegal activity. This has caused internal conflict as different sectors take sides and have their own perceptions about certain cases.

CHANGING POLITICAL CLIMATE FOR ANCESTRAL DOMAIN RECOGNITION

The restoration of a democratic spirit with the revolution of 1986 made way for new opportunities in protected area management, the promotion of indigenous peoples rights, and the recognition of the potential of community forestry schemes. Collaboration in management and the devolution of power to local governments, inclusion of community viewpoints, and multisectoral consultations indicate the importance placed on participation and collaboration mechanisms in the management of natural resources in the Philippines.

The early to mid-1990s ushered in progressive government support for community-based forest management, and the ancestral domain delineation and management program was a flagship program under the DENR, along with community-based forestry management. As of June 6, 1998, 181 CADCs covering 2,546,035 ha* had been awarded nationwide (see Chapter 11 by Luna). Theoretically, the DENR must turn over the operations in regard to the delineation and monitoring of ancestral domains in the country to the National Commission on Indigenous Peoples. The commission is the governing structure for the implementation of the ancestral domain delineation and titling legislation, Indigenous Peoples Rights Act (IPRA).

To date, the commission has had its share of problems. The constitutionality of the IPRA continues to be questioned in the Philippines' highest court; the release of government funds to support the commission has been suspended; and the commission is not fully staffed and does not have the technical capacity to oversee or fulfill its mandate. It is also currently highly vulnerable to political pressures in the selection of commissioners and of the executive staff. Many indigenous peoples' organizations and assisting organizations also question the credibility, competency, and sincerity of the commission.**

To make the situation more challenging for the indigenous peoples, the DENR, under new leadership, has shifted its perspectives. DENR Secretary Antonio Cerilles, appointed to the post by President Estrada in 1998, intends to keep upland dwellers

* Although these have all been issued, not all have been "awarded" (awarded is the DENR gatekeeper's term before a CADC can actually be enforced at the provincial level) and not all are recognized by provincial DENRs regardless of issuance as is the case for the Tagbanua of San Viente and Irawan, Puerto Princesa. Other CADCs were issued even when they had not gone through the proper consultations. This is the case for the indigenous peoples of the northern Sierra Madre and the Dumagat of Quezon province (see Chapter 11 by Luna).

** Many of these commissioners (1) were not chosen by the IPs themselves, but appointed by the president, without consultations, (2) have existing court cases against them, and (3) they have been said to support the interests of the corporate sector and not the IP sector whose interests they are meant to defend (Ballesteros, 1999).

out of the forests, particularly out of protected areas (Luna, 2001). He has also allotted a meager budget to community-based forest management programs as opposed to commercial forestry and plantation establishment program (LRC-KSK, 1999). This comes at an inopportune time for indigenous groups to proceed or expand their activities for ancestral domain delineation and management when the implementation of the IPRA is also currently stalled.

In Pala'wan, policy implementation has been volatile since the ancestral land program and the Pala'wan Plan were launched in 1992, and when the joint council and DENR task force was formed the year after. The ancestral domain program generated a diversity of positions, ideas, and interpretations of the rules and regulations regarding resource access and use within the delineated boundaries of the ancestral domains. The recurring debate is on indigenous peoples' rights, privileges, and limitations of extraction of forest resources and of actual land use, and the standards that must be set to regulate other people's resource access and use within the ancestral domains, for example, non-Pala'wan forest concession holders, farmers, and forest gatherers.

The effect of the long interruption of the distribution of ancestral domains has been quite a difficult experience for communities that were not able to obtain certificates before the commencement of the new administration of President Estrada in June 1998. The policy scenario that is currently shifting away from community management has left old, landed gentry* at the helm of forest concessions. Renegotiation and transactions with long-time powerholders then became necessary, as resource rights were not transferred. The animosity engendered by the radical spirit of community-based laws continues to strain the relationship with these, often powerful, political figures.

ADAPTIVE COLLABORATIVE MANAGEMENT IN ANCESTRAL DOMAINS

The Pala'wan case study underscored the importance of the cooperative effort between the Pala'wan and the provincial task force toward the delineation and actual survey of ancestral domains. Collaboration was key among the community members; between the communities and other stakeholders; and among the assisting organizations (e.g., NATRIPAL), relevant local authorities, and community DENR offices in the planning process. Collaboration and amicable relationships among these parties continue to be catalytical characteristics of the natural adaptation of the Pala'wan's management plans and of the actual community-based ancestral domain management in Campung-Ulay and Punta-Baja.

Aside from the importance of institutional support, the policy climate, provincially and nationally, promoted the conditions that were conducive for adaptive and collaborative processes. As a policy and management instrument, the Pala'wan CADCs and management plans spurred a heightened sense of solidarity and organized action among the indigenous residents in Campung-Ulay and Punta-Baja. The processes associated with it, from boundary negotiation and delineation, actual

* Many of them prominent government officials.

survey, to the formulation, enforcement, and evolution of the management plan proceeded in an adaptive and collaborative mode. This was not necessarily by design but was a natural consequence of the opportunities presented by a national and provincial policy framework, and the challenges presented by various constraints and changing circumstances. These processes have also set an encouraging tone to continue to leverage support for the other nine Pala'wan communities and other indigenous groups in the province, with standing ancestral domain claims.

Table 22.2 summarizes the mechanisms available within three national and provincial policy frameworks for ancestral domain delineation and management to apply ACM. In the Rizal case, the mechanisms available within the context of the Pala'wan Plan and the DENR administrative orders supported multisectoral collaboration, indigenous community engagement, and participatory processes from delineation to community-based management and monitoring of ancestral domains.

The case, more specifically, contributes the following perspectives about ACM within the context of the implementation of the Pala'wan Plan and the DENR CADC frameworks:

1. It promoted the Pala'wan's constructive engagement with government authorities from the outset of the delineation process, and continued, although less intensely, through to the management planning activities. Collaboration can be burdensome; yet, most often, it is timely and provides the necessary process to inform and draw acceptance of ancestral domain claims from other sectors.
2. Adaptation, although implicit in government-required annual work plans, was not necessarily prescribed. The amendment process that followed the initial drafting resulted from field-testing community policies or adapting them to respond to the policy climate, evolving community structures (e.g., emerging enterprise functions) and capacities, and other external events (e.g., calamities).
3. ACM served as a catalyst for community engagement. It allowed the Pala'wan to participate in a meaningful way in the CADC program from mere attendance in input sessions during earlier years to a wider scope of involvement in a range of community activities. The Pala'wan have been active discussants and primary actors in boundary conflict negotiation and resolution, actual survey and boundary demarcation of their ancestral domains, community planning and decision making on rules and regulations (and their amendments) for resource use, sanctions, and development of community structures that would oversee community livelihoods and resource management. Community engagement has been substantial and has evolved parallel to the development of local capacity, community organization, strengthening of intra- and extra-relationships based on trust and reciprocal benefits and agreements.
4. ACM created the opportunities for informed decision making. The nature of the implementation of the CADC program, being multisectoral and guidelines are provided for information dissemination, consultations, documentation, and other validation activities, in themselves provided the

means for information generation, exchange, and use of the information generated as basis for proceeding to the next stage of the CADC process. In the development of the ancestral domain plans, the planning process was participatory and transparent in terms of being open for scrutiny and advice from external actors such as the DENR or other collaborating organizations like NATRIPAL.

5. ACM allowed the development of local capacity and of community institutions. This was supported by the availability of technical and financial support for planning, building intra- and extra-community cooperation and coordination, and facilitating the development of the plan, the amendments, and in activating the community structures set up for livelihood and organizational development.
6. The purpose for collaboration is also adaptive. Upon the acquisition of the CADC and the submission of the management plan, collaboration was necessary for coordination and enforcement of ancestral domain laws; however, collaboration was no longer utilized for the direct management of and policy making within the ancestral domain. Collaboration became more of a mechanism for transparency with a clear delineation and respect of each other's roles: the Pala'wan as lead implementor and policy maker; the collaborating groups as observers and periodic advisors to the Pala'wan. This arrangement has inspired the Pala'wan to further recognize and act on their primary responsibility of forest protection. It has been observed that illegal activities were closely monitored and pro-environment policies (e.g., low-impact harvesting, minimal slash and burn) were effectively set in place.

The third policy framework described here, the 1997 IPRA, provides potentially expanded mechanisms for community-led adaptive planning and management and intracommunity collaboration.

As an adaptive strategy while the IPRA is currently suspended and contentious, Pala'wan's indigenous communities and their partners are trying to invoke other laws and make stronger alliances with diverse groups to secure the protection of their rights and the resources that lie within their zones. The Pala'wan Plan is still a functional provincial policy framework that mandates the demarcation of tribal ancestral zones as environmentally critical areas. NGOs and indigenous people's organizations are actively lobbying in the newly established Pala'wan Special Committee on Tribal Ancestral Zones, the multisectoral body tasked to process ancestral zones in the province, to influence the preparation of the guidelines of these zones. Once established, however, these zones may not be strong enough to grant resource tenure to indigenous peoples within the ancestral domains (Alisuag, 2000).* They are likely to be forced into another form of bureaucracy with no guarantee of securing economically viable instruments either to develop or to sustain community livelihoods.

* PCSD staff have stated that no instruments will be released to harvest natural resources. These would still have to go through the DENR.

TABLE 22.2
Available Mechanisms for Adaptive Collaborative Management in Policy Frameworks

Implementing Rules/Guidelines	Strategic Environmental Plan for Pala'wan	DENR Administrative Orders	IPRA
1. Governance/ organizational structure	• Pala'wan Council for Sustainable Development • Pala'wan Committee on Tribal Affairs with Provincial Special Task Force on Ancestral Domain • Pala'wan Special Committee on Tribal Ancestral Zones	• Pala'wan special task force on ancestral domain	• National commission on indigenous peoples • Ancestral domain offices (provincial level)
2. Boundary delineation and survey of ancestral domains	• Public information campaign • Documentation of boundary delineation (e.g., community mapping) • Gathering and submission of proofs of claim • Ocular inspection and negotiation of boundaries • Actual survey and mapping	• Public information campaign • Documentation of boundary delineation (e.g., community mapping) • Gathering and submission of proofs of claim • Ocular inspection and negotiation of boundaries • Actual survey and mapping	• Titling of ancestral domains delineated under DENR administrative orders • Documentation of boundary delineation (e.g., community mapping including census) • Gathering and submission of proofs of claim • Negotiation of conflicts • Actual survey and mapping • Registration of ancestral domain titles
3. Management of ancestral domains	• Management by indigenous community/community organization • Preparation and implementation of ancestral domain management plans • Availment of technical assistance from assisting organizations and the palawan council staff	• Management by indigenous community/community organization • Preparation and implementation of ancestral domain management plans • Preparation of indicative resource use plans and/or detailed workplans • Availment of technical assistance from assisting organizations and the provincial special task force on ancestral domain	• Management by indigenous community/community organization • Preparation and implementation of ancestral domain sustainable development and protection plans • Availment of technical assistance from assisting organizations and the ancestral domain office

TABLE 22.2 (continued)
Available Mechanisms for Adaptive Collaborative Management in Policy Frameworks

Implementing Rules/Guidelines	Strategic Environmental Plan for Pala'wan	DENR Administrative Orders	IPRA
4. Monitoring ancestral domain management	• Enforcement of rights and customary laws of the indigenous community • Enforcement of the ancestral domain management plan (i.e., community resource use plans; rules and regulations; penalties and sanctions)	• Enforcement of rights and customary laws of the indigenous community • Enforcement of the ancestral domain management plan (i.e., community resource use plans; rules and regulations; penalties and sanctions)	• Enforcement of rights and customary laws of the indigenous community • Enforcement of free and prior informed consent[a]

[a] As defined in the IPRA (1997), free and prior informed consent is the "consensus of all members of the indigenous communities to be determined in accordance with their respective customary laws and practices, free from any external manipulation, interference and coercion, and obtained after fully disclosing the intent and scope of a proposed activity, development programs and projects or other forms of intervention], in a language and process understandable to the community."

Sources: Republic Act 7611, 1992; DENR Administrative Order 2, 1993; DENR Administrative Order 34, 1996, Republic Act 8371, 1997.

CONCLUSIONS

ACM allowed the Pala'wan to meet three objectives: to promote the recognition of their rights to their ancestral domains; to meet their livelihood needs within the government's ancestral domain policy framework; and to maintain options and the flexibility to evolve as more information and resources become available. The participation of the Pala'wan in the multisectoral implementation of the CADC program and the development of their management plans in stages fulfilled the basic government requirements up to the level where their ancestral domain management plans are deemed as an acceptable basis to grant them permits to harvest and transport forest products from their ancestral domains. The Pala'wan's working relationship with government authorities brought legitimacy to their ancestral domain claims and viability in their organized action to assert management control within their bounded territory.

The flow of processes and activities that led to the current stage of actual community-based ancestral domain management was supported by available information and resources. The information campaign conducted by assisting NGOs and the DENR initiated the dialogue between these institutions and the Pala'wan in Rizal. The Pala'wan in Campung-Ulay and Punta-Baja were able to resubmit their CADC application because, unlike the other communities, they had the available resources (from NATRIPAL) to support the field activities such as meetings, ocular inspection, and validation activities of the task force, the actual survey, mapping activities, planning workshops, and the like.

At this juncture of actual community-based management and community interactions with outside forces, CAMPAL and PINPAL require substantial material and human resource investment (e.g., funds, technical assistance, provision of tools, facilitator assistance, and others) to continue to advance adaptive and collaborative processes within their Pala'wan ancestral domains. In the case study, the availability of resource support provided the communities the means to leverage opportunities for capacity building in terms of support for training, coaching, providing access to capital and markets, management of community livelihoods by the enterprise teams that they formed, sustaining subsistence livelihoods, building effective command of community leaders and responsible teams in enforcing ancestral domain regulations and protection.

In addition, with support for comprehensive local capacity building, the communities will have the means to negotiate persistent unequal power arrangements. Given the temporal quality and vulnerability to donor-driven demands of assisting organizations, however, other sources of support, such as government agencies, must be explored. Local government (e.g., the council, municipal government) and government line agencies (e.g., DENR), that also have a significant stake in forest protection and utilization, must be open and willing to contribute resources to communities that will enable them to manage their ancestral domains effectively (e.g., provide budget for paralegals, forest guards, funds for planning and management). However, this is potentially fraught with difficulties given the persistent collusion of vested interests (e.g., government officials and private sector groups).

The current Pala'wan scenario is precarious given that a self-sustaining mechanism to generate resources for capacity building and overall institutional support is still not fully in place in both communities. New policy frameworks must continue to reinforce an adaptive and collaborative approach to maintain the momentum for dynamic community-based processes in Campung-Ulay and Punta-Baja, as well as to stimulate a timely discussion and strategic community response on the issue of sustainability.

Finally, the larger application of the CADC experience in Pala'wan to other areas in the Philippines is complex, especially in the current mood of the national government to support industrial plantations and programs instead of community-based activities. The focus at present is to implement and improve operations in granted CADC areas around the country. Unfortunately, the implementation of policy appears to be different around the country and, as is the case of the Mangyan Alangan, not all government agencies are implementing the law as intended (Balmes, 1999).*

As progressive laws for indigenous people's rights and tenurial instruments remain in limbo, indigenous people and assisting organizations are trying to use other laws to continue to support community-based natural resource management

---

* In Paitan, Mindoro, Mangyan Alangan claim that provincial environment officials are forcing communities to apply for concessions even after the submission of management plans, thus adding to the bureacracy.

efforts. The same movement is also trying to reeducate government on the merits of these programs and share positive experiences with other indigenous groups. The performance of the two communities in Rizal and of other indigenous communities have extended implications for ACM and future delineation of other ancestral domains in Pala'wan and the rest of the Philippines. However, in a context where the policy scenario is volatile, it is the internal and external relationships that communities build and negotiate that will help determine if these will indeed go forward in Pala'wan or elsewhere in the country.

ACKNOWLEDGMENTS

The authors thank the board and staff of NATRIPAL, who shared their experiences and insights; the officers, members, and elders of the Pala'wan communities of Punta Baja and Campung Ulay, Rizal, who have kindly shared their challenges in ancestral domain management; Loreta Alsa, NATRIPAL Resource Management Coordinator, who conducted many of the interviews; and Melanie McDermott, Charles Geisler, and Louise Buck for conceptualization, guidance, and support.

REFERENCES

Alisuag, J., 2000. Notes from interview of Palawan Council Director, 12 July, Puerto Princesa City, Palawan.

Ballesteros, A. G., 1999. NCIP Administrative Order No. 3: Validating founded Fears on the IPRA, Tana-wan 2, No. 1 (a publication of LRC-KSK).

Balmes, B., 1999. Personal communication with Mangyan Mission staff at the conference on Southeast Asian Conference on Non-Timber Forest Products, 14 December, Puerto Princes City, Palawan.

CAMPAL, 1997. Ancestral Domain Management Plan.

Dalabajan, D., 1999. In Defense of Ancestral Homelands, *Bandillo Palawan Mag.,* 6(10).

Department of Environment and Natural Resources, 1993. DENR Administrative 02-93. Rules and Regulations for the Identification, Delineation of Ancestral Land and Domain Claims, Quezon City, The Philippines.

Eder, J. F. and Fernandez, J. O., 1996. Palawan, a last frontier, in *Palawan at the Crossroads: Development and the Environment on a Philippine Frontier*, Eder, J. F. and Fernandez, J. O., Eds., Ateneo de Manila University Press, Quezon City, Philippines.

Lania, R., 2000. Notes from interview of Council Elder by Loreta Alsa, NATRIPAL Resource Management Coordinator, 22 July, Punta-Baja, Rizal, Palawan.

LRC-KSK, 1999. No More Surprises from Cerilles, *Cebu Daily News,* 21 November, Legal Rights and Natural Resources Center-Kasama sa Kalikasan.

Limsa, R., 2000. Notes from interview of NATRIPAL Community Organizer, 7 July, Puerto Princesa City, Palawan.

Nasura, N., 26 July 2000. Notes from interview of PINPAL President by Loreta Alsa, NATRIPAL Resource Management Coordinator. Punta-Baja, Rizal, Palawan.

NATRIPAL, 1998. Biological Inventory Report Covering Sitio Kayasan, BarangayTagabenit, Puerto Princesa City and Campung-Ulay, Rizal, Palawan, report submitted by R. Melchor for the BCN Project, Puerto Princesa City, Palawan.

PINPAL, 1997. Ancestral Domain Management Plan.

Republic Act. No. 7611, 1992. The Strategic Environmental Plan for Palawan Act, Philippines.
Republic Act. 8371, 1997. The Indigenous Peoples' Rights Act of 1997, Philippines.
Taom, CAMPAL President, 2000. Notes from interview by Loreta Alsa NATRIPAL Resource Management Coordinator, 20 July, Campung-Ulay, Rizal, Palawan.
Terong, Community Elder, 2000. Notes from interview of Loreta Alsa, NATRIPAL Resource Management Coordinator, 26 July, Punta-Baja, Rizal, Palawan.
Warner, K., 1979. Walking on Two Feet: Tagbanwa Adaptation to Philippine Society, Ph.D. dissertation, University of Hawaii.
Yupracio, M., 2000. Notes from interview of NATRIPAL paralegal, 6 July, Puerto Princesa City, Palawan.

Index

A

C

D

E

G

H

I

J

K

L

M

N

O

P

Q

R

S

T

U

V

W

Y